Genetics and development

# Genetics and development

**J. H. Sang**
*School of Biological Sciences,*
*University of Sussex*

Longman
London and New York

**Longman Group Limited**
Longman House, Burnt Mill, Harlow
Essex CM20 2JE, England
*Associated companies throughout the world*



First published 1984
Second impression 1986



**British Library Cataloguing in Publication Data**
Sang, J. H.
Genetics and development.
1. Genetics
I. Title
575.1 QH430

ISBN 0-582-44681-3

**Library of Congress Cataloging in Publication Data**
Sang, J. H. (James H.), 1912–
Genetics and development.

Bibliography: p.
Includes index.
1. Developmental genetics. I. Title.
[DNLM: 1. Embryology. 2. Genetics. 3. Mutation.
QH 453 S225g]
QH453.S25 1984 574.87′328 83-13536
ISBN 0-582-44681-3

Set in 10/12 pt Linotron 202 Times Roman
Printed in Great Britain by The Bath Press, Avon

# Contents

# Preface

It may seem the wrong time to write about genetics and development when approximately two papers relevant to the subject are being published daily. Still, students have to be taught! But there are more important reasons for a new book. The past decade has seen a most remarkable convergence of interests on the problems of gene expression, as summarized by Lewin (1980) for example, and this has resulted in new ideas about the regulation of genes during development. Concurrently, geneticists have been identifying and exploring the actions of genes which directly affect development. Together these studies have raised the subject to a new level of understanding where we are at last beginning to appreciate the complex subtleties of genome organization. There is therefore a new framework of established facts and of tentative explanations which students must know if they are to follow research papers as they come out. This book attempts to provide that background.

Much of the new information comes from disciplines once considered separate from embryology and genetics, particularly from molecular and cell biology. This presents two problems. We cannot now assume that students wishing to specialize in the subject will all have the same elementary training. We have tried to cater for this welcome diversity of intake into the subject by assuming a minimal background, in the confident assumption that students who know more will skip what they have already learnt. The second problem is where to begin, what to emphasize and what to leave out, and here we have been guided by two principles. We have taken it that conventional and molecular genetics are now one, and that this combined approach will provide the most exciting future developments. Since developmental genetics is above all an experimental subject, we have also tried to provide sufficient technical information for the described experiments to be understood. The shape of the book has been dictated by these two prejudices. Unfortunately this has sometimes meant that history, particularly of ideas, has been given less attention than it deserves. The quotations which head the chapters may make up for this in some small way, just as the bibliography should allow the reader access to topics which have been omitted, or dealt with only briefly. Many of these papers have been chosen for their comprehensive introductory sections, for that reason.

Like many books, this one grew out of a course given to third year, and some graduate, students. The content and emphasis of that course changed annually as new work was published, and this book is quite different, too, for the same reason. Nevertheless, it owes a debt to my fellow teachers, Dr. J. I. Collett, Dr. J. R. S. Whittle and, too briefly, Dr. B. Lewin, and it is a pleasure to acknowledge their many helpful discussions. Thanks are also due to Sheila Buckingham, Sally Byatt, Jill Davey, Nicola Ford and Jo Harper for secretarial help, and to Colin Atherton for the original photographs. Permission to use published material is gratefully acknowledged here, and again specifically in the legends. The enthusiasm of Michael Rodgers, and the editorial help of Eleanor Lawrence and Gillian Holmes greatly, and expeditiously, eased the transition from manuscript to print.

# I Introduction

Our problem is to identify, and define in molecular-genetic terms, the mechanisms that underlie the orderly, sequential changes in gene expression during development. It is this harmonious growth process that accounts for the variety and organization of cell types in higher organisms, and the various phenomena which occur during growth have been explored in many of their descriptive details by experimental embryologists. Our first task is to describe these phenomena, for they provide a set of terms of reference into which genetic explanations must fit, and sometimes they may suggest a starting-point for molecular-genetic analysis. Unfortunately we know the embryology of too few organisms in the kind of detail needed, and the genetics of even fewer, so all generalizations must be suspect. Here we emphasize the main points discussed in the following chapter.

Cell differentiation is not generally due to loss of genes not required to make a particular cell type: the nuclei of all cells are at least pluripotent. Differences between cells must therefore result from differential gene activation during development. A number of strategies have been adopted to achieve this end. At one end of the spectrum are those organisms in which the deposition of maternally-formed substances in the egg produces a *mosaic* egg, whose early division products are immediately restricted in their developmental potential according to the substances they contain. Cells are committed at a very early stage to particular paths of development. This 'invisible differentiation' (Weiss, 1939) or *determination* is inherited by the cell's progeny, and the genes for the molecules that effect determination will be of great interest to us.

Mosaic eggs also illustrate another and less well understood, phenomenon. The determined mitotic progeny of the early division products can complete their differentiation independently of other cells. In some way they are preprogrammed *ab initio*, and overt differentiation follows only after a number of cell divisions. How particular genes are activated during this process is not at all obvious.

At the other extreme are eggs in which the initial division products are totipotent and can *regulate* to form whole organisms. In such embryos the determined state may not be established until the broad mass of the embryo is laid down. Determination must in this case depend on the relations of cells to one another, and possibly on the establishment of metabolic differences between them which by contact, or other mechanisms, direct different cell groups along divergent developmental paths. In some cases, the establishment of these cell states seems to depend on gradients of molecules, and in others on localized differences such that one cell group may induce changes in other, competent cells. The final determined state in such regulative embryos is the outcome of progressive stepwise restriction of a cell's developmental potential. We have to identify genes which disturb this progression if we are to understand it.

The distinction between mosaic and regulative eggs is only a rough one and there is a continuous series in between, for not only do some organisms adopt partial strategies but they may also use one strategy for one feature of development and the other for another. To an extent, the complexities of cell interactions reflect the elaboration of tissues to form organs containing many different cell types. The genetics of this aspect of development has been largely ignored.

Once determined, many cells follow regular division patterns before achieving their final differentiated state. Others differentiate only after exposure to a morphogenetic hormone, though plant hormones affect determination too. Morphogenetic hormones are therefore powerful tools for studying gene activation, and for exploring the molecular organization of the genes they affect.This particular morphogenetic strategy has not yet been studied by embryologists in all its detail, but we probably know enough to guide us through its complexities.

Experimental manipulations of embryos have been widely used for studying the interactions between different parts of the embryo during development. We might expect the consequences of experimental manipulations to be copied by some gene mutations and in a few cases this will direct our attention to what the genes do. However, we shall see in subsequent chapters that mutant genes are often more precise tools than the embryologist's scalpel, exposing new phenomena.

The embryology that we shall now consider poses three problems in molecular-genetic terms: (1) which genes make products (RNA and protein) that regulate other genes; (2) how is the DNA organized to allow some genes, but not others, to interact with these activator (and repressor) molecules; (3) what genetic mechanisms ensure the proper, orderly, distribution of gene products. Since development is a dynamic, epigenetic process, these general questions have many aspects, and we shall examine them in many different contexts later. For now the question is: what do embryos tell us?

# 1 Development and differentiation

. . .the fundamental problem of biological specificity can be viewed as an embryological one: by what chain of mechanisms does the action of the genes specify a cell or group of cells to become different from others.

Oppenheimer, 1967

Developmental genetics is a relatively new subject: the XIth International Genetics Congress of 1963 was the first to have a session with that name. Of course, geneticists have always found it interesting to ask how the particular abnormalities of mutants come about, how a mutation alters normal development; and they have had some success in answering such questions. In particular, when mutations affecting the relatively simple pigment systems of plants, of the mammalian coat, and of *Drosophila* eyes were studied, they showed that genes were involved at each step of these pathways of biochemical synthesis; a discovery that led more or less directly to the 'one-gene-one-enzyme' hypothesis (Beadle, 1958). Studies of more complex characters were less successful. This 'physiological genetics' (Goldschmidt, 1938) sought to elucidate the mechanisms of dominance, pleiotropy, position effect, gene dosage and gene interaction, and was set in the context of studies of inheritance. Its primary object was to explain genetic phenomena, not development; and we shall ignore much of it for that reason. More important, physiological genetics had little to say about the problems being uncovered concurrently by embryologists. Not only were the two disciplines concerned with different questions, they were using different organisms: the geneticists working mostly with plants, mammals and *Drosophila*, all then difficult to manipulate by the embryologist's microsurgery; and the embryologists using sea urchins, snails and newts whose genetics were not known. Waddington, an embryologist, appreciated the significance of this dichotomy, and his *Organisers and Genes* (1940) attempted to integrate the two subjects on the following argument: "A coherent theory of development cannot be founded on the known properties of genes; in fact, it seems much more hopeful to try to fit our somewhat scanty knowledge of the developmental actions of genes into a framework founded in the first instance on the direct experimental study of development." But this emphasis on the ignorance of both sides had little impact on developmental genetics, or on embryology (Waddington, 1956). Neverthless, work continued on the developmental consequences of mutations and environmental insults, as a means of elucidating gene action by comparing the abnormal with the normal. Only a few laboratories were involved (Wright and Hillman, 1977), but their studies established many of the observations and manipulative techniques which were to be built on subsequently

The situation is now very different. The problems of how the genetic information in the fertilized egg is used to produce a complex multicellular organism, with well-defined temporal and spatial patterns of cell differentiation, have come to the forefront of genetic work. New experimental systems have been developed employing both familiar and new organisms, and new ideas and new methods have been introduced, most deriving directly from the recent advances in molecular biology. We now know more about the embryology of the geneticists' organisms and the molecular biology of the embryologists' organisms, and many ideas concerning the regulation of gene activity are coming from work with prokaryotes and metabolic systems. However, this recent information does not yet add up to an orderly picture showing "by what chain of mechanisms . . . the genes specify a cell" (Oppenheimer, 1967). In fact, we are only at the beginning of the research programme which will fill in the blank spaces in that picture, and a great variety of different experimental systems are being explored with that end in view.

This diversity of approach (from the molecular to the mathematical) and the many different organisms being studied (from moulds to man) make it difficult to follow Waddington's precept and place current genetic findings in their embryological context. As we shall see when we examine some of the experimental systems, genetic studies have also exposed features of development not anticipated by embryology. Neverthless, in this chapter we shall look at some well-established aspects of embryology which define the essential terms of reference within which the genetic findings must eventually take their place. Before doing this we must ask a prior question: is the genome the same and complete in all cells whatever their cytoplasmic differentiation?

## Nuclear totipotence

Because mitotic metaphase chromosomes of different tissues of an organism look the same, it has long been assumed that they have identical genetic complements, and are all totipotent and capable of providing the genetic information for all cell types. The complexity of chromosome organization (Ch. 2) is now known to be too great for this argument from microscopic appearance to carry much weight; but the conclusion may nevertheless be correct. The proper experiment to test totipotence (Spemann, 1938) is to transplant nuclei from differentiated cells into non-nucleated eggs, whose subsequent development should reveal the nature of that nucleus. This technical feat was first accomplished using the frog, *Rana pipiens*, whose eggs can be activated to divide by pricking, prior to enucleation (Briggs and King, 1952). Nuclei from blastulae, gastrulae, and from the anterior mid-gut regions of neurulae transplanted into such prepared eggs gave suggestive, but somewhat ambiguous, initial results. Only a proportion of the eggs developed normally and grew into tadpoles, while the remainder were arrested at the gastrula stage, or later in development. This second class was later found to result from chromosomal deficiencies caused by the inevitable lack of synchrony between the division of some of the transplanted nuclei and the division of the ovum, and was therefore a technical artefact (Briggs *et al.*, 1964).

Gurdon's group (Gurdon, 1974) has since pursued this approach to the limit, using the clawed toad, *Xenopus laevis*. They have successfully transplanted nuclei from cultured adult skin cells, unambiguously identified as containing the skin protein keratin, and genetically marked as heterozygotes for the mutation anucleolate (*nu*) (Fig. 1.1). In this case, none of the first transfer embryos developed into tadpoles, for the technical reason just mentioned. First transfer partial blastulae were therefore used to provide nuclei (serial transplantation) for 11 second generation clones of embryos, of which 6 developed into complete swimming tadpoles, which were *1-nu* diploids. The mitotic descendants of the keratinized skin cell nuclei therefore gave rise to such specialized cells as nerve, muscles, melanophore, blood, lens and so forth of the swimming tadpole. Nuclei of other differentiated cells have similarly been shown to be totipotent (gut epithelium and lymphocytes of *Xenopus*), but fertile adults have been obtained only from nuclei of the tadpole's intestinal epithelium. It is hard to know if this failure to obtain adults from nuclei of adult, differentiated cells is significant, as King (1979) thinks, or reflects only the technical difficulties of this delicate experimental manipulation.

Similar tests have been done using *Drosophila melanogaster*, which has the added advantages of a short life cycle and where both host and donor can be genetically marked. The technical problems are even greater due to the small size of egg and nucleus and, especially, to the rapid division rate of the cleavage nuclei (once every 9 min, according to Zalokar and Erk, 1976). Not surprisingly, nuclear transplants into unfertilized eggs again result in partial blastoderms and defective embryos; but these embryos can be further cultured in the abdomens of adult females and metamorphosed by transplantation into pupating larvae (p. 221), when an array of normal

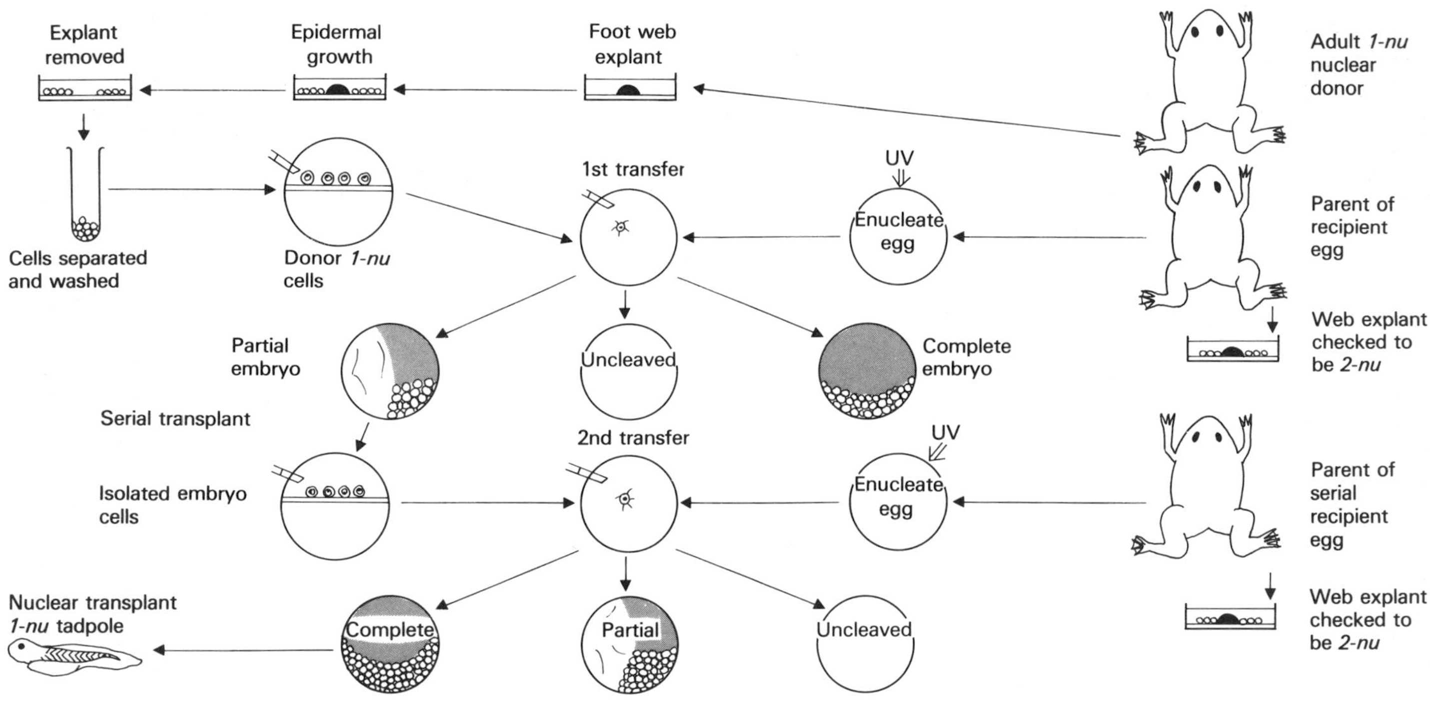
Explant removed
Epidermal growth
Foot web explant
Adult 1-nu nuclear donor
1st transfer
UV
Cells separated and washed
Donor 1-nu cells
Enucleate egg
Parent of recipient egg
Web explant checked to be 2-nu
Partial embryo
Uncleaved
Complete embryo
Serial transplant
2nd transfer
UV
Isolated embryo cells
Enucleate egg
Parent of serial recipient egg
Nuclear transplant 1-nu tadpole
Complete
Partial
Uncleaved
Web explant checked to be 2-nu

adult structures are found (see Illmensee, 1976). In other experiments, genetically marked nuclei transplanted into differently marked fertile eggs result in mosaic organisms made up of the two identifiable kinds of cells. Nuclei from blastoderm, from various parts of the gastrula stage, and from forebrain cells of the 6 h embryo (Zalokar, 1977) are all capable of giving rise to adult structures, and in a few cases they are incorporated into the gonads and become progenitors of a further, normal generation. Nuclei from some cultured cell lines, derived from embryos, are also capable of forming mosaic areas in larvae and in adults (Illmensee, *loc. cit.*), and not only in one tissue but in a variety (gut, fat body, Malpighian tubule, thorax, abdomen etc.). Nuclei from adult cells have not yet been successfully transplanted. The *Drosophila* work therefore corroborates the finding with frogs, and together they strongly suggest that most, if not all, nuclei are genetically intact (see Di Barardino, 1980), but we await a final proof of this proposition.

Experiments with plants have taken a different direction due to the difficulty of working with the seed. Steward (1970) showed that individual carrot cells are capable of forming complete plants when properly cultured (Fig. 1.2). And this procedure of making plants from cells has been successfully used for improving a great variety of crop and ornamental plants (Reinert and Bajaj, 1977,) but not all. Of course this does not prove the totipotence of all cells, only of some.

Nuclear totipotence is not universal. Boveri (1887) was the first to show that chromatin was eliminated from the presumptive somatic cells, but not from the germline, during embryogenesis of the nematode, *Ascaris*. In the common gut-worm, *Ascaris lumbricoides,* 27 percent of the DNA is eliminated from the somatic cells (Tobler *et al.*, 1972), and this removes DNA sequences found in sperm. A similar, but more dramatic, elimination has been reported for some species of Cecodomydae (gall midges). In this case

---

**Fig. 1.1** Serial transfer of nuclei from *Xenopus* adult skin cells into enucleate eggs, after Gurdon *et al.* (1975). Skin cells are obtained by culturing an explant from the foot web of an adult which is genetically marked as heterozygous for the anucleolate mutation (*1-nu*). After growth the cells are confirmed as adult skin—99 percent contain keratin. These cells are separated by trypsinization and washed prior to injection of their nuclei into eggs from normal *2-nu* females. The eggs are 'enucleated' by ultraviolet (UV) irradiation prior to the nuclear transplant. Seventy percent of the eggs fail to cleave, and 25 percent show partial cleavage. These latter are used for a second, serial nuclear transfer since earlier work shows that nuclei from partial embryos generally support more normal development than those from completely cleaved embryos. So the cells from partial embryos are dissociated and their nuclei transplanted into eggs from normal females, giving clones from each initial partial embryo. Thirty percent of these eggs give complete cleavage and about a third of the clones give rise to viable tadpoles, which are *1-nu* diploids. None of these tadpoles metamorphoses successfully and we can conclude only that adult skin cells are pluripotent. The reasons for this, and other similar results, are discussed by Di Barardino (1980).

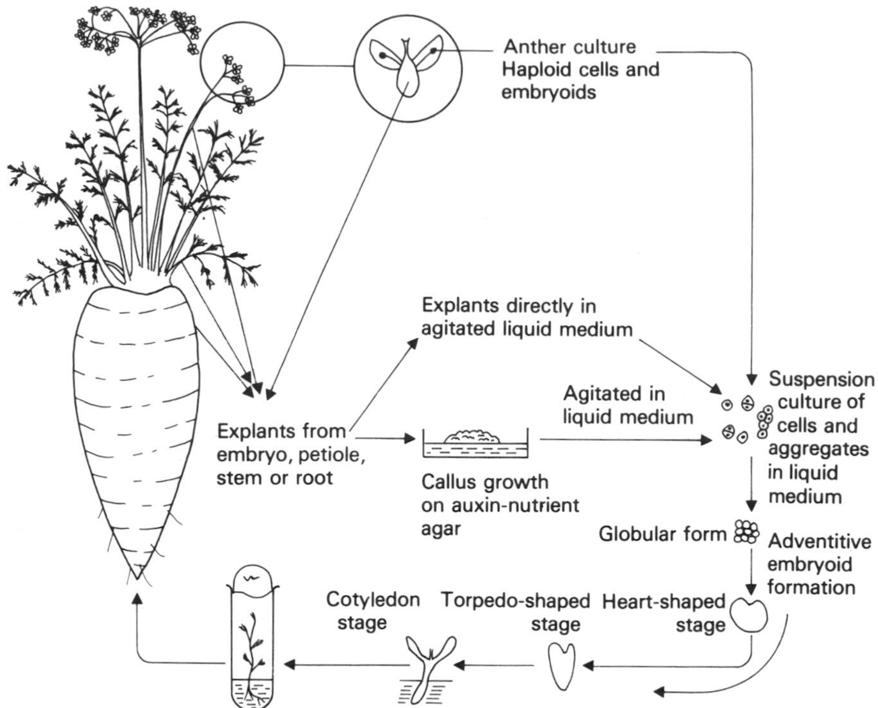


**Fig. 1.2** Regeneration of the carrot plant from tissue explants, according to the work of Steward and others. Explants from embryos, stem, petiole or root, or of pollen grains (anther culture), may be the starting material. The explants may be grown in liquid culture (nutrient medium with auxin) which is agitated to give a cell suspension, or the explant may be grown in nutrient agar with auxin to form an undifferentiated (usually friable) callus, which is then transferred to agitated liquid medium. The cell suspension medium contains nutrients and hormones, and the original experiments exploited the liquid endosperm of coconut milk as a source of undefined growth factors. The embryoids which differentiate are transferred to agar medium with appropriate hormone supplements and grow into proper plantlets, and then into normal plants. These plants flower and set seed; and the plant-cell-plant cycle can be repeated as often as desired. Whatever the source of the explanted cells, they carry the entire genome of the plant.

the full complement of 40 chromosomes is retained in the germ cell nuclei, but only 8 of them in the somatic cell nuclei. By removing the 32 chromosomes from all cells (Fig. 1.3), sterile, but otherwise normal, adults can be produced, suggesting strongly that the full genetic complement is necessary only for gametogenesis. Both kinds of example suggest that elimination of DNA, where it occurs in one fashion or another, will create a difference just between gonad and somatic cells. But at present we cannot discount other, and more subtle, mechanisms for developmentally organized gene loss.

Conversely, there is also evidence of gene amplification, again as a special case. The genes coding for ribosomal RNA (rDNA) of some vertebrates,

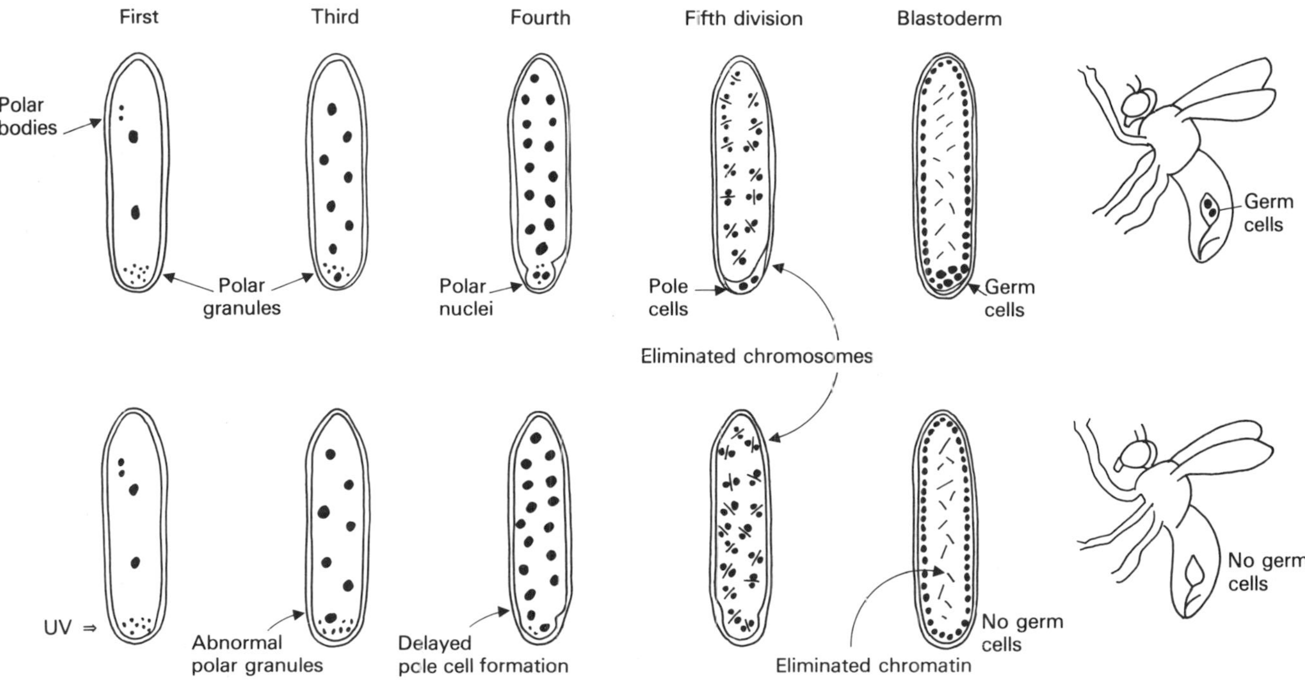


**Fig. 1.3** Normal germ cell formation in the gall midge *Mayetiola* (upper row) and its failure after irradiation of polar granules (lower row). The syncytial nuclei associated with the polar granules during the fourth normal division are protected from the chromosome elimination which occurs during the fifth division, but if the polar granules are damaged by UV irradiation (lower row) pole cell formation is delayed and chromosome elimination occurs as with the somatic cells. The UV irradiated egg results in otherwise normal adults lacking germ cells. The blastoderm first forms as a syncytium and the polar nuclei are the first to become cellularized. After Fischberg and Blackler (1961).

invertebrates and protozoa are found to be selectively amplified as extrachromosomal DNA sequences; for instance, in the oocyte of *Xenopus laevis*. These sequences arise very early in oogenesis, but not in spermatogenesis, and appear necessary to meet the heavy demands made on protein synthesis during early embryogenesis (see Bird, 1980). Again, the puffs in the salivary gland chromosomes of sciarid insects result from increased DNA synthesis, probably within the chromosomes themselves (Pavan and DaCunha, 1969).

Gene amplification and chromosome loss suggest that we must remain open-minded about the kinds of mechanisms involved in cell differentiation. There is another reason for this attitude, too: the repertoire of differentiated cells varies greatly among organisms. There are two cell types in the slime mould (*Dictyostelium discoideum*), a handful or so in *Hydra*, around two dozen in a plant, and up to about 200 in man. Depending on their functional specialization, the structural differences between these cells may be small, or they may cover a phenotypic range as great as that found within the phylum Protozoa. In view of this diversity (and of our lack of information about most organisms), we cannot conclude that all nuclei are totipotent. We can say from these transplantation tests that nuclei are at least pluripotent, and carry the information for many cell types. Cells must then differentiate by expressing some genes and not others.

## Developmental gene regulation

The classic example of gene regulation during development is the sequence of expression of L-lactate dehydrogenase (LDH) enzymes during embryogenesis and subsequent growth of the mouse (Fig. 1.4). Three separate loci code for LDH enzyme subunits, one type of which (A) is functional in heart muscle 9 days before birth. It is replaced by the B type in the adult heart. A is then found in, and is typical of, adult body muscle. During spermatogenesis, both A and B are found in spermatogonia, but a third gene (C) is uniquely expressed in the primary spermatocytes derived from them. This metabolic enzyme shows that gene regulation is a feature of development and that, as a consequence, different tissues may express different genes. There are many examples which show that one can generalize from this case (see Markert, 1975) and we shall note some in the following pages.

An even more important example is the proof that nuclei from cultured kidney cells of the toad, *Xenopus*, are reprogrammed by exposure to oocyte cytoplasm, in this case by being injected into enucleate oocytes of the newt, *Pleurodeles* (DeRobertis and Gurdon, 1977). Three days after transplantation, proteins specifically formed by the normal *Xenopus* oocyte are synthesized, and the proteins found in the cultured kidney cells are no longer made, as judged by two-dimensional gel electrophoresis. If nuclear

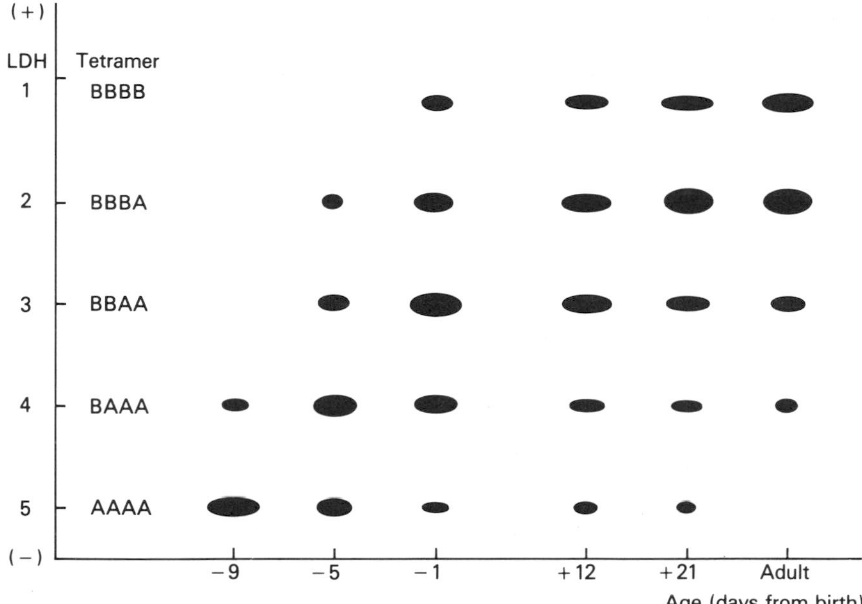


**Fig. 1.4** The changing isozyme pattern of L-lactate dehydrogenase (LDH) in heart muscle of the developing and adult mouse (after Markert and Ursprung, 1962). The enzyme is a tetramer formed by combinations of two subunits, A and B, each coded by a separate gene. The different tetramers can be readily separated by electrophoresis, and when this is done the embryonic heart (9 days before birth) is found to contain predominantly the LDH 5 combination; the adult heart contains mainly LDH 1 and 2. The data show the progressive change in gene expression during development. There is a specific foetal heart LDH in humans, and considerable variation in the presence or absence of these multiple loci among vertebrates generally. In some older papers A is designated as M (for muscle) and B as H (for heart).

transcription is blocked with the drug α-amanitin, the oocyte proteins are not found, showing that transcription (not translation) has been reprogrammed by specific, but unknown, gene-activating substances present in the egg cytoplasm. The previously differentiated nucleus is capable of responding to these substances, and the chromosomes still carry the relevant genes. Since the egg does not divide, this reprogramming does not require nuclear division, which has sometimes been postulated as a necessary mechanism for getting 'clean' genes.

One of our problems is identifying the particular genes which specify a cell type (sometimes called 'luxury' genes, as opposed to 'housekeeping' genes which are required for the general metabolic processes of all cells). This is more complex than is sometimes appreciated. For example, a HeLa (human fibrocarcinoma) cell contains 35,000 mRNAs, and a *Drosophila* cell about 4000 (Bishop *et al.*, 1974); specifying many more gene products than have yet been identified by biochemists. So we simplify the problem by looking

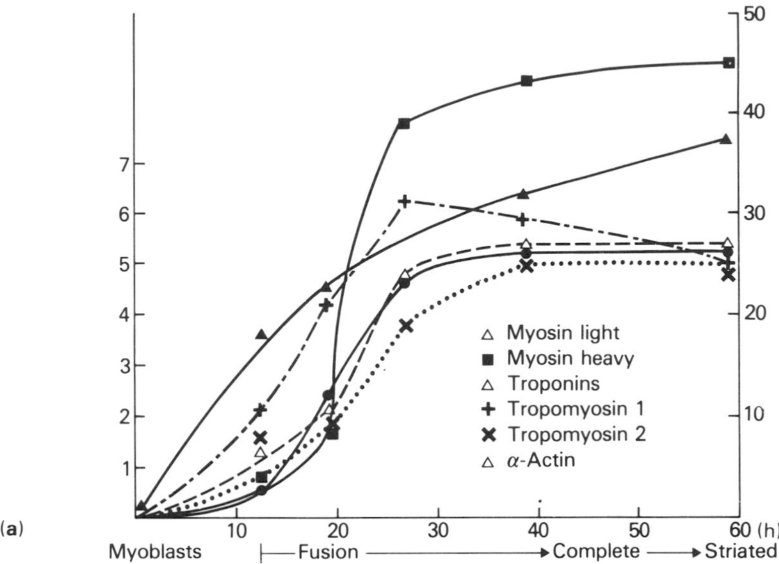


**Fig. 1.5** Regulation of protein synthesis during myoblast differentiation, according to Devlin and Emerson (1978). New protein synthesis is assessed by pulse-labelling quail myoblasts at the indicated times after transfer into culture media lacking growth factors. Fusion starts 12 h later and is completed by 40 h, and the myotubes are fully striated by 60 h. The scales measure the rate of new protein synthesis (uptake of radioactivity on the left and, for myosin heavy chain and α- and β-actin, by measuring specific immunoprecipitations, on the right). (**a**) shows that the synthesis of the major muscle proteins is initiated at fusion and increases rapidly in a highly coordinated fashion with about the same kinetics for each. Muscle-specific

at the proteins which are characteristic of the cell type: for example myosin, actin, tropomyosin, tropin, the α-and β-actinins, creatine kinase and phosphorylase *a* in the case of muscle. However, description of the time sequence of the appearance of these (or other relevant) proteins tells us only when they are translated, or cease being translated, as Fig. 1.5 shows. It tells us nothing about the mechanisms which regulate gene action in the orderly fashion proper to embryogenesis. This is the main problem, and we must now ask what experimental embryology tells us about it.

## Embryonic development

### *Mosaic and regulative development*

To begin at the beginning. Most fertilized eggs are enclosed by a membrane, jelly coat or shell which separates them from the environment, except for the exchange of gas and water. All the developmental information is there, enclosed. The mitotic products of the zygote would then necessarily be as

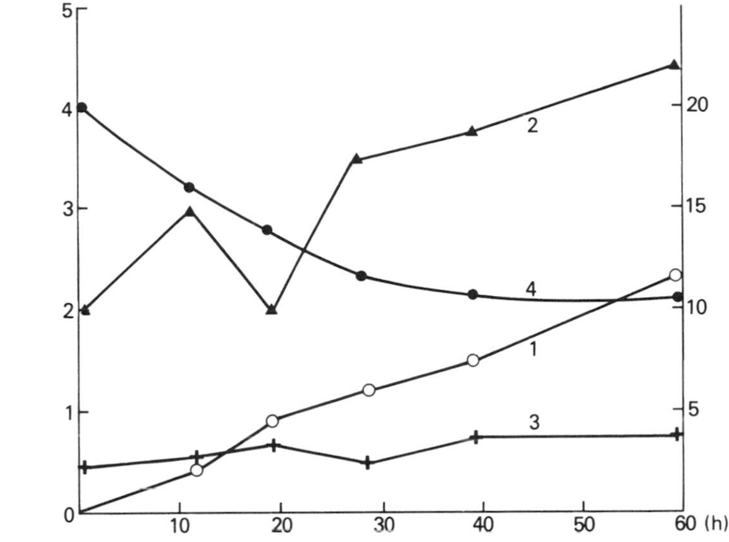


α-actin increases at about three times the rate of the others. (**b**) shows a sample of the changes which are found in 30 other proteins. The first class (1) also increases after fusion but only at about a hundredth of the rates in (**a**). The rate of synthesis of the second class (2) also increases after fusion, but its members are also being synthesized by the dividing myoblasts. The third class (3) is synthesized at a low, and relatively constant, rate, and the fourth class (4) is typified by β-actin (non-muscle) whose rate of synthesis declines after fusion. These data show that there are regulatory processes other than the activation of contractile protein synthesis, and it is still uncertain if these involve control of transcription or of later events.

identical as members of a cell clone unless the egg is in some way heterogeneous, or unless differences subsequently arise between cells because their relationships to one another change as they multiply.

Differences between the initial division products (blastomeres) of an egg can be tested by finding if the individual blastomeres are capable of forming the whole organism when separately cultured; that is, by finding if the cells are totipotent. In many instances, totipotence is lost at the first division (e.g. Fig. 1.6); in other cases, both division products are totipotent (e.g. frog), or the first four blastomeres may each be capable of making a complete, but small, embryo (e.g. sea urchin). But after these early divisions, blastomere totipotence is usually lost when the cytoplasm is no longer symmetrically apportioned (Wilson, 1925). Mammalian eggs (and turbellarian eggs) are the exceptions which prove this rule since, at least in mouse and rabbit, cells from the 8-cell stage can contribute to both the inner cell mass (which forms the embryo proper) or to the extraembryonic trophoblast (which will attach the embryo to the uterus), as Hillman *et al.* (1972) have shown. In this case, only the cells on the outside of the morula, or solid ball of cells

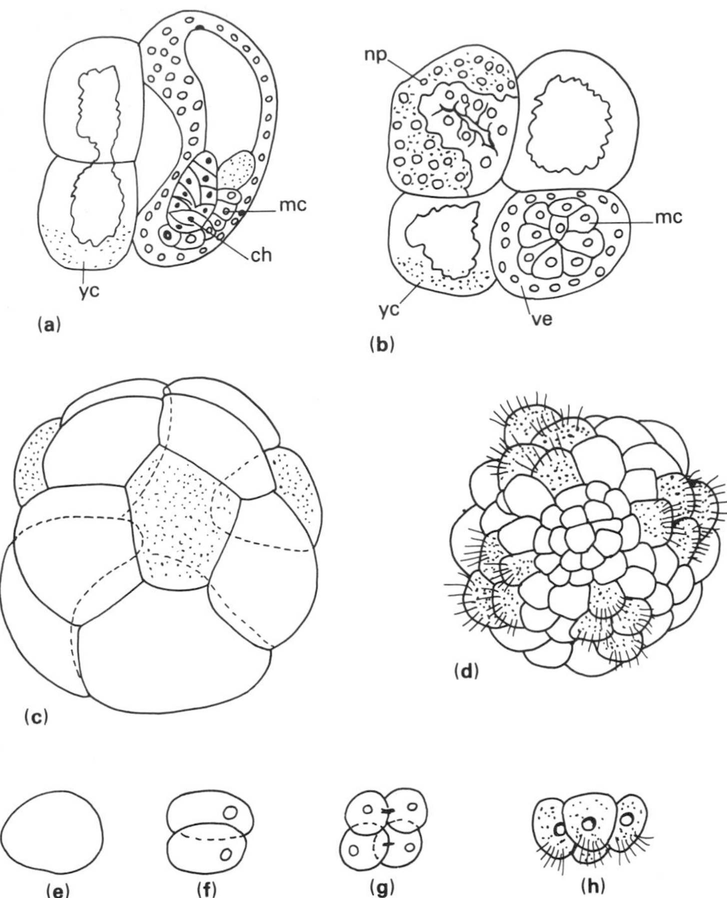


**Fig. 1.6** (**a**), (**b**), Partial embryos of the tunicate *Styela partita* prepared by ejecting 4-cell stage embryos through a pipette (Conklin, 1905). In (**a**), both cells on one side are damaged and do not develop and the remnants of the yellow crescent (yc) accumulate in the lower cell. The living half of the embyro continued to differentiate, and chorda (ch) and muscle cells (mc) could be identified. In (**b**), diagonal cells were destroyed and the others continued to divide and differentiate to form neural plate (np), muscle (mc) and ventral endoderm (ve). The 16-cell stage (**c**) of the mollusc *Patella coerulea* differentiates primary trochoblasts (shaded) which at the 10 h ctenophore stage become ciliated after division into 4 cells in each group (**d**). Wilson (1904) isolated primary trochoblasts by exposing the embryos to Ca-free sea water (**e**) and followed their division (**f**) and (**g**) and differentiation (**h**) as an independent group of ciliated cells. As Wilson concluded: "The history of these cells gives indubitable evidence that they possess within themselves all the factors that determine the form and rhythm of cleavage, and the complex differentiation they undergo, wholly independently of their relation to the rest of the embryo".

formed by the blastomeres, make the trophoblast; and their differentiation is determined by an unequal division into large polar peripheral cells and apolar central cells (Johnson and Ziomek, 1981). The inner cell mass (ICM) of later blastocysts fails to differentiate as trophoblast when implanted in a uterus, although ICMs from early blastocysts do (Rossant and Lis, 1979). The polarized division causes the cells to lose their totipotence, and the inner cells then follow the pathway of embryonic development.

We have then, a hierarchy of kinds of egg as judged by the loss of totipotence of blastomeres. At one extreme, the developmental fate of the initial pair of cells is programmed in different directions as a consequence of some heterogeneity within the egg itself. At the other extreme, differences between cells are established later and depend, apparently, on the relationships between cells. Provided there is communication between cells, any such early cellular difference may generate further differences, at least in principle. So we have three issues to look at in greater detail. First, what is packaged in eggs so that its unequal distribution to the blastomeres results in differences in their development? Second, are there a number of different substances which are selectively partitioned during many subsequent divisions to produce different cell types? Or, third, is development largely epigenetic, and thus dependent on interactions between cells, either early and/or late in embryogenesis? Of course, the facts discussed in the previous paragraph imply that we may get different (and unfortunately incomplete) answers to these questions when we consider different organisms.

## Mosaic development

Nematode worms have been classic subjects for the study of cell lineages (Boveri, 1899) and one species, the small (1 mm), free living *Caenorhabditis elegans*, has recently been developed for genetic work too (Brenner, 1974). Its embryonic lineage, which is invariant, has been described up to the 182-cell stage (Deppe *et al.*, 1978), which is only a little more than a cell cycle short of the 550 cells of the hatched juvenile worm. The first, unequal, divisions separate five somatic blast cells and the germline precursor, as shown in Fig. 1.7. Each of the somatic blast cells (AB,C,D,E and MSt of the figure) subsequently divides equally and synchronously, but at different rates for each precursor lineage, to form the equivalent of the germ layer tissues; ectoderm, endoderm, mesoderm and stomodeum, and germ cells, as indicated. The developmental fate of the cells depends, therefore, on the initial distribution of the cellular material of the egg to the half-dozen blast cells, each of which is different.

All of the 32 gut cells of the worm contain a fluorescent pigment, cytoplasmic 'rhabditin granules' derived from tryptophan, and these are first apparent when the E blastomere has divided to form 8–16 gut precursor cells at 4–6 h. If a cellular egg is broken so that the freed cells can multiply in a culture medium, fluorescent gut cells differentiate and twitching muscle is seen. The cells follow their determined fate although no longer having their normal contacts with other cells (Laufer *et al.*, 1980). These primary culture cells can have their cleavage blocked with cytochalasin B, and their nuclear division reduced with colchicine. The original 2, 4 or 8 cells, depending on starting age, then remain undivided, although they eventually contain many genome equivalents of DNA per blastomere. Some of these cells contain fluorescent rhabditin granules, and it can easily be shown that P1 cells fluoresce while ABs do not, that EMSt cells fluoresce and P2s do not, and that Es fluoresce while MSt cells do not. Cleavage-blocked embryos

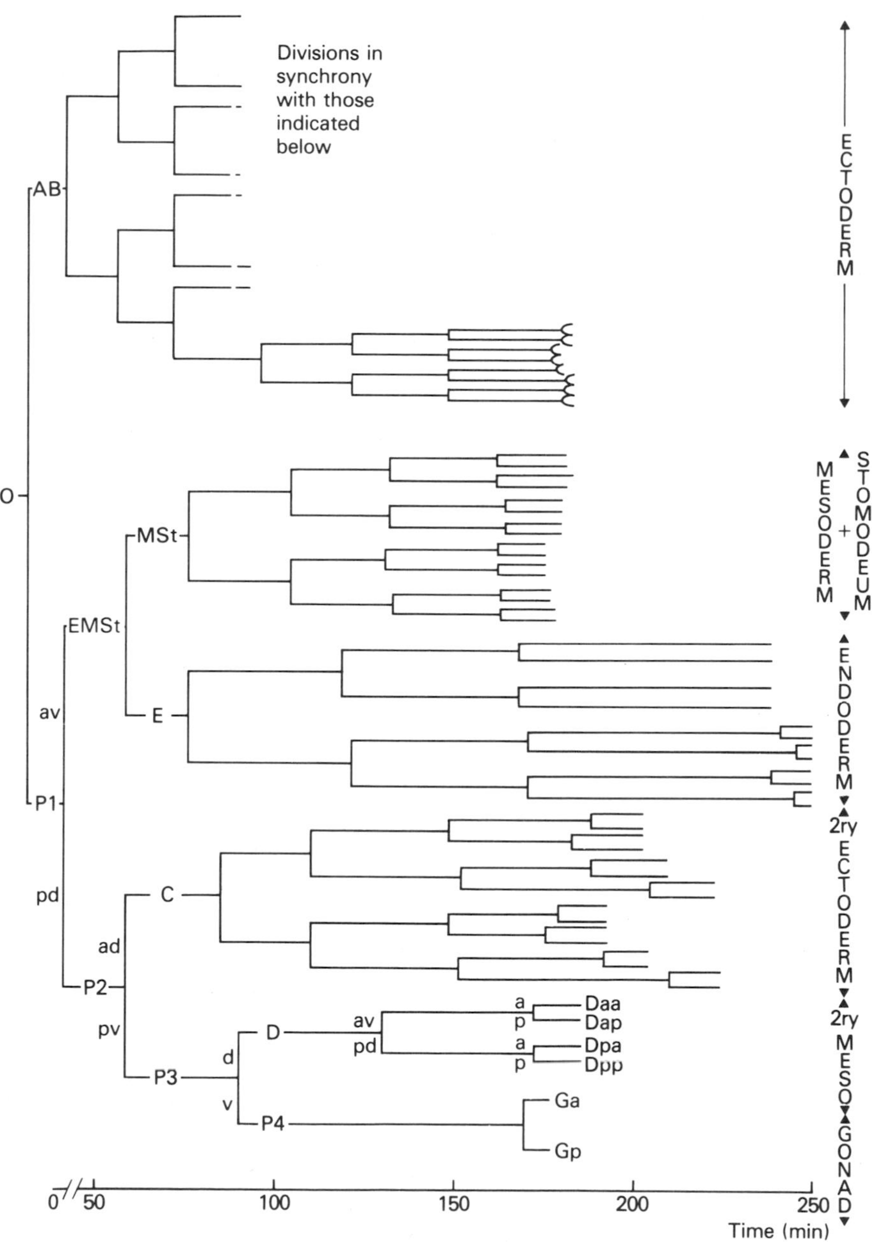

demonstrate the "asymmetric positioning of cellular components during the blast cell-generating divisions" (Laufer *et al.*, *loc. cit.*), and they also show that cell–cell interactions, and cell division, are unnecessary for this characteristic gut cell differentiation.

We do not yet know if the determinants for blast cell differentiation are activators of specific gene transcription, or maternal mRNAs which are translated after segregation, or activators of specific embryonic nuclear mRNA translation. We shall have to consider another system for information about this question. However, embryonic arrest mutants have been identified in *C. elegans* strains (Miwa *et al.*, 1980), and most show maternal effects. That is, mothers homozygous for the mutations form eggs that are incapable of normal differentiation even when the embryo is heterozygous for the gene: the female therefore places substances essential for embryonic development in the eggs she forms. Unfortunately, none of the mutants described so far affects the gut cells we have considered (except for regulation of their multiplication).

The classical material used for studies of cytoplasmic localization is the embryo of the ascidian, *Cynthia* (*Styela*), where there is visible segregation of cytoplasmic pigments to the blastomeres which can be followed through to the mesenchyme, muscle cells etc. (Conklin, 1905). Again we have a strictly determinate cleavage pattern, and an invariant cell lineage which allocates particular regions of the cytoplasm to particular tissue lines (Fig. 1.8). If particular blastomeres are removed the expected tissue is not formed. For example, destruction of the B4.1 pair results in loss of larval muscle tissue: and, conversely, the B4.1 pair divide and differentiate into larval muscle if cultured apart from the embryo (Whittaker *et al.*, 1977), Despite the many cytoplasmic movements following fertilization, the egg is a mosaic of localized tissue determinants, and one can draw a 'fate map' on the egg surface, showing which tissue a region will give rise to.

Three tissue-specific enzymes have been studied in a related ascidian, *Ciona intestinalis*, and indeed, the technique of cleavage arrest with cytochalasin B was first used on this species (Whittaker, 1973). The three enzymes are acetylcholinesterase (AChE), which is histospecific for larval

**Fig. 1.7** The invariant cell lineages of *Caenorhabditis elegans*, according to Deppe *et al.* (1978). The first zygote division uniquely separates the embryo into an anterior (AB) and a posterior (P1) cell. The AB cell is the stem cell of the primary ectoderm (128 cells): only one example of the lineage is shown for simplicity, and the other seven clones divide in synchrony with it. Five further stem lines arise during the next divisions of the P1 lineage, and the smaller of these unequal divisions (P2–P4) gives rise to the germline (G). The other P1 product forms the EMSt cell which divides into the separate stem cells for endoderm (E) and for mesoderm and stomodeum (MSt). The P2 and P3 divisions form the stem lines for secondary ectoderm (C) and for secondary mesoderm (D). Each stem line divides at a different, characteristic rate. The entire embryo has about 550 cells at hatching. Each cell can be defined by its division pattern, shown here only for the D cell line, where a/p is an anterior/posterior division and d/v is dorso/ventral division.

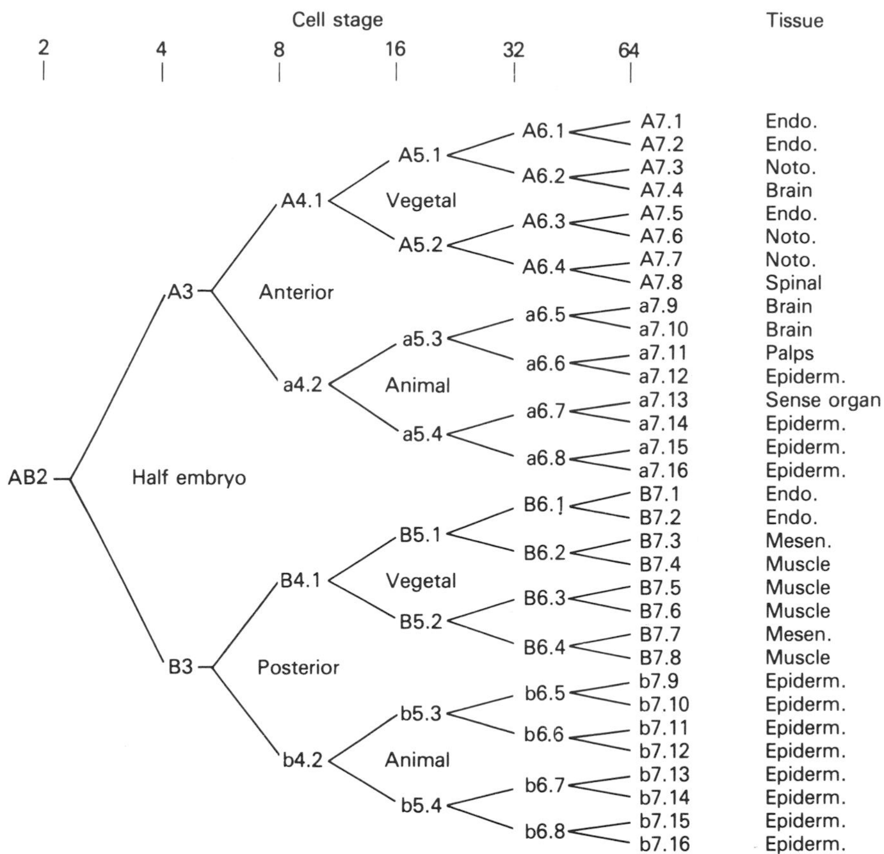


**Fig. 1.8** The first division of the egg of the ascidian, *Styela* (now called *Cynthia*) forms the two halves of the embryo (the CD half is not shown). This is followed by a further vertical division separating each into anterior and posterior cells, and then by an equatorial division separating the upper (animal pole) cells from the lower vegetal cells. Pigmented materials are segregated in the egg cytoplasm shortly after fertilization, and Conklin (1905) showed that these were partitioned during cell division such that the tissue formed depended on the material they contained, a process completed by the 64-cell stage. A transparent cytoplasm went only to ectodermal cells, a light yellow to the coelomic mesoderm, dark yellow to the tail muscles, light grey to the notochord and neural plate, and a dark grey pigment to endoderm cells. This is a different pattern of cell inheritance from Fig. 1.7, but is similarly determinate.

muscle, alkaline phosphatase localized in endoderm, and tyrosinase which is associated with two giant, black-melanocytes in the brain. As before, blocking cleavage at different stages and allowing the non-dividing embryos to age, shows that the partitioning of determinants proceeds along the expected cell lineage pathway, through cytoplasmic segregation. If mRNA synthesis is blocked by exposing embryos of different ages to actinomycin

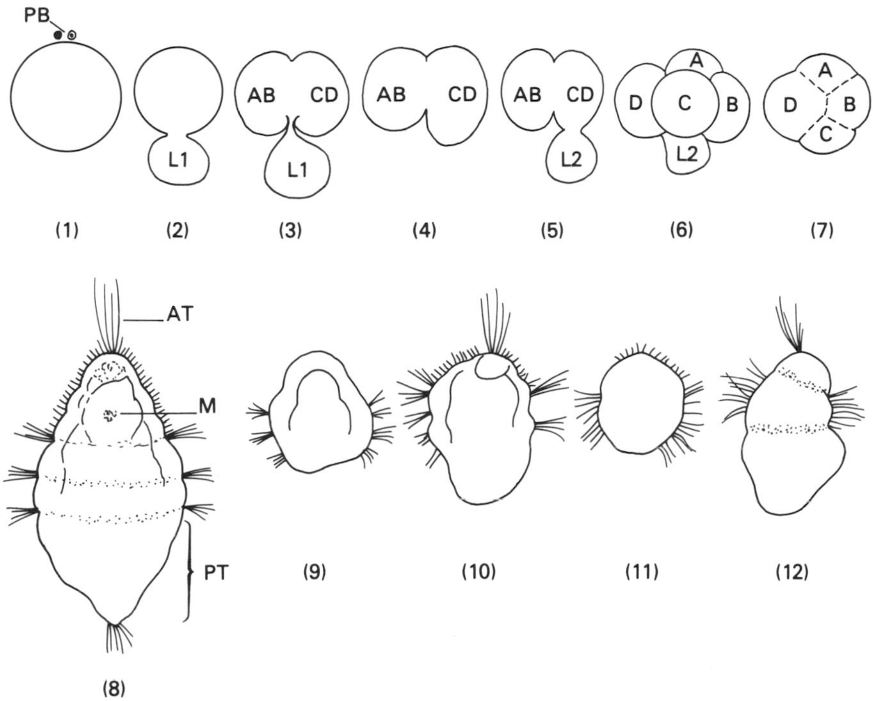


**Fig. 1.9** Early cleavage and fate of the first blastomeres of the mollusc, *Dentalium*, after Wilson (1904). The mature egg (1) with polar bodies (PB) extrudes cytoplasm from the vegetal pole (2) as a polar lobe (L1), prior to cleavage (3). This is the trefoil stage, and the two blastomeres, are named AB and CD by convention. The lobe material is predetermined to flow back into the CD blastomere to give an unequal cleavage (4). The second cleavage (5–6) also involves polar lobe formation (L2) and this lobe flows back into the D cell again to give an unequal division. The AB cell divides to give two cells similar in size to C (7). The normal trochophore larva (8) has an apical tuft (AT) and a ring of cilia around the mid-region, behind the mouth (M) and below this a post-trochal region (PT). Differentiation continues as a free-living form. Isolated blastomeres continue to cleave, the AB blastomere (lacking L1) makes no post-trochal structures and lacks the apical tuft (9). The CD blastomere (10) has a relatively large post-trochal region and a proportionately small pre-trochal region, but is otherwise 'normal'. Comparison of the C(11) and D(12) blastomere development shows that the lobe material is necessary for apical tuft formation and for the post-trochal apparatus. This early example of the cytoplasmic determination of morphogenetic potential is considered further in the text.

D, the alkaline phosphatase differentiation is not affected at any age. AChE synthesis is blocked at 5–6 h of embryogenesis, and tyrosinase synthesis at 6–7 h (Whittaker, 1977; 1979). These results argue that alkaline phosphatase synthesis depends on preformed mRNA passed into the egg from the mother, while the AChE and tyrosinase mRNAs are synthesized by the embryo genome. The AChE and tyrosinase determinants, whatever they may be, must be specific gene activators, as Morgan (1934) postulated long ago. And they must also be deposited in the egg by the mother, like the alkaline phosphatase mRNA.

Evidence from molluscs suggests that determinants may sometimes be associated with the egg cortex. In *Dentalium*, the material of the vegetal pole of the egg is shunted into a large polar lobe (Fig. 1.9) which is associated with the CD blastomere after the first division. A lobe is again formed at the second cleavage, and is partitioned to the D blastomere of the 4-cell stage, which becomes the dorsal quadrant of the embryo. Removal of the polar lobe has a dramatic effect on development: the adult shell, foot, operculum, statocysts, eyes, tentacles and heart are absent. Not all these structures originate from the D quadrant, which must be indirectly involved in the differentiation of other cells (Cather, 1971), but the polar lobe must carry the determinants for those that are (heart, intestine etc.). If the cytoplasm that might carry these determinants is sucked out from the polar lobe, development proceeds normally. Conversely, if the contents of the second polar lobe are injected into a B blastomere, development is also normal (van den Biggelaar, reported in Dohmen and Verdonk, 1979). Yet if the first cleavage is equalized by treatment with cytochalasin B, lobe dependent structures are duplicated in the embryo. The obvious conclusion is that the determinants are not free in the cytoplasm but are bound to, or are incorporated in, the plasma membrane of the vegetal hemisphere of the egg. What these cortical localizations might be is quite unknown.

On the basis of these manipulations of mosaic eggs, we can briefly answer the questions raised earlier (p. 13). There are, indeed, substances made during ovogenesis and laid down in the egg which are selectively partitioned among the blastomeres and which determine their differentiation. These substances include mRNAs (maternal message) which are subsequently translated in some cells only, and unidentified activator substances which cause specific gene transcription in the cells which carry them. Some of these activators may be associated with, or be part of, the egg cortex. Perhaps the most striking proof of this is that when zygotes of the annelid, *Chaetopterus*, are treated with isotonic KCl they differentiate without cell division, become ciliated and have the form of the trochophore larva (Lillie, 1902). Differentiation is independent of nuclear cleavage, but dependent on the topographical distribution of cytoplasmic 'determinants' in the egg. It should be possible to find mutants, maternal-effect mutants, which alter both the character and distribution of such determinants.

## Regulative development

Sea urchin eggs (e.g. *Paracentrotus lividus*) show mosaic characteristics, but are also capable of regulation. The first two meridional divisions give totipotent blastomeres, but the third equatorial division does not. Separation of the upper quartette (animal hemisphere) results in their differentiation into an ectodermal vesicle after division, whereas the lower quartette (vegetal hemisphere) forms a large endodermal gut, lacking a mouth (Horstadius, 1935). Cells of the two hemispheres are therefore fated to develop in separate ways at this 8-cell stage, due to the unequal partition of the egg components.

At the 16-cell stage, eight mesomeres occupy the animal hemisphere, and four macromeres and below them four micromeres, make up the vegetal hemisphere, and these easily identified groups can be separated and recombined. The animal (mesomeres) and vegetal halves (macro- plus micromeres) develop and differentiate, as just described. However, mesomeres plus macromeres produce a virtually complete larva, with mesenchyme and skeleton which normally derive from the micromeres (Fig. 1.10). Mesomeres plus micromeres can also develop into a nearly normal larva; although this combination lacks the macromeres which should provide the endoderm. The last two combinations are therefore capable of regulating their development by reprogramming cells to function like micromeres, or macromeres, respectively. Thus: (1) the cytoplasmic substances which determine this simple differentiation do not do so directly, otherwise mesenchyme could not be formed without micromeres and endoderm could not develop without macromeres; (2) regulation must depend on the cells being capable of differentiating in a number of ways at this stage, and on the mass of cells re-establishing the normal pattern of development as they multiply. In this particular example, there is evidence that the pattern depends on two chemical gradients, one centred on the animal pole of the egg and the other on the vegetal pole (Runnström, 1928); but the substances involved are not known. Only combinations which have cells from both the animal and vegetal hemispheres would be expected, on this assumption, to form complete larvae—as is the case.

RNA, and that mostly maternal RNA, is not distributed homogeneously between the mesomeres, macromeres and micromeres (Rodgers and Gross, 1978), but attempts to find the RNAs responsible for the different characteristics of these cell groups have been frustrated by their number and complexity. At present there is therefore no evidence that regional localization of sets of maternal messages can account either for the gradient system or for the way the cells respond to any gradient (Ernst *et al.*, 1980). If echinoderm differentiation is not dependent on the unequal distribution of qualitatively different determinants to parts of the egg, we are confronted with another kind of gene regulation. Cell responses would then result from different reactions to the amount of some substance (or substances) present in a gradient throughout the egg. Such a system is compatible with regulation following cell removal (see Wolpert, 1969). In short, we cannot always assume the action of specific determinants: the cells may sometimes respond, and respond differently, to the amount of some substance passed from cell to cell.

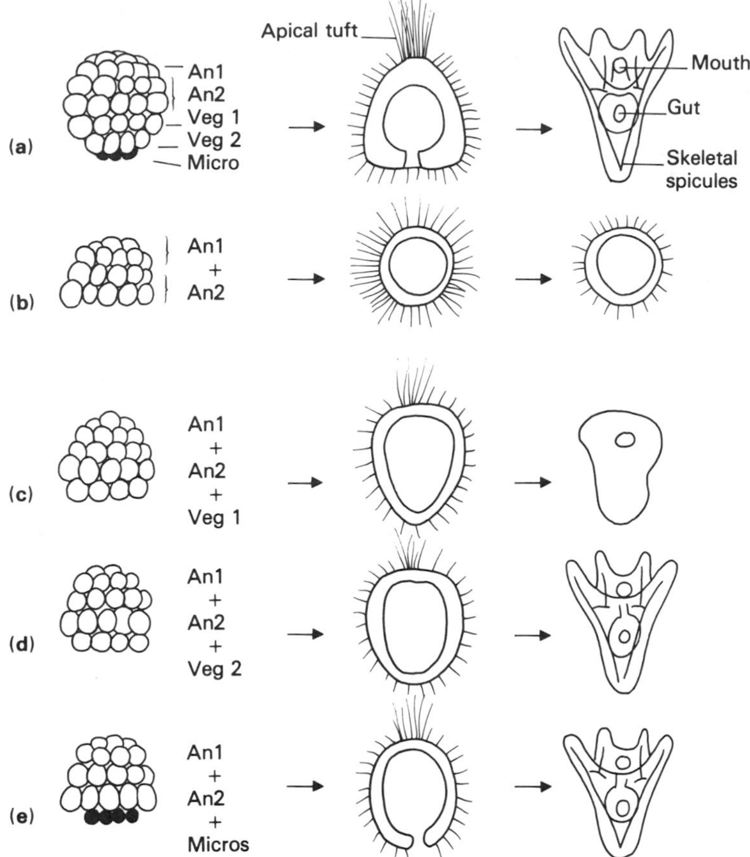


**Fig. 1.10** The development of *Paracentrotus lividus*, and of combinations of mesomeres with macromeres and micromeres, after Horstadius (1939). The normal embryo (**a**), forms a ciliated gastrula with an apical tuft which grows into a free-living pluteus larva with a mouth, oesophagus, stomach and intestine, having its form supported by a skeleton of calcareous spicules. By following the cells marked by dyes, the animal 1 mesomeres (An1) are found to form the apical cilia, some of the epithelium and the mouth. The animal 2 mesomeres form epithelia as do the descendants of the vegetal 1 macromeres (Veg1). The vegetal 2 macromeres (Veg2) form the gut system and the archenteron; and the micromeres (Micro) make the spicule system which supports the larva. The An1 + An2 cells alone form only a ciliated ball of cells, (**b**), with the tuft spreading widely over the surface. The same happens when an egg is physically divided into two and the animal half allowed to differentiate after it is fertilized. This might suggest that the egg is a mosaic. However, when Veg1 cells are added to the animal group, (**c**), the apical tuft is restricted to its normal position and size, although a proper pluteus larva does not form. On the other hand, with added Veg2 cells (**d**), development is more or less normal and skeletal spicules are laid down, although often incompletely. Thus Veg2 cells form skeleton which they never do during normal development. Again, (**e**), the addition of micromeres to the animal cells is sufficient to permit harmonious development of the larva, even when micromeres are combined with only An1 or only An2 cells. The egg is therefore regulative, not mosaic.

### *Induction*

Diffusible molecules have a major role in the differentiation of the amphibian egg. The animal–vegetal axis of the egg has a dorso–ventral polarity superimposed on it after fertilization. Opposite the point of sperm entry, and following movements of the cortical cytoplasm, a 'grey crescent' area appears subequatorially, to give the egg a bilateral symmetry. The grey crescent becomes the site of blastopore formation at the beginning of gastrulation. Cells are totipotent up to this early gastrula stage: prospective epidermis (as determined from the fate map) may become endoderm or mesoderm when transplanted into the appropriate region of the embryo (Mangold, 1923). The cells regulate to conform to their position. This situation changes after gastrulation. If cells from the dorsal lip of the blastopore (e.g. of the newt, *Triturus*) are transplanted to the prospective belly ectoderm region of another gastrula, a secondary set of axial structures (neural tube, notochord, somites) is formed (Spemann and Mangold, 1924). The blastopore lip is the *primary organizer* region which induces embryonic development, and equivalent regions have been identified in many vertebrates, from *Amphioxus* to Aves. A heat-killed organizer region still functions in transplants, and culture medium in which blastopore cells have been kept will cause small fragments of ectoderm to differentiate into nerve (Niu, 1956). The inducing substance, the inductor, is clearly a chemical molecule, and Løvtrup *et al.* (1978) have evidence which suggests it is heparan sulphate. Cells from the animal hemisphere of late blastulae of *Xenopus* do not differentiate in culture, but in the presence of heparan they become mesenchyme cells, nerve cells etc. Further, the cells have to be of the right developmental age to respond to the heparan, as is true for experimental inductions in whole embryos (Holtfreter and Hamburger, 1956). So primary induction in amphibian embryos probably depends on the synthesis of a simple chemical by a small group of localized cells and its interaction with other cells competent to respond by differentiating. We do not know how the secreting cells arrive at that state, or what the condition of competence implies for the molecular organization of responding cells.

Many secondary, tertiary, and higher level inductions have been identified during vertebrate embryogenesis, but we shall look at only two examples. The first concerns the induction of the balancers found in most newts and salamanders. These are tentacle-like outgrowths which lie behind the mouth angle, below the eye; and they are often induced when the blastopore lip is transplanted. If epidermis from any part of the neurula stage of a species with a balancer (such as *Triturus taeniatus*) is transplanted to the side of the head of a species without balancers (*Ambystoma mexicanum*) then the transplanted epidermis forms a normal balancer in the right place. The reciprocal experiment results in the absence of a balancer in the transplanted tissue (Mangold, 1931; Rotmann, 1935). These experiments show that: (1) the inductive stimulus for balancer development is still present in a species which has been without these organs for a very long time; and (2) it can turn on the 'balancer genes' (which have presumably been lost by *Ambystoma*) when presented with them.

The second example concerns the differentiation of feathers. Birds' feathers are ectodermal structures which have different shapes, sizes and pigmentation patterns in the various regions of the body, and the great variety of poultry breeds shows that each of these characteristics is genetically controlled. Feather differentiation is induced by the underlying mesoderm, and it is the nature of this mesoderm that determines the kind of feather formed. A piece of thigh mesoderm transplanted under the ectoderm of the wing bud of a 3 day chick embryo results in thigh feathers forming on the wing (and vice versa in the reciprocal transplant). Similarly, foot mesoderm will cause scales to develop on the wing. The mesoderm has a regionalized, specific inductive capacity to which a large part of the embryonic flank ectoderm is competent to respond, and the organization of these inductors must be under genetic control. Conversely, the mesoderm has its responses to its inducing tissue, the ectoderm. The developing wing bud ectoderm forms an apical thickening, the apical ridge, and if this is excised the underlying mesoderm does not differentiate into cartilage and muscle. The wingless mutation of the fowl blocks apical ridge formation and, hence, the development of the wing (Zwilling, 1956).

These examples of late inductions show that specific chemical signals, themselves genetically determined in time and space, pass from one tissue to another and activate particular genes in the responsive tissue. Of course, this reciprocal relationship is restricted to particular tissues, and even to regions of tissues, with respect to any one inductive event. During development there is also a sequence of inductions which cause less obvious changes in cells. These limit their potential for futher development, and are defined as determinative events.

### *Determination*

The most comprehensive analysis of determination has used the newt (*Triturus* species) in which, as we have already seen, small tissue fragments can be transplanted to foreign locations. After gastrulation, such transplants no longer regulate. Prospective neural tissue becomes brain or spinal cord wherever it is grafted, and epidermis remains epidermis when placed in the neural area. Both cell types have become determined at this stage, and their developmental potential has become restricted in a precise fashion. Determination is "a process which initiates a specific pathway of development by singling it out from among various possibilities for which a cellular system is competent" (Hadorn, 1965). Or, we might say, determination blanks out part of a cell's total potential, and also that of its progeny since it is a cell inherited characteristic.

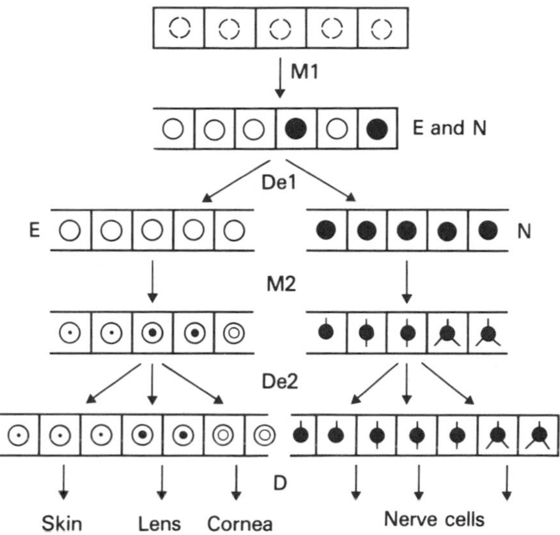


**Fig. 1.11** The formal sequence of steps required for the differentiation of amphibian ectodermal derivatives, according to Hadorn (1965). The gastrula ectoderm has to undergo a process of maturation (M1) before it is competent to become determined as either epidermis (E) or nerve (N) precursors. The first determinative event (De1) separates the cells into these two pathways. Further maturation (M2) is required before the epidermal cells lying over the inducing eye cup are capable of responding by becoming determined (De2) for skin, lens or cornea. They, or more accurately, their progeny, then differentiate (D) as the appropriate cell types. The same progression occurs in the nerve pathway, ending with a series of different nerve cell types. Differentiation results from a restriction of cell potential from general properties to successively more specific ones, in a stepwise fashion. It is uncertain if cell division is a necessary condition for these quantal changes, or if there is a real molecular distinction between determination and differentiation (Holtzer *et al.*, 1975).

Figure 1.11 summarizes a number of experiments which take this analysis of a cell's potential further. It shows, by the test of what differentiates, which is the only assay we have, that determination is progressive, and successively limits the developmental potential of the descendant cells. In normal development, the determinative steps result in the branching of cell lineages (Fig. 1.11) and are of major importance in shaping form and pattern, but we know nothing about the nuclear changes involved. It may be that the primary determinative events are different in kind from those that follow, e.g. as in the establishment of the six blast cell types of *Caenorhabditis*. This seems also to be the case with respect to determination of the sex cells of some insects and amphibia. Figure 1.3 shows that only the multiplying nuclei which reach the polar area of a *Mayetiola* egg become pole cells. These cells are then carried into the larval body during embryogenesis, and there become the primary sex cells. This early differentiation depends on the presence of particular ribonucleoprotein granules (polar granules) formed in the pole region late in ovogenesis. If the polar granules are made inactive by UV irradiation, pole cells are not formed and the emerging adults are normal but sterile. Transfer of pole cytoplasm (in *Drosophila*) confirms this conclusion since pole cells develop in the abnormal locations where the polar plasm has been injected (Illmensee *et al.*, 1976) and, they can be shown (again by transplantation) to produce viable germ cells. Thus, these maternally inherited polar granules are primary germ cell determinants. The vegetal plasm of some anuran eggs has similar properties (Smith and Williams, 1975). In both cases, of course, the cells undergo many changes prior to their final differentiation into sperm and ova. Very little has been done to see if there is a true distinction between primary and subsequent determinative events.

## Autodifferentiation and teratomas

Although induction plays an important role in organizing the normal sequence of determinative steps during embryogenesis, it is not so certain that it is essential for differentiation as such. If an embryo is disorganized by homogenization so that the cells are separated, and then cultured, they will form differentiated structures. This has been tested using cells from an early chick embryo (Andres, 1953). Injecting them into the veins of another chick embryo results in their aggregation into clusters at various places in the embryo, or on the chorioallantois. There they differentiate into nerve, muscle, cartilage, epidermis etc., forming teratomas; disorganized mixtures of differentiated tissues (Fig. 1.12). Teratomas occur spontaneously in the gonads of some mouse strains, where they derive from malignant primitive germ cells; or they can be made by transplanting the genital ridge of 12-day embryos (or whole 7-day embryos) into the testes of syngenic males where subsequent growth results in differentiation of tissues from all three embryonic germ layers, but in a disorganized fashion (Graham, 1977). That is, embryonic cells differentiate even when they are not subject to the normal sequence of tissue relationships and of inductive stimuli, and the result is developmental disorder. Or again, cell suspensions made from early gastrulae of *Drosophila* differentiate when cultured *in vitro* to give nerve, muscle, fat body, chitin-secreting capsules, trachea and the adult structures called imaginal discs (Shields *et al.*, 1975). All these cells, and the heparan-treated amphibian cells described earlier (p. 23), have undergone their primary determination. This suggests that they, or more exactly their progeny, are then preprogrammed to complete their differentiation, and so may autodifferentiate under appropriate conditions. The inductive interactions which we have noted may then give temporal and spatial order to differentiation. If this is so, then teratoma cell lines, which multiply in the undifferentiated state but then spontaneously differentiate into all possible

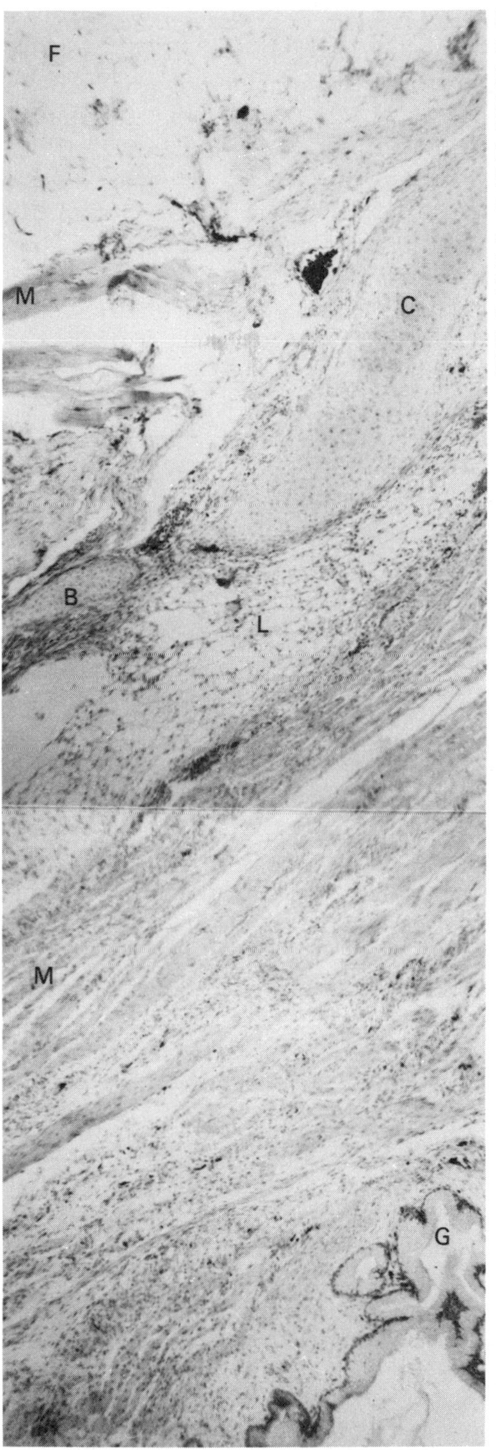


**Fig. 1.12** Not all teratomas derive from gonadal tissue, and this photograph is of part of a rare pinealoma of a chick, described by Campbell (1962). The section shows fat (F), smooth and striated muscle (M), cartilage (C), bone (B), lung (L) and intestinal epithelium (G). All three primitive germ layers are represented by these differentiated tissues, but they are organized in the completely chaotic manner typical of teratomas. Slide kindly provided by Dr. J. G. Campbell.

final forms in a disorganized, random fashion, may provide material for exploring these secondary cell–cell interactions. Primary determination is clearly a complex, and still little understood phenomenon.

## Morphogenetic hormones

The only defined chemicals which are known to cause differentiation are the morphogenetic hormones. And anyone who has fed thyroid tablets to tadpoles will be aware of the dramatic changes produced by this, the oldest and simplest, of these hormones. Tail cells are caused to die, limb and other cells to grow and differentiate, and the fundamental pattern of nitrogen excretion is changed. Many of the nuclear events preceding and accompanying this metamorphosis have been described, including the synthesis of new mRNAs, but their complexity leaves us quite uncertain as to how the hormone acts (Beckingham-Smith and Tata,1976). The pupation hormone of insects (the cholesterol derivative, 20-hydroxyecdysone) has the same range of effects, and we shall consider it as an example of this class of hormone since it will be referred to later.

Figure 1.13 shows that the hormone is made by the developing embryo, but to what end is not known, since the cultured cells mentioned above differentiate in its absence. There are two hormone pulses prior to the 1st and 2nd instar moults, and then a large pulse two days later which initiates pupariation. During pupation there is a further pulse, and this is accompanied by degeneration of larval tissues (except nervous tissue and Malpighian tubules), the differentiation of the exoskeleton from cells of the imaginal discs, and of other structures from nests of imaginal cells present in the larval organs (Fig. 1.14). Figure 1.13 also shows that the enzyme dopa decarboxylase, which is involved in chitin sclerotization, is induced by the hormone at the times when new chitin is laid down. Ecdysones are also made in the adult fly, in which they activate the synthesis of yolk proteins by fat body cells (p. 62). Like thyroxine, the pure hormone can be used to induce these changes in organisms, tissues and cells (Cherbas and Cherbas, 1981).

Wigglesworth (1935) had earlier identified a hormone secreted by the insect corpus allatum (juvenile hormone, or methyl 10,11-epoxy-3,11-dimethyl-7-ethyl-*trans*-2,6 tridecadienoate) which inhibited metamorphosis and stimulated ovarian development in some insects. Essentially, in moths like *Hyalophora cecropia*, this hormone blocks the metamorphosis induced by ecdysones, causing the organism to repeat the *status quo*, possibly by preventing a reprogramming of its chromatin (see Willis, 1981). This is almost certainly an oversimplification since there is evidence that while some cells (e.g. imaginal disc cells in *Drosophila* are blocked in differentiation, as the *status quo* hypothesis implies, other tissues (e.g. muscle) differentiate

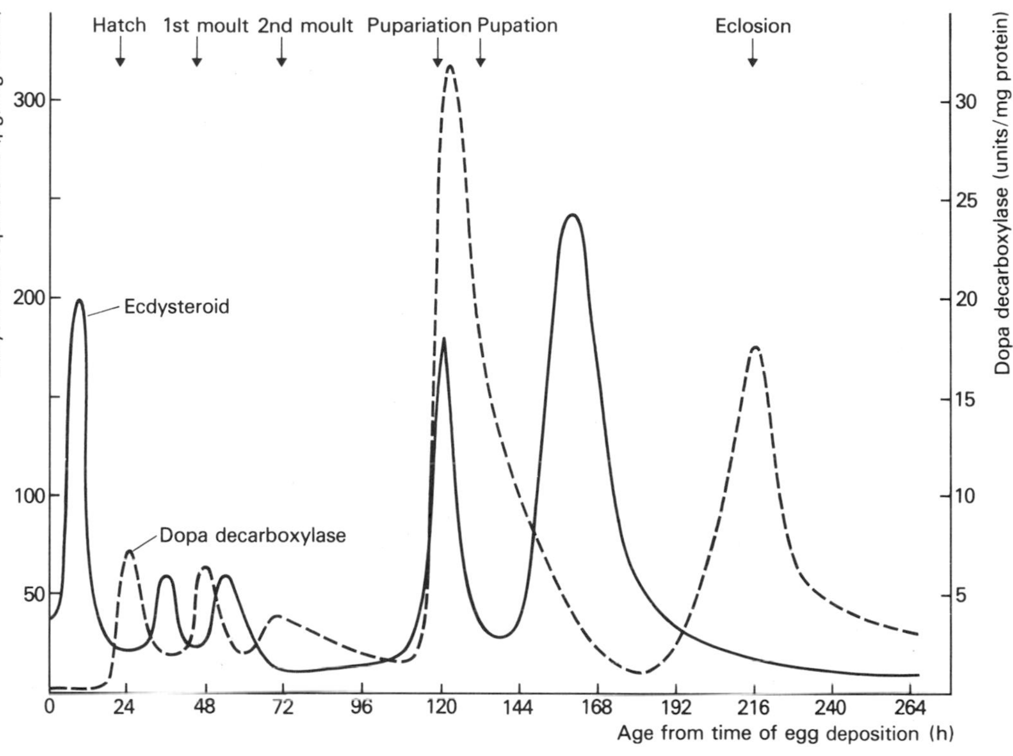


**Fig. 1.13** The cyclical changes of ecdysteroid levels during *Drosophila* development, according to Kraminsky *et al.* (1980). Richards (1981) discusses the different measures of these changes reported in the literature, and Handler (1982) provides more detail of the pulses found during pupal and adult stages. The hormone is first found in the 6-h embryo, before the secretory ring gland is formed. Subsequent peaks precede the larval moults and pupariation, but the major peak during the pupal stage surprisingly antedates eclosion by 60 h. The cyclical pattern of hormone levels implies its destruction by an ecdysonase cycling out of phase. The hormone-induced changes in dopa decarboxylase levels (DDC) are considered in Ch. 6.

normally (Milner and Dübendorfer, 1982). What juvenile hormone does, and it may do many things since it is a metamorphic hormone for barnacles (cirripeds) according to Ramenofsky *et al.* (1974), remains to be resolved (Richards, 1981).

This description identifies four classes of developmental problem that can be studied using morphogenetic hormones. The first centres on the timing of synthesis of particular molecules. Ecdysones are made by the insect ring gland, secreted into the haemolymph, and there transported to the tissues. This process is cyclical and may be regulated by the insect's circadian 'clock', via the neurosecretion of prothoracotropin (Doane, 1973). The control of this activation hormone is unknown in fact, but a sequence of critically timed events (activation of enzymes for ecdysone synthesis, and the secretion processes) before the appearance of the effective hormone is implied.

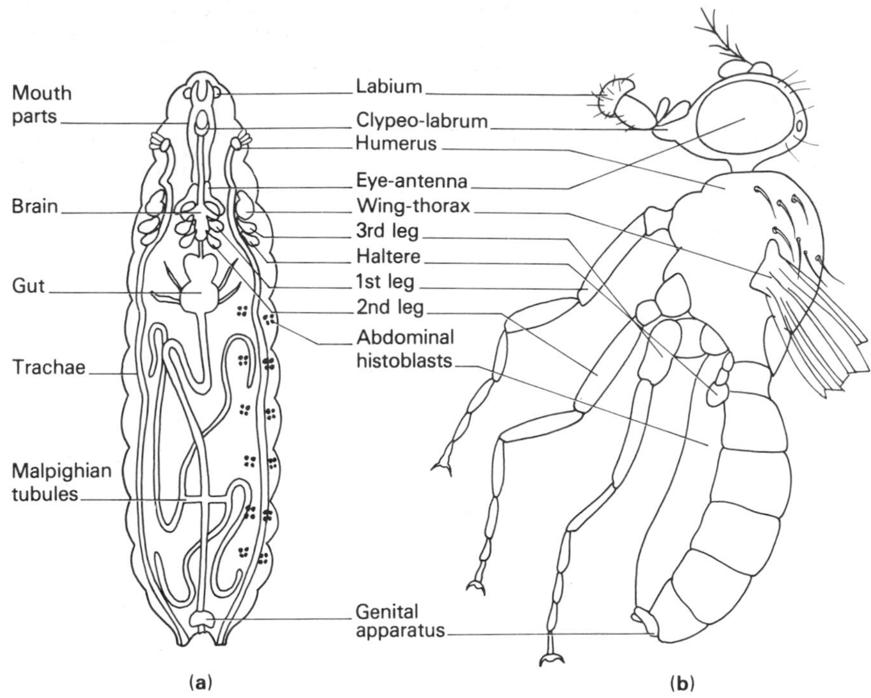


**Fig. 1.14** The larval imaginal discs and imaginal precursor cells (**a**), and the adult structures which they form during the pupal metamorphosis (**b**), after Wildermuth (1970a). The differentiation, and in some cases multiplication, of these cells is induced by the ecdysterones, and can be manipulated *in vitro* (Dübendorfer and Eichenberger-Glinz, 1980).

Further enzymes which break down the hormones are secreted out of phase, as Fig. 1.13 indicates.

The second set of problems concerns the acquisition by the tissue of a state in which it can respond to the hormone: early 3rd instar larvae cannot be made to pupate even if exposed to excess ecdysone. It may be that they have not yet synthesized the receptor proteins that are required to transport the hormone to the nucleus (Stevens *et al.*, 1980), or there may be some other cause. Whatever the explanation, there remains the problem of regulating the synthesis of such key proteins as these receptors.

The third question concerns the different responses of different tissues and cells to any individual hormone (and possibly to different levels of that hormone). This implies that their DNA must be in a new guise, and hormones can be used to explore the involvement of DNA organization in the patterns of cell differentiation.

Fourth, the hormones may provide the easiest approach to studying the regulated control of transcription of such an array of genes, and we might expect each array to be distinguished by identifying elements in the molecular organization of the DNA itself. In the simpler cases, where the hormone induces the synthesis of only a few mRNAs, hormones may provide the easiest approach to studying this important aspect of DNA organization.

Morphogenetic hormones are powerful tools for probing many differentiation events. However, we cannot assume that inducing agents will act in the same way, and it seems unlikely that determinants such as polar granules will do so, either. Plant hormones are less precise in their actions and interactions, and have to be considered separately.

## Plant development

We shall have little to say about plants, mainly because the genetic and molecular analysis of their development is some way behind studies with animals, although this is not always the case. Since plant cells are totipotent—a plant can be grown from a leaf cutting as well as from cell culture—the great wealth of mutants affecting leaf pattern, flower-shape and colour, growth rate and so forth, shows that genes are selectively activated during differentiation. It is therefore assumed (e.g. Wareing and Phillips, 1978) that the processes involved are analogous to those described for animals, and that the differences are only in the details. Two obvious divergencies derive from the nature of the plant cell: (1) the rigid cell walls prevent cell movement and hence the interactions so typical of animal gastrulae and; (2) as a corollary, only the local 'embryonic groups' of the meristem multiply during plant growth in contrast to general cell proliferation during animal development. Growth from the apical meristems of root and shoot is therefore accretionary and indeterminate. Futher, determinate meristems of leaves, flowers and fruits form during stem growth, usually in regular patterns and sequences, and proceed to their final differentiation. The total range of tissue and cell types is more restricted than in higher animals, and diversity (e.g. of leaf shape) arises from the patterns of cell multiplication.

The distinction between root and shoot arises with the first division of the zygote which is transverse, making two unequal cells (Fig. 1.15). The larger, basal cell continues to divide transversely to form the suspensor (p. 74) while the smaller, terminal cell divides twice longitudinally, and then transversely to give eight cells which give rise to the embryo proper. These cells then divide to form the progenitors of the epidermis on the outside, and of the cotyledons and hypocotyl on the inside. The suspensor cell next to the embryo divides to make the root cap and the rest of the radicle. The embryo is polarized as root and shoot from the beginning. Although there is preformed mRNA in the long first cell, it seems unlikely that its unequal partition is the cause of this first determinative step. Individual cultured cells taken from the shoot (Fig. 1.2) are similarly polarized, divide unequally and form embryoids. The root pathway seems stably determined, nevertheless,

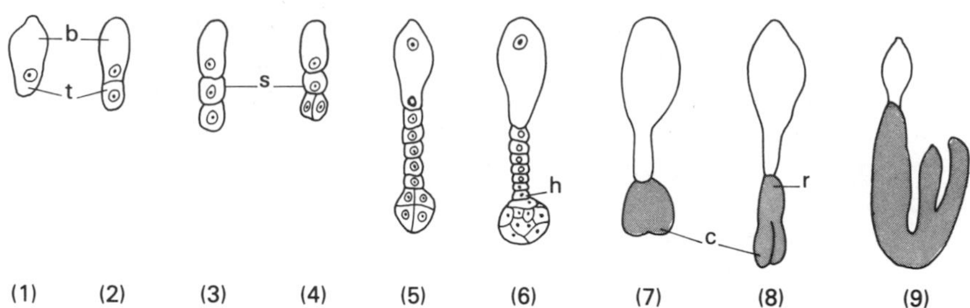


**Fig. 1.15** The development of *Capsella bursa-pastoris* (shepherd's purse), after various authors. The egg is polarized and the first division is at right angles to this polarity (1) and (2) to form a larger basal cell (b) and a small terminal cell (t). The basal cell divides transversely to form the suspensor cell (s) (3), and these two cells subsequently divide to form a suspensor of up to 10 cells ((5) and (6) and then shown only in outline in the following figures). The terminal cell divides to give 2, 4, 8, 16–32-cell stages of the embryo (4)–(6). The globular embryo forms the primordia of the cotyledons (heart-shape stage in (7)) and elongates to give the torpedo stage (8), of Fig. 1.2. The distal cell of the suspensor, the hypophysis (h), divides and its daughter cells contribute to the embryonic root apex and root cap (r). The suspensor cells have begun to degenerate by the heart-shaped stage, and they are crushed by the fully formed embryo (9), which folds upon itself.

since root fragments in culture usually make roots, not shoots. On the other hand, shoots will root, but in this respect they are polarized since if a shoot fragment is inverted, roots form at the upper end. Plant cells are polarized characteristically, but the mechanism is unknown.

We shall look later at the clonal history of the cells of a shoot apex (primordium); here we want to note that a leaf primordium first appears at the flank of such an apex, and it is then not determined. If a cut is made between the primordium and the apical cell of the shoot it develops as a bud, i.e. radially symmetrically and not flattened dorso–ventrally as a leaf. Later primordia are determined and continue to develop as leaves. Excised and cultured primordia behave in the same way. Thus, as described for animal development, determination is a progressive process, and is stable (cell inherited) once established. Unfortunately, it is not clear if this is an intrinsic property of the cells which they achieve through multiplication, or if it depends on association with other cells and is due to extrinsic factors. The change from the vegetable to the reproductive phase (flowering) of growth depends on day length, and is certainly extrinsically determined.

Plant hormones (such as indole-3-acetic acid, gibberellins and cytokinins) certainly regulate RNA metabolism in plant cells, and affect their differentiation (Jacobsen, 1977). The question is: does the response to a hormone depend on the determined state of the cell itself, or is it caused by the hormone? While it is difficult to generalize about a great variety of hormone effects, the answer seems to be that the specificity of the hormone action resides in the determined state of the cell itself, as with the morphogenetic hormones of animals. For example, indole acetic acid stimulates cell division in the cambium, but causes vacuolation in growing internodes. And so on. Thus, in plants we have some processes analogous to those found during animal development, and other, as yet unexplained, events which seem to be characteristic of the plant cell.

## The problems

If we accept Oppenheimer's proposition at the head of this chapter, we have to recognize that no-one has yet elucidated the chain of events which results in only one particular gene array being regularly expressed in a single differentiated cell type. The great variety of embryological processes, briefly outlined above, may seem to imply that this task is too formidable to be accomplished even in the simplest of cases. Such an analysis appears equally daunting at the molecular level. For instance, even in the relatively simple situation when the vegetative cells (carrying 6000 poly(A) RNAs, i.e. functional mRNAs) of the fungus *Aspergillus nidulans* are caused to sporulate, about 1300 new structural genes are then expressed (Timberlake, 1980). It is hard to see a way through this complexity. For now we must therefore formulate our problems more generally, and leave the detailed pursuit of individual differentiation events to future developmental geneticists. Oppenheimer's formulation defines a long-term goal.

Our first conclusion above was that cells of an organism are pluripotent (p. 10), and we shall see that the molecular analysis of the genome, as far as it goes, confirms the generalization that differentiation does not generally result from organized gene loss but from selective gene activation/inactivation. The exceptions prove this rule. Thus our central problem is: how is eukaryotic gene expression regulated? Essentially, this is the problem of the regulation of transcription in eukaryotic cells, but we must recognize that it may also involve all the steps prior to translation, including the processing of the RNA transcript. It seems likely that subsequent translational control, where it exists, only modulates and refines these earlier activities. In theory, we are not concerned with the activation of all genes but only with those involved in the production of the 'luxury proteins' which distinguish one cell type from another. In practice, we know so little about the regulation of genes for 'housekeeping' or for secretory proteins that we shall consider them, too. At the very least, we may expect their organization to throw light on gene regulation generally, and at most we shall recognize that both these classes of gene vary in their expression in different tissues and at different times. Some of the housekeeping genes (p. 175) may indeed be involved in regulating the production of luxury proteins.

Our central problem can be put in other terms: what is the molecular organization of the DNA which allows an 'activator' substance to cause the initiation of transcription of particular luxury protein genes? Or, conversely, how is the DNA organized so that the majority of these genes is not transcribed in most tissues? Since we have seen that many different molecules (maternally deposited substances, inducers, hormones etc.) may be involved as activators, each may require some specific, corresponding organization in the DNA itself: specificity of action may lie with the DNA–activator complex. Granted this specificity, the remainder of the transcriptional organization may be common to all genes; or this, too, may be organized in different ways, perhaps corresponding to the hierarchical organization of developmental events. However, as both determination and differentiation are clonally inherited cell states, it is difficult to imagine an activator mechanism which would provide this stability. It therefore seems possible that there is some further organization that 'blanks out' parts of the genome so that transcription of these regions is no longer possible. This could occur at the level of DNA itself, or it may involve some developmental reorganization of the DNA in its chromosomes such that only limited regions are accessible to transcription. It follows that our first task is to look at the molecular organization of DNA and chromosomes, and this we shall do in the next three chapters.

# II The organization of the genome

Conventional and molecular genetics are now one. While a gene may be studied with respect to its chromosomal location, or the nature of its product, or the role and function of that product during development, we shall find ourselves continually returning to the problems of gene regulation and, thus, to the molecular organization of the genome and to genes as DNA. Although we do not yet know how the expression of a single eukaryotic gene is controlled, what we do know about the molecular genetics of eukaryotes provides the second, essential, frame of reference for development studies. This section therefore proceeds from molecular descriptions and experiments to what we know about active genes, and then to models of the functioning genome. This overtly reductionist pursuit, the explanation of development in molecular terms, is at present the most active approach to these studies.

The organization of eukaryotic genes is more complex than that of prokaryotes, and gene-sized segments of DNA are found interspersed between stretches of DNA carrying hundreds or thousands of repeated sequences; or, they may lie in long uninterrupted stretches. Two broad types of arrangement are found: the '*Xenopus* type' where single copy sequences (~1000 nucleotides long) alternate with short repeated sequences (300 nucleotides) and a '*Drosophila* type' where long single copy sequences are interspersed with long (5000–6000 nucleotide) repeats. These two arrangements account for about half the DNA in the genome, and the significance of this organizational difference, which is found in other plants and animals, is unclear.

Another major difference between eukaryotic and prokaryotic genetic organization concerns the presence of 'introns'—long, non-coding and untranslated stretches of DNA within individual genes. When eukaryotic genes were isolated and analysed in fine detail by genetic engineering techniques, some, but not all, were found to be 'split' by (sometimes many) intron sequences. These are transcribed into RNA but are 'spliced out' of the RNA before it leaves the nucleus. Unlike the prokaryote genes therefore, the eukaryote gene may be variously divided into 'exons' which combine to form functional mRNA, and introns which are removed in the nucleus.

A third difference is the existence of families of dispersed transposable elements, some of which may be proviruses. This DNA has been classed as 'selfish' DNA, which is parasitic and has no function in the organism. It may account for most of the middle repetitive DNA, and it adds a further, unexpected complexity which will make it difficult to sort out the relevant functional organization of DNA at this general level.

Cloned genes and their flanking sequences can now be cut and manipulated *in vitro*, allowing the sequences necessary for transcription to be identified. Such experiments have found that the genes transcribed by RNA polymerase III (tRNA, 5S RNA) have their control regions within the gene, whereas polymerase II (which transcribes mRNA precursors) requires control regions upstream from the initiation site. A fourth difference in eukaryotes therefore derives from the characteristics of their DNA-dependent RNA polymerases, and these have still to be worked out in detail.

The final difference, which is most important to us, is the organization of eukaryote DNA into chromatin and then further into nucleosomes. A variety of techniques, including the susceptibility of actively transcribed genes to preferential digestion by DNase I, show that the conformation of the DNA of active genes within chromatin is altered, so making them accessible to transcription. The details of this change, its possible dependence on an association with high mobility group proteins, on the methylation of the DNA and a range of other changes known to occur in histones and in the DNA itself, have still to be elucidated. Transcriptional activity is also related in some way to the organization of the nucleus itself, to the structure of the 'nuclear cage'. There are nuclear changes which regulate the genes active in chromosomes.

In polytene chromosomes the DNA appears to be organized into bands (chromomeres) and interbands, but the apparent correlation of one gene with one chromomere is no longer supportable. This correlation is, however, still of value for the cytological location of genes to bands. Active genes are expressed as 'puffed' bands (but not invariably) and the changing patterns of puffs during development (or after experimental manipulation) show that genes whose coordinate function is necessary are not grouped together in sets. Some gene products may activate (or inactivate) other genes, sequentially; and others may act directly between the paired chromomeres. The changes visible at the chromosome level are too coarse to inform us about the structural changes in the active genes but, by combining cytological and gene cloning techniques, progress is being made towards the understanding of the activation of heat shock proteins, which are found throughout the living kingdom, from bacteria to man.

For lack of comprehensive facts, we can only speculate about how the eukaryotic genome is regulated during development. These speculations start conventionally from the prokaryote operon model, but have to take account of diploidy and of the coordinate activation of metabolically unrelated genes. There are possible transcriptional control models, translational control models, and models which imply successive inactivation of chromatin regions. All these models may explain too much to be open to critical tests, and there is no good reason for not assuming that different systems (or organisms) may employ a variety of these various possible strategies. Models of the active genome therefore provide a final frame of reference for the genetic studies of subsequent chapters.

# 2 The molecular organization of the genome

Since the genes form a chain, the chemical alteration of a gene might at times be a matter entirely confined to the individual links, and it might, on other occasions, by destroying or breaking the link or its connection with an adjacent link, result in breaking of the whole chain.

H. J. Muller, 1932

As shown by Jacob and Monod (1961) in bacteria, there are also parts of the gene material that appear not to synthesise protein, yet to act as regulators of other gene material which does so. There should be far more, and far more varied, systems of processes of this kind, and of gene material underlying them, in higher forms.

H. J. Muller, 1966

As these quotations show, a generation of research changed our understanding of the molecular organization of the genome. The static, mechanistic, string of beads hypothesis suggested by the genetic study of mutations was replaced by the knowledge that the structural genes of prokaryotes were separated by untranscribed sequences, and regulated by other genes. The DNA was functionally organized. It was therefore anticipated that higher forms should have an even more elaborate DNA organization to cope with the precise and dynamic regulation of gene activity during their development and differentiation. Our concern now is to ask if subsequent research has shown that this is so, and what this molecular organization might be. If we had the full answer to this question we should be well on the way to understanding development. As it is, we only have many partial answers; but the gaps in our knowledge are being filled very rapidly in this expanding field of study. We shall pay some attention not just to results but also to the methods used to get them, since these techniques have still much to offer.

## Sequence organization of nuclear DNA

Studies of the general organization of the DNA extracted from chromosomes take advantage of the physical properties of this polymeric molecule.

The DNA double helix falls apart when it is heated to 80–90 °C in a solution of physiological ionic strength, or at half that temperature in the presence of a denaturing agent like formaldehyde. This denaturation, or melting, can be measured by following UV absorbance at 260 μm, which increases with melting (or by measuring other physical properties). On cooling, the denatured DNA renatures, and the UV absorbance decreases again (to about 10 percent above the original) showing that slow cooling allows the original duplex structure to reform, as a native-like DNA. Rapid cooling gives results which suggest a random aggregation. Renaturation depends on DNA strand complementarity and does not occur between molecules with different sequences, such as DNAs from unrelated organisms. The same sequence complementarity relationship allows hybridization of RNAs to the sense strand of the DNA.

Elaborations of this methodology enable separation of renaturation products. By trapping single-strand DNA in agar gel and incubating sheared complementary radioactive DNA (or RNA) with it, and then eluting the non-complementary sequences, the extent of hybridization can be measured by the retained (duplex) radioactivity. Or nitrocellulose filters can be employed instead of agar-immobilization, since they absorb single-strand DNA (but not RNA). Treatment of the filters with albumin after they are loaded with single-strand DNA, permits annealing with a second, labelled DNA as single strands are not then retained unless they combine with the first DNA. Hydroxyapatite columns permit another manoeuvre for they retain duplex DNA at the low ionic strengths (0.12 M phosphate buffer) which elute single strands. Increased ionic strength (0.27 M phosphate buffer) is then used to remove the double-stranded DNA. These techniques allow not only the study of the whole DNAs and RNAs, but also of DNA fragments, when we can ask about the arrangement of their nucleotide sequences.

Renaturation of the complementary strands of a DNA involves a random collision process, and depends on the initial concentration ($C_0$) and on time ($t$). This is measured, as decreased optical absorbance, in terms of the $C_0t$ which is expressed as moles of nucleotide × second/litre. $C_0t$ curves for simple DNAs have the same, second order reaction, form (Fig. 2.1); and $C_0t_{\frac{1}{2}}$ values, at which half the DNA is again double stranded, increase with the sequence complexity of the DNA. The $C_0t_{\frac{1}{2}}$ is proportional to the length of the DNA (Fig. 2.1), and this relationship can be relied on to calculate the repeating length of an unknown DNA when its $C_0t_{\frac{1}{2}}$ is measured under the same conditions as that of a 'standard' DNA. For example, in Fig. 2.1, *Escherichia coli* DNA has a length of $4.6 \times 10^6$ base pairs (bp) and a $C_0t_{\frac{1}{2}}$ of about 6, while the $C_0t_{\frac{1}{2}}$ of calf thymus (non-repetitive) DNA is about 3000. By simple proportion, the calf thymus DNA must have a basic unit of $2.3 \times 10^9$ bp. In this example, we know that the DNAs contain single sequences and are free of repeated sequences (or relatively free), and the length of the DNA unit therefore measures its 'sequence complexity'. For

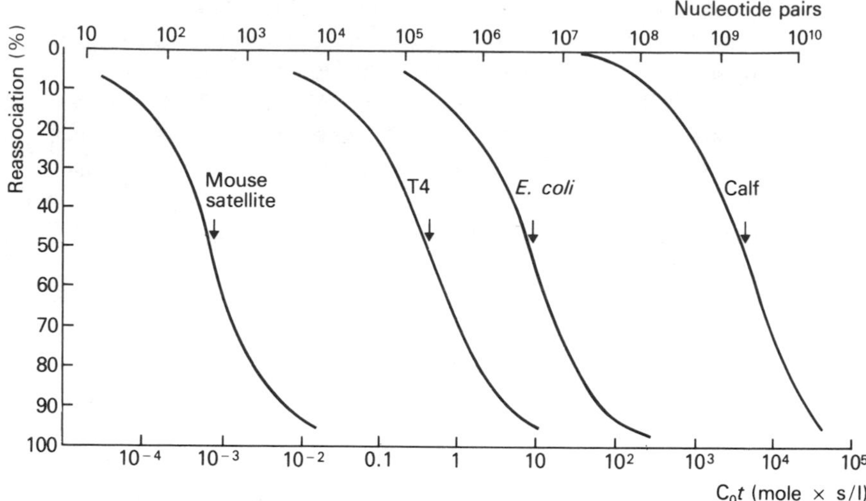


**Fig. 2.1** The reassociation curves of different DNAs in 0.12 M phosphate buffer, after Britten and Kohne (1968). Higher $C_0t$ values are found with DNAs of increasing sequence complexity; from mouse satellite through phage T4 and *E.coli* to the non-repetitive fraction of calf thymus DNA. The $C_0t_{\frac{1}{2}}$ values estimate the length of the basic repeating unit of the haploid genome in base pairs (bp) as indicated by the upper axis. Note that the scales are logarithmic.

a given DNA concentration, when the sequence complexity is low, as with the bacteriophages, there will be many like molecules present, they will be repeated many times, and complementary strands will come together quickly (low $C_0t$): a high $C_0t$ will be measured when the sequence complexity is great, as with the calf DNA.

The renaturation curves of DNAs of eukaryotes are more complex, and they differ from one another in detail (Fig. 2.2). However, they show three general characteristics. First, there is a level of spontaneous, zero time, renaturation of a few percent of the DNA. This is due to inverted, repeat sequences (palindromes) in the DNA chain, and since this renaturation occurs in very dilute DNA solutions it is likely that the single strands containing palindromes form hairpin loops a few hundred nucleotides long (foldback DNA), quite spontaneously. These palindromes may provide protein–nucleic acid recognition sites, if one can argue from the association of the *lac* repressor, or of restriction enzymes (p. 47), with similar short DNA palindromes in prokaryotes. Second, true renaturation takes place in three phases: very rapidly, with $C_0t_{\frac{1}{2}}$ values of ~ 0.005; rapidly with $C_0t_{\frac{1}{2}}$ values between 0–1, and slowly with $C_0t_{\frac{1}{2}}$ values measured in thousands. By the argument of the preceding paragraph, the three fractions must represent *highly-repeated* short sequences; longer *moderately* or *intermediately-repeated* sequences; and *single copy* DNA. Third, the proportions of these three classes differ among metazoa (Table 2.1). The distinctive difference

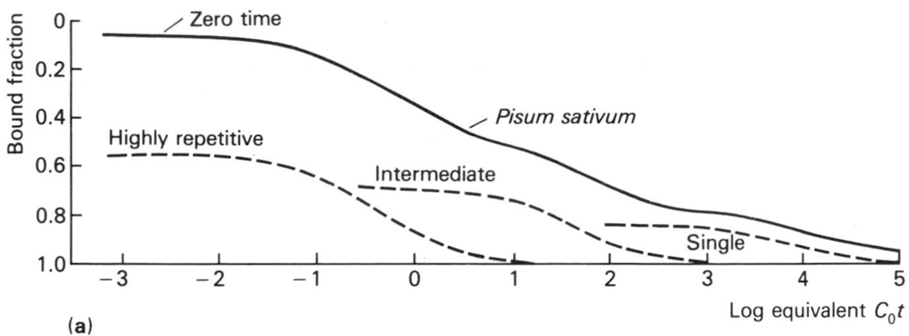


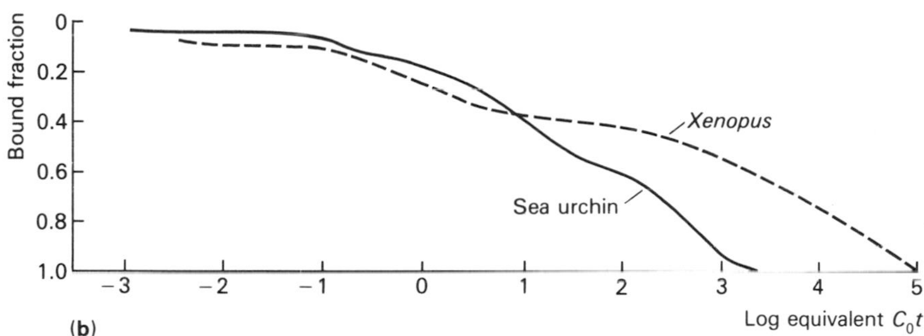


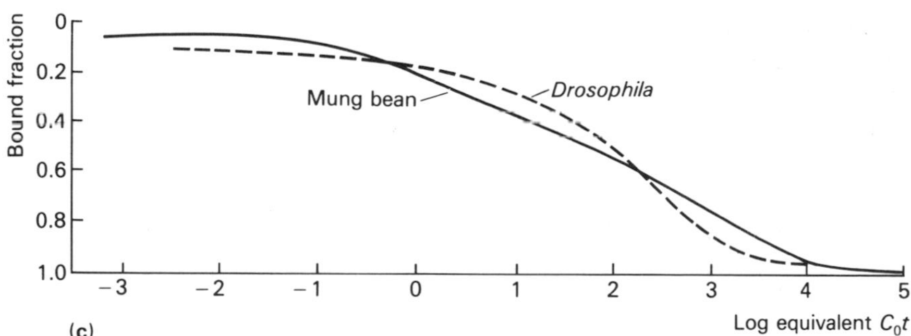


**Fig. 2.2** The reassociation kinetics of short (300–450 bp) nucleotide fragments: some typical cases. The fraction of the fragments binding to hydroxyapatite (i.e. containing duplexed regions) is plotted as a function of the $C_0t$. (**a**), Data for the pea, *Pisum sativum*, together with the least-squares fitted three-component model (broken lines) to match the overall curve. This, and the data for mung bean (**c**), are taken from Thompson *et al.* (1980). The *Xenopus* and sea urchin data (**b**) are from Davidson *et al.* (1973) and Graham *et al.* (1974), respectively. The *Drosophila* data (**c**) are from Manning *et al.* (1975).

**Table 2.1** Sequence organization of the DNAs of some eukaryotes

| Organisms | Fast and zero time | Highly repetitive | Intermediate | Single copy |
|---|---|---|---|---|
| *Strongylocentrotus purpuratus* ($8.6 \times 10^8$ bp) | | | | |
| % of DNA | 3 | 7 | 27 | 63 |
| Reiteration | – | 6000 | 250 | 1+ |
| Complexity (bp) | – | $1.3 \times 10^4$ | $1 \times 10^6$ | $4.6 \times 10^8$ |
| *Xenopus laevis* ($2.8 \times 10^9$ bp) | | | | |
| % of DNA | 3 | 6 | 31 | 60 |
| Reiteration | – | 32,000 | 1600 | 1+ |
| Complexity | – | $6 \times 10^3$ | $6 \times 10^5$ | $1.6 \times 10^9$ |
| *Drosophila melanogaster* ($1.3 \times 10^8$ bp) | | | | |
| % of DNA | 5 | 12 | 12 | 70 |
| Reiteration | – | 24,000 | 70 | 1 |
| Complexity | – | $7.2 \times 10^2$ | $2.5 \times 10^5$ | $1 \times 10^8$ |
| *Mus musculus* ($2.7 \times 10^9$ bp) | | | | |
| % of DNA | 10 | 14 | 11 | 58 |
| Reiteration | – | 7500 | 150 | 1.6 |
| Complexity | – | $2.9 \times 10^4$ | $1.1 \times 10^6$ | $9.3 \times 10^8$ |

Data from: *S. purpuratus*, Graham *et al.* (1974); *X. laevis*, Davidson *et al.* (1973); *D. melanogaster*, Manning *et al.* (1975); *M. musculus*, Laird (1971).

of eukaryote DNA is that it contains repeated sequences as well as single copy sequences (Britten and Kohne, 1968).

### *Arrangement of single copy and repeated sequences*

When eukaryotic DNA is sedimented in a CaCl density gradient, satellite bands are found associated with the main DNA band, either above or below it, and constituting less than 10 percent of the total. This DNA can be separated, and it renatures at the very low $C_0t$ value of a highly repeated DNA fraction which is characteristic of the species involved. Analysis shows that this satellite DNA is made up of short units (100–400 bp) of simple composition and repeated many times (Sutton and McCallum, 1971). The location of this DNA in chromosomes has been determined using an important elaboration of the DNA hybridization technique (Gall and Pardue, 1969; John *et al.*, 1969).

When cells are squashed on a slide to release their chromosomes, then dehydrated and dried, their chromosomal DNA can be denatured by alkali (NaOH) and a radioactive complementary DNA (or RNA) will then hybridize to it. After hybridization is completed (at low $C_0t$) and the excess radioactive DNA washed off, the slides are dipped in a photographic emul-

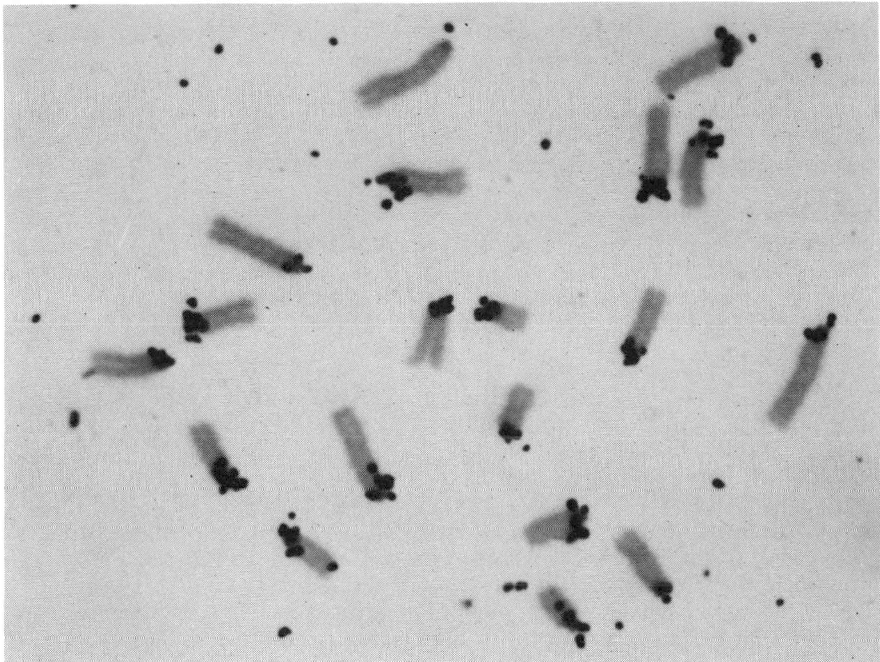

**Fig. 2.3** Mouse satellite DNA was copied *in vitro* to make a radioactive RNA, which was then hybridized to the complementary sequences in the DNA of the chromosome of a cultured mouse cell line. The satellite sequences are localized in the centromeric heterochromatin exclusively. From Pardue and Gall (1972) with permission.

sion and stored while the radioactive decay causes the silver of the emulsion to be deposited as grains. The emulsion is developed, and the slide then stained so that the chromosomes can be located (Fig. 2.3). Using this method, radioactive satellite DNA is found to hybridize to the centromeric, constitutive heterochromatin (Gall and Pardue, 1969). This heterochromatin is not transcribed, so it is not surprising that satellite DNA does not hybridize with cell RNA. This highly repetitive DNA may have some role in chromosome pairing (Peacock *et al.*, 1978), and there is no evidence that it functions in development. This technique does not tell us if there is any similar DNA present elsewhere on the chromosome in minute amounts, or how it is organized in the heterochromatin, it only demonstrates where most such sequences are found.

The question of the organization of repetitive and single copy DNA is obviously important, and it, too, can be explored using renaturation. The approach is to provide a great excess of short (~ 400–500 nucleotides) pieces of 'driver' DNA, and hence of short repetitive sequences. Labelled DNA is then sheared, and chromatographed to particular lengths which are renatured to these repetitive sequences in small amounts and at low $C_0t$, as 'tracer' sequences. The duplexes then formed are bound on hydroxyapatite

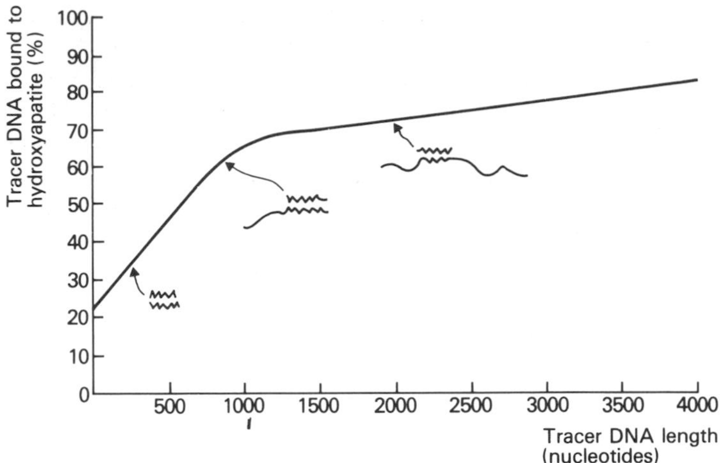


**Fig. 2.4** Proof of the interspersion of repetitive and non-repetitive sequences in *Xenopus* DNA (data of Davidson *et al.*, 1973). In this experiment, different tracer (labelled) lengths of DNA are annealed to a great excess of short (≏ 450 nucleotides) unlabelled fragments at low $C_ot$ so that only repetitive sequences form duplexes. The proportion of duplexed tracer is then determined and plotted for each DNA length. The ordinate intercept shows that about a quarter of the DNA is repetitive. The increase of binding to hydroxyapatite with increased tracer length is due to repetitive sequences covalently linked to non-repetitive sequences, as illustrated diagramatically below the curve (and as confirmed by electronmicrography of these structures by Chamberlin *et al.*, 1975). The obvious inflection of the curve at DNA lengths of 800–1000 nucleotides indicates a large class of single copy sequences of this size linked to repetitive sequences. These, and the longer single copies found with larger tracers, show that repetitive and non-repetitive sequences are interspersed in the *Xenopus* genome.

columns and their radioactivity measured. Short tracer DNA lengths that renature under these conditions will generally carry repetitive sequences, while short single copy sequences will not. The zero intercept (Fig. 2.4) thus measures the proportion of repetitive sequences in the whole DNA, usually 25–30 percent. Longer tracer DNA may contain both repetitive and single copy, but the proportion bound to the driver DNA will depend on the pattern of interspersion of the two classes under these low $C_0t$ conditions. If the repetitive DNA is all, or nearly all, in one block, and the single copy sequences quite separate in another, the proportion of duplexes formed would be nearly the same as with short DNA lengths. If repetitive and single copy sequences alternate, the proportion of duplexed DNA will increase with tracer length since the bound repetitive sequences will carry longer and longer single copy tails. If the interspersion has a long period, the proportion of DNA renatured will fall between the two classes: short tracers will be either repetitive or single copy and long tracers carrying repetitive sequences will have even longer tails than with short period interspersion.

The results for *Xenopus* or for sea urchin DNA are typical of many of the organisms studied. With long (4000 nucleotide) fragments, 80 percent of the

DNA is bound and must therefore contain repetitive sequences (Fig. 2.4), although these sequences constitute only about a quarter of the DNA. Consequently, there must be a close interspersion of repetitive and non-repetitive sequences. The remaining 20 percent of this DNA, not bound to the hydroxyapatite, cannot have repetitive sequences within their 4000 nucleotide lengths, or perhaps for even greater lengths, and this fraction must therefore be classed as long period interspersed DNA. The inflection of the curve at 700–1200 nucleotides implies that the average length of a repetitive plus non-repetitive sequence must be about this size. The presence of a second repetitive sequence in the fragment would not increase the binding, and the curve would plateau at this point if the fragments were 'repetitive-single copy-repetitive'. Since the curve continues upwards there must be fragments with longer non-repetitive 'tails' of greater length, up to, or more than, the tested 4000 nucleotides. The size of the repetitive sequences can be measured by denaturation, by removing the single strand tails with nuclease, and by viewing the duplexes in the electron microscope (see Davidson and Britten, 1973). Their modal length is 300 ± 150 nucleotides.

Single copy and repeated sequences are not randomly distributed in these DNAs, but ordered. A small fraction of the DNA (< 10 percent), is made up of small, clustered, repeated sequences. Larger repeated sequences (25 percent of the DNA) are interspersed with gene-sized sequences of 700–1200 nucleotide lengths in the short period pattern (35–40 percent of the genome), and more than 20 percent of the genome is in the long period interspersion pattern where the single copy sequences are at least 4000 nucleotides long between the repeated sequences. Within this general pattern (and that described in the next paragraph) are four groups of structural genes which are themselves repeated and separated from one another by 'spacer' sequences, in tandem arrays (Table 2.2). They account for only a small fraction of the interspersed patterns just described.

It was thought at first that the pattern of DNA organization just described was universal, but a different organization was found in *Drosophila* DNA (Manning *et al.*, 1975; Crain *et al.*, 1976). Using the methods just described, *Drosophila* is found to lack the 300 nucleotide repetitive elements, and to have much larger repetitive sequences of about 6000 nucleotides. Short

**Table 2.2** The numbers of some repeated genes

| Organism | Xenopus laevis | Drosophila melanogaster | Yeast | HeLa cell |
|---|---|---|---|---|
| rRNA | 450 | 250♀ (150♂) | 140 | 280 |
| 5S RNA | 24,000 | 165 | 140 | 2000 |
| tRNA | 1150 | 860 | 250 | 130 |
| Histones | 40 | 100 | – | 40 |

These estimates (from various sources) are based on saturation hybridization averages, and are therefore approximate.

period interspersion is not found, and single copy sequences extend to 10,000 or so bases before they are interrupted by a repeated sequence. So there is a *Drosophila* pattern of organization quite distinct from the *Xenopus* pattern. Surprisingly, some insects (e.g. *Apis* and *Chironomus*) follow the *Drosophila* pattern while others (e.g. *Musca* and *Antharea*) do not. Since *Drosophila* and *Musca* have quite similar embryologies, it must follow that the features of the DNA organizations just described are not essential for development as such. And the same is true for plants, where the related pea (*Pisum sativum*) and mung bean (*Vigna radiata*) have, respectively, *Xenopus* and *Drosophila*-type DNA organizations, but notably similar developments (Thompson *et al.*, 1980). In this latter case, there is also a dramatic difference in the amounts of the haploid genomic DNAs—4.6 pg in the pea and 0.5 pg in the bean. Similarly, the dipteran, *Sarcophaga bullata* has between three and four times (0.61 pg) the DNA content of *Drosophila* (0.18 pg), but a *Drosophila*-type DNA organization (Samols and Swift, 1979). One must conclude that a great part of the DNA of some species is non-informational.

While DNA hybridization has given us an interesting, and unexpected, picture of the organization of DNAs, it is too broad a technique to tell us about the molecular arrangement around the informational sequences. It suggests that much of the DNA may be without a developmental function, but the limitations of the techniques do not allow us to conclude that all of the sequence arrangements exposed by such analyses are unimportant. To find out more, we have to turn to the finer analyses coming from the explosion of work on 'genetic engineering'.

## Gene organization

### *DNA cloning techniques*

DNA cloning involves the insertion of a length of foreign DNA into the circular chromosome of a plasmid (or of a bacteriophage), and the infection of a bacterial cell with this hybrid (or chimeric) chromosome so that the foreign DNA is multiplied within the bacterium to provide sufficient copies of the segment for analysis. Particular cloning vectors have been developed for this purpose, particularly the plasmid Col E1, and smaller derivatives made from it, and safe variants of phage lambda (e.g. phage Charon). These carry resistance genes (for antibiotic or colicin resistance) so that uninfected bacteria can be eliminated, and bacteria carrying the plasmid can be selected and multiplied (see Glover, 1980).

Both the foreign DNA and the plasmid DNA have to be broken before joining together, and this is done in a variety of ways. Restriction endonucleases, such as *Eco*R1 and *Hin*dIII (Table 2.3), are commonly used since

**Table 2.3** The cleavage sites of some commonly used restriction endonucleases

| Enzyme code | Source | Cleavage site |
|---|---|---|
| *Eco*R1 | *Escherichia coli* | 5′–G↓AATT–C–3′<br>C TTAA↓G |
| *Bam*H1 | *Bacillus amyloliquefaciens* H | G↓GATCC |
| *Hin*dIII | *Haemophilus influenzae* | A↓AGCTT |
| *Mbo*1 | *Moraxella bovis* | ↓GATC |
| *Pst*1 | *Providencia stuartii* | CTGCA↓G |
| *Sal*1 | *Streptomyces albus* | G↓TCGAC |
| *Alu*1 | *Arthrobacter luteus* | AG↓CT |
| *Hae*III | *Haemophilus aegyptius* | CG↓CC |

For *Eco*R1 the arrowed cleavage sites are shown for both DNA strands, and the others should be read similarly. This enzyme generates mutually cohesive 5′ ends, but *Pst*1 makes 3′ ends and *Alu*1 and *Hae*III cut blunt ends. The enzymes are coded according to their source of origin, and almost 200 are now available.

they cut the DNA at specific sites (*Eco*R1 at G ↓ AATTC, for example) and leave cohesive terminal tails, as indicated in outline in Fig. 2.5. Alternatively, the DNA may be sheared physically, or cut with a nuclease which gives blunt-ended preparations; or a complementary DNA (cDNA) may be made from an mRNA template using a reverse transcriptase. These blunt-ended preparations are joined using T4 DNA ligase, but usually a terminal transferase is used to attach deoxyadenylic acid 'tails' to the 3′OH ends of, say, the foreign DNA, and reciprocal thymidylic acid 'tails' to the ends of the plasmid DNA (Fig. 2.5). The foreign DNA is then annealed into the broken plasmid chromosome to re-establish the plasmid (or phage) circle, and used to transform (or transfect) a bacterium, usually *E. coli*, and is there multiplied. Many variants of this methodology are being developed to improve its efficiency (see Old and Primrose, 1980; Maniatis *et al.*, 1982).

The nature of the eukaryote DNA insert is known, of course, when it is, for example, a satellite DNA, or an rDNA, which has been separated by density gradient centrifugation or by some other procedure, or if it is a cDNA made from a known mRNA using reverse transcriptase (see Glover *et al.* (1975) for some of these applications). But most studies involve the use of a gene 'library' (or 'bank'), which is a set of plasmids (or of phage) carrying the complete gene complement of the organism distributed in fragments across the plasmid collection. Usually, the genome DNA sample is sheared to fragments of 10 or 20 kb (kb = thousand bases) lengths, and fractionated by electrophoresis to give a population of random samples of the genome of this size. Relatively large (20,000–50,000) plasmid populations have to be made to ensure that all the genome is in the library, the actual numbers depending on genome and fragment sizes (see Maniatis *et al.* (1978) for examples). Libraries exist for all the important organisms, and there are many separate libraries of the genomes of the commonest experimental animals and unicells. The difficulty of making pure DNA limited work with plants, but plant genes can now be cloned like any other DNA.

As with many libraries, the problem is then to find what one wants; which plasmids contain the DNA of interest, in whole or in part. When a particular

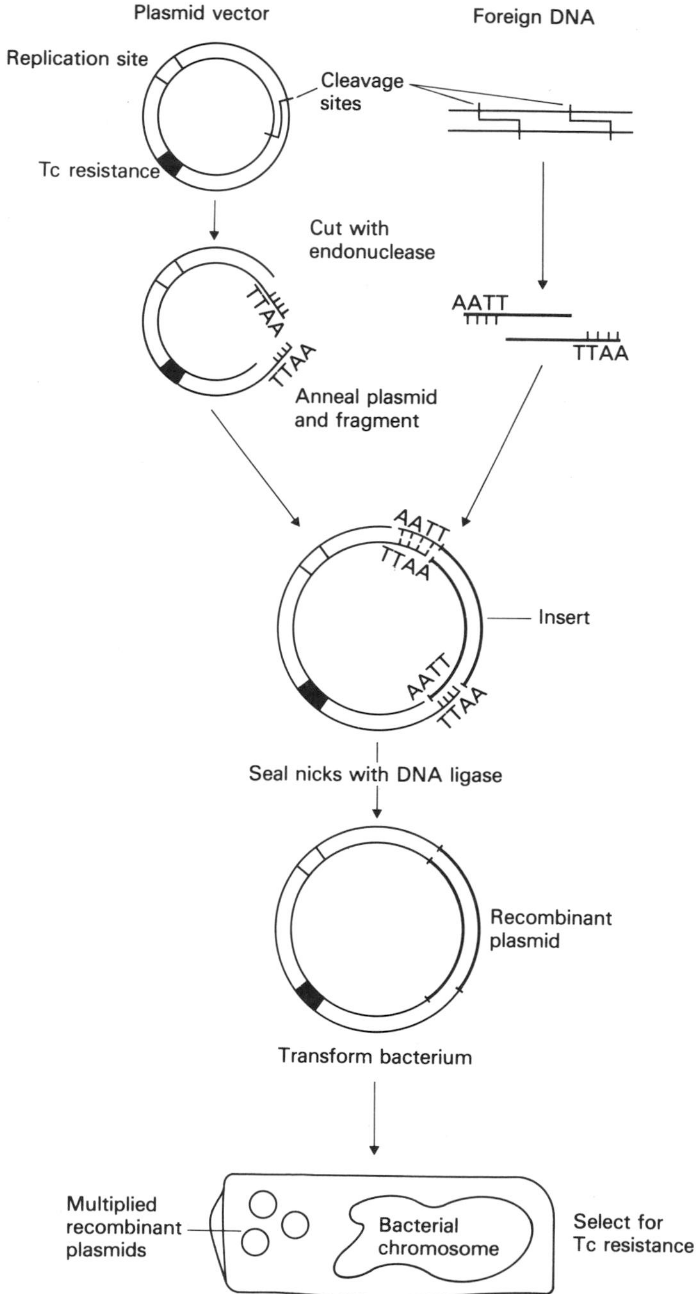

mRNA is predominant in a cell, or tissue, like the mRNAs for globin, or ovalbumin, or silk fibroin, that mRNA can be isolated, purified, and made radioactive *in vitro* using polynucleotide kinase. The mRNA can then be used as a probe to search among the plasmids (or viral plaques) for those carrying DNA sequences which anneal with it. One general technique (Grunstein and Hogness, 1975) is to grow the bacterial colonies on nitrocellulose discs placed on the surface of the usual agar plates. Replica plates of the colonies are made and stored, and the original colonies lysed, their DNA denatured with NaOH and fixed as a DNA 'print' on the nitrocellulose filter. The defining radioactive 'hot' RNA probe is then hybridized to any complementary DNA present, and identified by autoradiography. In this way colonies carrying the relevant DNAs are located, and their identified replicas on the stored plate can then be multiplied. A similar procedure can be used for recognizing relevant phage plaques (Maniatis *et al.*, 1978).

The DNA fragment in the clone has to be excised from the plasmid circle by a restriction enzyme, and because it is still 10–20 times longer than an average gene it is cut again with another restriction enzyme, or with endonuclease S1. These DNA fragments are separated by size on an agarose gel, and the alkali-denatured DNA of the gel strip is then drawn through onto a nitrocellulose filter strip (by placing a buffer solution below and a filter paper above it) where it attaches and forms a 'Southern blot' (Southern, 1975). This array of DNA fragments on the nitrocellulose paper is then probed with the radioactive mRNA, as before, to identify, by autoradiography or fluorography, those fragments carrying the complementary sequences of the gene involved. In their turn, the positive fragments can be further amplified by another round of gene cloning. Thus, when an mRNA probe is available, it is possible to find the complementary DNA clone in the gene library, and to make further clones limited to the gene in question and to its adjacent DNA sequences. (This procedure can be inverted to make a 'Northern blot'. RNAs are separated on agarose and blot-transferred onto diazotized filter paper to which they bind covalently, and where they can be probed with a radiolabelled DNA (Alwine *et al.*, 1977)).

Since the library is made up of relatively large DNA fragments, a number of clones will carry the required gene at different places along their length.

**Fig. 2.5** Construction of a recombinant plasmid. In this example the restriction enzyme *Eco*R1 is used to cut both the vector and the foreign DNA to provide the sticky ends which allow the two to be annealed. The strand nicks are sealed *in vitro* using DNA ligase, and the complete recombinant circles are used to transform *E.coli*. The concentrations of the chimeric plasmid and of bacteria are chosen to average one plasmid per cell, but some bacteria will then, by chance, not be transformed. These bacteria are eliminated by exposing them to the antibiotic tetracycline since only the plasmid-carrying bacteria have the tetracycline (Tc) resistance gene brought in with the recombinant DNA. A great many other ways of engineering recombinant DNA vectors are summarized by Glover (1980) and by Maniatis *et al.* (1982).

They will also carry adjacent sequences common to other clones. Some such clones may bear a second, neighbouring gene, but not the first, and it is then possible, by DNA–DNA hybridization, to use the first clone to probe for clones carrying genes contiguous with the first gene. By repeating this process, it becomes possible to 'walk' along the genome from a known starting gene locus to other gene loci.

Because we know so much about the location of *Drosophila* genes, it was to be expected that the first 'walks' would exploit *Drosophila* gene libraries. *Drosophila* has a further advantage: the cytological DNA–DNA hybridization technique (p. 43) can be applied to the giant salivary gland chromosomes, and the chromosomal location of a 'hot', cloned DNA fragment recognized as a radioactive spot somewhere on the chromosomes (see Fig. 3.5, p. 81). Using this method, the direction of a 'walk' can be determined, too, and a gene whose mRNA is not known can be identified in a clone, according to its chromosomal location.

### *Analysis of gene structure*

A major objective of cloning a gene is to define its molecular structure, and to relate this to its function. The first task involves determining its base sequence and comparing it to the sequences of other known genes, and the second often involves returning the gene to a eukaryotic cell to see what it does (see Ch. 12), or transcribing its message *in vitro* to identify the product. Since sequencing kilobases of DNA is expensive and not always

---

**Fig. 2.6** Preparation of a restriction (enzyme site) map of a DNA fragment (in this case containing histone genes). The fragment is excised from the plasmids (e.g. by *Eco*R1 digestion of the chimeric plasmid of Fig. 2.5), separated by electrophoresis and then extracted from the gel. The 5 ′-phosphoryl groups are removed with a phosphatase and replaced by radioactive phosphate, using [$^{32}$P]ATP and T4 polynucleotide kinase. The two ends are then labelled, and the fragment is then cut in two with another nuclease and the two subfragments separated and sized. The two fragments each now carrying label at one end, are analysed separately. Under conditions which allow only partial digestion by the endonucleases chosen for mapping, a series of overlapping digest products will be formed, each identifiable by its radioactive terminus (R), i.e. fragments RA, RAB, RABC, etc. These fragments are size separated by electrophoresis on agarose gels, and identified by autoradiography. The figure shows the autoradiographs of gel tracks of a digested 5.1 kb fragment from the sea urchin histone gene complex, simplified from the data of Smith and Birnstiel (1976). Standards are run in the first track, but are here represented arbitrarily to emphasize the logarithmic scale of the size separation. The sites of action of the three endonucleases in the succeeding tracks are read from the scale and mapped to the DNA, as shown at the foot of the diagram. The endonuclease sites are not randomly arranged either within the genes (blocks) or their spacers (lines), and this was confirmed using other endonucleases and fragments. Restriction maps provide a rapid identification of sequence arrangements (Table 2.3) which allow comparisons, for example, of the genetic organization of the units of the histone complex, and the like.

necessary, the organization of the DNA is often 'mapped' by finding the locations along it which are broken by different restriction enzymes—i.e. making a DNA restriction site map.

The cloned eukaryotic DNA is often engineered into its vector so that the whole fragment can be excised by a single restriction endonuclease. The fragment can then be separated from the rest of the plasmid (which may also be broken) by electrophoresis on agarose slab gels; and so recovered. Using Smith and Birnstiel's (1976) method, the 5′ end of the DNA is then labelled with $^{32}P$, using polynucleotide kinase, to give the primary orien-

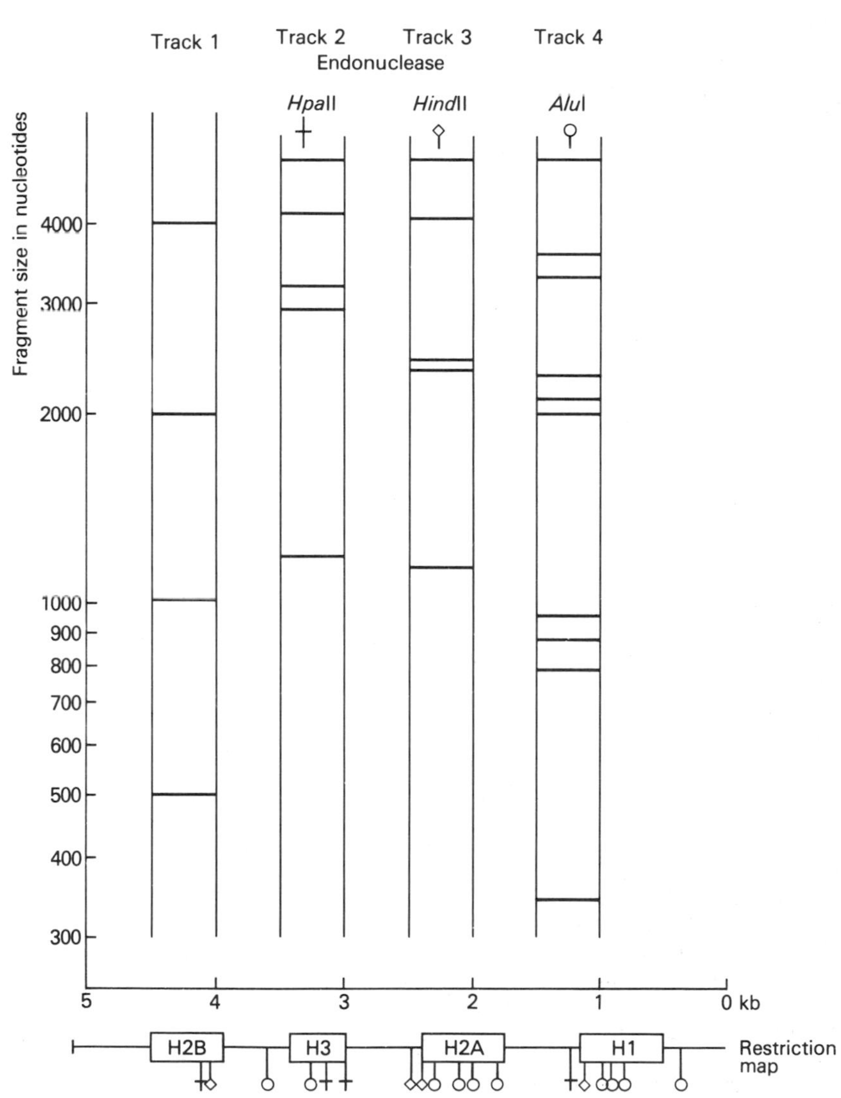

tation of the molecule. Complete digestion with a restriction enzyme will then cut the DNA into fragments at its target sites, and these can then be sized on agarose (Fig. 2.6). Since the enzyme attacks its sites at random, a partial digest will make an overlapping series of fragments, and those with the $^{32}P$ label will all have a common terminus. A restriction map can then be made from analysis of the size distribution of these partial digest products as separated in the agarose gel. Usually a number of digests using different enzymes are run together to map the distribution of their sites of attack along the DNA (Fig. 2.6), but sometimes two enzymes may be combined so that the site of action of the second may be identified within the small DNA fragment made by the first, or the fragments made by one endonuclease may be further digested using a second (e.g. Saint and Egan, 1979).

Restriction maps permit a more or less detailed comparison of genes which are repeated in the genome, and identification of any differences which may have evolved in them. And where the map suggests that the fragment contains repeated sequences (i.e. there are repeats of the patterns of endonuclease attack), these fragments can be isolated, cloned and used to probe other parts of the genome for the same repeats. Or, again, the restriction fragments may provide convenient material for sequencing, using one or other of the available rapid sequencing techniques (Air, 1979) which would be difficult to apply to large, uncut DNA fragments. Thus cloned DNA can be analysed in detailed ways which are not possible when whole chromosomal DNA is being handled. We shall now look at some of the results coming from the molecular analysis of DNA fragments.

### *Organization of some cloned genes*

The first genes to be isolated and identified in clones were those whose DNA, or mRNA, could be readily separated and identified, like the genes for histones, rRNA, 5S RNA and tRNA; or genes whose product was the predominant RNA of a particular cell, such as globin in erythroid cells, or immunoglobulin in mouse myeloma cells, where the mRNA could be separated and used to probe the DNA library. Advantage was similarly taken of tissues or organs which responded by making a specific protein, and thus mRNA for a probe, when exposed to a hormone, as in the production of ovalbumin by the estrogen-stimulated chick oviduct. It is only recently that genes, and in some cases mutant genes, have been identified by 'walking'. It is already obvious that the more genes identified in a library the more rapidly will further genes be found, and we must expect a logarithmic expansion of information on the detail of gene organization. Here we can only sample existing studies, taken from the three groups of work just indicated, to look at gene organization. We shall pay some attention to the hormone-induced systems since they suggest how this developmental problem may be explored.

## Universally required genes

All cells must have the genes for cell division, transcription, translation, etc. and for 'housekeeping' metabolism, and it is reasonable to expect that these genes will be similarly, but not identically, organized among the eukaryotes. The histone genes are among the most extensively characterized of the genes coding for structural proteins, and they are found in clusters, each cluster unit being repeated hundreds of times. In sea urchins, the five major kinds of histone are encoded in a 6–7 kb repeated unit (Fig. 2.7) in which the coding sequences are separated from one another by 'spacer' sequences which account for ≏ 60 percent of the DNA (Schaffner *et al.*, 1978). These spacers are AT-rich and often contain short runs of homopolymers, and they are frequently different in length and sequence pattern among individuals of the species. The coding regions (and flanking sequences?) are conserved between species, but the spacers may show little homology (Cohn and Kedes, 1979). Studies of the histones made during development show that at least three subtypes of H2A, H2B, and H1 are made (Weinberg *et al.*, 1978), and in *Lytechinus pictus* two 'non-allelic' cluster units have been distinguished, showing that one can have a large number of copies of a limited number of unit types. *Drosophila* also has two types of unit, differing only in the size of their H1 region, and both types show a difference in the pattern of unit transcription (Fig. 2.7) when compared with the sea urchin units (Lifton *et al.*, 1977).

The cloned histone gene clusters show that coding sequences are separated by non-coding spacer sequences of about the same size as the structural genes themselves. Nothing is known about regulatory sequences within these spacers, but the *Drosophila* organization implies that these may be differently organized in different groups of organisms. 5S RNA and rRNA genes are also redundant, and the units are again separated by spacers. In *Xenopus* there is a single, autosomal nucleolus organizer of 400–500 copies

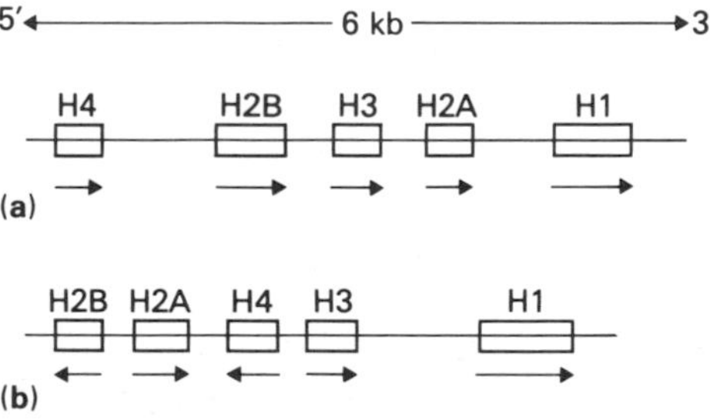


**Fig. 2.7** Organization of histone gene repeat units (**a**) of the sea urchin (see preceding figure and Schaffner *et al.*, 1978) and (**b**) of *Drosophila* (Lifton *et al.*, 1977). There is evidence that some *Drosophila* histone gene clones carry an insert between H3 and H1. The direction of transcription (arrows) is from the 5′ end for all the sea urchin genes, but from opposite strands in *Drosophila*. This argues against a polycistronic (multigene) transcript which is subsequently processed to make the five mature histone mRNAs.

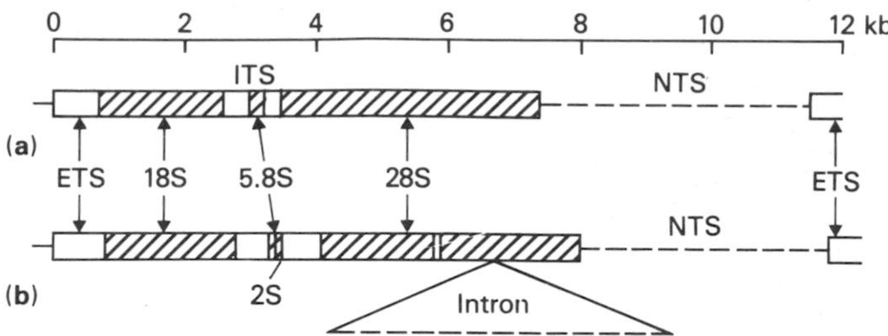


**Fig. 2.8** The genomic organization of the ribosomal gene repeat units of (**a**) *Xenopus* and (**b**) *Drosophila*, after Dawid *et al.* (1978) and Jordan and Glover (1977). Transcription is from left to right in both. It starts from external transcribed sequences (ETS) and runs through internal transcribed sequences (ITS), which are not conserved in rRNA and are shown as open blocks. The non-transcribed spacers (NTS and dotted line) vary in length, as does the inserted sequence in the *Drosophila* 28S DNA. This intron, which does not coincide with the interruption of the normal 28S sequence, is found in only some DNA clones, and the 'normal' arrangement lacks it. The 2S gene is found only in the *Drosophila* unit, adjacent to the 5S sequence.

of rDNA units, and the spacers may be transcribed, or not (Fig. 2.8). The non-transcribed spacer varies greatly in length (2.7–9 kb) and is composed of similar, short, 15 nucleotide, repeats (Botchan *et al.*, 1977). The transcription units, on the other hand, are virtually constant in length and sequence composition. *Drosophila* has two nucleolus organizer regions, one on the X and the other on the Y chromosome, each containing ≏ 250 repeat units. However, *Drosophila* has two types of repeat unit (Glover and Hogness, 1977): the first being like the *Xenopus* unit, and the second being interrupted within the 28S segment (Fig. 2.8). The mature 28S rRNA lacks this insertion and is continuous at this site, and the insertion can be identified as a single strand DNA loop in rRNA–rDNA duplexes. This was the first, surprising, evidence that coding sequences (exons) could be interrupted by insertions (introns), and we shall see that the intron–exon arrangement (Gilbert, 1978) is common among, and typical of, eukaryote genes. We should also note that two types of intron are found in this second class of rDNA, type 1 being absent from the Y chromosome and type 2 making about a sixth of both X and Y rDNAs (Wellauer *et al.*, 1978). Sequences homologous to these introns are found throughout the *Drosophila* genome (Dawid and Botchan, 1977; Kidd and Glover, 1980).

5S DNA is differently organized in *Drosophila* and *Xenopus*. *Drosophila* has a single cluster of 160 repeat units on chromosome 2, whereas *Xenopus* 5S genes are found at the telomeres of all chromosomes (Fig. 2.9). In both cases the structural genes are 120 bp long, and the difference in the repeating unit lies in spacer organization; 250 bp (AT rich) at the 5′ end of the *Drosophila* unit, and a small 15 bp unit at the 3′ end (Artavanis-Tsakonas *et al.*, 1977). In *Xenopus* there are two types, one transcribed during oogenesis and the other during post-embryonic development (Ford

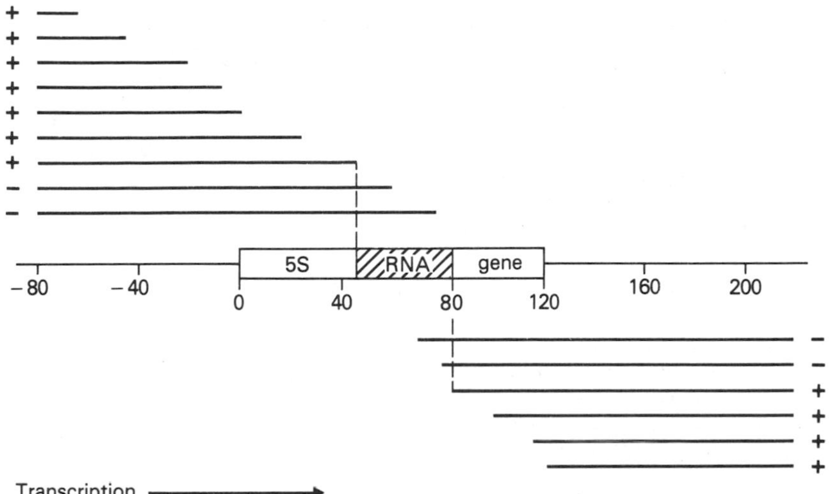


**Fig.2.9** Identification of the control region of the 5S RNA gene by deletion analysis *in vitro* (see text). The deletions made from the 5′ end of the plasmid are shown by the lines above the gene, and those from the 3′ end by the lines below it. Mutations which support transcription are indicated by +, and those which do not by –. All the mutations which allow transcription leave the centre (shaded) of the gene undamaged, indicating that the transcription control region is intragenic and can be defined by this technique.

and Southern, 1973). The oocyte family of 24,000 copies has a variable (360–570 bp) spacer, and the structural gene is followed by a small spacer and a sequence-related, GC-rich untranscribed 'pseudogene'. None of these genes has introns.

The somatic 5S gene of *X. borealis*, which has an untranscribed 750 bp, GC-rich spacer, and a 120 bp gene region, is particularly interesting. This somatic 5S repeating unit must contain the signals for initiation and termination of transcription since the cloned unit directs accurate 5S RNA synthesis when either injected into oocytes (Brown and Gurdon, 1978), or processed in an oocyte nuclear extract where polymerase III is the active enzyme (Brown and Jordan, 1978). The latter assay system has allowed Sakonju *et al.* (1980) and Bogenhagen *et al.* (1980) to ask which region of the unit is essential for transcription, by using exonuclease III to digest away parts of the unit from either the 5′ or the 3′ end, after its linearization by excision from the plasmid carrying it. The fragments obtained are recloned, and then tested for transcription in the nuclear extract system. The results of these 'genetics *in vitro*' experiments are summarized in Fig. 2.9, and lead to the surprising conclusion that the control region for transcription lies in the centre of the gene, between bases 41–87. This fragment itself has been cloned and tested, and it is clear that the surrounding regions are not required for transcription.

The experiments show two further points of interest. In the complete repeating unit, transcription starts at the first nucleotide of the gene itself, of course, and we might anticipate that removal of this residue and its neighbours would block initiation. On the contrary, when the gene is cut into from the 5′ end, linking and plasmid sequences replace the excised gene regions when the DNA is cloned, and transcription starts within these foreign sequences to give transcripts between 116 and 121 nucleotides long, compared with the normal 120. Hence, initiation is not necessarily sequence dependent, but is started at a more or less set distance from the control region (Sakonju *et al.*, 1980). Note, too, that these RNAs are fused transcripts, initiated in the vector or linking sequences, transcribed through the remnant of the 5S gene and terminated at its end. On the other hand, removal of the final sequences from the 3′ end (TTTT on the non-coding strand) stops termination. Consequently, studies of 'deletions' of the 3′ end of the molecule involved adding cordycepin triphosphate to the assay system, since this analogue (of 3′-deoxyadenosine) terminates transcription artificially whenever it is incorporated in place of ATP.

RNA polymerase III is a large (700,000 molecular weight) multimeric enzyme, so it may be that it contacts a large part of this small gene and has recognition, binding and initiation properties, but this is uncertain. What is important is that this first major exercise in genetics *in vitro* has demonstrated the power of genetic engineering techniques for the analysis of gene structure and organization.

The genes for tRNAs are also redundant: in *Drosophila* there are 90 or so different 'species' each represented by about a dozen copies. They are scattered throughout the genome, as shown by hybridization to salivary gland chromosomes (Hayashi *et al.*, 1980), but only a few have been cloned and their organization is still uncertain (Dudler *et al.*, 1980). We shall make only one point about a cloned tRNA. Yeast tyrosine tDNA has a short 14 bp intron near its 3′ end, and this is spliced out when the RNA is made. DeRobertis and Olson (1979) have tested the generality of this splicing mechanism by microinjecting the cloned complete gene into a *Xenopus* oocyte nucleus, and find that frog nuclei will process the introns out of yeast tRNA. This shows, too, that frog polymerase III can transcribe the yeast RNA faithfully. The same oocyte system will process rRNA (requiring polymerase I) and structural genes (requiring polymerase II), and this provides a rapid technique for *in vivo* analysis, although polymerase I and II are there less efficient in their action than polymerase III (see also p. 331).

These results with universally required genes provide a detailed picture of genome organization which could not readily have been deduced from the DNA hybridization studies described at the beginning of the chapter. The interspersion of genes with repetitive sequences turns out to involve the separation of genes by spacer regions, some of which are frequently redundant and which are not transcribed. More surprising is the discovery of introns which are processed out of the RNA, as are transcribed spacer sequences. Each gene, or repeated gene unit, it would seem, has its own arrangement of flanking sequences and introns. We must now ask if the same is true of structural genes.

**Structural genes**

We shall consider the molecular organization of some structural genes in later chapters. Here we shall look at the well-studied globin genes for what they can tell us about genome organization.

Haemoglobins have been studied intensively during this century because of their medical importance, and their amino acid sequences are known. It is therefore not surprising that the genes responsible for the pairs of globin units, which constitute the final haemoglobin tetramer, were among the first to be studied using gene cloning techniques (for a review see Leder (1980)). The ready availability of globin mRNA in red cells provided the necessary probe, and the globins from human, mouse, rabbit and chick gene libraries have all been identified. Most work has been done using the human genes, which are remarkably similar in organization to those of other mammals.

The human α-like and β-like globin genes have many homologies, indicating a common origin, but they are now found on separate chromosomes and exist as clusters (Fig. 2.10). Both α-like and β-like genes contain pairs of introns, and while there are sequence differences in the exons (which, of course, identify the globins) and in the introns, the intron insertion sites are regularly placed and characteristic of the α-like and β-like types (Fig. 2.10). The same insertion sites are found in mouse and rabbit, and each individual exon-intron complex is transcribed as a unit, as an hnRNA (heterogeneous

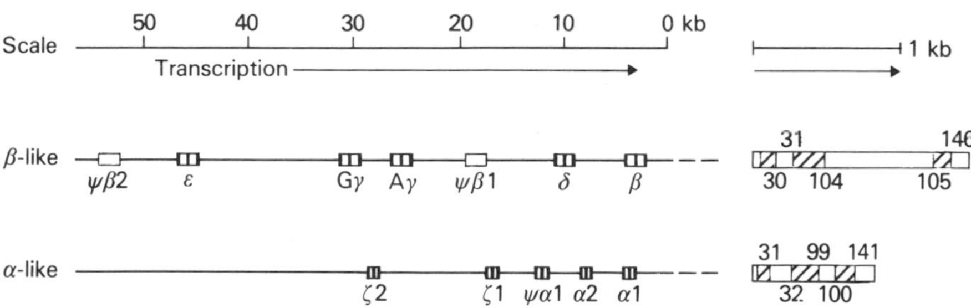


**Fig. 2.10** The chromosomal organization of human α-like and β-like globins (left) and the detailed organization of these genes (right),after Proudfoot *et al.* (1980). The untranscribed α-like pseudogene (Ψα1) carries two introns, but the two untranscribed β-like pseudogenes (Ψβ1 and Ψβ2) do not. The α-like genes have a ±95 bp insertion (unshaded) between coding sequences 31 and 32, and a ± 125 bp insertion between codons 99 and 100. The β-like genes have 122–130 bp inserts between codons 30–31 and 850–900 bp inserts between codons 104–105. The haemoglobin tetramers generally have two each of an α-like and a β-like polypeptide—the embryonic being $\varepsilon_2\zeta_2$, $\alpha_2\varepsilon_2$, $\zeta_2,\gamma_2$, $\alpha_2\zeta_2$; the foetal $\alpha_2\gamma_2$, and the adult $\alpha_2\delta_2$, $\alpha_2\beta_2$, again illustrating the regulation of transcription during development.

nuclear RNA) which is then processed to form the shorter, intron-lacking, mRNA. Each cluster also contains 'silent' untranscribed genes. These pseudogenes have a 75–80 percent homology with their normal confreres, but they carry insertions or deletions which act as frameshift mutations and cause the premature termination of their transcription.

The members of each cluster are spaced widely apart, and the spacer sequences contain members of the *Alu* family of interspersed, repeated sequences which constitute ~ 5 percent of the human genome (and of the genome of other mammals surveyed, according to Jelinek *et al.* (1980)). *Alu* is a 300 bp sequence which is cut in half by the restriction enzyme *Alu* 1 (hence its designation) and is repeated some hundreds of thousands of times in the genome. Although it is transcribed and found among the hnRNAs, its function is unknown. The *Alu* family almost certainly constitutes part of the short, highly repeated DNA complement noted earlier.

Rather surprisingly, the globin genes are ordered on the chromosomes: the embryonic lying at the 5′ end, the adult at the 3′ end with the foetal in between. Nothing is yet known of the regulation of this developmental progression. However, studies of deletions which cause the hereditary persistence of foetal haemoglobin and β-thalassaemia suggest that (deleted) sequences near the 5′ end of the δ gene may be involved in the *cis* suppression of the distant foetal γ-globin genes in adults (Fritsch *et al.*, 1979). This might imply that the complex is regulated as a whole, or that the conformation of the chromatin, and with it the spatial relations of the DNA sequences concerned, is important for gene regulation.

Studies of DNA methylation, which is a post-replicative event converting cytosine to 5-methylcytosine, have shown that inactive members of the γ-δ-β globin cluster are highly methylated, whereas the active members are less so (Van der Ploeg and Flavell, 1980). The technique for showing this is to use the restriction enzymes *Hpa*II and *Msp*1; the former of which cuts CCGG but not $C^{m}CGG$, whereas the latter cuts both. Thus methylated sites can be identified by comparison; and the DNAs of different cell and tissue types can be probed using cloned DNA of the β-like globin cluster using the Southern blot technique. The results (Fig. 2.11) suggest that active globin genes are less methylated than inactive ones.

It has been suggested that DNA methylation may be a general mechanism for gene inactivation (Razin and Riggs, 1980). The methylase acts preferentially on half-methylated strand pairs, i.e. on $\begin{smallmatrix}-C^{m}G-\\-G\ C-\end{smallmatrix}$ as opposed to $\begin{smallmatrix}-CG-\\-GC-\end{smallmatrix}$, so that methylation would be clonally inherited as required for its cell-by-cell perpetuation. However, since sperm DNA, and the DNA of early embryos, is highly methylated (Fig. 2.11), gene activation would involve the regulated activity of a demethylase which has not yet been identified. Still, growing mouse cells with 5-azacytidine, which blocks methylation, does indeed result in their differentiation into a variety of forms (Taylor and Jones, 1979). Even genes on the inactive X chromosome (p. 125)

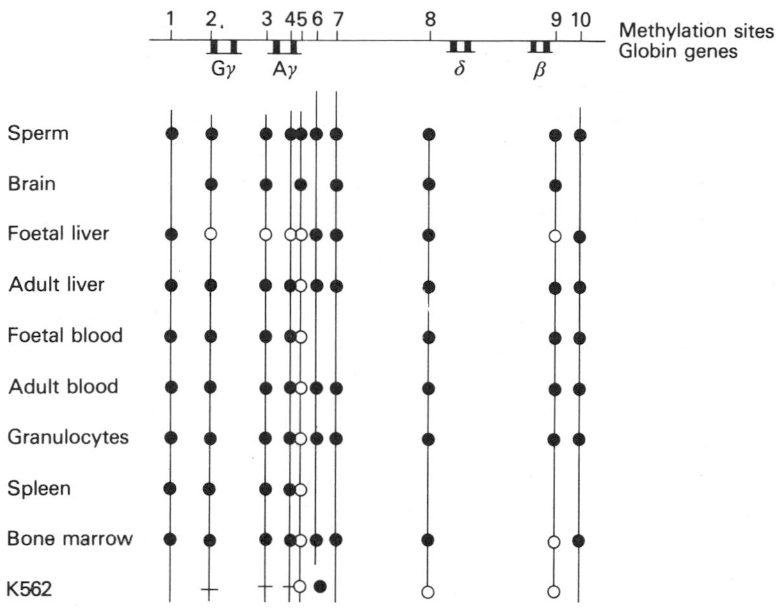


**Fig. 2.11** The relationship between methylation and gene expression as seen in the β-like globin gene array, simplified from Van der Ploeg and Flavell (1980). The methylation sites, as identified using *Msp*1/*Hap*II nucleases, are numbered above the line, and the location of the globin exons below it. Full methylation is shown by closed circles, part methylation by open circles and unmethylated sites by a dash. Blanks were not tested. The sites are totally methylated in sperm and foetal brain where the genes are not expressed, and the γ-like genes are methylated in the K562 leukaemic cell line which expresses them. Unfortunately, the other tissues represent mixed cell samples where there is evidence that some cells are expressing the genes (e.g. γ-globin in foetal liver) but others are not. Methylation may be a necessary, but not sufficient, condition for suppression of gene activity; it may also involve spacer sequences adjacent to the genes. See also Shen and Maniatis (1980).

are re-expressed when 5-azacytidine is added to the culture medium of a hybrid mouse cell line in which the reactivated gene can be detected by its particular electrophoretic properties (Graves, 1982). The fact that *Drosophila* has little 5-methylcytosine in its DNA cannot be taken to exclude this possible mechanism of gene regulation since it is not claimed that this is an exclusive system; it is only one way in which some genes may be inactivated.

Finally, the globin genes are all transcribed in one direction from one strand of the DNA. Scrutiny of their flanking sequences (Nishioka and Leder, 1979) shows some suggestive similarities. There is a TATAA$^{A}_{G}$ hexanucleotide regularly sited 30 nucleotides upstream from the 5′ putative initiation site. This TATAA, or 'Hogness box' is also found in the same position in other eukaryotic genes such as the histone gene array and the ovalbumin gene. Its likeness to the 'Pribnow box' of prokaryotes suggests it may be the signal for transcription initiation. Two other 'boxes' appear to be associated with the sites of post-transcriptional modification of the mRNAs. The cap (7′-methylguanosine added to the first nucleotide of eukaryote messages) is possibly signalled by TTGCT, and there are regions of homology about 25 nucleotides before the poly(A) addition site, centred on the hexanucleotide AATAAA. However, there is no evidence of like sites which might be involved with the coordinate control of the paired genes that are simultaneously transcribed.

It is not known if the embryonic, foetal and adult haemoglobins are made by successive members of the same cell family at different stages during the differentiation of the clone, or by cells of different origins. In either event, the production of the different haemoglobins reflects different states of cell differentiation, and this is generally the case for families of related, but non-identical sequences. Such multigene families need not be clustered, and in the case of the actin genes (which have been so highly conserved during evolution that a chick actin gene can be used to probe for *Drosophila* loci) six different actin genes are widely dispersed in the *Drosophila* genome (Fryberg *et al.*, 1980). Most vertebrates also have six actin genes, but *Dictyostelium* has about fifteen. The individual *Drosophila* (or chick) actins are not all found in one cell type, and muscle-specific actins have been identified and cloned (the *Drosophila* actin has an intron). Members of multigene families therefore tend to be expressed in particular cells. The mutation of any one may cause specific defects which cannot be covered by the activity of the others, and the existence of these families of variants does not negate the finding from mutation studies, namely, that structural genes exist as single functioning copies. The differential activation of members of a multigene set exposes an important aspect of differentiation, and we can expect more attention to be paid to these and other repeated genes.

## Hormone-activated structural genes

One set of hormone activated genes is of particular interest: those specifying steroid hormone-induced proteins. We have a number of different systems which, together, should give us the information needed for understanding the molecular basis of hormone action, and hence of hormone-induced differentiation.

***Chick ovalbumin and related genes*** When estrogens are applied to the immature chick oviduct it secretes the four egg white proteins: ovalbumin, ovomucoid, conalbumin and lyzozyme. The mRNAs for these proteins then become abundant and can be isolated, providing a system which can be regulated at the experimenter's discretion and involving the coordinate control of four genes. As a bonus, conalbumin (transferrin) is also made in the liver where it is not so strictly under hormone control and must therefore be differently organized. So far most work has been done with the first three proteins, and their organization has proved to be exceedingly interesting.

When the cloned ovalbumin gene is compared with the cDNA made from the ovalbumin message, it is found to contain *Eco*R1 and *Hin*dIII sites absent from cDNA. This can be explained only if the chromosomal gene is interrupted, or split, by DNA sequences not mirrored in the cytoplasmic mRNA (Breathnach *et al.*, 1977), as we have seen. Subsequent work shows that the gene is made up of eight blocks of structural sequences (exons) separated by seven intervening sequences (introns). The map of the gene is given in Fig. 2.12. If denatured DNA is hybridized with pure mRNA we see in the electron microscope that gene and message are indeed not colinear, they form R loops. The seven introns form loops, and their sizes can easily be related to the restriction map. About 7.6 kb of DNA are needed to code for the 1859 nucleotides of the mRNA (Dugaiczyk *et al.*, 1979; Chambon *et al.*, 1979). Since a heterogeneous nuclear RNA (hnRNA) of the right size and character is found as a precursor in the nuclei of oviduct cells, the total gene must be transcribed, introns and all (Roop *et al.*, 1978). During the processing of the hnRNA, the introns must be excised and the exons spliced together, as for other structural genes. Sequencing of intron–exon boundaries suggests that there are directly repeated sequences which must in some way be involved in this unexpected process (Lewin, 1980a, b). DNA from a non-secretory tissue such as the erythrocyte, which never makes ovalbumin, shows the same intron–exon organization, as we should expect from previous considerations of nuclear totipotency.

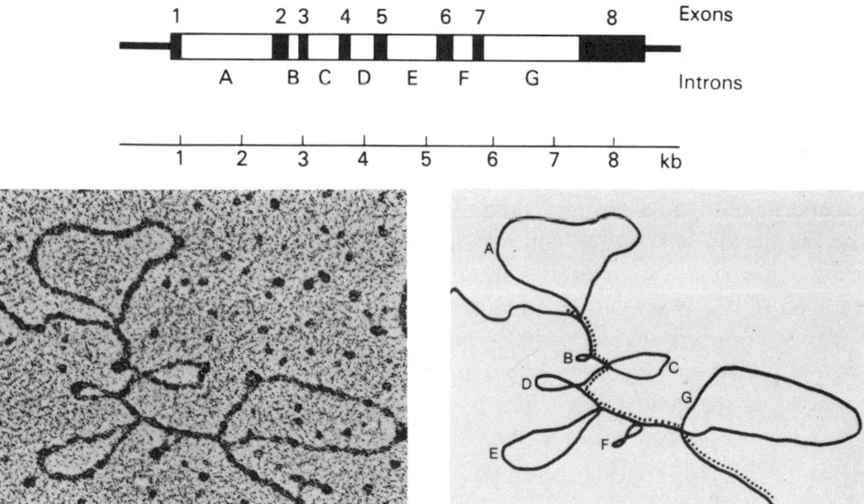


**Fig. 2.12** Organization of the natural ovalbumin gene (Dugaiczyk *et al.*, 1979). The flanking sequences are indicated by the thin line, the eight exons by the thick closed lines, and the seven introns by the thick open lines. The electron micrograph shows a hybrid molecule between the total ovalbumin gene region DNA and ovalbumin mRNA; the line drawing indicates the intron loops which are then excluded. By courtesy of Dr. A. Dugaiczyk.

Ovalbumin was studied first because its mRNA makes up half the total in oviduct tubular gland cells. It is more difficult to isolate ovomucoid mRNA, but with this as a probe, clones of the ovomucoid gene have been identified in a chick 'library' (O'Malley *et al.*, 1979). The gene again bears six, probably seven, introns, and the exons are very short since this mRNA is only about 850 nucleotides long. Similarly conalbumin gene clones suggest that 17 very short coding sequences (75–200 bp) may be interrupted by 16 introns. The conalbumin mRNA is about 2400 nucleotides long compared with a 'gene' length of 10.3 kb (Cocker *et al.*, 1979). No differences have been found in the conalbumin genes isolated from oviduct or from liver (Chambon *et al.*, 1979).

The flanking sequences of each of these proteins have been determined and they show two homologies, which have also been found for other genes. The TATA box is found in the same situation as before. In addition, seven nucleotides after the 3′ end of the coding sequence there is a tetranucleotide (TTGT), also found in other sequences, which is taken to be the termination signal. As yet, there is no allocation of sequences to hormone activation functions.

***Vitellogenins*** Estrogens also induce the synthesis of yolk precursors (vitellogenins) by liver. This is being studied using *X. laevis* where cDNAs have been used to identify four vitellogenin genes which are simultaneously expressed. These fall into two classes, genes A1 and A2 and genes B1 and B2, each of which is sequence homologous within 5 percent as judged by heteroduplex annealing, whereas the A and B classes differ by about 20 percent. A1 and A2 have been cloned, and R loop analysis shows that each carries 33 introns, which have different lengths and sequences in the two genes. In other words, the introns show divergence, as we might expect with non-functional sequences (Wahli and Dawid, 1980). These, and the preceding intron-interrupted genes, raise interesting questions concerning the regulation of splicing, since exon 1 must join exon 2, and not exon 3 or 4 etc., implying a segmental, directional process (Crick 1979; Lewin 1980a, b). The flanking sequences of the A genes have not yet been analysed.

*Drosophila* provides a parallel system of vitellogenin synthesis (Bownes, 1982). Three yolk proteins are synthesized by the fat body of the mature female, and transferred via the haemolymph to the developing ova. This synthesis is apparently induced by the pupation hormone, ecdysterone, and can be manipulated using cultured ovaries. As with the estrogens, intact males can be caused to synthesize vitellogenins when they are exposed to hormone although male and female fat bodies differ in that the latter alone continue to make yolk after the ecdysterone stimulus is withdrawn. Ovaries transplanted into males make yolk, and this comes not from the fat body but from the ovary itself, which secretes the same three vitellogenins (as judged by electrophoretic behaviour) and then transfers them to the ova (Postlethwait *et al.*, 1980). The stimulus for this secretion appears to be juvenile hormone, and in females we have the curious situation of two organs, fat body and ovary, making the same specific proteins under the control of two, usually antagonistic, hormones—ecdysterone and juvenile hormone. The *Drosophila* yolk protein genes have been cloned, and sequenced (Hovemann and Galler, 1982; Hung and Wensink, 1981).

We have digressed from gene structure studies since it seems likely that the three systems just considered may provide the detailed information we need concerning the sequence organization involved in hormone regulation of gene activity. However, the situation may be more complex that we have so far implied. These steroid hormones are transported in the cell on specific receptor proteins which themselves have a developmental history (e.g. estradiol receptors appear only after 10–14 days of chick embryogenesis) and we need to know more about them and how, in this case, they are stimulated to increase five-fold by the hormone itself. Further, cDNA clones identify a low level (4 and 30 molecules per cell) of ovalbumin and conalbumin mRNA production in the unstimulated oviduct (Moen and Palmiter, 1980), and it remains to be seen if this is a consequence of earlier embryonic hormone production. Nothing is yet known about the development of the *Drosophila* receptors, but it is interesting that the hormone induces the production of an entirely different set of proteins in the larval fat body, larval serum proteins 1–3, (Roberts and Brock, 1981). The fat body thus provides a second system for study.

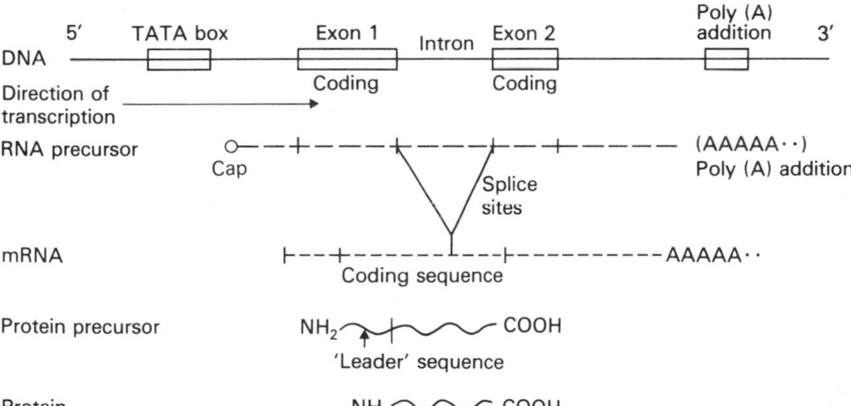


**Fig. 2.13** The relation between the DNA sequences of a typical gene and its protein product. The TATA box (promoter) and the poly(A) addition site are indicated as well as two exons separated by a single intron. As noted in the text the number of introns may range from none to many. The precursor RNA is shown as capped (i.e. carrying an additional 7-methylguanine (or similar modification (see Lewin, 1980c) at the 5′ end) and with its added poly(A) tail. The intron is spliced out of the mRNA. The precursor protein is shown as having a short leader, or 'signal' sequence (usually of 15–30 amino acids) which distinguishes secretory from other proteins. This allows insertion of the growing protein into the cell membrane, after which the sequence is cut out by a protease of that membrane to give the final protein product.

Figure 2.13 summarizes what we know about gene structure and organization from these and other studies. Although the TATA box is usually found 30 or more base pairs upstream from the starting point of transcription, its deletion, or chemical alteration, does not always cause transcription failure, although it may do so (Corden *et al.*, 1980). The experimental demonstration of polymerase II binding to this site is still lacking. Similarly, although sequence comparisons suggest that we know something about the signals for cap sites and for poly(A) addition, these suggestions still require experimental verification. Detailed sequence data are not as easy to interpret as might have been expected, and they have yet to provide guidance concerning the regulatory sequences which must be involved in the differential control of gene activation. Contrary to the finding with 5S RNA, where the regulatory sequences appear to lie within the gene itself, the many interruptions in the vitellogenin genes might suggest that regulatory sequences lie in the flanking DNA. We will therefore look more closely at this aspect of gene structure, in Ch. 13.

## Transposable elements

The discovery of introns was quickly followed by the equally unexpected finding that the DNA of some, perhaps all, eukaryotes contains many dispersed, repeated sequences which are nomadic. These sequences are transcribed and at least some are also translated. They are defined as elements because they are of some complexity, and they may inactivate genes with which they become associated. We shall consider this mutator role in Ch. 5 and look here only at the molecular organization of these elements.

When the most copious poly(A) RNA of *Drosophila* cultured cells (3–4 percent of the total) is used to probe a genomic library, the identified segment (called copia) is found to hybridize with 30 or so widely scattered sites on the salivary chromosomes (Finnegan *et al.*, 1978). Copia is repeated, but unlike the tandemly repeated rDNA, it is also dispersed. This dispersal is not to fixed chromosomal locations, since hybridization of copia clones occurs at different sites in the chromosomes of different *Drosophila* strains, and sometimes even between chromosomes of individuals of the same strain (Young, 1979). Copia can apparently move around the chromosome, and in cultured cells the number of copia elements increases as might be expected if they could multiply more readily in this less stringent environment (Potter *et al.*, 1979).

The organization of the 5 kb copia element is also significant (Fig. 2.14). It carries a 276 bp direct terminal repeat, and each of these has a short inverted terminal repeat of 17 bp (Levis *et al.*, 1980). In addition examination of genomic fragments with and without inserted copia sequences

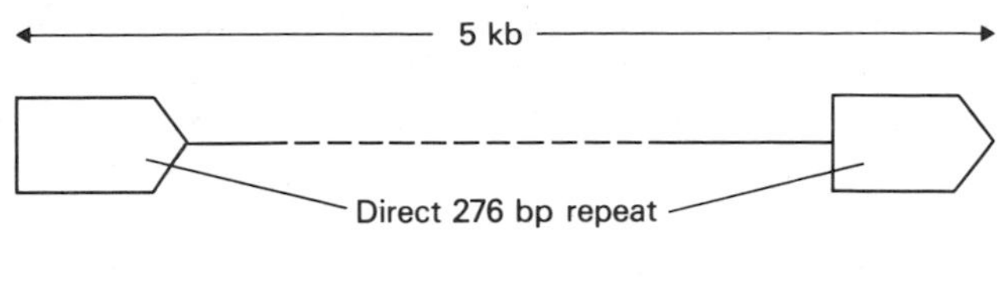


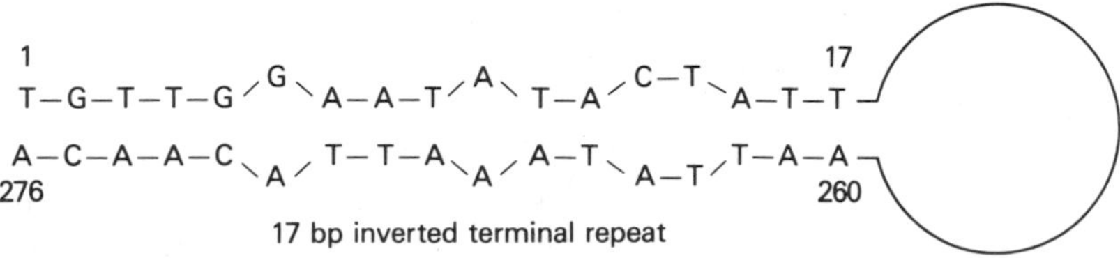


**Fig. 2.14** The organization of the copia transposable element, showing the direct repeat at the two ends of the 5 kb sequence and, below, the 17 nucleotide sequences of the imperfect, inverted terminal repeat found at the ends of the direct repeats. (Based on Levis *et al.*, 1980).

shows that a 5 bp sequence present at the target site is found at both ends of the copia element after insertion. There is no sequence homology (i.e. no preferred sites of insertion) since the duplicated 5 bp sequences are different, at least in the limited cases so far examined (Dunsmuir *et al.*, 1980). These molecular characteristics are also found in the previously identified transposable element (Ty1) of yeast (*S. cerevisiae*), which is 5.6 kb long and is repeated 35 times, and which similarly generates a 5 bp duplication (Gafner and Philippsen, 1980); and in the integrated proviruses of two retroviruses (Shimotohno *et al.*, 1980). Eukaryote transposable elements therefore have virus-like characteristics, and Young's attempt (reported in Ashburner, (1980)) to breed copia out of *Drosophila* by taking advantage of location differences in different strains, should tell us if it is truly parasitic, or has some role in *Drosophila* development. Since *D. simulans*, which was originally confused with *D. melanogaster* in Morgan's laboratory, has only three copia elements, and *D. mauritiana* has apparently none, the likely answer is that this element carries no crucial information for making the fly, despite apparent developmental regulation of transcription (Flavell *et al.*, 1980). This very abundant mRNA can be transcribed *in vitro* (Falkenthal and Lengyel, 1980; Falkenthal *et al.*, 1982), and it was a surprise when it was shown that the protein made corresponded to the protein, a reverse transcriptase, present in virus-like particles present in flies and cells (Shiba and Saigo, 1983). Although these RNA particles have not yet been shown to be infective, it is probable that chromosmal copia is the provirus of a copia retrovirus of *Drosophila*.

Just as there are other repeated elements in yeast, there are many, perhaps 20–30, families of nomadic, copia-like sequences in *Drosophila*. Elements 412 has no sequence homology with copia, for example, and element mdg3 appears to carry an intron (Ilyin *et al.*, 1980) and has both strands transcribed. The element families are therefore extremely varied in organization, but together they must constitute about 5 percent of the *Drosophila* DNA, which is about a third of the DNA which we defined earlier (p. 40) as middle repetitive. Since there are many known *Drosophila* RNA viruses (Brun and Plus, 1980), it will be interesting to learn if any correspond to members of this copia-like group of elements.

The polymorphic localization of copia (and copia-like) sequences provides a new approach for identifying particular structural genes in a library. The strategy is to look for and identify a strain with a copia sequence adjacent to the relevant structural gene. Copia DNA is then used to probe a library from this strain to collate segments carrying copia and joined flanking sequences, some of which will contain the relevant gene. The sample of copia-containing clones is then hybridized individually to the salivary chromosomes of a strain which does not have copia adjacent to this gene (e.g. *D. simulans*). Only the clone carrying flanking sequences of the gene in question will then have the information allowing it to hybridize to the gene locus. This is how the white locus was cloned (p. 131).

Two further characteristics of *Drosophila* DNA organization have been found by examining the structure of cloned DNA segments. By looking for clones with inverted repeats (foldback DNA), Potter *et al.* (1980) found a great variety of segments (size range from 0.8–4.25 kb) which, unlike the copia-class, carry inverted terminal repeats. These also seem to be transposable (possibly by similar mechanisms to the bacterial transpons which

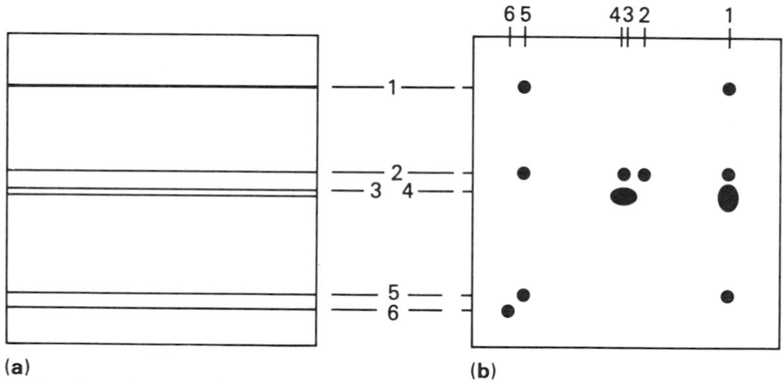


**Fig. 2.15** 'Southern cross' demonstrating the scrambled and clustered arrangement of moderately repetitive elements in *Drosophila* DNA, simplified from Wensink *et al.* (1979). A 22 kb plasmid, the sequences of which were repeated about 90 times in the *Drosophila* genome, was digested with *Eco*R1 and *Bam*H3, and run on a gel to give the six fragments shown in (**a**). This was then transferred to nitrocellulose paper *à la* Southern. The same plasmid was made radioactive and treated identically. The second 'blot' was then laid over the first, but at 90° to it, and annealed (**b**). If the fragments had no homologies, radioactivity would be found only on the diagonal (as with fraction 6). Instead, 5 has homologies with 1 and 2, and so on. Tests with other plasmids show similar repeated clusters of sequences.

also have inverted terminal repeats). How far this group is included in the next one we shall discuss is uncertain.

The second question was whether a cloned segment contained repetitive elements, and if these repetitive elements were also found in other clones. The DNA from a clone was fragmented with restriction enzymes, the fragments separated on a gel and transferred to nitrocellulose paper by the Southern technique. An identical clone was then labelled with $^{32}P$ by nick translation, and the same restriction fragments separated. The first gel was then placed at right angles across the second, and the second set of labelled fragments transferred through the first under conditions which allowed the labelled DNA to remain on the paper only if it hybridized with unlabelled fragments of the first gel. This two-dimensional matrix, or 'Southern cross', should show radioactivity only along the diagonal if there is no homology between fragments; fragment 1 hybridizing only with the fragment 1 below it, and not with fragments 2, 3 and so on. In the case studied, hybridization is found off the diagonal (Fig. 2.15). For example, fragment 1 hybridizes with fragments 1 and 5 (and vice versa) showing that there are elements common to both, i.e. there are repetitive elements within the DNA segment. Different clones can be tested in the second dimension to identify their elements common to the first, and these can be used in their turn in all combinations. Such a study (Wensink *et al.*, 1979) shows that these moderately repeated elements may be in large, dense clusters in some DNA segments, or scrambled in different arrangement within other segments, or associated with clones carrying particular genes, or even within the transposable element, 412. It is not yet clear if these elements are themselves transposable, or associated with any transpositions. These two DNA classes must, together with the copia-like sequences, account for most of the *Drosophila* middle repetitive DNA.

## Selfish or parasitic DNA

The discovery of these apparently non-genetic DNAs resolves the old C-value paradox: the paradox that the haploid DNA content of a cell (its C value) exceeds the estimates of gene numbers gained from mutation studies by one or two orders of magnitude, thus implying that most of the genetic material is not genetic, in the usual sense. The dramatic disparity between the C-values of many closely related species (p. 46) may also be explained by their differing contents of these non-specific DNAs, but we are still without the detailed data needed to prove this for any particular case.

The evolution of these DNAs lacking phenotypic expression has been discussed by Ford Doolittle and Sapienza (1980) and by Orgel and Crick (1980). Both argue that once this DNA arises it will spread by making copies of itself within the genome (as copia does), provided that multiplication has only trivial selective disadvantages for the organism. They become 'selfish DNAs', existing in a parasitic fashion within the DNA molecule, and their only 'function' is their own survival. From our point of view, we must ask if any of this selfish DNA might nevertheless evolve a specific function, and become part of the developmental genetic system?

As Orgel and Crick (1980) argue, we have no good theoretical explanation for the evolution of the coordinate control of sets of genes, previously regulated individually. The scattering of many kinds of selfish DNA throughout the genome might therefore allow, by chance, one particular repeated element to be used for turning on (or off) a set of genes. The difficulties with this hypothesis are clearly formidable, particularly in assessing how the element might evolve so as to control a subset, initially small, of the overall genome so that its coordinate activity would then have a selective advantage. We have no evidence that this has occurred, but then the problem is too new to have been properly explored. What is clear is that the apparently random distribution of selfish DNAs, and their different distributions in lines and species, implies that most of them will have no special function. Selfish DNA, at this time, adds to our difficulties in understanding the detailed organization of this polymeric molecule (Spradling and Rubin, 1981).

## Chromosomes and chromatin

The integrity of the DNA in chromosomes has been studied by Kavenoff and Zimm (1973), using rearrangements. An inversion which displaces the centromere makes no difference to the length of the DNA which can be isolated from the chromosome: the centromere does not interrupt the polymeric molecule. A translocation or a deletion produces an appropriately longer or a shorter DNA molecule than the standard. DNA molecules are organized without interruption or break in each chromosome. The DNA is also packaged in the same way throughout the length of the chromosome, and we must examine this since some higher order organization may be relevant to specific gene transcription.

The chromatin filament which can be isolated from nuclei is composed of about equal amounts of DNA and histones, together with non-histone proteins in amounts averaging about one third that of DNA, and a few percent of RNA. Four of the small globular histones are regularly organized as a protein octamer core (2(H2A, H2B, H3 and H4)), and a 195±5 bp length of DNA is coiled twice around this to form the basic repeat unit of chromatin, the nucleosome (Fig. 2.16*a*). The nucleosome is the essential, basic arrangement of the eukaryote chromosome, and details of its organization are given by Klug *et al.* (1980) and by Mirzabekov (1980). The remaining histone (H1) lies outside the nucleosome, and the central globular portion

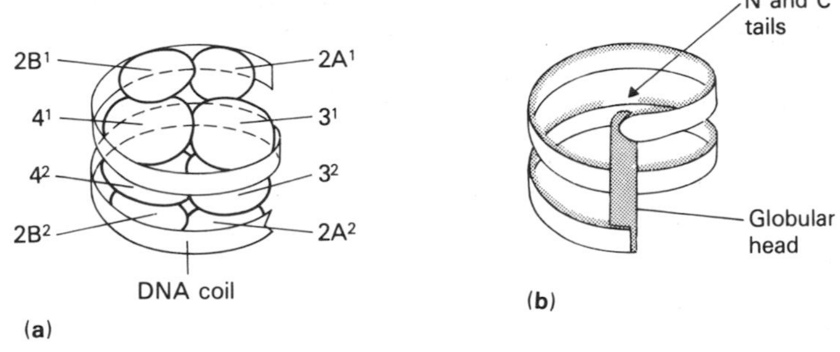


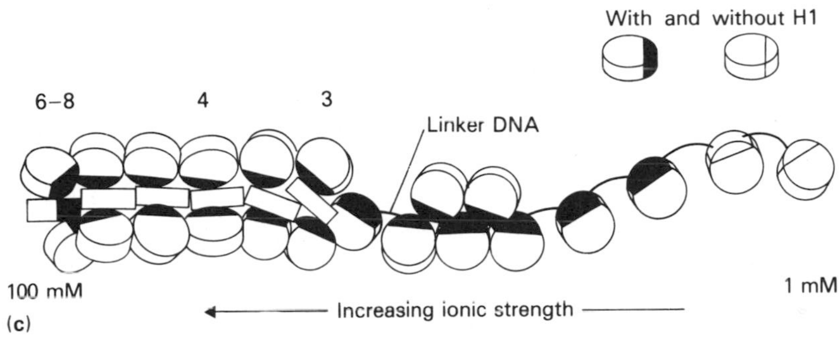


**Fig. 2.16** The organization of DNA and histones. (**a**), The double tetramer of the four core histones (H2A, H2B, H3 and H4) is shown diagrammatically (for structural details see Klug *et al.*, (1980)). The globular parts of the histones have the property of cross-linking in the arrangement shown, in mirror image apposition. The $NH_3$ terminal 'tails' and parts of the globular proteins attach to the DNA which is arranged as a 140 bp double coil around the nucleosome protein core. (**b**), Histone H1 lies across this coil, spanning its entry and exit points from the nucleosome, and the H1 N and C tails attach along part of the DNA, penetrating between gaps in the core proteins, according to Mirzabekov (1980). The DNA filament links nucleosomes together in a regular fashion. (**c**), Histone H1 also has the property of cross-linking nucleosomes under appropriate conditions (Thomas *et al.*, 1979). When H1 is absent no higher order structures are found in electron micrographs, and at low ionic strengths the nucleosomes (with H1) are well separated (right). At higher ionic strengths the nucleosome filament folds to form a fibre, and at still higher strengths the nucleosomes become arranged in a close zig-zag of 6–8 units, and the 'linker' DNA is no longer visible (left). This fibre, stably cross-linked by the H1 proteins, is the first higher-order structural arrangement of nucleosomes.

of this molecule spans the points of entry and exit of the DNA coil, protecting 15 base pairs on either side of the core-associated coils (Fig. 2.16*b*), but not the 'linker' DNA which runs from nucleosome to nucleosome (Allan *et al.*, 1980). Removal of H1 causes the nucleosomes to separate (Fig. 2.16*c*), and H1 must be involved in maintaining the higher order structure of chromatin. But it is not yet clear if the disc-shaped nucleosomes are then stacked above one another or, more probably, are packed more tightly to give the 250±50 Å fibrils seen in electron micrographs of chromosomes (Du Praw, 1970). The histones, which differ little between tissues, therefore provide a packing structure for the DNA, to the extent that the length of the coiled and folded DNA is reduced to one fifth its fully extended length. Further folding reduces this length by an order of magnitude.

Apart from H1, whose presence clearly affects the accessibility of the DNA, the histones apparently have a scaffolding function. So it is not surprising that nucleosomes seem randomly located with respect to DNA sequences (Prunell and Kornberg, 1978), and that there is no fixed relationship between sequence and nucleosome. An average gene must spiral around five nucleosomes, and an average 13 kb transcript around 65 nucleosomes. It seems unlikely therefore, that chromatin is organized to facilitate the accessibility of promoter and other regulatory sequences through simple architectural arrangements of the DNA which might bring them into a regular working relationship with activator molecules. Nevertheless, there is evidence that different regions of the DNA are selectively accessible to nucleases.

Staphylococcal nuclease digests chromatin at the linker region breaking it into its nucleosome units, showing that H1 protects the 165 bp non-linker DNA from this digestion. It does not affect functioning genes differently from the rest. On the other hand, pancreatic DNase preferentially digests active genes and not the linker. And the gene need not be being actively transcribed for this to happen (Weintraub and Grandine, 1976). Active, or recently active, genes must differ from untranscribed genes in their structural arrangement in chromatin, and this must involve some decompaction of the nucleosome so that the DNA is extended to a transcriptionally available state (Franke *et al.*, 1978). We do not yet know if this altered arrangement of DNA in the nucleosome is a cause, or a consequence, of transcription. Nor is it clear how the acetylation, phosphorylation or methylation of the histones affect their binding properties, as they certainly do, with respect to transcription (Johnson and Allfrey, 1978). Exposure to low levels of sodium butyrate, which blocks deacetylation, prevents the induction of egg white genes by steroids (McKnight *et al.*, 1980), for example, and causes Friend erythroleukaemic cells to differentiate into non-dividing, haemoglobin-synthesizing, normoblast like cells (Leder and Leder, 1975). Post-synthetic modification of chromosomal proteins therefore appears to play some part in regulating transcription, but it is not yet clear how these somewhat general processes can act in a specific way. This suggests that non-histone proteins (NHPs) may be regulatory.

About 450 NHPs can be identified from a single cell type (HeLa). By analogy with the bacterial *lac* repressor, between $10^3 - 10^4$ molecules might be required to regulate a gene in a cell, and these small numbers would generally escape detection and would be difficult to separate from the commoner NHPs of this large and varied group. One class, the high (electrophoretic) mobility group (HMG), seemed to play a role in active genes, but as they are found in all tissues and are associated with satellite DNA, as are most of the NHPs, it now seems more likely that many NHPs have some function in chromosome organization and its changes during the cell cycle (MacGillivray, 1977). Nevertheless, the circumstantial evidence suggests that protein components of chromatin are involved in gene regulation (Paul, 1978), and we now await the development of simple and more exact experimental systems which will enable their precise definition. Studies of this sort have just started. For example, Howard *et al.* (1981) have used the monoclonal antibody technique (p. 317) to identify an NHP apparently associated with active genes. Or, again, the gene product which blocks transcription when organisms are heat shocked (p. 91) is being studied. But the nature and characteristics of these proteins are still not known.

There is evidence that NHPs form a chromosome scaffolding, which shows banding patterns. However, this apparently regular feature may prove to be an artefact since the scaffold is lost under some conditions of nuclear preparation (Hadlaczky *et al.*, 1981). Nevertheless, it seems clear that nuclear DNA is organized in loops (Paulson and Laemmli, 1977) and is probably attached to subnuclear structures, the nuclear matrix or cage. DNA replication is associated with this matrix material since DNA pulse labelled with [$^3$H]thymidine, is preferentially associated with the cage (McCready *et al.*, 1980). More important, the evidence also suggests that specific sequences are attached to the cage, and the DNA is not randomly organized with respect to the cage structures (Cook and Brazell, 1980).

When the intact, histone-free DNA is isolated in the membrane-free nuclear cage, as 'nucleoids', the DNA loops can be digested to various extents with restriction endonucleases, and the cages sedimented free of the released, no longer attached, DNA. The protected, cage-associated DNA can then be purified and fragmented completely with the same endonuclease. If a gene is randomly situated in the DNA with respect to the cage, we should expect that, in equal aliquots, there should be no enrichment for a particular gene in the cage-associated DNA compared with that of the initial digest. Probing nucleoids from HeLa cells with cloned α- and β- and γ-globin genes after this treatment shows that the α-genes are enriched and must therefore resist detachment from the cage, and so lie close to the point of attachment between DNA and cage. The β and γ sequences are not so associated (Cook and Brazell, 1980). DNA is therefore organized in the nucleus in a regular fashion related to the protein structures of the cage. It may be that this protein–DNA organization also plays some part in the control of gene transcription, although the experiments to test this have still to be done. Thus, it is possible that the way DNA is organized in its physical relation to nuclear proteins may be significant for the determined state of a cell.

## Conclusions

At the beginning of this chapter we asked if recent studies had exposed the complexity of DNA organization which Muller (1966) expected would be characteristic of the eukaryotic genome. The identification of highly repetitive and middle repetitive sequences, of introns and transposable elements, and the identification of some of these as selfish DNA, presents us with an overall structure many times more complex than anything Muller might have imagined. Some of the general characteristics of the DNA, such as the repetitive sequences, can now be attributed to transposable elements and other identified sequences, but there is much DNA yet to be accounted for (e.g. the spacer DNA between the globin genes). There are still difficulties too, in the way of disentangling the aspects of DNA organization relevant to gene regulation, but it seems that cloned genes, and *in vitro* genetics, will provide the tools for doing this, as exemplified by the work with 5S DNA. Genetic engineering is now giving us a picture of the sequence organization necessary for transcription, but we still lack an insight into the molecular detail which would explain the coordinate activation of gene sets during development. Similarly, the details of chromatin organization need further clarification before we can understand how inducers and hormones change the states of cells so that they become competent, or determined, or differentiated. Since there are other data still to be considered which reflect back on the organization of the DNA, we shall return to these matters in the last chapter. There we shall see that a gene, defined as a transcription unit, is even more complex than shown in Fig. 2.13.

# 3 Active chromosomes

Ever since the formulation of the chromosome theory of heredity cytologists and geneticists alike have dreamed of the day when someone would find somewhere an organism in which the chromosomes were so large that it would be possible to see qualitative differences along their lengths corresponding to the different genes which we know must reside there. . . .
. . . With these four discoveries before us (i.e., constant and distinctive patterns, somatic synapsis, the behaviour of active and inactive regions and the separation of the arms of the large autosomes) it was clear that we had within our grasp the material of which everyone had been dreaming. We found ourselves out of the woods and upon a plainly marked highway with by-paths stretching in every direction. It was clear that the highway led to the lair of the gene.

T. S. Painter, 1934

When Painter (1933, 1934) discovered that the nuclei of *Drosophila* larval salivary glands contained giant chromosomes, he started a new area of cytogenetic studies which has been actively pursued ever since (reviewed by Ashburner and Berendes, 1978). We shall only sample the information coming from this work which tells us how the functioning genome is organized. At this cytological level we are necessarily looking at less detailed facts than in the last chapter, by about three orders of magnitude.

## Polytene chromosomes

The *Drosophila* giant chromosomes were polytene, and this differentiated nuclear state is not uncommon. It arises through repeated duplication of the chromosomes within the intact nuclear membrane (endoreduplication), without spindle formation. Often the multiple chromonemata are only loosely associated, and the extent of this reduplication, where it occurs, may vary considerably among the different tissues of the organism. Cytological studies have therefore concentrated on the favourable material, where alignment is exact and the level of reduplication is high. In plants there is the rare instance of the embryo suspensor cells of the bean (*Phaseolus vulgaris*) with a DNA content of 8192 C, and the 2*n* chromosomes condensed and banded in a regular fashion (see Nagl (1981) for others). Diptera have the additional advantage of somatic pairing (p. 00) which doubles the width of the single (paired) chromosomes seen (Fig. 3.1). In *D. melanogaster*, the salivary gland chromosomes have undergone 8 cycles of endoreduplication by pupariation: *Chironomus tentans* achieves 13 cycles. This replication does not include the heterochromatic regions, which are associated in a chromocentre (Fig. 3.1). The bands or discs (chromomeres) which stain with acetocarmine make up a non-repetitive, aperiodic sequence, clearly separated by low staining regions (interbands), but there is some ambiguity in the definition of this pattern at the electron microscopic level (Saura and Sorsa, 1979).This biological magnification of chromosome organization apparently shows, too, that the same chromomere pattern,

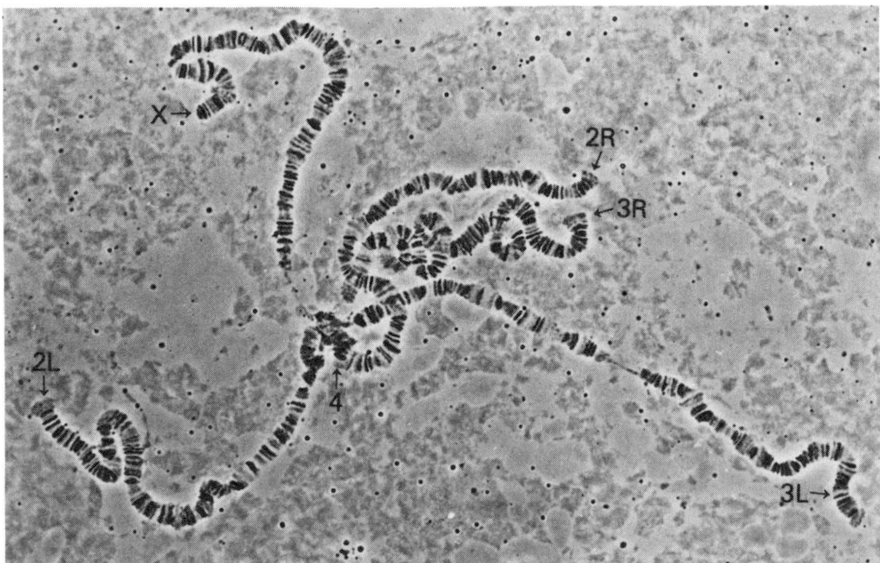


**Fig. 3.1** The polytene chromosomes from *Drosophila* salivary glands showing the banding patterns and puffs, by courtesy of Dr. S. Pinchin. For reference, the chromosomes are divided into 102 numbered groups, each of which is further subdivided into lettered groups (A–F) starting from a sharp band, and the bands are then numbered in each subdivision, as shown in Fig. 3.3. The division numbers start from the distal end of the X chromosome (1–20), then along the left arm (21–40), and the right arm (41–60) of the second chromosome, and correspondingly on the third chromosome (61–80 and 81–100), with two divisions (101–102) for the small fourth chromosome.LeFevre (1976) has published a full photographic sequence of the chromosome complement showing the band coding, or salivary gland map. Bands can be dissected from the chromosomes to provide DNA for gene cloning (Scalenghe *et al.*, 1981).

number and sequence is expressed in all tissues in which it can be identified, presumably reflecting the constancy of the DNA molecules and their general interactions with nuclear proteins, as seen at this level (Beerman, 1972).

Study of the banding patterns in transpositions, duplications and deletions showed correlated genetic changes: a translocated gene was always accompanied by a translocated band. This suggested to Painter (1934) that he was on the highway that "led to the lair of the gene"; or as Bridges (1935) put it more exactly: "each of the faint crossbands . . . corresponds to one locus". We must therefore ask if modern data support this one-band-one-gene hypothesis, for the problem of gene number might then be resolved by counting chromomeres.

Bridges also noted that some bands became dispersed, or puffed, during development, but this phenomenon was largely neglected until Beerman (1952), working with *Chironomus tentans*, showed that the puffing patterns differed within four different tissues of an individual. In this favourable material he was able to show that a puff arose from a single band (or interband), that adjacent bands puffed and declined independently, and that tissue specificity could be attributed not to single puffs but to the total puffing pattern. That is, he demonstrated morphologically that giant chromosomes consisted of linearly arranged, independently reactive, genetic units of the order of magnitude of the bands; and deduced that puffing indicated changes, "most probably increases, in the activity of gene loci" (Beerman, 1956). If this is correct, gene activity can be directly visualized in polytene chromosomes, which makes them very attractive to study. And this can now be further refined by applying nucleic acid hybridization techniques (p. 43) to this favourable material.

### One gene, one chromomere?

Our understanding of the molecular organization of polytene chromosomes is too fragmentary for us to deduce whether genes correspond to chromomeres. The description (p. 69) of chromatin organization suggests this to be unlikely, and the fact that the transcription unit (embryonic) in *Drosophila* is half the size (average 18 kb) of a chromomere (average 30 kb) makes the conclusion improbable. But the issue is complicated by the evidence that bands and interbands display different levels of polyteny, and of folding (Fig. 3.2). The hypothesis has therefore been tested, in true Popperian fashion, by attempts to find if the number of genes caused to mutate within a defined chromosomal region corresponds exactly to the number of chromomeres there, or not.

Initially, detailed analysis of the well-studied zeste ($z$)–white ($w$) region (Judd and Young, 1973) which spans 0.75 map units between 3A2 and 3C2 of the X-chromosome salivary map of *D. melanogaster*, produced a remarkable correspondence: an impressive 344 independent mutations could be

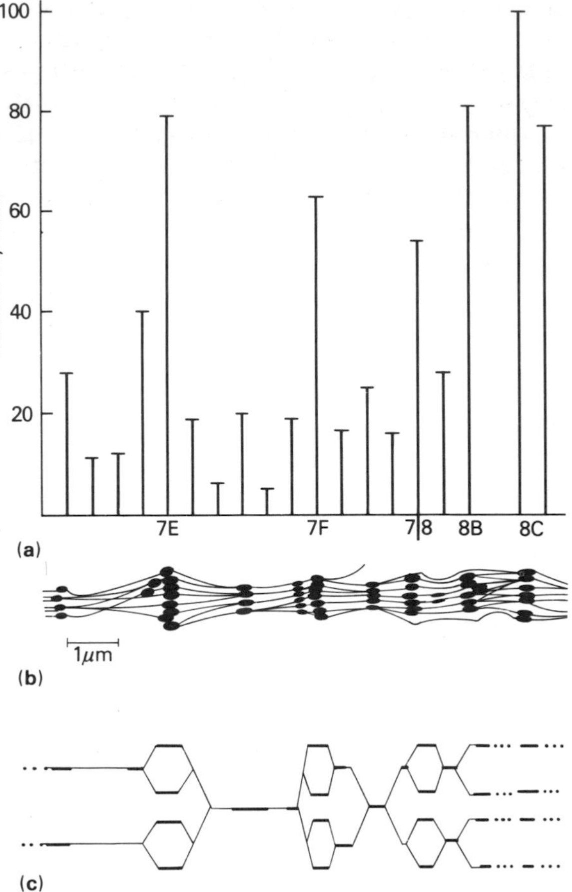


**Fig. 3.2** Analysis of the dry mass of part of a male X chromosome from *Drosophila*, simplified from Laird (1980). The relative cross-sectional dry mass (**a**) is indicated directly above the drawing of the electron micrograph of the salivary gland fragment analysed (**b**). The model (**c**) below assumes that the different polytenys arise at the DNA replication forks. Bands and interbands are indicated by thick and thin lines. The surprising conclusion is that the mass of bands and interbands are in the ratio of 4:1, whereas it has generally been taken as 20:1, or more. It is therefore unlikely that interbands are restricted in their coding potential (Crick, 1971a), although the discovery of Z DNA in interbands (see text) may mean that some are.

grouped to give 15 recessive lethal, or semi-lethal, and 2 visible, unambiguously ordered sites in a region with 15 bands (Liu and Lim, 1975). Similarly, the 50 bands of the small fourth chromosome corresponded to 37 essential genes (the complementation groups mentioned on p. 125) and 6 visibles, and the repeat mutation frequency suggested that the total map should not exceed 60 genes, and might be lower (Hochman, 1976). The correspondence was too good to be true! The mutations being isolated were either essential genes (and therefore lethal when mutant) or visibles, and O'Brien (1973) pointed out that known biochemical mutants are generally viable and not scored. It is easy to imagine that many such non-vital loci (behavioural as well as metabolic) may by concealed by the homeostatic systems of the organism, which would invalidate this approach.

A second, and theoretically more powerful, methodology was adopted by LeFevre (1973), using irradiation-induced transpositions in *Drosophila*. His argument was that rearrangements affecting a *single* band should be different in their expression if the band coded for two or more independent genetic functions; in other words, band breakage should separate the genes. Unfortunately the uncertainties of cytology make the identification of break-points very dificult, so that the firm data are limited. Breaks affecting known sites of visible mutants (e.g. band 1B3 = scute (*sc*), 3C2 = white (*w*), 3C5 = roughest (*rst*)) produce the expected mutant phenotypes, and nothing else. Similarly, breaks at 1A5–6, for instance, produce lethals which are lethal in heterozygotes. This limited information suggests that generally, 'all rearrangements affecting a given band produce the same genetic consequences'. More interestingly, a number of breaks were without effect; the flies were normal. Of 51 breakpoints distal to 20A3 (on the X chromosome), 17 were non-lethal; precisely one third. LeFevre (1973) suggests from this and more extended evidence that about half the euchromatic rearrangement break-points are not associated with mutant effects.

More recently, Young and Judd (1978) used deficiencies and transloca-tions to re-examine the *z–w* and associated regions (3A–3C) for non-essen-tial sequences. Like LeFevre, they find that there are such sequences (Fig. 3.3): in particular, the division 3B (six bands) is found to have nine complemention groups, while 3C has more bands than complementation groups. Of course there is the problem of gene duplication (and epistasis) to confuse the issue, but the general conclusion is surely that genes are not exclusively organized in chromomeres. This does not resolve the relation of genes to bands and to interbands since these techniques treat the band–interband as a unit.

An alternative approach has been used by Jamrich *et al.* (1977) to look at what is being transcribed in *Drosophila* salivary glands. By making anti-bodies to *Drosophila* RNA polymerase II and linking them to fluorescene it is possible to locate the polymerase on salivaries. The label is found to coincide with all the interbands (Fig. 3.4). Conversely, antibodies to histone H1 are found exclusively in the bands. When particular puffs induced by heat shock are looked at (p. 90), it is interesting that the converse picture is then found—no H1 can be detected and the polymerase is present in the puffed chromomere. In short, condensed, inactive chromatin is not tran-scribed, while uncondensed chromatin is. And this uncondensed chromatin, which can include band and interband, is seen as an active puff, as we shall see. Zhimulev *et al.* (1980) suggest that the transcription of interbands, noted by Jamrich *et al.* (1978), is a general phenomenon, implying that housekeeping (or metabolic) genes are located there (and are active in all

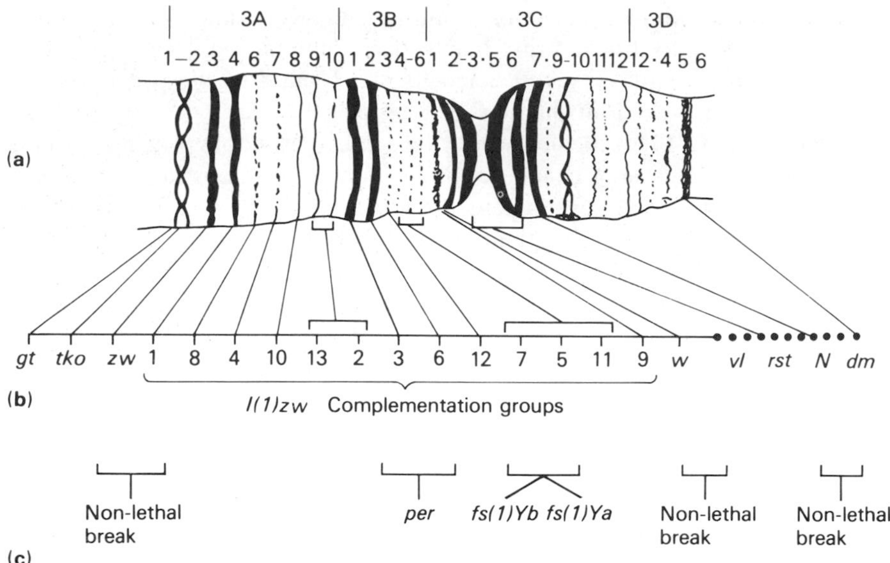


**Fig. 3.3** The 3A–3D region of the salivary gland X chromosome and the relation of its banding pattern to known complementation groups (see text). The band pattern and its cytological map divisions (**a**) are related to known genes and to the zeste-white lethal series obtained by saturating the region with mutations (**b**). The notable correlation between bands and mutable loci suggested the one-band-one-gene hypothesis. However, a search for non-lethal sites (**c**), using breaks and translocations (Young and Judd, 1978) showed three regions where loss had no functional significance, and exposed two female-sterile loci (*fs*(*1*) *Ya* and *fs*(*1*) *Yb*). The *per* locus which when mutated disturbs the eclosion rhythm of flies, their circadian rhythm, had been found earlier (Konopka and Benzer, 1971) and is located to the 3B1–2 region. Thus, the 3B region contains more complementation groups than polytene bands, and parts of 3A and of 3C apparently have no functional significance. The other genes shown are: giant (*gt*), technical knock out (*tko*), zeste (*z*), white (*w*), verticals (*vl*), roughest (*rst*) and diminutive (*dm*). *N* is the dominant Notch locus (p. 175).

tissues), whereas more specialized (or luxury protein) genes are sited in bands, and specifically activated when a band puffs. If correct, this rescues the one-gene-one-chromomere hypothesis by locating only the genes for luxury proteins exclusively in the bands. It also implies that chromosomes are differently organized with respect to the two classes of gene. This revised hypothesis may be more difficult to disprove, due to the technical problem of discriminating between band and interband at the puff level.

Interband regions have recently been shown to have an unexpected DNA organization (Nordheim *et al.*, 1981). We have assumed so far that chromosomal DNA has the usual right-handed, stable β-helical structure, but alternating purine–pyrimidine sequences (CpGpCpGp . . .) may hydrogen bond to form a left-handed zig-zag (Z) conformation. Z-conformation

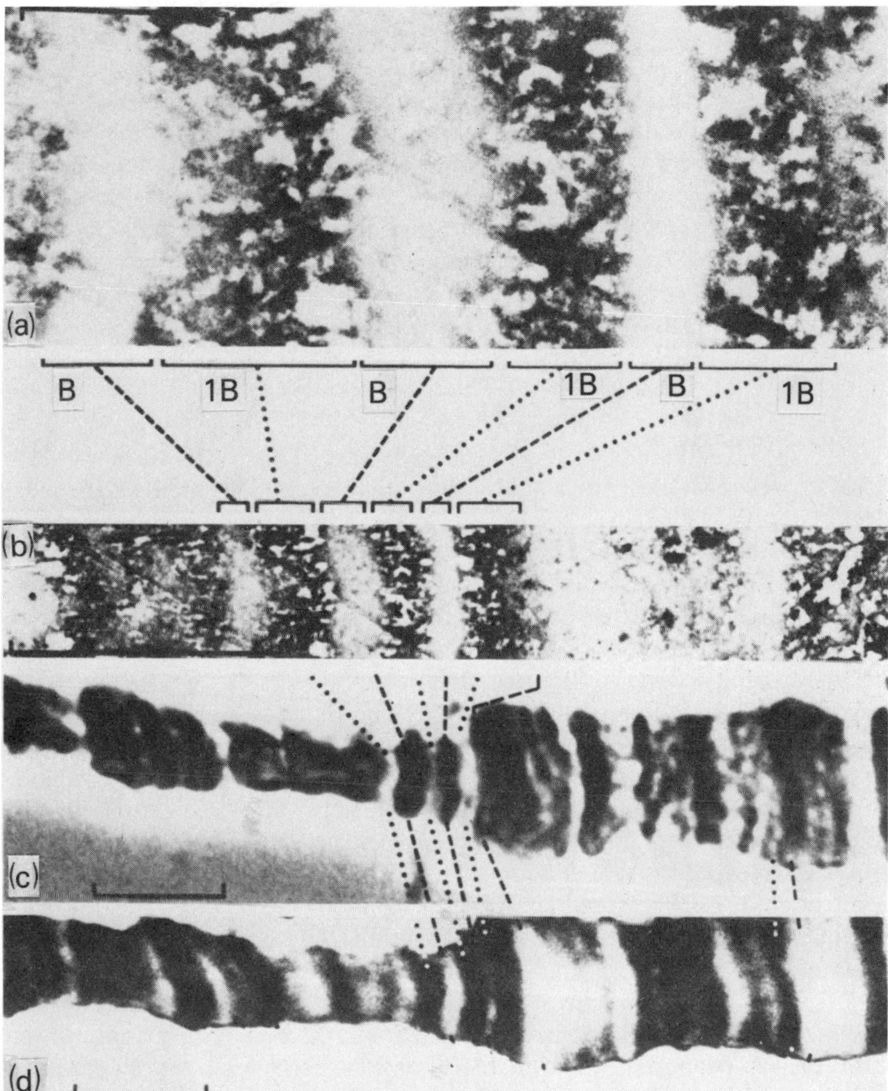


**Fig. 3.4** Sass and Bautz (1982) have identified the locations of polymerase B on *Chironomus* salivary glands by attaching rabbit anti-polymerase antibody to these, and using peroxidase conjugated to goat anti-rabbit immunoglobulin to visualize them. (**c**) is the unstained, phase contrast view of a stretched segment of chromosome III, and below it is the same chromosome stained with immunoperoxidase (**d**). The bands are connected by dashes and the interbands by dots. Ultrathin sections are shown at (**b**) and (**a**), where the homogeneous labelling (dark) of the interbands is in clear contrast to the unstained, condensed bands. The enzymatic labelling is always bound to the chromatids which traverse an interband. Puffed bands (not shown) also contain high levels of polymerase B (polymerase II) which is therefore present in puffs and interbands, but not in condensed bands. By permission of Dr. Sass and Springer-Verlag.

polymers are strong immunogens, unlike B-conformation polymers, and if *Drosophila* salivary glands are exposed to anti-Z antibody it reacts exclusively with interband regions. It is possible that any Z-DNA in bands is inaccessible to the antibody, due to chromatin conformation, but it seems unlikely that the result merely reflects this. Again, the antibody may merely pick out the AT runs noted earlier, when interbands would only reflect this aspect of sequence organization. But there is also the possibility that we are observing a structural change in the DNA, perhaps a switch from the B to the Z form which might be stabilized by a non-histone protein, and this may reflect a functional difference between bands and interbands. We have still much to learn about organization in chromosomes.

## Puffed chromomeres

The experiments described above show that decondensation of chromomeres occurs when at least some of the genes in that section of the chromatin are activated, as Beerman supposed. Although salivary glands are differentiated, specialized secretory organs with inherent physiological limitations, many studies have been based on the assumption that puffing reflects gene transcription, and we shall consider some of them for what they tell us about gene activity in different situations.

Puffing is first detected as the accumulation of stainable non-histone proteins at or near the band, which becomes decondensed and extended. The protein concentration increases and spreads throughout the puff region. Isolated nuclei do not puff, so the proteins derive from the cytoplasm and puffing depends on cell integrity. RNA synthesis can be detected in the puff soon after protein accumulation starts by following the incorporation of [$^3$H]-uridine, using autoradiography. There is some evidence that this transcription is confined to part of the extended DNA. Generally, the size of a puff as a measured by autoradiography or by its physical extent, depends on the strength of the stimulus, but it is not clear if this reflects a variation in the number of transcribed chromonemata, or in the rate of their transcription (i.e. on the proportion of available initiation factors). Withdrawal of the stimulus causes puff regression; but some puffs normally regress in the presence of the stimulus (p. 87) after a short period of activity. Puff regions translocated to new chromosomal sites behave normally, showing that the control is local. Autoradiography also exposes RNA synthesis in unpuffed bands (micropuffs or housekeeping genes?). We have still much to learn about the biochemistry of this sequence of events.

Radioactive label is also used in a powerful extension of the RNA–DNA hybridization technique. The squashed chromosome preparation is treated to remove RNA and histones, and the DNA denatured by heat, low pH or formamide. Purified, highly radioactive RNA is then placed over the cytological specimen at 10–20 °C below the melting temperature of the DNA,

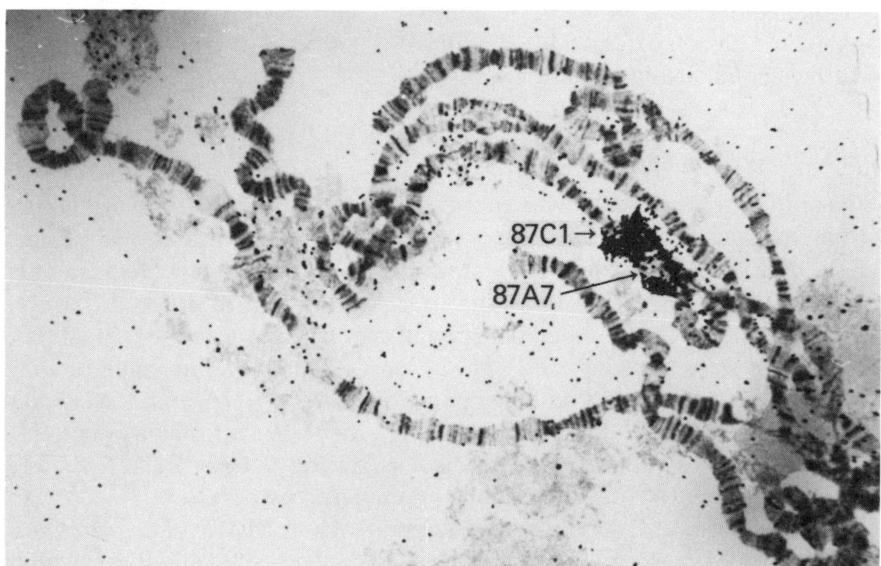


**Fig. 3.5** In this autoradiograph a radioactively labelled, cloned, heat shock protein DNA sequence has been used to probe for the mRNA heat induced at the 87C and 87A loci (see Fig. 3.10). The conditions are therefore chosen to allow DNA-RNA hybridization, and not DNA–DNA hybridization. The salivary gland was heat shocked 10 min earlier, and the probe identifies the mRNA subsequently produced. A control treated with RNase shows none of the label above the bands. Photograph of this preparation kindly provided by Dr. S. Pinchin.

and annealed to its complementary sequences. The hybrid regions are then identified by autoradiography (Fig. 3.5). In principle, this method allows the localization of any RNA to its chromosomal site of origin and permits molecular gene mapping (see Wimber and Steffensen (1973) for some limitations). This technique was first used for redundant genes (rRNAs, tRNAs, and histones). The last example is interesting because when mRNA from early sea urchin embryos, which form histone mRNA in quantity, was annealed to *Drosophila* polytenes, the region 39E–40A was found to be labelled (Pardue *et al.*, 1972): the urchin histone message is sufficiently like the *Drosophila* one to match its genes. Plasmid DNAs can be similarly exploited, as we have seen.

We might expect the pattern of uridine incorporation of a puffing chromosome, its transcription autoradiogram, to be the same as its hybridization autoradiogram when the RNAs in the cell are used to produce the latter. This is not the case (Bonner and Pardue, 1977): low (or no) incorporation in some bands of the transcription autoradiogram may match high levels in the hybridization autoradiogram, and vice versa. And micropuffs may incorporate and hybridize as well as large puffs. It follows that we have to interpret these forms of chromosome labelling with some caution: the reiteration of genes, the presence of sequences common to a number of genes, rates of transcription and message lifetimes, may all contribute to this apparent anomaly. The first example we shall look at has been so thoroughly studied that these qualifications do not apply.

**Balbiani Rings in chironomids**

Some of the puffs of polytene chromosomes ($8 \times 10^3$–$16 \times 10^3$ chromatids) of chironomids take a giant form (Fig. 3.6), and are named Balbiani Rings (BR) after their discoverer. They are found in small numbers (1–5) in only a few cell types, and the BR puff pattern is different in different tissues, and even in different functional regions of the same tissue (Beerman, 1952). BRs derive from a single band and, at least in larval salivary glands, they dominate RNA synthesis, making about 75 times the RNA found in smaller puffs (Lambert, 1973). It is taken that there they code for salivary proteins. Because they are so large, it is possible to dissect out particular BRs and to identify their RNA, relatively uncontaminated by others.

As shown by electrophoresis (Daneholt, 1972), BR2 of *Chironomus tentans* forms one main product, a 75S RNA, followed by a 'tail' of smaller molecules. These smaller molecules are not found in the nuclear sap. An inhibitor of mRNA synthesis progressively eliminates the 'tail' and the 75S molecules in 5 minutes (Egyhazi, 1976), and this, along with EM studies (Lamb and Daneholt, 1979), suggests that the tail represents nascent, incomplete 75S molecules. The completed molecules are not stored in the BR. After a delay, the 75S RNA is found in the cytoplasm, and can be hybridized back to BR2 (Lambert and Edstrom, 1974). Of course, this proves only that some of this RNA is not degraded or processed in the nucleus, but an RNA of this size has been isolated on polysomes, indicating that it is translated into polypeptides (Daneholt *et al.*, 1978). 75S RNA isolated from BRs 1 and 2 (both make this product), and then translated *in vitro* using the reticulocyte lysate system, does indeed make the major salivary secretory protein component—component I of 500,000 or greater molecular weight (Rydlander *et al.*, 1980). This component is immunologically distinct from the other two major secretory polypeptides, strongly suggesting that component I is, or is related to, the primary translation product of the 75S RNA. However, the DNA in the BR2 chromomere ($1 \times 10^5$ bp) greatly exceeds that of the transcription unit ($37 \times 10^3$ bp) for the 75S RNA, which, in its turn, is larger than the adjacent interchromomeres ($5 \times 10^3$ bp) The greater part, or possibly all, of the transcription unit must therefore be accommodated in the chromomere, which may contain other units (Case and Daneholt, 1978).

Little is known about the control of these puffs, although they can be manipulated (Beerman, 1973). In mid-4th instar larvae, BR1 is collapsed while BR2 is puffed (Fig. 3.6*a*). This state is not changed if the larvae are exposed to protein synthesis inhibitors. However, if the larvae are exposed to 1% galactose the situation is reversed: BR1 expands and BR2

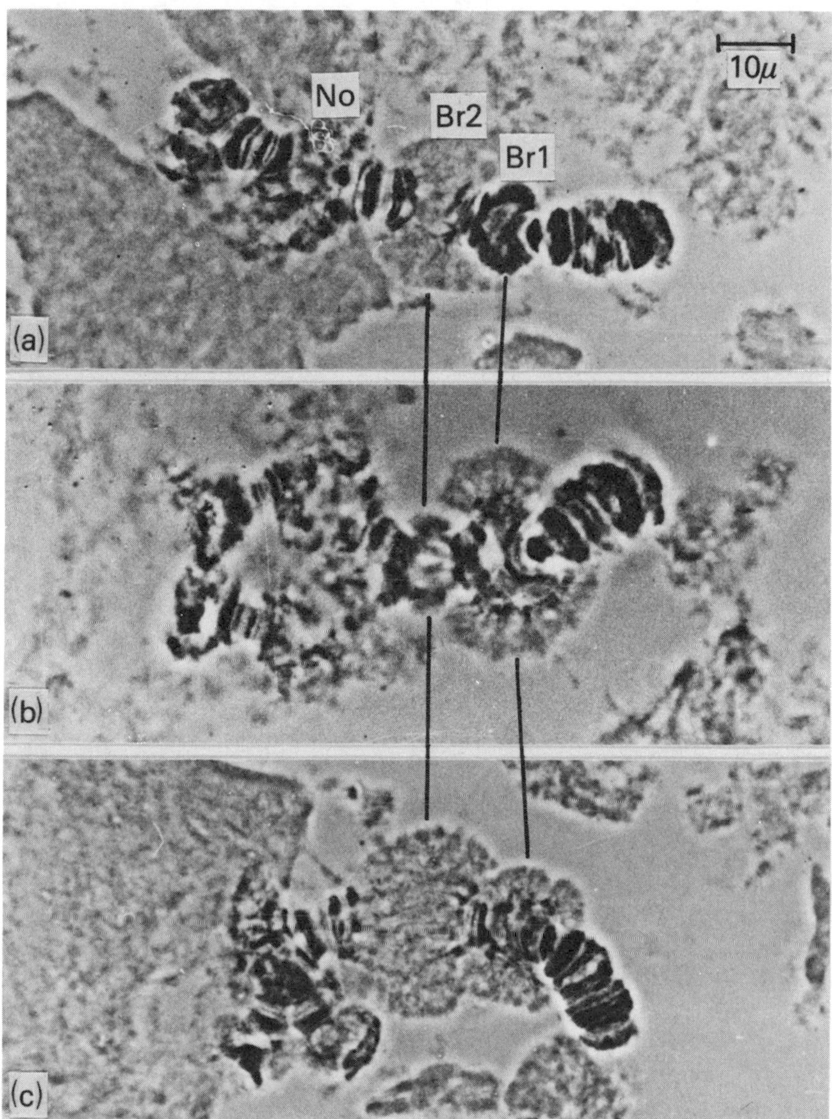


**Fig. 3.6** Balbiani rings (BR) of chromosome IV of middle 4th instar larvae of *C.thummi*. In (**a**) BR1 is collapsed and BR2 is expanded in this control. In (**b**), pretreatment of the larvae with galactose reverses this situation; BR1 expands and BR2 collapses. In (**c**), inhibition of protein synthesis with cycloheximide in galactose pretreated larvae causes selective reactivation of BR2. Photographs by courtesy of Dr. J. L. Diez.

collapses (Fig. 3.6*b*). This process, contrariwise, is blocked by protein synthesis inhibitors, and BR1 then tends to contract and BR2 expands (Fig. 3.6*c*) and synthesizes RNA (Diez *et al.*, 1980). This implies that galactose induces the synthesis of a protein which inhibits BR2 puffing, and that by blocking that protein synthesis BR2's normal puffed state is retained. The data for BR1 are less secure, but at least they imply that these two puffs, making the same 75S RNA, come under quite separate control systems, possibly one positive and the other negative.

**Ecdysterone-regulated puffs**

There is a cycle of puff formation in the salivary glands of *Drosophila* which can be directly associated with changes in the level of the pupation hormone, 20-hydroxyecdysone (Fig. 1.13). Because earlier stages are difficult to study, most work relates to 3rd instar larvae, and to the prepupal and pupal stages preceding the breakdown and autolysis of the salivary gland. Becker (1959) was the first to show, by ligaturing larvae so that the hormone was excluded from the gland cells, that the puff cycle was hormone dependent, and this was subsequently confirmed by direct study of the effects of the hormone on glands cultured *in vitro* (Ashburner, 1971). In general, the pattern of hormone effects is that (1) intermoult puffs are repressed by the hormone, (2) several new (early) puffs are briefly induced into activity and (3) a third series of (late) puffs appears after some delay. The appearance of the late puffs depends on protein synthesis, since the chromomeres do not puff if larvae are treated with cycloheximide, and the late puffs are therefore assumed to depend on factors produced by the early puffs. The natural puffing succession suggests sequential gene activation (Clever, 1964). We shall consider the intermoult, and the early and late puffs separately.

***Glue proteins in Drosophila.*** About the middle of the 3rd larval instar (86 h after egg laying), *Drosophila* salivary glands synthesize a saliva which they discharge just prior to pupation. This synthesis then stops. The saliva hardens into a glue that sticks the pupa to its substrate, and Korge (1977a) has shown that it contains four proteins, two of which are glycosylated. (Although it does not affect what follows, we should note that Bekendorf and Kafatos (1976) find six proteins, four being glycosylated). Only two major puffs (at 3C and 68C) are active during the period of saliva synthesis and these regress when it ceases, probably under the action of the hormone ecdysone which is released at that time. Actinomycin treatment of larvae in the 75–80 h period blocks glue synthesis and the appearance of the two associated puffs (Zhimulev and Kolesnikov, 1975), but is without effect at 90 h, suggesting that the message has about a 10 h life. Of course, correlation of activities is no proof of causal relationships, but in this case two proteins (salivary gland secretion proteins, Sgs3 and Sgs4) have been independently associated with puffs: fraction 3 with the puff at 68A8–68C11 (Akam *et al.*, 1978), and fraction 4 with the 3C puff (Korge, 1977a). The proof of these relationships is genetic, and we shall look at Sgs4, as an example.

Electrophoretic variants of protein 4 are found in the Oregon-R wild-type stock (variants 4a, b and c); in heterozygotes both variants are made and can be followed—the genes are co-dominant. These genes were shown to be X-linked and, by crossover analysis, to be located in the 3C section of this chromosome. This is the Notch region for which a number of deficiencies are available (p. 176); and deficiency analysis (i.e. identification of females producing half the amount of protein 4 and all of it the electrophoretic variant on the wild-type chromosome) located the gene to between bands 3C10 and 3D1. This corresponds to the known puff at 3C11–12, which is not formed in the relevant deletions. Korge (1977a) and Berendes (1970) believe there is , in fact, only one fine band (3C11) at this point along the chromosome, and that Bridges was mistaken in drawing two.

Another wild-type fly strain, Hikone-R, produces little or no protein 4; it apparently carries a null mutation of *Sgs4*. Although no cytological deficiency can be seen, homozygous Hikone-R chromosomes do not puff at 3C11. On the other hand, when such a chromosome is heterozygous with an Oregon or Berlin wild-type a new, and faster migrating, Hikone-specific protein (4h) is found on electrophoresis, and the 3C-region is puffed, if somewhat asymmetrically (Fig. 3.7). This change occurs only when the two chromosomes are synapsed; and when the Hikone-R is heterozygous with the inversion-containing FM1 chromosome, which suppresses pairing, no 4h protein is made and only the FM1 wild-type gene puffs (Korge, 1977b). There must therefore be an interaction between some product of the wild-type 3C locus and the Hikone-R locus which activates the Hikone-R 3C structural gene when the two are paired. This relationship is called a transvection effect, and we shall see that it was earlier deduced from genetic experiments (Lewis, 1955). Transvection implies that some product diffuses locally from one chromosome to the other, and is capable of activating the structural gene at the defective locus.

Since glue proteins are the major salivary gland product at this time, it is possible to isolate their mRNAs, and to use them to identify the relevant gene sequences in a *Drosophila* library (Muskavitch and Hogness, 1980). The size of the gene and flanking sequences, determined using deletions, is between 16–19 kb (McGinnis *et al.*, 1980), and the structural gene is very much smaller at 0.9 kb. The locus is transcribed distal to proximal, but details of possible control regions have not yet been identified but are being pursued (Korge, 1981). An interesting feature of the structural gene is that it contains tandemly repeated 21 bp sequences which vary in number between fly strains, and which account for the protein variants found on electrophoresis.

Finally, we must note that this part of the salivary gland secretion programme is determined early, as are other larval cell characteristics. If anterior halves of 10–20 h embryos, containing the gland precursor cells, are cultured in the abdomens of adult flies (p. 221) they grow and secrete saliva, more or less normally. Their activity programme is in some way built

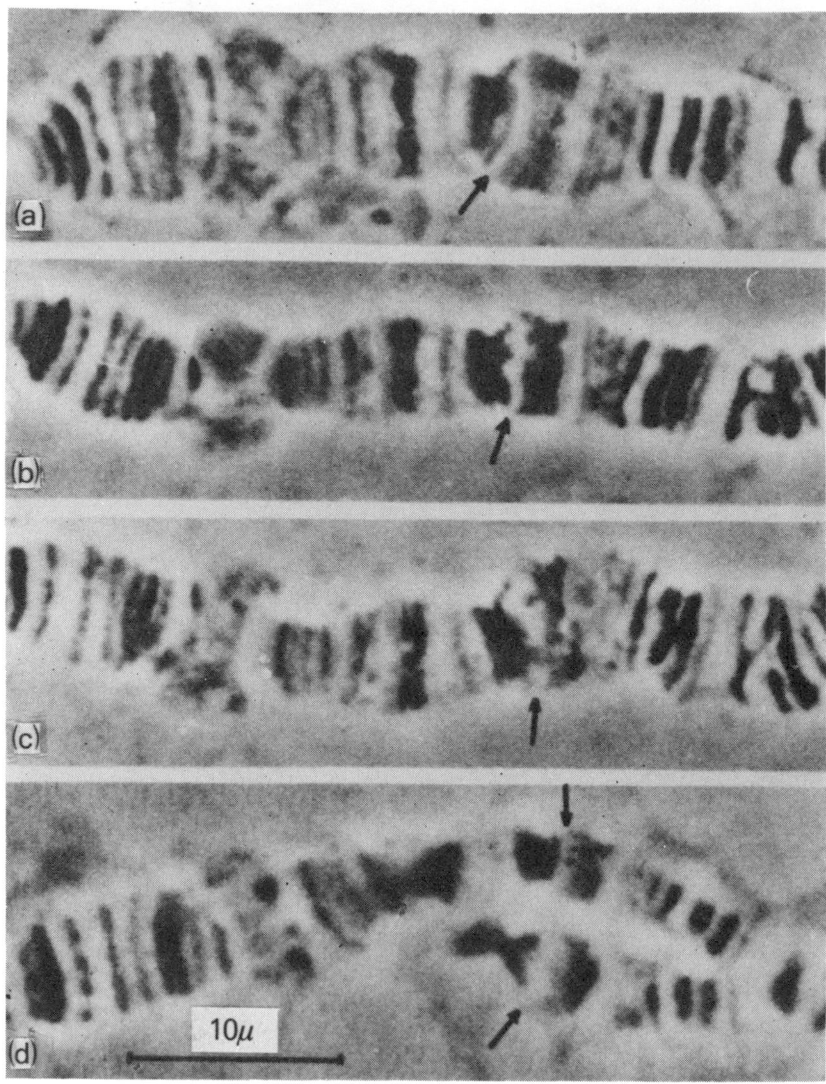


**Fig. 3.7** Photographs of the left end of the X chromosome of female 3rd instar larvae of *Drosophila*. (**a**) is from the Hikone-T strain showing the unpuffed 3C11-12 band (arrow). (**b**) and (**c**) are Hikone/Berlin heterozygotes with strongly paired chromosomes displaying asymmetrical puffing, and (**d**) is the same combination but with the chromosomes unpaired and only one chromosome puffed. See text for details. Photographs by courtesy of Prof. G. Korge.

into the cells (Korge, 1980). Cessation of function, on the other hand, is caused by the secretion of ecdysterones, which causes the activation of still other puffs.

***Ecdysone-induced early and late puffs.*** Puffing of salivary gland bands follows a well-established pattern during the later larval and early pupal stages of both *Drosophila* and of *Chironomus* (Ashburner and Berendes, 1978; Richards, 1980a). Just prior to puparium formation in *D. melanogaster*, the few larval intermoult puffs regress and an elaborate pattern, invoving over 100 puffs, is then initiated. This change depends on the normal secretion of ecdysone prior to pupariation, which it induces (Fig. 1.13). It can be copied by adding 20-hydroxyecdysone to the defined culture medium (Grace's medium) in which explanted glands are handled *in vitro*. The pattern of regression and induction is complex (Fig. 3.8)—some puffs regressing, others appearing, regressing and reappearing, and still others arising late in the cycle. In all cases the duration of a puff is short (3–6 h), and appears to be accompanied by, or associated with, the production of new protein by the gland. What these proteins are, what their physiological function might be, and whether they account for all the puffs, is unknown (but see Kress, 1981). The system is especially interesting because the dynamic relationships of gene activation by the hormone can be followed visually as the appearance and disappearance of puffs.

When isolated mid-3rd instar larval glands are exposed to $10^{-4}$ – $10^{-6}$ M 20-hydroxyecdysone in organ culture, puffs at 74EF, 75B and 23E appear

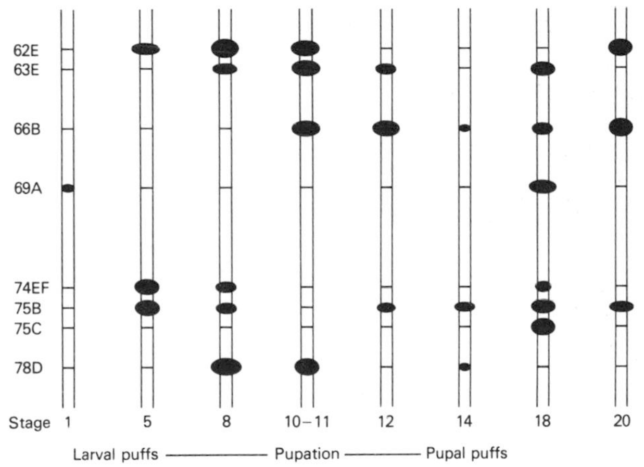


**Fig. 3.8** The sequence of appearance and regression of eight of the puffs on the left arm of chromosome 3 of *Drosophila*, after Ashburner (1975). The extent of puffing is shown approximately by puff size, and there are gradual changes between the stages shown. Twenty-one puff stages have been defined: 1 corresponds to the feeding 3rd instar larva, 5 to migrating larvae whose salivary glands have begun to fill, 8 to the stage when the gland is bloated, 10–11 is the emptying of the gland and puparium formation, 12 is about an hour later, 14 is 4 h after pupation, 18 about 10 h after pupation and 20–21 is 12 h after puparium formation. After this the glands degenerate. Of the four additional puffs mentioned in the text, 23E behaves much like 63E, 52A and 93F like 69A, and 82F puffs just prior to pupation and again at stage 18.

within minutes. They reach their maximum size in 2–3 h and then regress even in the presence of hormone. Withdrawal of hormone prematurely causes regression, but this is blocked by cycloheximide. The hormone is therefore an immediate cause of puffing, and regression depends on protein synthesis. Late puffs, at 62E, 78D and 82F, appear only after some hours, and their expression requires the normal, *in vivo* level of hormone ($10^{-7}$ M). This puffing is inhibited by cycloheximide, and is presumably dependent on protein synthesis. Sustained hormone levels are not necessary for puff maintenance, and hormone withdrawal induces premature puffing (Ashburner, 1973; 1974). Late puffs are not directly induced by the hormone.

A similar, related sequence of puffing events occurs between pupariation and pupation (the prepupal puffs) involving some, but not all, of the puffs just considered. By adding ecdysone at hourly intervals to cultured prepupal gands these puffs can be induced, but only after a refractory period of $4\frac{1}{2}$–6 h post-pupation (i.e. they do not respond to ecdysone as in the larval stage, but nevertheless puff earlier than they would do *in vivo*). Indeed, some larval puffs (78D) cannot be reinduced, and other new puffs appear (e.g. 93F). Culture with cycloheximide prevents the achievement of competence to respond to ecdysone, showing that it is a positive process, but the protein synthesis inhibitor has no effect once the nuclei are competent. These results imply that competence depends on the absence of ecdysone, and 3 h culture without the hormone is found to be the necessary minimum to allow subsequent induction of the late prepupal puffs (62E, 74EF, 93E). Ecdysone inhibits the acquisition of competence if present at levels higher than $0.5 \times 19^{-9}$ M. Presence of ecdysone inhibits, and its absence induces, the puffs (63E, 69A, 75CD and 52A) which appear briefly 6–8 h after puparium formation, during the period of competence acquisition. Cycling of the ecdysone level is therefore an essential aspect of the regulation of the normal puffing sequence. The correctness of this conclusion has been confirmed by taking 3rd instar larval glands, culturing them for 6 h with ecdysone, then for 3 h without, before returning them to ecdysone treatment. This regime induces the entire puffing sequence including that normally occurring during the prepupal stage (Ashburner and Richards, 1976; Richards, 1976).

Formally at least, these two early/late puff cycles can be explained by the same model (Fig. 3.9). This assumes that there are two qualitatively different kinds of puffs; early puffs which are directly induced by ecdysone, and late puffs which are induced, directly or indirectly, by products (P) of the early puffs. Conversely, ecdysone is taken to inhibit late puffs, and the product (P) to inhibit early puffs. So there is a reciprocal interaction between ecdysone and puff product, the two being supposed to compete for a chromosomal site. Prepupal puffs can be brought into this scheme if the stability of the product P is taken to depend on the presence of ecdysone (at least after the larval sequence); and its disappearance when the ecdysone

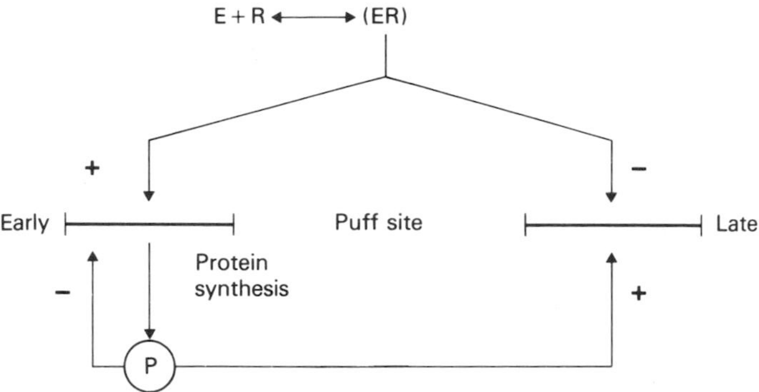


**Fig. 3.9** Model for the control of early and late puffs in *Drosophila* larval salivary glands. Ecdysone (E) combines reversibly with receptor (R) protein(s) to form the active ecdysone complex (ER). The early puff responds quantitatively to this (hence to the amount of E) by producing a product P, which is either a protein or requires a protein for its formation. P inhibits the early puff, and induces the late puff. By blocking the synthesis of P, cycloheximide will prevent regression of the early puff and inhibit the late puff. ER inhibits the late puff; in both situations ER is competitively replaced by P. Early removal of ecdysone causes regression of the early puff, independently of P and of protein synthesis, and correspondingly results in induction of the late puff to an extent dependent on the amounts of P previously synthesized (Ashburner *et al.*, 1974).

titre falls 3 h after pupariation can be equated with the 'acquisition of competence' stage, necessary for subsequent puff activation by ecdysone. Kress (1981) also has a model which involves two qualitatively different puffs, the first cycle, early ecdysone-induced, puffs are assumed to produce protein activators, as before, but also other proteins which inactivate the product of the intermoult puffs required for glycosylation. These different hypothetical proteins have yet to be identified.

Although the assumptions in these models have still to be proven, the data strongly suggest that ecdysones cause the expression of some genes (early puffs) and suppress others (intermoult puffs), and that the puff products themselves may similarly act positively or negatively. The hormone effects are orderly and involve groups of genes, presumably reflecting the developmental state (competence) of the tissues affected. We should then expect the puffing patterns to be different in different tissues, but the evidence for this is still ambiguous. Berendes (1967) found that the same puffs formed in gut, Malpighian tubule and salivary glands of *D. hydei*, and Richards (1980b) notes that salivary gland and fat body chromosomes exhibit the same puffs, with minor exceptions. On the other hand, Bonner and Pardue (1977) find that RNAs isolated from imaginal discs do not hybridize to the puffed bands of salivary nuclei prepared from larvae of the same state. Unfortunately the activity pattern, the puffs, of disc cell nuclei cannot be discerned, and the reciprocal test is not possible. The problems of the specificity of puffing patterns in different tissues have still to be resolved. This is not the case, so far as we know, for the general response of cells to heat shock.

## Heat shock puffs

Ritossa (1962) found that a specific puff pattern is induced when *Drosophila* larvae are exposed to 37 °C for 20 min: nine puffs in *D. melanogaster* and six in *D. hydei*. Within an hour the previously active 25 °C puffs regress leaving this new 'gene family' as a distinct pattern of puffs dispersed through the genome (Table 3.1). Other insults, such as exposure to ammonia, to anaerobiosis, to inhibitors of electron transport and to uncouplers of oxidative phosphorylation, induce the same puffs, or some of them. Puff size is proportional to temperature, to duration of exposure (to a limit, for they eventually regress), and heat shock plus anaerobiosis is roughly additive. Embryos, specific tissues and cell lines respond in the same way, so it is likely that we are concerned with a general, rapid cell response to extreme environmental conditions, reflecting a homeostatic mechanism not yet understood. Analogous responses have been found by heat shocking *E. coli.*, maize, sea urchin embryos, chick, mammal and *Dictyostelium* cells (Schlesinger *et al.*, 1982), but the most detailed work has employed *Drosophila*, particularly *D. melanogaster* salivary glands and cell lines, as recently reviewed by Ashburner and Bonner (1979). The system is an attractive and simple one to manipulate, and we can ask what it tells us about gene organization and regulation.

**Table 3.1** The heat shock genes of *D. melanogaster*: their puffs, messages and proteins

| Puff locus | Puff size | mRNA | | Protein |
|---|---|---|---|---|
| | | Size | Amount | |
| 33B | Minor | Minor | 3% | Unknown |
| 63C | Minor | 20S | 6% | 84K |
| 64F | Minor | Minor | 3% | Unknown |
| 67B | Major | 12S | 5% | 22, 23, 26 and 27K |
| 70A | Minor | Minor | 2% | Unknown |
| 87A | Major | 20S | 10% | 70K |
| 87C | Major | 20S<br>12S | 34% | |
| 93D | Major | No poly(A) | 35% | Unknown |
| 95D | Major | 20S | 4% | 68K |

Puff are classed only as minor or major in size. The heat shock RNAs sediment at 12S or 20S, but can be further fractionated by electrophoresis. All except 93D produce poly(A) messages.

Heat shock provides a direct proof of the relationship between puffing and protein synthesis. Tissières *et al.* (1974) demonstrated that the great array of protein bands displayed by gel electrophoresis of normal tissues is then replaced by eight major heat shock polypetides (hsps); and this new synthesis, like the puffs, can be blocked by actinomycin D (Lewis *et al.*, 1975). The pattern of hsp synthesis is the same in diploid or polytene, and larval or adult tissues, so tissue culture cells can be used to provide the quantities of material necessary to explore the molecular events resulting from heat shock.

The immediate event, seen as a puff within 1 min, is the synthesis of the hsp mRNAs. At the same time pre-existing polysomes disappear and are replaced by a new group carring 20–30 ribosomes per message. Heat shock therefore affects translation as well as transcription. Nothing is known of the mechanism involved at either level, although it is generally assumed that anaerobiosis (the common denominator of effective treatments) is the immediate signal for these changes. Cytoplasm from heat shocked cells induces heat shock puffs when injected into salivary glands, suggesting that some specific activator may be produced (Compton and McCarthy, 1978), although we should note that cell damage may mimic heat shock. Nor do we know how the transcription of previously active genes is turned off, though this occurs immediately (Findly and Pederson, 1981). The evidence is against the coordinate control of a gene battery, for the six proteins induced in *D. hydei* appear sequentially within 20–90 min of exposure to 37 °C, and only two of the *D. melanogaster* proteins are made at the intermediate temperature of 29 °C (Lewis *et al.*, 1975). This does not exclude some coordination since, by using cloned genes as probes for RNA synthesis, Findly and Pederson (*loc. cit.*) find that mRNAs for the 70,000 and 26,000 molecular weight proteins are made synchronously. Both are found at low levels of synthesis prior to heat shock, showing that we are dealing not with new gene activation but with a magnification of transcription already under way (cf. ovalbumin induction, p. 63).

Since the polysomes of heat shocked cells can be separated, the poly(A)mRNAs they carry can be isolated on denaturing gradients and further separated by electrophoresis. If these mRNAs are hybridized to salivary chromosomes they find the locations recorded in Table 3.1. Only puff 93D is not accounted for. It presumably does not make a poly(A) message, and is also peculiar in that it is specifically induced by benzamide (Lakhotia, 1971) and by old culture media (Bonner and Pardue, 1976). It may not be a true member of the heat shock gene family. The most interesting result, however, is that the major (20S) mRNA hybridizes to puffs at both 87A and 87C, implying gene duplication at these loci (Henikoff and Meselson, 1977).

These poly(A)RNAs can also be translated *in vitro* producing a predominant [35S]methionine-labelled protein of 70,000 molecular weight corresponding to the 87A and 87C puff products (Spradling *et al.*, 1977; McKenzie and Meselson, 1977; Mirault *et al.*, 1978). Since clean separation of the mRNAs is difficult, further protein identification has involved the technique of hybrid arrest (Paterson *et al.*, 1977). In this, cloned DNA for the locus is hybridized to the mRNA mixture, inactivating the complementary mRNA. Comparison with the unhybridized control mixture allows identification of the protein which is not made (e.g. Livak *et al.*, 1978; Holmgren *et al.*, 1979). Electrophoretic variants of the hsps have also been used to confirm the relation of puffs to protein (Peterson *et al.*, 1979). This gives the allocation of proteins to puffs of Table 3.1. Three of the minor mRNA classes have not yet been identified as protein. The molecular organization of some heat shock genes, particularly the interesting 87A and 87C puffs, has been explored both genetically and by using their abundant mRNAs to identify DNA clones complementary to them (Fig. 3.10). Deleting the 87C locus does nothing to the heat shock protein patterns, whereas deleting both 87A and 87C removes the 70,000 molecular weight (hsp70) protein from the gel array, implying that like genes are present at both sites (Ish-Horowicz *et al.*, 1977). Restriction maps of cloned DNAs show that 87A (actually at 87A7) carries two hsp70 coding sequences lying back to back, and separated by about 1.5 kb. The 87C region (87C1 to be precise) is more complex. It has one hsp70 gene separated from two tandem hsp70 genes by 38 kb (Livak *et al., loc. cit.*; Craig *et al.*, 1979; Mirault *et al.*, 1979; Ish-Horowicz and

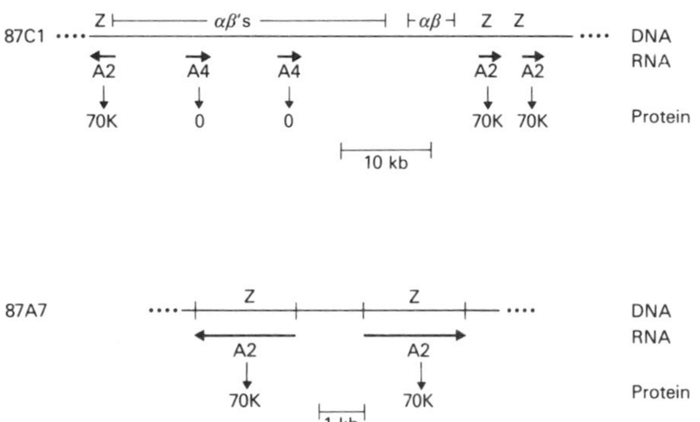


**Fig. 3.10** The organization of the 87C1 and 87A7 heat shock loci of *D.melanogaster*, after Ish-Horowicz and Pinchin (1980). The repeat units (Z) code for an A2 mRNA which corresponds to the 70,000 molecular weight heat shock proteins. In 87C1 there is an isolated hsp70 gene separated from a pair of tandem repeat genes by a 38,000 bp region containing α-β sequences which make A4 transcripts which are not translated. The Z repeat units are also found at 87A7, where they are transcribed in opposite, back-to-back, orientation. The relationships between Z and α-β sequences vary between strains and species, and there are γ sequences (not shown) of uncertain function associated with the Z regions. A single band may therefore carry sequences for two different RNAs (A2 and A4); a single gene (A2) may be present in more than one band; and multiple copies of a gene may be found at one locus. How far this complex gene organization is unique to these hsp sequences is uncertain.

Pinchin, 1980). As judged by restriction maps, there are minor differences between the genes at the 87A and 87C sites, but in both cases the 2.5 kb segment carrying the hsp70 coding regions has a 5′ non-coding region (0.3 kb) associated with it. The 2.5 kb unit has been called the Z element, and the Z non-coding region ($Z_{nc}$) may be involved in the regulation of the Z coding ($Z_c$) region (Mirault *et al.*, *loc. cit.*). The large 38 kb insertion in the 87C1 locus carries a secondary set of sequences which are transcribed after prolonged heat shock (Lis *et al.*, 1978). These α–β sequences are not found at 87A7, but are present in the chromocentre where they are not transcribed. The α–β sequences are present in tandem arrays, together with a third γ element, making α γ units. They are not transcribed and have yet to be allocated a function, if any.

The puff at 67B has also been analysed by finding cloned DNA segments corresponding to the puff site (Corces *et al.*, 1980). The four small heat shock polypeptides are encoded as a cluster within an 11 kb segment of DNA and separated from one another by 1.0–4.7 kb spacers. Since three of the genes exhibit alternating polarities they are probably organized as individual transcription units without introns. Each of the four contains TATA sequences in accord with this conclusion, but the only other common feature shown by sequence analysis is an ACTTTNA run situated $195 \pm 12$ bp upstream from the initiation site. This suggests to Ingolia and Craig (1981) that this sequence may be exposed on the nucleosome in alignment with the initiation site, and outside the core particle.

Sequence data are as yet too incomplete to suggest what may be involved in the regulation of these genes, but the hsp loci are prime structures for this type of analysis. It is therefore particularly interesting that the chromatin of the active hsp70 genes is sensitive to DNase I digestion (p. 70) and that there are sites at the 5′ end of these genes which are hypersensitive to the enzyme (Wu, 1980). And the same is true for the 67B gene cluster (Keene *et al.*, 1981). This may imply that there are recognition elements at the 5′ end of the genes which function in their regulation. There must also be some regulation of translation (p. 109). The reticulocyte lysate and wheat germ extract systems used for translation *in vitro*, as noted above, tell us nothing about this. This problem has been studied by Storti *et al.* (1980) using lysates of *Drosophila* cells, either cultured at 25 °C or shocked at 37 °C. The normal cell lysate translates the 37 °C mRNAs and also the 25 °C mRNAs from what must be a mixture extracted from heat shocked cells. Hence, the array of mRNAs being made at 25 °C is not broken down or inactivated by heat shock, but is stored and available to the cell when it is returned to the lower temperature. Lysates from heat shocked cells, on the other hand, preferentially translate heat shock messages. This discrimination must result from more or less stable alteration to the components of the heat shocked cell, and from the presence of specific information encoded in the heat shock mRNAs. Although the heat shock system may be unique in these respects, its ease of manipulation should permit the detailed examination of these interesting aspects of genome organization (Scott and Pardue, 1981).

These studies, which combine cytogenetic and molecular techniques, have demonstrated that: (1) two separate bands may contain the same, or nearly the same, genetic information, and (2) one band may possess two or more genes, including duplicates and triplicates of one gene. Our original question about the relation of chromomeres to genes is therefore categorically answered in these specific cases, confirming the earlier but cruder genetic analysis (p. 77). This does not tell us if the chromosome folding pattern, identified as a chromomere, is related to gene activity (or, more probably, inactivity). So it is interesting that King *et al.* (1981) have recently confirmed Ribbert's (1979) observation that the banding pattern of oocyte nurse cells differs from that of pupal bristle-forming cells in *Calliphora*, by showing the same to be true for nurse cells and salivary glands of *Drosophila*. Tissue differences in polytene banding present a new, and still to be explored, aspect of chromosome organization which may be very relevant to understanding gene activity, should it be confirmed.

## Lampbrush chromosomes

During mid-prophase of the first oocyte meiotic division, the separated chromosomes of many organisms attain the appearance of a *Lampencylinderputzer*. This formation is unique to the germline, is a 'hairy' lampbrush in many invertebrates, but appears as large loops in amphibia where they have been especially studied (Fig. 3.11). In *Triturus* there may be 5000 or so loops of an average 100 μM length in the haploid set, calculated as some 50 cm of DNA in the total of 10 m (Callan, 1963). Each homologous chromosomal pair consists of two fine chromatids, and the loops are unit lateral extensions arising from the chromomeres which carry transcribed RNP particles (Fig. 3.12). The loops are an intregal part of the chromatid since they can be pulled apart without breaking its linear continuity (Callan and Lloyd, 1960). These lateral loops contract and are drawn into the chromatid as the oocyte reaches maturity. Since the loops have distinctive morphologies, heterozygous pairs can be followed in inheritance and are found to obey mendelian rules.

The detailed molecular organization of chromomere and loop is not known (see Macgregor (1980) for a critical discussion of this and other aspects of lampbrushology) and we cannot properly draw parallels between these structures and the polytene organization considered above. However, since amphibians have a wide range of genome sizes, even within a genus, we can ask what is the relationship between chromosome length and chromomere number? The data of Vlad and Macgregor (1975), based on careful

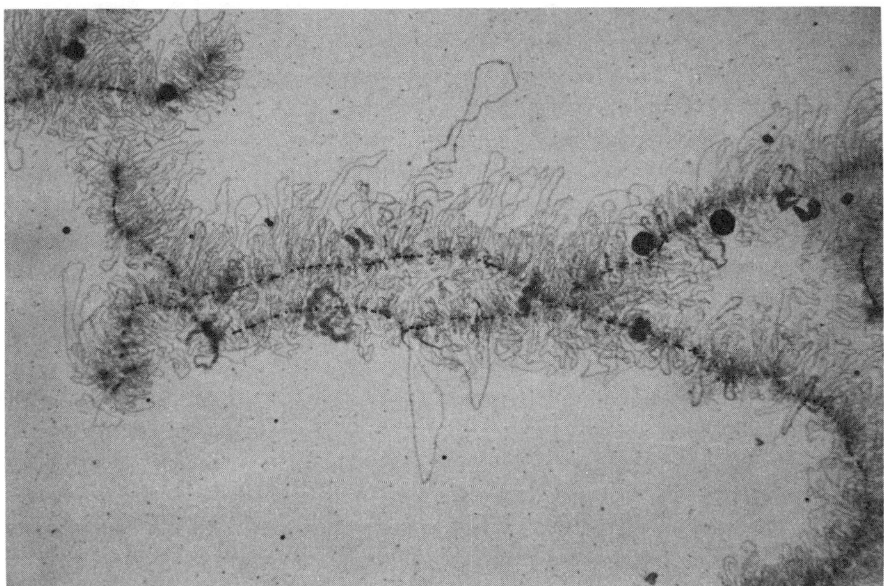

**Fig. 3.11** Part of the meiotic diplotene bivalent of chromosome X of *Triturus cristatus carnifex*, showing all of the shorter 'right' arms. The interchromomeric main axis contains two DNA duplexes, whereas the loop axis contains only one and is part of a single chromatid. The DNA duplex is continuous throughout the length of the chromatid, and the DNA/histone complex is reflected back on itself at intervals to form the lateral lampbrush loops (see Fig. 3.12). Photograph by courtesy of Prof. H. G. Callan.

measurements of five selected lampbrush chromosome segments, show that the number of chromomeres per unit chromosome length, or per pg of DNA, is constant (Table 3.2). Put another way, *Plethedon vehiculum* and *P. dunni* with about twice the C values of *P. cinereus*, have 60–70 percent more chromomeres at the lampbrush stage than this last species, although the development of all three is notably similar. As might be expected, more of the DNA of the first two species is moderately repetitive and the repeti-

**Table 3.2** Estimates of chromomere numbers in *Plethodon* species, based on measurements of five lampbrush chromosome segments

| **Species** | **C value (pg)** | **Chromomere distance ($\mu$m)** | **Haploid length ($\mu$m)** | **Chromomeres per** | |
|---|---|---|---|---|---|
| | | | | **Haploid set** | **pg DNA** |
| *P. cinereus* | 20.0 | 2.06 | 6386 | *3100* | *155* |
| *P. vehiculum* | 36.8 | 1.96 | 9775 | 4987 | 178 |
| *P. dunni* | 36.8 | 2.13 | 10,680 | 5014 | 146 |

The chromomere distance was obtained by dividing the total length of segments measured by the total number of chromomeres counted. Haploid length is that of the entire genome. Data from Vlad and Macgregor (1975).

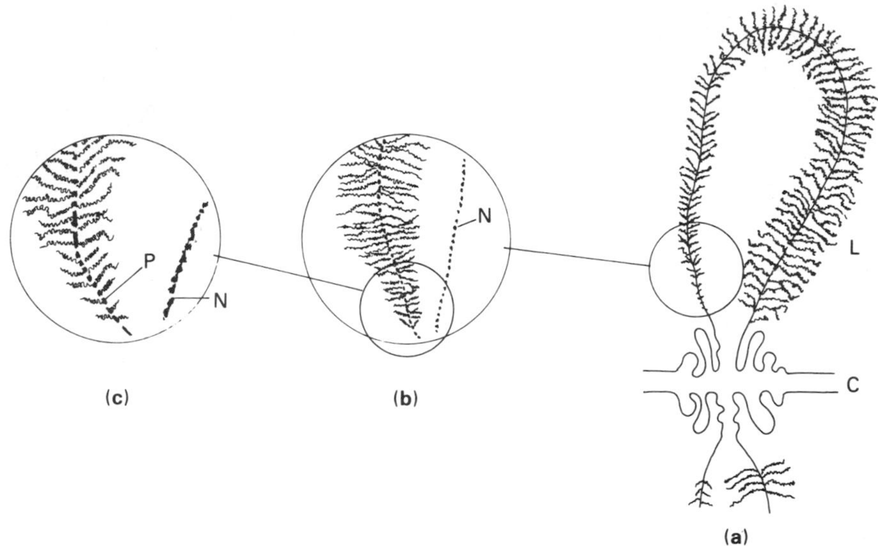


**Fig. 3.12** Diagrammatic representation of the ultrastructure of a lateral lampbrush loop, showing the detail of the transcription process. (**a**), The loop (L) is shown in its relationship to the chromomere (C). The detail of a 'Miller spread' (Miller and Beatty, 1969) is shown in enlargement (**b**) with the origin of transcription at the bottom and an increasing gradient of transcript lengths proceeding towards the top. Note that the chromatin before the origin of transcription is organized in nucleosomes, as is the untranscribed strand on its right (N). (**c**), shows the RNA polymerases (P) at the bases of the transcripts and their tight packing along the DNP strand being transcribed. Transcription at this lampbrush stage must account for the large amounts of maternal RNAs in amphibian eggs and the more or less complete inactivity of the genome during early embryogenesis.

tive sequences have little in common between the two species classes (Mizuno and Macgregor, 1974). These data then, provide no support for the one-chromomere-one-gene hypothesis, and they suggest that chromomeres *per se* are not organized in any simple way in relation to function. As a corollary, average loop length (i.e. the distance from the chromomere start to the inflexion of the loop) is greater in species with high C values: 17 μm in *P. dunni* and 8.3 μm in *P. cinereus*, for example. Thus, in general, there may be no regular relationship between the genetic information exposed on the loops of even related species.

Lampbrush loops are transcribed. An array of RNA fibres arise from polymerase molecules attached closely (~ 100 nucleotides apart) along the loop, and the length of these fibres increases around the loop (Fig. 3.12). It is reasonable to assume that the growing fibre reflects the progressive synthesis of an RNA molecule as transcription proceeds. RNA polymerase does not attach to the chromomeres or to the axis, which are not transcribed. The transcription unit is often the whole loop, as judged by the region between the shortest and longest fibres, but some loops are not completely transcribed. Other loops contain two or more transcript arrays with the same polarity, and some have two arrays with opposite polarity and a spacer between them (Angelier and Lacroix, 1975). Loops may therefore have single or multiple functions (genes?) exposed for transcription, but the situation is, quite literally, clouded by the aggregation of RNP particles formed around the loops, and these make it uncertain whether loops are always single, or composed of small multiple loops. Since poly(A)RNAs are made prior to the appearance of lampbrush loops (but not necessarily before they start to form) and can subsequently be identified in the egg (Darnborough and Ford, 1976) it is difficult to relate lampbrush transcripts to particular 'maternal message', i.e. to the mRNAs stored in the egg. The most attractive assumption, nevertheless, is that this massive activity provides a store of transcripts used early in development. Davidson (1976) speculates that this may include regulatory sequences needed by the developing oocyte.

Attempts to probe lampbrush function using purified or cloned DNA for cytological DNA–RNA hybridizations have been less than satisfactory (see Macgregor, 1980; Callan, 1981). The clones (or purified DNA) frequently contain repetitive sequences which hybridize to sites scattered throughout the lampbrush array, as in Pukkila's (1975) annealing of purified *Xenopus* 5S DNA to the lampbrush set of *Notophthalmus viridescens* which showed consistent labelling of loops, whereas we have already noted (p. 54) that 5S genes are telomeric. Or again, the work of Old *et al.* (1977), using purified sea urchin histone genes against *Triturus cristatus*, showed hybridization at eight loops on chromosome 1 and further 'histone loops' on three other chromosomes, whereas more detailed work using the salamander *Notophthalmus* shows that the repeated histone locus is at only two sites and that transcription on the loop runs through the entire set of repeats (p. 53) including the spacer sequences (Diaz *et al.*, 1981). The very large loop transcripts (150 kb average) must therefore contain repetitive sequences, spacer sequences and structural genes. Apparently transcription initiates at the histone promoter but fails to terminate and so runs through the repeated (satellite) sequences. This failure to recognize termination signals may explain the long transcripts of lampbrush chromosomes, but it gives no clue as to the organization which results in the formation of loops at particular places along the length of the chromosomal DNA.

Lampbrush chromosomes are also found in the spermatocytes of insects (e.g. grasshoppers, *Chironomus* and *Drosophila*) but, although transcription of the loops is known to occur, the details are not yet clear. In *Drosophila hydei* (Fig. 3.13), and probably all the Drosophilidae, loops are associated with the Y chromosome, and form a limited and distinctive set regularly associated with particular regions of this chromosome. Deletions of the Y covering any one of these regions cause infertility, as do mutants affecting the unfolding of the loop, or inhibition of RNA synthesis by the loops with actinomycin. Since only half the sperm contain a Y chromosome, these

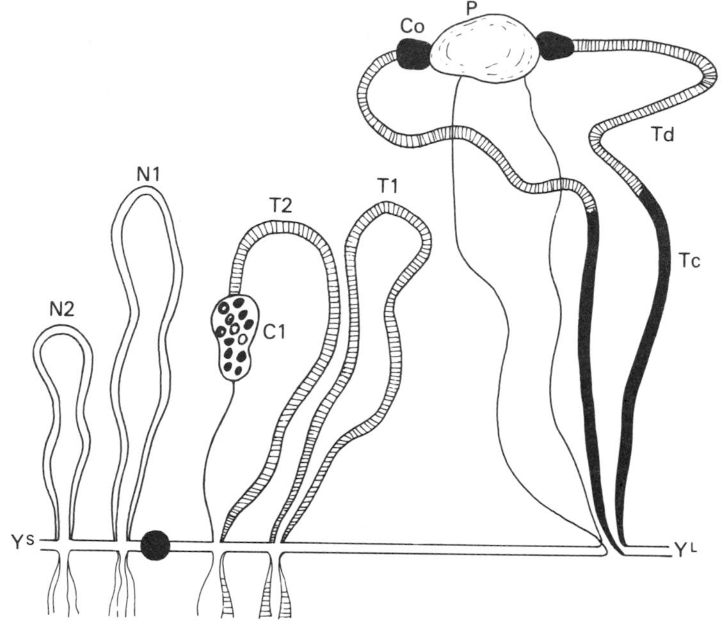


**Fig. 3.13** Lampbrush loops on the Y chromosome of *D.hydei*, after Hess (1975). The Y chromosome is heterochromatic and apparently largely untranscribed except during the diploid lampbrush stage of the growing primary spermatocytes. These strange loops are transcribed to give RNAs which are unique to the spermatocyte nuclei and are essential for male fertility. XO males are sterile, as are males deficient in any of the loop regions of the chromosome. There are three pairs of loops on the long arm of the Y ($Y^L$) and two pairs on the short arm ($Y^S$), here shown diagrammatically. These arms also each carry a nucleolus organizer region (not shown). N1 and N2 are noose loops, T1 and 2 are tubular ribbons, the second pair carrying club-like structures (C). The even more complex thread pair (both shown) are compact proximally (Tc) and diffuse distally (Td), and are associated through cone-like structures (Co) with a pseudonucleolus (P).

'fertility factors' must make their contribution prior to sperm formation, but there is no evidence that they contribute particular structural molecules needed for sperm differentiation, the usual complement being present, but disorganized, when a loop is absent (Hess, 1975). Stored message of as yet uncertain function clearly derives from the activity of these loops. In the female *Drosophila* on the other hand, ovogenesis is meroistic, and if we consider the role of the lampbrush to be the production of 'stored mRNAs' this function is taken over by the nuclei of the 15 nurse cells which, with the oocyte, make up the ovarian cyst (Fig. 3.14). These nurse cells are linked together by 'ring canals' through which they pass their products to the oocyte causing it to grow in size. As the cyst matures the nurse cell nuclei endoreduplicate (to 1024 C), and this expansion of the genome permits the rapid growth of the egg. RNA, particularly rRNA, is rapidly synthesized and passed to the oocyte, but the detailed nature of the other

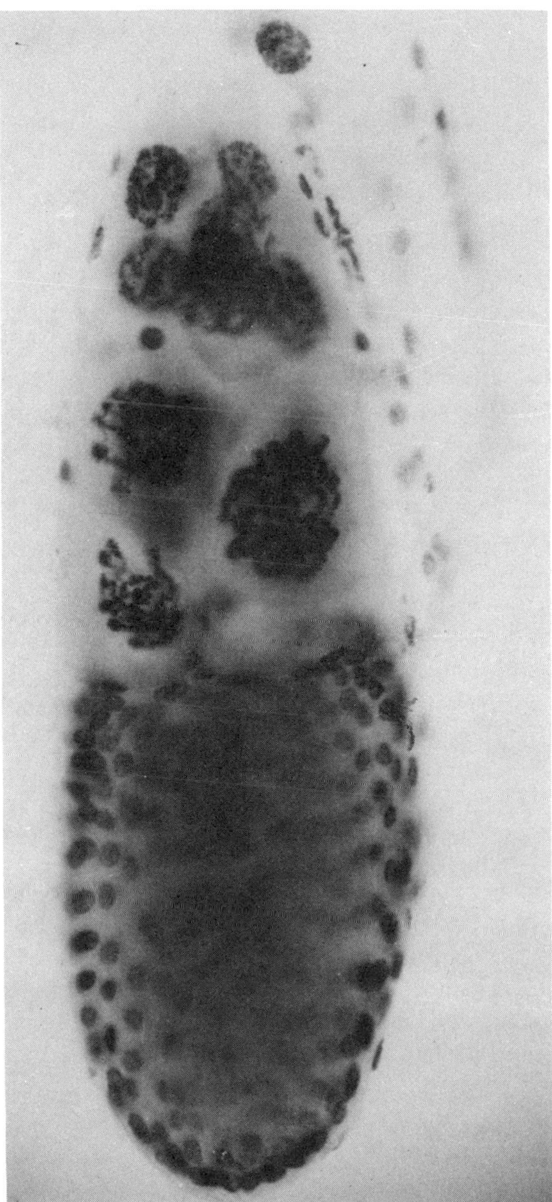

**Fig. 3.14** A Feulgen stained *Drosophila* ovarian chamber, at developmental stage 10 of Cummings and King's (1970) classification. The follicle cells surround the half-grown oocyte, and their densely staining nuclei mask the oocyte nucleus which lies close to the top of the growing oocyte. Ten of the fifteen nurse cell nuclei can be seen, and their polytene nature is obvious. The nuclei closest to the oocyte are 1024C, and those at the top of the chamber are half this level of ploidy. The nurse cell products pass into the oocyte cytoplasm through interconnecting ring canals (Fig. 7.3.) Shortly after this stage the nurse cell nuclei begin to degenerate; but the oocyte, surrounded by follicle cells, continues to grow until it fills the chamber.

products is unknown. Many mutations affecting all the distinguishable steps of ovogenesis have been listed by King and Mohler (1975), but it is only recently that mutations producing analysable ovarian polytene chromosomes have been isolated (King *et al.*, 1981).

## Conclusions

These microscope studies of active genes extend our understanding of the genome organization described in the previous chapter. In the first place, they show that genes which function together are not grouped but are spread around the chromosomes, as witnessed by puffing patterns and the identification of the loci for salivary proteins, heat shock proteins etc. Nor is there necessarily a gene per chromomere, though sometimes there is this association. Some genes, on the other hand, are duplicated both within and outwith a chromomere. When a gene lies in the condensed chromatin, which we define as a chromomere, its activation depends on this being unfolded, but we do not know the molecular arrangement of this transcriptionally accessible DNA in either puffs or lampbrush loops. This change, insofar as one can argue from puffing, depends on protein synthesis; so the apparent direct activation of some loci by hormones (or homone–receptor complexes) may involve many steps. Some active genes then make proteins which activate other genes; but there is also another class of gene product which functions as an inactivator, just as hormones also do when they directly cause the suppression of active puffs. These opposite responses probably depend on the state of the chromatin locally, as it affects the competence of genes to respond; and this condition depends on some prior developmental sequence of events. It is not certain if these changes of chromosomal organization are reflected in the banding differences seen in various tissues (but still to be confirmed), or if they are more subtle and so concealed by the regular folding of the chromatin which enables us to make cytological maps. Nor can we be sure, although we have assumed this, that genes are turned on from a null state, or if the manifest rise of RNA synthesis in puffs and loops reflects only an increase of already existing transcriptional activity. In lampbrush loops, this transcription reads through great lengths of DNA which include repetitive sequences, spacers etc., as well as structural genes. These long transcripts must carry information for their nuclear and cytoplasmic processing, but the sequences involved in these steps have not been identified.

The complexity of these phenomena make it difficult to construct models of developmental gene activation which are much more than redescriptions of events; except for models of limited phenomena like the induction of early and late puffs already described, which, in any case, are probably too simple. Nevertheless, it is useful to have models, and we shall consider some next. These models hark back to the molecular description of the genome and take account of some only of the higher order phenomena which we have just described. Although they are in this sense incomplete, they provide terms of reference which will allow us to place in context the experimental studies which we shall consider subsequently.

# 4 Models of the functioning genome

The discovery of units of coordinated genetic activity and of regulator genes which govern, negatively, the activity of structural (i.e. enzyme forming) genes via cytoplasmic repressors, and which are able to interact electively with exogenous or endogenous chemical agents, appear to offer precisely the type of elements needed to build the complex and precise chemical networks of information transfer, upon which must rest the development of the embryo, as well as the physiological functioning of the adult organism.

J. Monod, F. Jacob and F. Gross, 1962

The discovery of regulatory genes in prokaryotes resolved the problem of differential gene activation in principle, as Monod *et al.* (1962) were quick to see. It has conditioned all our subsequent thinking about the regulation of genes during development.

We shall look at this operon model first since it is the simplest system, and the elaborations on it which follow attempt to account more exactly for some of the developmental phenomena described in the first chapter. Unfortunately this increase in the complexity of the models (or hypotheses), which ensures that they fit the facts, makes it difficult to disprove them for that very reason: equations with more than three terms can be manipulated to fit any data! Nevertheless, the models which we shall examine focus attention on some of the key questions requiring answers, and they provide a framework on which we can hang new information. Some of the earlier models may now seem outdated, but they may in fact still be valid. We have three (or perhaps more) classes of gene activity in developing cells and tissues: the activation of 'housekeeping genes' which are responsible for general cell metabolism, but which may be qualitatively and quantitatively different between tissues with different functions (e.g. nerve and liver); the activation of genes making the 'luxury proteins' which typify particular cell types (e.g. muscle actin and myosin) and which concern us most; and hormone (and evocator?) induced proteins (e.g. the oviduct products already considered). At present there is no good reason to assume that these classes are regulated in precisely the same way, and the models remind us of this complication.

## The operon model

Beckwith and Rossow (1974) have emphasized the considerable differences of functional detail among the prokaryote gene control systems and we must here restrict ourselves, with that caution, to general principles. The simplest model (Fig. 4.1) has a *regulatory gene* coding for a regulatory protein which (1) interacts with an *operator* site adjacent to the structural gene (or genes), and which (2) has its affinity for that site modified by inducer or repressor substances (generally substrates or products). There are two classes of control system: positive systems where the regulatory gene product switches on the gene(s) when combined with the operator, and negative systems where this combination switches off the structural genes. The four possible types of control circuit are summarized in Table 4.1. Even this simple model is complex enough to cover everything we know about gene switching, and the question is: can we identify its components in diploids?

Regulatory genes should map separately from the gene(s) they control, and their mutation should have multiple effects in homozygotes, which

**Table 4.1** Prokaryote control systems

| | **Positive control** | **Negative control** |
|---|---|---|
| **Effect of regulatory gene product** | **Apo-inducer switches on the operon** | **Apo-repressor switches off the operon** |
| **Inducible systems** | | |
| Co-activator function | Co-inducer activates apo-inducer | Co-inducer inactivates apo-repressor |
| Regulatory gene deletion | Uninducible | Constitutive |
| Regulatory gene mutation* | Constitutive | Uninducible |
| Operator deletion | Uninducible | Constitutive |
| **Repressible systems** | | |
| Co-activator function | Co-repressor inactivates apo-inducer | Co-repressor activates apo-repressor |
| Regulatory gene deletion | Super-repressed | Derepressed |
| Regulatory gene mutation* | Derepressed | Super-repressed |
| Operator deletion | Super-repressed | Derepressed |

*The mutation prevents binding with the co-activator, but not with the operator. The presence of a normal regulatory gene cannot interfere with this so the mutant effect is dominant.

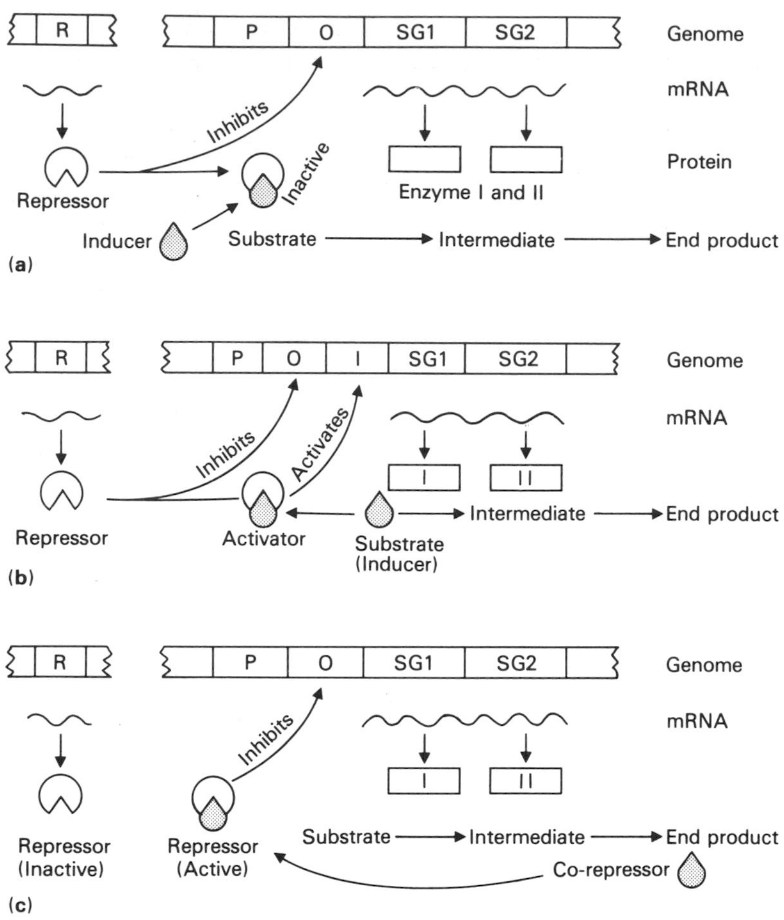


**Fig. 4.1** Operon organization. (**a**), Negatively controlled inducible system in which the regulatory (R) product combines with the operator (O) to prevent transcription of the two structural genes (SGI and SGII), but is inactivated by the inducer thus permitting transcription. (**b**), Positively controlled inducible system, in which the operon is activated by the regulatory gene product (apo-inducer) combining with the substrate (co-inducer) and acting at the initiator sequence to initiate transcription. (**c**), Negatively controlled repressible system, where the regulatory gene product is inactive until combined with the co-repressor (usually end product) which then inhibits the operator. P, promoter sequence; I, initiator sequence. Table 4.1 summarizes the details.

should distinguish them from mutations of the individual genes which they regulate (but see below). Table 4.1 indicates the consequences of two kinds of regulatory gene mutation; one a deletion (or equivalent null mutation) and the other preventing binding with the co-activator. These mutations have opposite effects within each system. Deletions will be recessive in heterozygotes since the wild-type chromosome will make enough apo-inducer (or apo-repressor) for the system to function. Binding-site mutations will inactivate the system, and so behave as dominants (or semi-dominants if incomplete in their effects). Deletions (or null mutations) of the operator locus will repress structural gene transcription in positive control systems, and cause constitutive functioning (depression) in negative systems (Table 4.1). Since operator sites are contiguous with, and control, the structural genes only on their own chromosome, operator mutations will be *cis*-dominant and *trans*-recessive in heterozygotes. In double heterozygotes also carrying alleles of the relevant structural gene(s) it should be possible to identify the operator mutations (e.g. by their effects on expression of allozyme mutants (p. 140)). Similarly, promoter (polymerase binding) site mutations will be *cis*-dominant and *trans*-recessive, but of course they will block transcription in both positive and negative systems.

These predictions can obviously be tested only by using the genetic fine structure analysis techniques developed with prokaryotes, and we have a dearth of mutants suitable for such studies in eukaryotes. The examples we have are described in Ch. 6. However, there is considerable evidence of enzyme induction in eukaryotes (see Fincham *et al.*, 1979), but the message is not then polycistronic. For example, the histidine (*his4*) locus in yeast codes for a large (95,000 molecular weight) polypeptide, which is a single, multifunctional, enzyme complex (Bigelis *et al.*, 1977). Thus the operon model may have 'eukaryote modifications'. Paul (1972) has proposed a further, but theoretical, modification: he adds an 'address locus', or specific DNA configuration, ahead of the promoter, to which a destabilizing, polyionic molecule, such as a non-histone protein, could attach and so reduce DNA coiling and allow transcription. This suggestion therefore effectively substitutes NHPs (or similar molecules) for activators in order to account for luxury gene transcription. While there is general evidence, then, that an operon-like system may regulate housekeeping genes, we have to make further assumptions, as Paul does, when we consider the biochemically unrelated syntheses of luxury proteins. This requires an even more complex model.

## The Britten and Davidson model

Davidson and Britten (1973) review some of the more plausible molecular models for the regulation of developmental events. The most comprehensive of these was proposed by Britten and Davidson (1969), and subsequently revised in the paper just quoted. It takes full account of the need to turn on genes in different biochemical pathways, for the coordinate production of myoglobin, myosin, actin, tropomyosin and creatine phosphate, for example, during the differentiation of muscle. And it relates this coordination to the interspersion of repetitive and non-repetitive sequences in the genome (p. 45) by attributing an integrating role to the repetitive

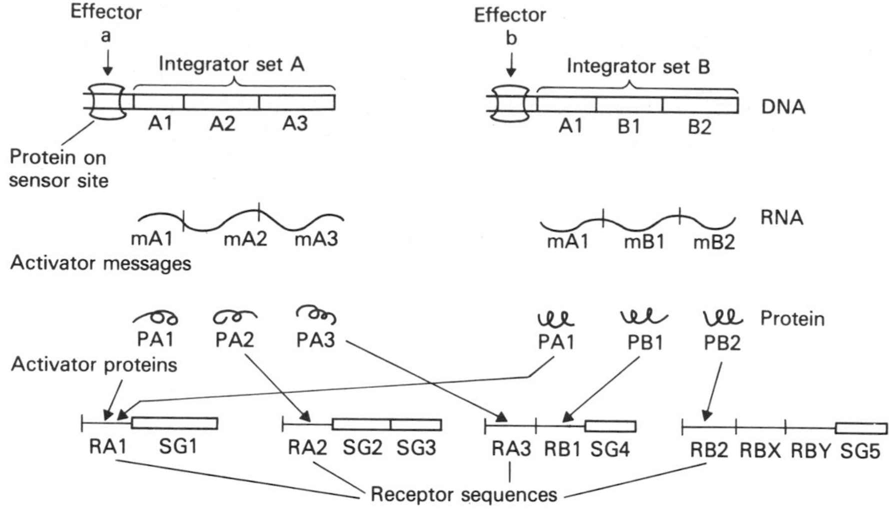


Effector a activates SGs 1, 2, 3 and 4

Effector b activates SGs 1, 4 and 5

**Fig. 4.2** The Britten and Davidson model. For simplicity only two integrator gene sets are illustrated. Each is activated by its specific effector, a or b, which combine specifically with the proteins of their sensor sites. This causes the coordinate transcription of the entire integrator gene set, and the formation of the activator mRNA (mA1 etc.) and of activator proteins (pA1 etc.). These activator proteins (or mRNAs) combine with specific receptor (RA1 etc.) sites (operators) adjacent to structural genes (SG1 etc.), which are then transcribed and translated. Structural gene transcription depends on the effector molecules at one remove, and the pattern of this transcription on (1) the organization of the integrator gene set and (2) the association of receptors with the structural genes, since a particular structural gene (e.g. SG4) may have more than one associated receptor.

sequences. The same repetitive sequence is taken to lie adjacent to structural genes which are activated together. It is an elaboration of the operon model, but one of some complexity.

The Britten–Davidson model of the organization of eukaryote DNA (Fig. 4.2) assumes that:

1. There are *effector substances*, which are signals external to the genome which interact with genome *sensors*.
2. The *sensor* is a sequence of DNA and a sequence-specific sensor protein which specifically binds its appropriate effector.
3. The effector–sensor interaction results in the transcription of an adjacent *integrator gene* set, which codes for sequence-specific *activator proteins* (i.e. it is a negative system).
4. The activator proteins recognize *receptor sequences*, again specifically, and the associated structural gene (or genes) is then transcribed.

5. Unlike the prokaryote model from which these ideas derive, a particular receptor sequence is not uniquely associated with a particular gene set, but may be contiguous with a number of physically separate sites as repeats of the same receptor sequence. There may be more than one receptor sequence associated with a structural gene (or gene set). Thus an activator may switch on more than one gene, and one gene may be switched on by more than one activator.
6. The characteristics of the integrator set determine the activator messages, which need not be exclusive since the same message may be repeated in different integrator sets.
7. It is recognized (Britten and Davidson, 1969) that the activator message set (RNA) may be directly involved with the selective transcription of structural genes without being translated into protein. This would allow the activation process to be intranuclear, and independent of protein synthesis. But there are arguments in favour of the alternative possibility.

The coordinate production of a functionally related set of enzymes (and structural proteins) is not necessarily dependent on a prokaryote operon-like cluster of structural genes. The essential coordination of development comes from the existence (proposed) of integrator gene sets which control the coordinated transcription of gene batteries. The integrator sets would not be expressed in the usual sense, except ephemerally as activator RNAs or activator proteins. A mutation at their sensor site would mimic a regulatory gene mutation of an inducible negative control system, and would have a similar inheritance pattern. Mutations of individual members of the integrator set would appear as structural gene mutations, except if the integrator gene was also repeated at another, activated set. Mutations of a receptor region would be like an operator mutation of an inducible positive control system (uninducible); but an associated, unmutated receptor region might still function when an appropriate, but different, activator combined with it.

Phenotypic analysis of this model is difficult, and the fact that mutations will be expressed only in some cells, and then only at certain times during development, complicates matters further. The model also assumes, by implication rather than directly, that effector substances are generated in some orderly fashion during development (and may be determinants in the egg). Holliday and Pugh (1975) have given greater precision to the timing problem by suggesting that methylation of palindromic control sequences (p. 40) may be used for counting cell divisions (Fig. 4.3). Indeed, this mechanism of cell division counting can be elaborated to cover whole sequences of gene activation, but we lack evidence for the necessary multiplicity of repeated sequences, as we have seen. This model cannot be a general one since DNA methylation is rare in the *Drosophila* genome, for example.

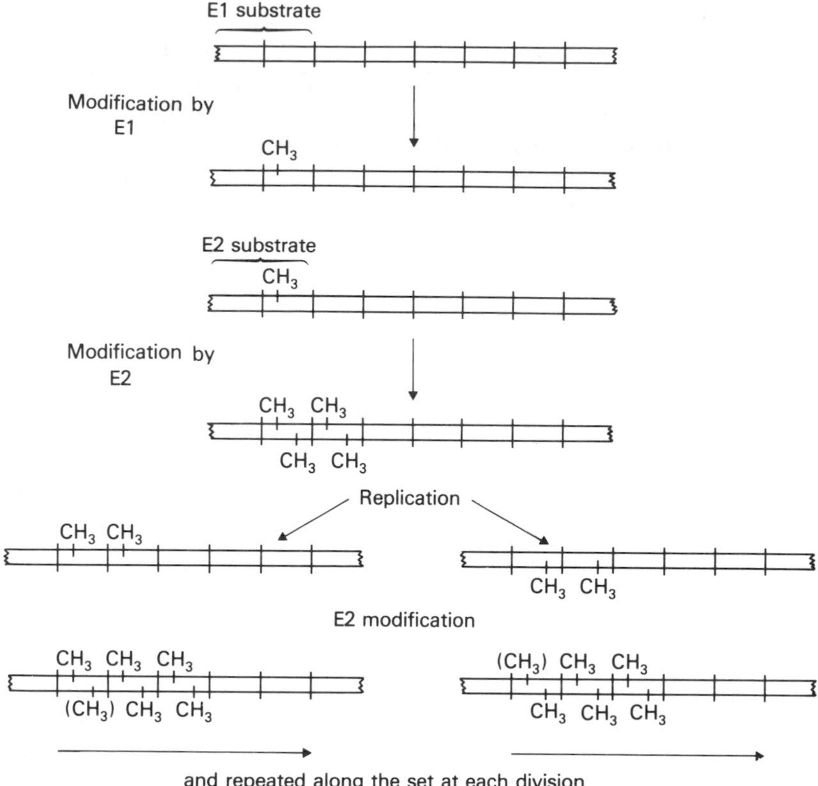


**Fig. 4.3** Counting cell divisions by methylation of palindromic controlling sequences (Holliday and Pugh, 1975). The enzyme E1 starts the methylation by recognizing a starter sequence at the left of the set of repeated sequences: it switches on the clock. This methylated strand is a substrate for a second enzyme E2, which methylates the opposite strand and both strands of the next repeated sequence. After division, E2 modifies this pair and the next adjacent pair of repeated sites, and so moves progressively at each division to the end of the palindromic repeat set. It is the counting enzyme, or 'clock' which ticks off divisions. The bracketed methyl groups need not necessarily be modified at each division.

The major difficulty with the Britten and Davidson model is that it was designed to fit all the developmental facts together with those then known about the molecular organization of DNA. The only valid analyses possible were of its components, and these were difficult to approach. However, its key assumption was that all important control was transcriptional (translational control merely modulating the transcription patterns), and this proved to be incorrect, as we shall see in the next section. It does not follow, however, that the model is meaningless in all situations, since particular effectors may act as postulated. The insect hormone, ecdysterone, may function in this way to cause the many different patterns of transcriptional change apparent during pupariation and ecdysis, although this is by no means proven.

## The Davidson and Britten model

For the most part we have ignored the post-transcriptional fate of RNAs, although we noted that a high proportion of nuclear RNA (nRNA) is not passed to the cytoplasm. The complexity of this nuclear population has long been a puzzle, and this has been resolved (if not in all its fine detail—see Darnell, 1979) by using complementary DNAs (cDNAs) made against particular mRNA populations to probe the nature of other RNA populations. $^3$H-labelled single copy cDNA made against sea urchin blastula mRNA hybridizes 100 percent with excess blastula cytoplasmic RNAs, but only 12–13 percent with RNAs from intestine or coelomocytes, confirming earlier studies showing that adult tissues carried only a fraction (10–20 percent) of the blastoderm mRNAs (Galau *et al.*, 1976). This reflects the tissue differentiation we should expect. However, if the blastula cDNA is then used to probe the nRNA it is found to hybridize 100 percent with the nRNAs from intestine or from coelomocytes (Wold *et al.*, 1978). Similar results have been found when mouse brain cDNA is used to probe mouse kidney cytoplasmic and nuclear RNAs (quoted by Davidson and Britten, 1979), and when tobacco leaf cDNA is used to explore stem cytoplasmic and nuclear RNAs (Kamalay and Goldberg, 1980). It is difficult to resist the conclusion that all single copy genes are continuously transcribed in cell nuclei, but only a specific proportion of them are processed to become the characteristic polysomal populations of each cell type. But the very large numbers of mRNAs found in all cells (Fig. 4.4) make it difficult to prove the constitutive transcription of the entire genome, particularly if transcription is regulated in amount.

Like the genome, most nRNAs contain interspersed repeated sequences. When these are identified using cloned repeats, some repeat families are found to be common and others rare according to the tissue examined, if one can judge from sea urchin data alone (Scheller *et al.*, 1978). That is, certain repeat sequence families are prevalent in gastrula nRNA transcripts, but rare in adult intestine nRNA, and conversely for other repeat families. This allocates a key function to the regulated transcription of interspersed repetitive sequences, instead of to structural gene transcription as in the previous model. However, since it is the structural gene mRNAs which are selectively processed we have further to assume an interaction between the structural gene nRNAs, and the transcriptionally-regulated nRNAs. The simple, but by no means certain, assumption is that the repetitive sequences result in an association between the two kinds of nRNA. Such double-stranded RNAs have already been implicated in the processing of mRNAs (Jelinek and Leinwand, 1978).

The Davidson and Britten (1979) model (Fig. 4.5) is built around the premises just outlined. The constitutive structural gene transcript (CT) carries the coding sequence flanked by short repetitive sequences, accounting for the complexity of this mRNA class in the nucleus. The inte-

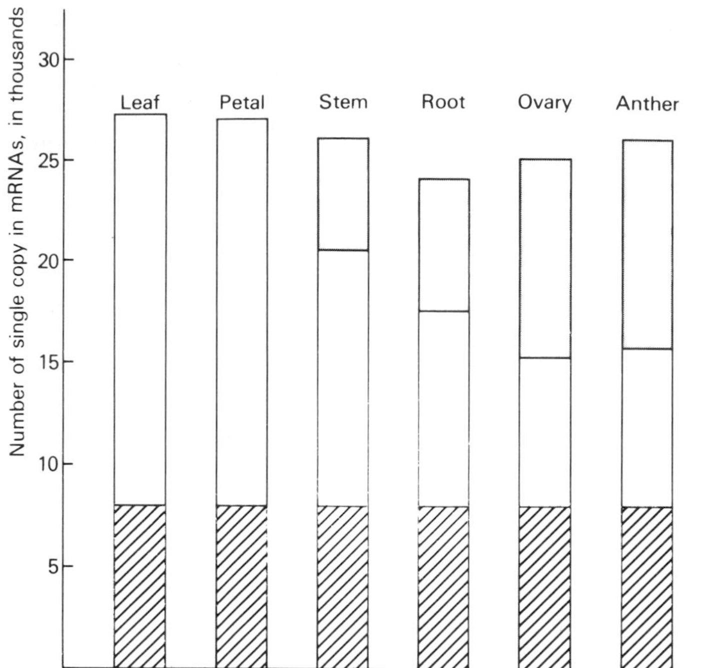


**Fig. 4.4** The polysomal mRNA sets of tobacco plant organs. Plants have the advantage of small numbers of cell types (*c.* two dozen), so that their organs are less heterogeneous than most animal organs. Total nuclear complexity is of the same order for both; in this case $3.3 \times 10^7$ nucleotides in the tobacco leaf, or $25 \times 10^3$–$27 \times 10^3$ mRNAs. There is a set (diagonal stripes) of mRNAs common to all organs. The leaf set (unshaded) is repeated in the petal, as we should expect; and is found in varying amounts in each of the other organs. These also have a unique set (shaded), not found in leaf. The sum of these unique sets together with the leaf complement suggests that about 60,000 structural genes are required to establish and maintain the mature plant (data of Kamalay and Goldberg, 1980).

grating, regulatory transcription unit (IRTU) is preceded by a nucleoprotein sensor which when activated by an external signal, results in selective transcription of the unit in a cell specific fashion. The IRTU carries interspersed repetitive and single copy (not structural gene) sequences. The IRTU transcript (IRT) is the cell specific nRNA which can combine only with structural gene transcripts carrying complementary, matching repeated sequences. Only this CT–IRT complex can be processed. Since the repetitive sequences of an IRT may be matched by the repetitive sequences of a number of CTs, the IRT will coordinately control the processing of a battery of CTs, in a fashion similar to the integrator gene sets of the previous model. The 'gene battery' of that model now becomes the set of structural genes bearing the same (or the same family of) repeated sequences. As before, gene expression during differentiation depends on the activator molecule–sensor response system, and, in this case, on the subsequent

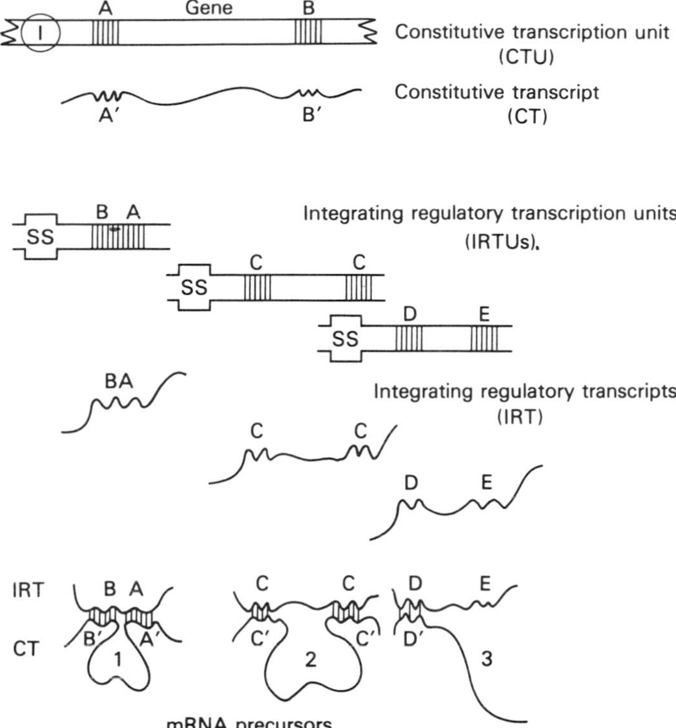


**Fig. 4.5** The Davidson and Britten (1979) gene regulation model. The constitutive transcription unit contains the structural gene, its introns and flanking sequences, including the initiation site (I) and the repetitive sequences A and B. This is constitutively transcribed into the constitutive transcript (nRNA). Three possible forms of integrating regulatory transcription units are shown, each with different arrangements of interspersed repetitive sequences (A, B, C, D and E). Their transcription is controlled by nucleoprotein sensors (SS) which respond to external signals and then make the integrating regulatory transcripts (nRNA). The IRTs duplex with the CTs, by sequence-dependent base pairing between the repeated elements. Three possible kinds of arrangement are shown (but the CTs for 2 and 3 are not included). These IRT-CT duplexes must form before the mRNA can be processed, and then translated.

pattern of IRTU transcription. As sensors may arise during development, their absence or presence might be involved in determination.

This ingenious model leaves transcriptional control in charge of differentiation, but indirectly, through a special class of RNAs which determine the selective processing of the constitutively transcribed total structural gene population. While this seems an uneconomic arrangement, it nevertheless fits available information. As with the first model, it is difficult to predict what mutations of the system's components might do. Of course, structural gene mutations would be expressed as such, conventionally. Mutations of their initiation sites or of their associated repetitive sequence(s) would appear like structural gene changes and would have to be identified by fine structure analysis of the region. A sensor mutation would again involve the coordinate loss of a number of gene products. And all these mutants would be recessive if the wild-type chromosome allowed normal heterozygote development. Mutations within the IRTU complex might also be recessive, except if the mutant IRT competes with the wild-type IRT during duplex formation: then it might be semi-dominant or dominant. The obvious test of the model is the clear identification of IRT–CT duplexes, and this has still to be done.

## The Caplan and Ordahl model

The above models are concerned with the differential regulation of transcription, and they assume that the organization of DNA in chromatin has no specific role in this respect. The opposite view would allocate a regulatory function to some of the chromatin proteins, particularly non-histone proteins. Unfortunately, such information as we have (p. 70) is too imprecise for the formulation of a detailed model, but the general characteristics of one such scheme can be briefly outlined. This is the irreversible gene repression model of Caplan and Ordahl (1978), which states that, initially, all genes are accessible to transcription, but that cells become restricted in transcription during development until each cell expresses one, and only one, phenotype. The pattern of cell lineages mirrors the different arrangements of gene inactivation, presumably sequentially caused by the induction of site specific NHPs, or other proteins (although the theory does not make this clear).

On the face of it, this model seems improbable, if only because we have already seen that some genes undergo cycles of activation and inactivation during development (p. 10). However, most of the information we have is for metabolic (housekeeping) genes, and the model assumes that these can be regulated according to need. The differentiation state-specific (luxury protein) genes are the class to which the theory applies, i.e. the genes coding for the particular products by which we identify a cell type. The argument is here supported by the finding, for example, that adult-type globin mRNA is found in *Xenopus* oocytes, although adult haemoglobin is not made in quantity until after metamorphosis (Perlman *et al.*, 1977). The model is summarized in Fig. 4.6.

Mutation could block the production of a particular inhibitory protein, and result in the continuous production of a luxury protein. In this respect it would not differ from an ordinary constitutive mutation, as judged by phenotype. The critical assessment of the model must therefore depend on showing that differentiation state-specific genes are irreversibly, and sequentially, repressed during normal development.

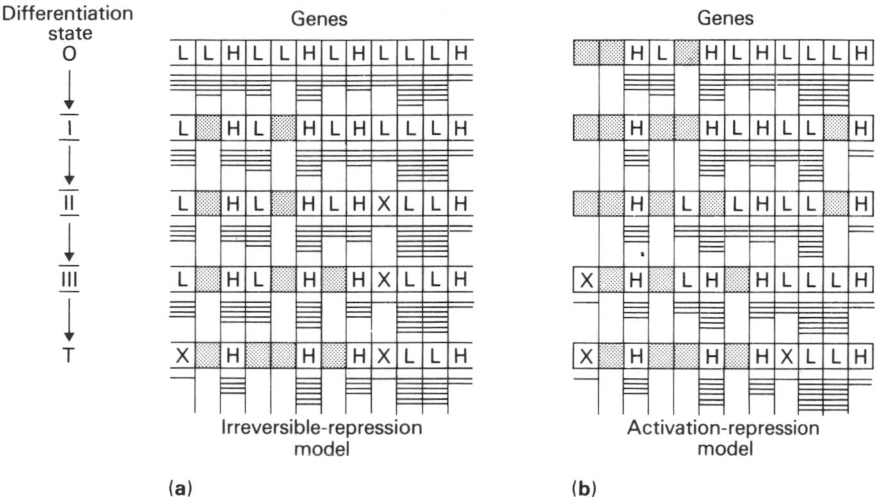


**Fig. 4.6** Comparison of the irreversible-repression (**a**) and activation-repression (**b**) models of the genome, after Caplan and Ordahl (1978). The patterns of RNAs (dashes below the genes) are the same at the terminal stage (T). Housekeeping genes (H) are transcribed throughout and luxury genes (L) are transcribed unless blocked-off (shaded) i.e. repressed. (**a**) starts with all genes active in the egg (0) and shows a stage by stage loss of L gene activity, including a change to a state (X) when transcription only ticks over. (**b**) starts with most L genes repressed, followed by a changing pattern of expression and repression during the developmental stages.

## Conclusions

The last model emphasizes the point made at the beginning of the chapter: we have to distinguish between the genes for housekeeping and for luxury proteins if they are regulated in quite different ways. Davidson and Britten (1979) also except secretory proteins from developmental control systems since, after stimulus, many thousands of copies of their mRNAs are made, as we have already seen for ovalbumin (p. 61). But it is not always easy to distinguish between these categories when the data (e.g. Fig. 4.4) show that there are apparently many thousands of members in each expected class, and the boundaries of the classes are smudged by the evident transcription of selfish, or junk, DNA (p. 67), often from both strands. Thus, it is at present very difficult to build a bridge between the reductionist, and very detailed, studies on the molecular organization of DNA described in Ch. 2, and the more general propositions outlined in this one. In part, at least, this derives from our lack of information concerning the differences among the proteins which typify particular cells, and when, and how, their synthesis is initiated and organized. Of course, certain proteins are of major importance in this context, just as they were for the exploration and manipulation of the prokaryote operon. Unfortunately, however, there is currently no obvious approach to the identification of the nucleoprotein sensors of the Davidson and Britten model, or to their activator proteins, if such there be, let alone to the batteries of proteins whose synthesis they are taken to regulate. And the same is true for the Caplan and Ordahl model.

Although the experiments considered in the following chapters were obviously not designed to test these models, we can reasonably ask if their results are compatible with, or exclude, any one model—or even if they eliminate them all. There is, indeed, no *a priori* reason for thinking that any single model applies to all the developmental changes which occur in all eukaryotes. Since the mechanisms regulating different systems may provide clues to the organization of other systems, we shall not ignore the arrangements relating to housekeeping or secretory proteins, in favour of those relating to luxury proteins. These classes may be interlinked physiologically in any event, since some metabolic changes, *per se*, may provide the extrinsic signals for the activation of structural gene batteries. We shall therefore start conventionally by considering the genetic analysis of the genome, and then explore the actions of genes with notable phenotypic effects, with these models in mind.

# III Mutations

We expect to find mutations affecting both regulatory and coding sequences. Changes in sequences, including deletions, are found, but a surprising proportion of spontaneous mutants, probably over one third in *Drosophila*, are due to the insertion of transposable elements. Whatever the cause of the mutation, most mutations identified are changes in the coding sequences of structural genes, not of regulatory genes. While these mutations can be classified according to their phenotypic effects, the complexities of metabolism usually make it difficult to deduce the changes occurring at the molecular level. Consequently, the classification of mutations as amorphs, hypomorphs, hypermorphs, antimorphs and neomorphs is only a convenient shorthand rather than a meaningful definition of altered gene function. Similarly, dominance (and recessivity) may reflect only the gene's role in a complex network of kinetic reactions where primary gene products are co-dominant. Overproducers (hypermorphs) may sometimes indicate mutations of regulatory sequences.

The identification of a structural gene by complementation tests (the *cis-trans* test defining the gene as a *cistron*) is an essential manipulation for separating genes with like effects. However, allelic complementation occurs when the products of the cistron make an active multimeric protein. This protein–protein interaction may be positive or negative, respectively correcting the genetic defects in heterozygotes, or exaggerating them to give incomplete dominance. Some loci are more complex, and the example of white eye shows that such loci may be functionally diverse. Molecular analysis confirms the genetic conclusion by showing that part of the locus is structural and part regulatory, but it also shows that the overall locus organization is even more elaborate than the breeding data suggest.

Genetic fine-structure analysis, particularly of the *Drosophila* rosy eye colour mutants, shows that there are *cis*-acting elements immediately adjacent to the structural gene, regulating the level of product made. Similar *cis*-acting elements are found contiguous to other structural genes, but it is not yet certain if these are the 'upstream sequences' previously described. This question will soon be answered using cloned mutant genes, and we can now anticipate that past genetic studies will provide the guide-posts to immediately profitable molecular pursuits.

There is also genetic evidence for *trans*-acting elements lying some distance from genes whose tissue-limited activity they regulate. However, the contrast between the regulation of *Drosophila* and mouse α-amylase genes cautions us against hasty generalization, for the first is *trans*-regulated and the second involves different tissue-specific transcription mechanisms. Or again, *Drosophila* chorion protein genes are amplified at a particular stage of oogenesis, presumably to match the required burst of chorion protein synthesis. None of these examples yet illustrates the coordinate regulation of many structural genes (the only known examples of that are among the metabolic genes of prokaryotes) and even the production of dopamine acetyltransferase is not coordinated with the hormonally induced cycles of dopa decarboxylase activity. However, the complexity of ecdysis implies that other genes must be activated then.

Mutants of the class rudimentary show that coordination may sometimes be achieved, as in prokaryotes, by the clustering of genes into a complex locus; in this case the first three genes of the pyrimidine synthesis pathway lie together, but the remainder are separate. Genetic analysis shows, too, that we may have clusters of multicopy genes—chorion and glue proteins are examples. Other multicopy genes (actins, heat shock proteins) are, however, dispersed throughout the genome. We have a diversity of gene arrangements to reckon with when we come to consider their tissue-by-tissue activation.

Lethal mutations provide a great wealth of material illustrating developmental anomalies, often stage and tissue specific in their effects. Some mutants affect gametogenesis and, among other things, show that one developmental sequence may be uncoupled from others, and proceed independently. Maternal-effect lethals disturb the primary determination of the egg, and pursuit of their mechanism of action opens up the possibility of understanding the molecular basis of this fundamental process. Later acting lethals may fail to interpret the information laid down in the egg, or fail to make a necessary product, causing abnormalities of cell and tissue differentiation. Still later acting lethals sometimes show complex pleiotropic effects, indicating that the gene product is essential for the differentiation of many tissues, but not for all. Lethal mutations identify a hierarchy of developmental problems, many of which await study with modern techniques.

# 5 The nature and significance of mutations

The effect of a change in a gene (which if recessive means, of course, a pair of like genes) frequently produces a more localized effect than a doubling or trebling of genes already present, because a change in one gene is more likely to upset the established relation between all genes than is an increase in the number of genes already present. By extension, this argument seems to mean that each gene has a specific effect on the course of development . . . all genes or many of them work together toward a definite and complicated end-product.

T. H. Morgan, 1929

Early work with mutants quickly led to the conclusion summarized above; namely, that each mutation has a specific developmental effect. (Some mutants are without apparent phenotypic effect, presumably due to homeostatic compensation.) One can therefore deduce the general function of the normal, wild-type gene from the developmental defect caused by the mutation: pigment loss means that the unmutated gene is necessary for pigment formation; and the like. It is more difficult to move from the general to the specific, however, and to infer which particular molecular step is involved when the end product is complex, as it generally is. For example, we still do not know what the white eye mutation of *Drosophila* does although it was the first *Drosophila* mutant found (Morgan, 1910). This difficulty is compounded when we consider 'developmental' genes, where we have no knowledge of the nature of the gene product, or of how this acts to cause specific developmental events. Most current studies therefore start from genes with known products and explore their regulation, as we have already seen for ovalbumin etc. and shall pursue further in Ch. 6. But before looking at specific genes we must examine more carefully what we mean by mutations and what we know about their general characteristics.

## Mutation

### *Molecular changes*

All mutations result from changes in the DNA sequence organization. Alterations to regulatory sequences will affect what is transcribed, but attention in the past has been directed to changes in structural genes, the protein coding sequences primarily, which may alter spontaneously or through treatment with mutagens. Addition or deletion of one or more base pairs which displaces the triplet reading frame (frameshift mutation), results in the production of an abnormal peptide from the site of the mutation onwards, whereas the deletion or addition of three base pairs deletes or adds an amino acid to the sequence. Base substitutions result in the replacement of one amino acid by another (missense mutation), or they may cause premature chain termination (nonsense mutation) when they result in UAG, UAA or UGA codons. We therefore anticipate, and do find, mutations which result in no identifiable gene product being made (early frameshift or nonsense mutations), or in a product which is nearly normal (insertion, deletion or missense mutations) and, depending on how the active centres of the peptide are affected, all grades in between. In short, mutations are found as allelic series, and when these mutant genes make allozymes, i.e. products which have different electrophoretic properties due to insertions, substitutions etc., we can be confident that we are dealing with mutations of structural genes (see O'Brien and Macintyre, 1978). Allozyme variants occur in about one third to one half of the loci examined in *Drosophila* and in man, and we shall see below that temperature-sensitive mutations and dosage dependence (i.e. the strict dependence of the number of molecules of a protein on the number of its structural gene copies in a cell) indicates that the great majority of identified mutants result from structural gene changes. However, another class of mutation has recently come to the fore which we must consider briefly.

### *Transposable elements*

It has long been known from the work of McClintock, Brink and others (reviewed by Fincham and Sastry, 1974) that there are elements in the genome of maize and other plants which modify the activity of genes and produce mutant phenotypes. McClintock named them 'controlling elements' for that reason, but as they also move from one part of the genome to another we shall call them transposable elements, or insertion sequences. As Temin (1980) notes, they have characteristics similar to the transposons and phage Mu of bacteria, the movable Ty1 plasmids of yeast, and the copia-like sequences of *Drosophila* (p. 65). Elements of this class are present in most, possibly all, eukaryotes, and they typically result in the inactivation of the gene with which they are associated. Some of the mutants we shall consider are certainly the result of the insertion of these elements (p. 262), and many others may be: cloned genes are required to identify these inserted sequences.

The general properties of transposable elements can be identified from one of McClintock's early experiments where an element caused pigment variegation of the maize kernel (and in other parts of the plant). The whole kernel became spotted, or streaked, with pigment (Fig. 5.1). The pigment gene was therefore present and normal, but suppressed in activity. Escape from this suppression was associated with chromosome breakage and removal of the element associated with the pigment locus ($C_1$). Since chromosome breakage was regularly associated with the element, it was therefore called Dissociation (*Ds*). After the break, *Ds* might be found close to another identifiable locus such as waxy (*Wx*) which was inhibited in its turn and failed to make the normal amylose starch when its homologue carried the recessive, *wx*. This new *Ds* location became the region of subsequent breaks causing waxy variegation, and further transpositions. If there were two *Ds* elements in the genome, say one associated with $C_1$ and the other with *Wx*, then the coloured patch overlapped the region of altered (amylopectin) starch, and there was an integration of *Ds* breakages (McClintock, 1949).

*Ds* is obviously a *cis*-acting element with respect to gene inactivation, but breakage occurs only when another transposable element is present to activate it—the Activator or *Ac* element. This element need not be on the same chromosome and is therefore *trans*-acting, and must produce an activating substance. Rather surprisingly, manipulating the dosage of *Ac* from 0 to 3 (the endosperm and aleurone layer of the kernel is triploid) shows that a single dose of *Ac* produces maximum effect, and that larger doses delay the onset of breaks giving smaller patches of variegation, or almost none at all. *Ac* elements also transpose, and they also inactivate loci. These two properties, activation of *Ds* breakage and gene inactivation, can be separated by chromosome breakage (McClintock, 1962) showing that this is a complex element. It is difficult to guess, at present, if the mutational effect of these elements results only from some interference with the tran-

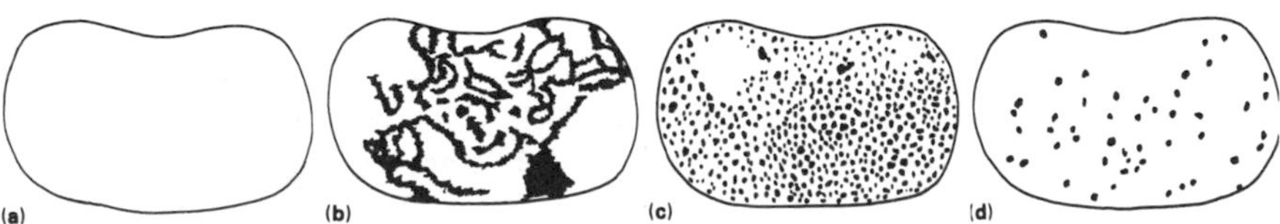


**Fig. 5.1** The effects of *Ds* and *Ac* elements. In (**a**) the maize kernel is colourless due to the association of *Ds* with the pigment genes whose action is suppressed. *Ac* is absent. In (**b**) there is one copy of *Ac* which causes early dissociation of *Ds* and produces large clones of pigmented cells. Kernels depicted in (**c**) and (**d**) carry two and three copies respectively of *Ac*. *Ds* breakage (transposition) is delayed and only small pigment spots form, fewer and smaller with increased *Ac* dosage. Similar transpositions occur in the germline. After McClintock (1951).

scription of the associated gene, or from more complex causes. Different *Ac*-inactivated *Wx* loci recombine among themselves and with other conventional, stable mutants, suggesting that *Ac* is not inserted at a special site within the locus or at a control region adjacent to it (Nelson, 1968). We await the molecular analysis of these cases for an understanding of this class of mutation.

The first clear evidence that transposable elements were present in the *Drosophila* genome was the observation that some second chromosomes from wild caught flies greatly increased the frequency of crossing-over in males (and females) and elevated the recessive mutation rate—the MR mutator system (Hiraizumi, 1971). This was an example of a syndrome of effects, now called 'hybrid dysgenesis' (Kidwell *et al.*, 1977; Engels, 1981), of which two systems have been described. If wild caught males are mated to long established laboratory strains, the hybrid progeny are sterile (gonads fail to develop at elevated temperatures), chromosomal inversions and translocations are found, and mutation frequency is increased. The reciprocal cross is normal. Genetic analysis of the most studied class shows that effective paternal chromosomes from wild flies carry P factors which are expressed only in the cytoplasm of laboratory strains—the M-cytotype, which also depends on chromosomally-linked factors. Unlike the *Ac–Ds* system, there is no expression of the interaction in somatic cells. We now have molecular proof that the P factor is a transposable element.

Since P factors induce mutation (and some mutations preferentially), Rubin *et al.* (1982) isolated P–M-induced white eye mutants and cloned the DNA of the locus so that it could be compared with the organization of the $w^+$ locus (see below). In a sample of five white mutants, they found that the loci were normal except for inserts which ranged in size from 0.5–1.4 kb. The insertions were homologous despite their size differences. Using these P sequences to probe salivary gland chromosomes, Bingham *et al.* (1982) found that M-cytotype strains carried no P elements, and that P strains carried 30–50 P elements, on all chromosomes. Also, in the $F_2$ of dysgenic hybrids about one P element could be identified on each X chromosome derived from the M mother. P elements transpose, and they do so cleanly since revertants of the P–M-induced white mutants have the normal wild-type sequence organization. A second, different, system similarly involves chromosomal inducing (I) factors and an appropriately reactive (R) cytoplasmic state (Bregliano *et al.*, 1980), and there is again good genetic evidence that in females heterozygous for I, homologous and non-homologous chromosomes lacking I may acquire I (Picard, 1976), proving transposition.

Several phenotypically identifiable transposable elements are also known in *Drosophila*, and that deriving from a white-apricot ($w^a$) X chromosome is perhaps the most interesting (Ising and Ramel, 1976; Ising and Block, 1981). This is a large element (TE1) covering four polytene bands (3C2–6) including the roughest ($rst^+$) and verticals ($vt^+$) loci as well as $w^a$. It has been found to transpose spontaneously to some 150 locations throughout the chromosomes (but has some preferred sites) and induced lethal mutations at about half of them. Like bacterial transposons (Bukhari *et al.*, 1977), this element can lose its associated mendelian gene (i.e. $w^a$, $rst^+$ or $vt^+$) and pick up new ones, suggesting that it is flanked by independently mobile insertion sequences. Gehring and Paro (1980) tested this assumption by isolating a recombinant plasmid containing the sequences present in this element, and they find it also contains a segment homologous to a member of the copia family (p. 65). Since copia is a mobile element, this association may be responsible for the transposition of TE1.

The discovery of transposable elements, and the variety, number and heterogeneous disposition of middle repetitive sequences in chromosomes, leaves us in some uncertainty concerning the precise nature of some of the mutants to be considered later. It was simpler when mutations could be attributed to the molecular changes described earlier in this chapter! Nevertheless, mutants can be given some systematic order by classifying them according to their phenotypic effects.

## Phenotypic classification of mutants

Muller (1932) used small deficiencies (Df) and duplications (Dp) to alter the numerical dosage of sex-linked genes in *Drosophila* to see if mutations could be categorized by their phenotypes. His classification has come back into use after a period of neglect, and it seems to be generally applicable to all diploid eukaryotes.

1. The simplest case is the *amorphic* mutation, where the dosage results imply that no active gene product is formed: it is a null mutation. For example, the lowest member of the white ($w$) eye colour allelic series (p. 129) can be present in haplo-, diplo- or triplo-dose in females (i.e. $w/Dfw = w/w = w/w/Dpw$), and no eye colour forms in each case. Amorphs are equivalent to, and difficult to separate from, small deletions; and it is only when the wild-type enzyme (or product) is identified that one can look for an immunologically related, but non-functional, protein (cross-reacting material, or CRM) which would identify a change within the gene which causes inactivation of the product. Amorphs are usually recessive.
2. *Hypomorphs* are mutations that are less effective than the wild-type gene; they are usually recessive and may be missense mutations. A great many examples will be given later: the classic instance is white-apricot ($w^a$) at the locus already considered, which makes the *Drosophila* eye apricot-coloured, not red. In females, the following eye colour relationships are found, ranked by increasing pigmentation from light apricot to wild-type:

$$w^a/Dfw^a < w^a/w^a < w^a/w^a/Dpw^a \leqslant +/w^a = +/+$$

That is, the eye pigmentation is graded according to gene dosage and the triplo-female approximates, like the heterozygote, to the wild-type. The additive phenotypic effect of gene dosage identifies the hypomorph, and can be explained on the assumption that the gene makes a less potent product than wild-type. *Conditional hypomorphic mutations* support this conclusion. For example, temperature-sensitive (ts) mutations are wild-type at the permissive temperature, but mutant at the restrictive temperature. The genetic lesion affects the configuration of the gene product in some environments but not in others, as we might expect from missense changes. Such changes may alter the charge on proteins so that variants may be identified by electrophoresis, to give a series of allozymes, as we have already noted. Some of these variants approximate to the wild-type and can be identified only in combination with other mutants or under environmental stress. Somewhat confusingly, they are then called isoalleles.

Hypomorphs and amorphs represent stages in the continuum of an allelic gene series reflecting graded damage to the structural gene, and affecting more or less significant sites in the enzyme or structural protein.

3. Muller extended his classification to *hypermorphic* mutations, citing a wild-type revertant as hypermorphic to the original mutation: a play on words. A true hypermorphic change has recently been shown when mouse or guinea pig cultured cells are exposed to methotrexate, which inhibits the enzyme dihydrofolate reductase and kills the cells. By progressively increasing the dose of this folic acid analogue, resistant cells were isolated (Shimke *et al.*, 1977), and this resistance increased in successive selection steps. When the cells were examined the dihydrofolate reductase genes were found to have multiplied to the point that the extra DNA could be seen in the chromosome. This gene amplification, by unequal crossing-over, may be the main explanation of hypermorphic mutations in Muller's sense (but see below).
4. *Antimorphs* have 'an opposite action to the normal allelomorph, competing with the latter when both are present'. In this situation a heterozygote is mutant, the heterozygote for a deficiency of the locus is not, and a double dose of the mutation with the normal allele is more mutant than the heterozygote, i.e. $x/x{+}>x/{+}>Dfx/{+}={+}/{+}$. Portin (1977) finds that this relationship applies exactly to some mutations at the Abruptex (*Ax*) locus of *Drosophila*, a duplication that reduces wing venation etc., and he has proposed a molecular mechanism which could explain the phenomenon. He assumes that the mutant product combines with its substrate but cannot convert it, thus competing with the + gene product. Homozygous mutants should then be lethal, as is the case; and gene expression may be affected by

temperature, which is also true. Some *Ax* mutants are hypomorphs, as we might expect if they caused other kinds of change to the polypeptide product. If Portin's deduction is correct, dominant antimorphs are due to a particular class of structural gene change.

5. Finally, a *neomorph* is a mutant which results in the appearance of a new character. Addition of the + allele has no effect on its expression since the + allele then behaves as an amorph. The neomorph should therefore be dominant (or semi-dominant), and a double dose should increase the expression of the character. Muller cites Hairy wing (*Hw*) as an example since it behaves in this fashion, i.e. $Hw/Hw = Hw/Hw/Dp+ > Hw/+ = Hw/+/Dp+$, with the genotypes to the left of $>$ more abnormal than those to the right. However, like Bar eye (*B*) which also behaves in this way, *Hw* is due to a duplication, and we do not know what this implies at the gene level. Denell (1973) has adduced evidence that the gene mutation Nasobemia (*Ns*) is a neomorph which transforms the *Drosophila* antenna to leg. We shall consider this class of novelty-producing genes in some detail in Ch. 10, for they apparently involve the regulation of numbers of genes.

Muller's classification is a convenient shorthand, and we shall use it. However, not every mutant classed as, say, a hypomorph has been properly assessed by Muller's rigorous criteria, and we must remember this reservation.

## Dominance

Dominance and recessivity are not properties of genes or of characters, but descriptions of the relationships of phenotypes to their respective genotypes; the two homozygotes and the heterozygote. Most mutants are recessive because the change they make to the total output of a system, in which they are linked kinetically through their substrates and products, is small. Even with a null mutant, where a measure of the gene product in a homozygote shows that it is 50 percent the homozygous wild-type amount (co-dominance at this level), it is frequently difficult to detect a phenotypic change, although this may be exposed by stress of one kind or another (Fig. 5.2). The wild-type phenotype of the many haplo-sufficient loci (i.e. heterozygotes for a deficiency of a locus) found by Lindsley *et al.* (1972) well illustrates that mutant recessivity is the "inevitable consequence of the kinetic structure of enzyme networks" (Kacser and Burns, 1981), and, hence, that + genes are usually dominant. Intermediate dominance may be found if the metabolic pathway is short, or the mutation affects its last steps and there are few kinetic interactions to compensate for the mutation-caused lesion in the pathway.

Dominant mutations are rare and they may be due to different causes, two of which have been suggested as explanations of antimorphs and

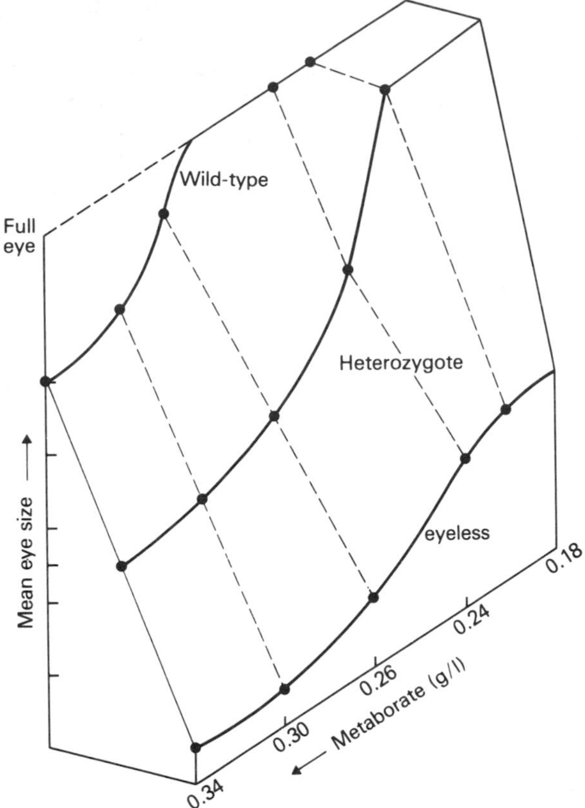


**Fig. 5.2** Feeding sodium metaborate to *Drosophila* larvae produces phenocopies of the eyeless (*ey*) mutation in suitable strains. This morphological change is not inherited, and the extent of phenocopy expression is proportional to amount of metaborate fed. The expression of eyeless in *ey* mutants is increased additively by this agent, and heterozygotes fall intermediately between the two homozygotes, as illustrated in this experiment by Sang and McDonald (1954). The treatment exposes the relationship of the amounts of 'eye-forming substance' in the three genotypes, and suggests that heterozygote makes enough eye material to form normal eyes when metaborate is not fed. The *ey* mutation is therefore recessive. Eyeless is a mutation which shows variable penetrance and expressivity, the former being the proportion of flies which are eyeless to any degree, and the latter measuring the extent of eyelessness in flies showing the character. These proportions can be converted to normal statistical measures, as here (Sang, 1963).

neomorphs. Overproducing mutants (hypermorphs) are also possible dominants, as has been proposed for the Beadex (*Bx*) mutant of *Drosophila* which makes the wing narrow and scalloped (beaded). Four doses of the wild-type gene ($Bx^+ \ldots Bx^+/Bx^+ \ldots Bx^+ = Bx/Bx^+$) produce the phenotype, showing that overproduction is involved, and *Bx*/*Bx* has a stronger phenotype than the heterozygote. *DfBx*/+ and *DfBx*/*Bx* are wild-type, as we should expect of an overproducer. Rather surprisingly, all 13 X-ray induced

*Bx* revertants carry their wings held up (*hdp*), like an already identified recessive of that name. These mutations map only 0.0045 units from *Bx*, so Lifschytz and Green (1979) postulate that *Bx* is a regulatory element for *hdp*, making a bipartite genetic unit. Although not yet proven for this example, overproduction due to mutation of a controlling element may explain some dominant mutants (but see p. 129). Finally, allelic complementation in the polymeric polypeptides, described below may give us another dominance mechanism when the heteropolymer is less efficient than either homopolymer (negative complementation).

Dominance and recessivity, *per se*, tell us very little about the nature of a mutation or of its role in the making of an end product. As we saw earlier when we considered the operon model (p. 103), these problems have to be explored genetically and by following the molecular products involved. But we can say, at very least, that dominant mutants are a class which merits special attention since they may include regulatory genes.

### Dosage compensation

Although females have two copies and males just one copy of any X-borne gene, the phenotype determined by it is usually the same in both species. There is a compensation for the difference in X gene dosage, and measures of mRNA production show a near equality in the two sexes. In mammals, the mechanism which brings this about is heterochromatization of one of the X chromosomes very early in female embryogenesis; electively of the paternal X in marsupials, for example, and randomly for either X in eutherians (Lyon, 1972). This state is cell inherited so that every female of the latter class is a mosaic of cells each expressing a single dose of the genes from one or other X chromosome. We shall note the significance of this mosaicism in Ch. 8. *Drosophila* adopts the alternative strategy of equalizing the output of the pair of female X genes and the single male X gene, with some exceptions. The sizes of salivary gland puffs are about the same, and measures of mRNA or its products are similarly nearly equal in the two sexes. A segment of X chromosome translocated to an autosome is still dosage compensated, and an autosome fragment translocated to the X is not (see Stewart and Merriam, 1980). This suggests some local form of interesting gene regulation, and it is only recently that it has been found to be related to sex determination, as we shall see in Ch. 11. What matters in the present context is that both genes are equally expressed in females, as we implied above, and that the dosage compensation mechanisms prevent any imbalance of gene products between the sexes.

## Complementation

We have defined mutants largely in terms of their phenotypes, but many mutations may cause the same phenotypic change. If two mutants with similar effects map to different chromosomal locations there is no problem in deciding their separate identities. If they map together we have to use the complementation test to see if they are distinct mutations or alleles of one locus.

Mutations are *complementary* when the two mutants ($\frac{a}{a}$ and $\frac{b}{b}$) each produce the same, or closely similar, phenotype, and when the double *repulsion*, or *trans* heterozygote ($\frac{a}{+}$; $\frac{+}{b}$) is not mutant. Such mutants affect sequential steps of a metabolic pathway $X \xrightarrow{a} Y \xrightarrow{b} Z$, and the activity of one gene complements that of the other (but see *ma-l* and *ry* in the next chapter for another kind of relationship). Although they have the same effect (failure to make Z), the mutations are non-allelic, or complementary. Allelic mutations ($\frac{xa}{xa}$ and $\frac{xb}{xb}$) would give the mutant phenotype ($\frac{xa}{xb}$) in heterozygotes. This would also be so if either mutation was dominant ($\frac{A}{+}$ ; $\frac{+}{b}$), so the mutations should also be tested together in the *cis* configuration ($\frac{a\,b}{+\,+}$) when one +-chromosome would make the wild-type protein in the absence of dominance. This *cis/trans* test therefore defines non-complementing allelic mutations when the *trans* heterozygote is mutant, and the *cis* heterozygote is wild-type. It delimits the genetic functional unit, the *cistron*, which may mutate to a number of allelic states.

True alleles sometimes show complementation when their polypeptide chains form a dimeric (or higher order) functional product. This *allelic complementation* (sometimes called *intracistronic complementation*) arises because the hybrid polypeptide complex ($\frac{a\;+}{+\;b}$) functions more effectively than either faulty homodimer ($\frac{a\;+}{a\;+}$ or $\frac{+\;b}{+\;b}$), due to the mutual correction of their different configurational defects which allows expression of the wild-type phenotype. Enzymes of this sort are necessarily less effective than the wild-type enzyme since their 'correction' is incomplete; and by random association half the enzyme molecules formed in the heterozygote will anyway be the ineffective homodimers. Conversely, the abnormal heterodimers may be less effective than the homodimers giving more extreme mutant expression (the negative complementation noted under the last but one heading). Monomeric enzymes do not show this form of molecular correction since the *trans* heterozygote will make two, separately active but faulty, polypeptides.

Allelic complementation requires us to extend our definition of a cistron to: "a segment of the DNA within which pairs of mutations in the *trans* configuration are either deficient for a particular enzyme, or give rise to that enzyme in a structurally abnormal form" (Whitehouse, 1965).

Complementation tests can be used to make a *complementation ,nap* of a locus which tells us about the sites of the mutational changes within the cistron, at least in simple cases. The alleles are mated in all combinations and scored for complementation, as in Fig. 5.3. The map is prepared by showing non-complementation as a straight line and complementation by the absence of overlap of the lines. In the example, mutation group 1 fails

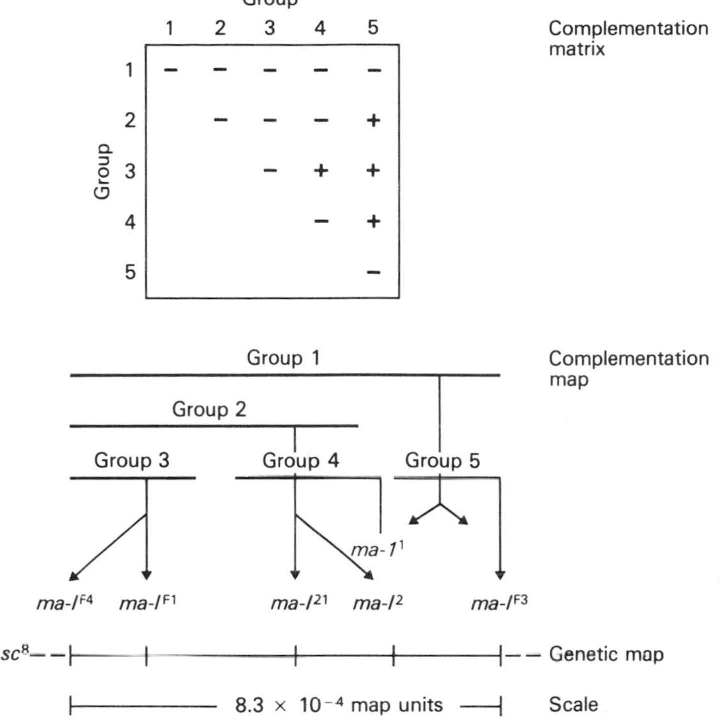


**Fig. 5.3** When mated *inter se*, thirteen viable maroon-like (*ma-l*) *Drosophila* mutants are found to fall into five complementation groups. In the complementation matrix shown in the figure, + indicates complementation between group members. We derive the complementation map by following the rule that complementing mutations are indicated by non-overlapping lines, and non-complementing mutations by overlapping lines. This map is roughly co-linear with the genetic map below it, derived from recombination tests between the mutations. Group 1 contains seven mutations (not shown) three of which (*ma-l*[29], *ma-l*[30], *ma-l*[6]) map with *ma-l*[F3], and four (*ma-l*[23], *ma-l*[25], *ma-l*[28] and *ma-l*[F2]) which map with *ma-l*[2]. Simplified from the data of Duck and Chovnick (1975).

to complement with any of the other mutants and mutation group 2 complements only with group 5. The mutation of group 1 must therefore affect most of the transcript (early frameshift or nonsense mutation?), and 2 and 5 must change different regions of the transcript. And so on. This map might represent three co-transcribed cistrons (groups 3, 4 and 5), but Duck and Chovnick (1975) have shown this is not so. Heterozygotes for the double mutant, *ma-l*[F4] and *ma-l*[F3], with *ma-l*[1] produce no enzyme activity (except for the rare crossover) showing that *ma-l* is a single genetic element exhibiting allele complementation. In general, *ma-l* shows there is co-linearity between the genetic and complementation maps. This simple co-linearity is lost when the functional gene products are highly folded structures, because the complementation map reflects the organization of the protein complex.

## Complex loci

Not all loci are simple. Some mutants behave as alleles by the *cis-trans* test, but can be separated by crossing-over. Such loci are sometimes called pseudoallelic for that reason. For example, Green and Green (1949) analysed the relations of 18 pseudoallelic lozenge (*lz*) mutants which change the *Drosophila* eye shape, and found that they could be allocated to four contiguous subloci. No mutant involved more than one sublocus (Fig. 5.4). The apparent single unit of function was complex and divisible. Since the subloci must be concerned in the making of the same final product, the locus may (or may not) be a complex of cistrons for the enzymes successively involved in the (unknown) pathway, like the operons described for prokaryotes. Complex loci therefore suggest a genetic organization of function, and they have been found in a wide range of organisms including mammals, maize, and fungi. They are uncommon since, in eukaryotes, most genes for particular pathways, are dispersed throughout the genome and are not clustered.

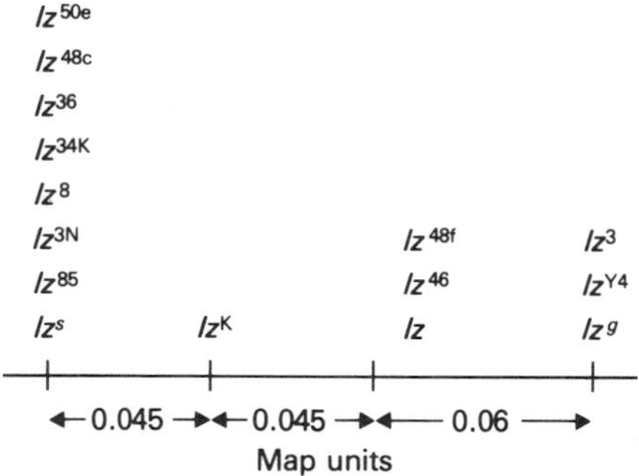


**Fig. 5.4** The lozenge complex locus of *Drosophila*, after Green and Green (1949). The crossover map of lozenge (*lz*) alleles identifies four contiguous subloci. The leftmost group of alleles are named after lozenge-spectacled ($lz^s$), the next has a single representative ($lz^K$), the third group are alleles of the original *lz* mutant and the fourth is the lozenge-glossy ($lz^g$) series. All double mutants show the extreme $lz^s$ phenotype. There is some evidence that $lz^+$ is the structural gene for monophenol oxidase, but this is not conclusive (O'Brien and Macintyre, 1978).

The *lz* locus suggests a simple organization, but other complex loci are more intricate. For example, the white (*w*) locus in *Drosophila* is among the most thoroughly characterized genetically of all eukaryote genes (reviewed by Judd, 1976) and it displays some regulatory features which shall be our concern here. The locus is made up of seven sites separable by recombination (Fig. 5.5). Deletion of the locus shows that it is not vital, nor does it affect fertility. Deletion, and extreme mutant alleles, cause almost complete loss of pigment from the eyes, ocelli, larval Malpighian tubules and adult testis sheath. Since these pigments involve two pathways (Fig. 5.6), the red pteridines deriving from guanylic acid and the brown

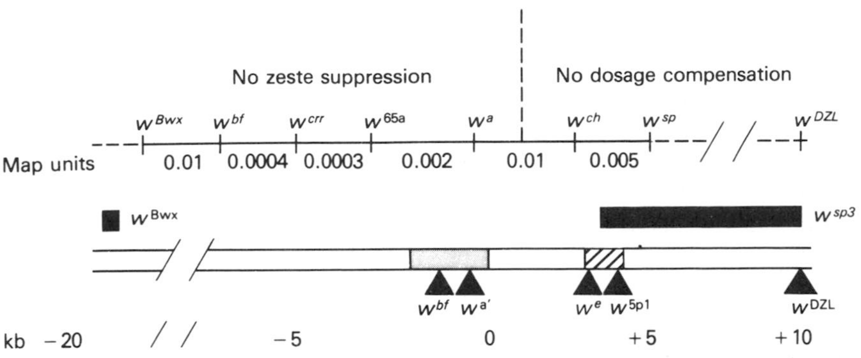


**Fig. 5.5** The genetic fine structure map of the white complex locus of *Drosophila*, according to LeFever (1973). The seven mutant sites which are separable by recombination are indicated by the type mutants above the line: $w^{Bw}$ gives a brown eye in heterozygotes and homozygotes, $w^{bf}$ gives a buff eye, and half a dozen alleles have been identified; $w^{crr}$ (carrot) makes a brown coloured eye, as does $w^{65a25}$, and other alleles known; $w^{a}$ is an apricot eye and a series of identifiable alleles have been described. This distal group is dosage compensated and does not suppress zeste. The $w^{ch}$ allelic series gives cherry through eosin to white eyes; and $w^{sp}$ series cause a fine mottling of the eyes with yellow-brown spots. Members of both these loci are dominant suppressors of zeste and are not dosage compensated. The map distances below the line are approximate. This locus has been cloned (Levis *et al.*, 1982a), and some of Zachar and Bingham's (1982) data from analysing the clones are given below the gene map. Some mutants, like $w^{Bwx}$ and $w^{sp3}$ are due to deletions (solid bars), others to insertions (▲) and no structural changes have been found for others (e.g. $w^{crr}$). The stippled segment covers the left-hand, and the hatched segment the right-hand, regions of the locus (see text) as found by deletion analysis.

ommochromes from tryptophan, we have the apparent coordinate regulation of both. This pleiotropic effect of the locus may be due, however, to a single metabolic lesion in the pteridine pathway resulting in a cofactor deficiency affecting the other pathway, or from a modification of the protein component of the pigment granules: the biochemistry is not yet conclusive either way (Philips and Forrest, 1980). The *w* locus also interacts with zeste (*z*), half a map unit distal to it, such that $z\ w^{+}/z\ w^{+}$ females have lemon-yellow eyes, while $z\ w^{+}/Y$ males are wild-type. Deletion of one *w* in females ($z\ w^{+}/z\ w^{-}$) restores wild-type (suppresses *z*) and, conversely, $z\ w^{+}/Dpw^{+}/Y$ males are zeste in phenotype (Gans, 1953). The effect of zeste depends on the number of $w^{+}$ genes in the individual cell. Not all the *w* subloci behave in the same way; the two loci at the right of the locus (proximal) $w^{prx}$ behave as zeste suppressors, (like $w^{-}$) whereas the five subloci distal to the centromere ($w^{dst}$) permit the co-expression of zeste, even when mutant. The *w* locus is functionally heterogeneous.

A great many different *w* mutants, deletions and duplications have been used for the further analysis of the *w* locus and they all confirm its internal differentiation (see Judd, 1976) and suggest that the subloci in the right-most portion are regulatory. These latter upset the pattern and timing of pigment deposition (causing spotting), interact with *z*, and are not dosage compensated (see p. 125), Judd (1974) therefore suggested that they are regulatory sites (*cis*-dominant since there is no exchange of signals between alleles in homologous chromosomes) which also affect $z$. $w^{dst}$ would then contain structural gene sequences.

This hypothesis has recently been inverted after further studies of $z-w$ interactions, and following the discovery of a new dominant white allele which produces a zeste-like phenotype ($w^{DZL}$). As with $z$, homozygous $w^{DZL}$ females have lemon-yellow eyes and males red-brown eyes; but unlike $z$, heterozygous females ($w^{DZL}/w^{+}$) have lemon eyes (Bingham, 1980). This

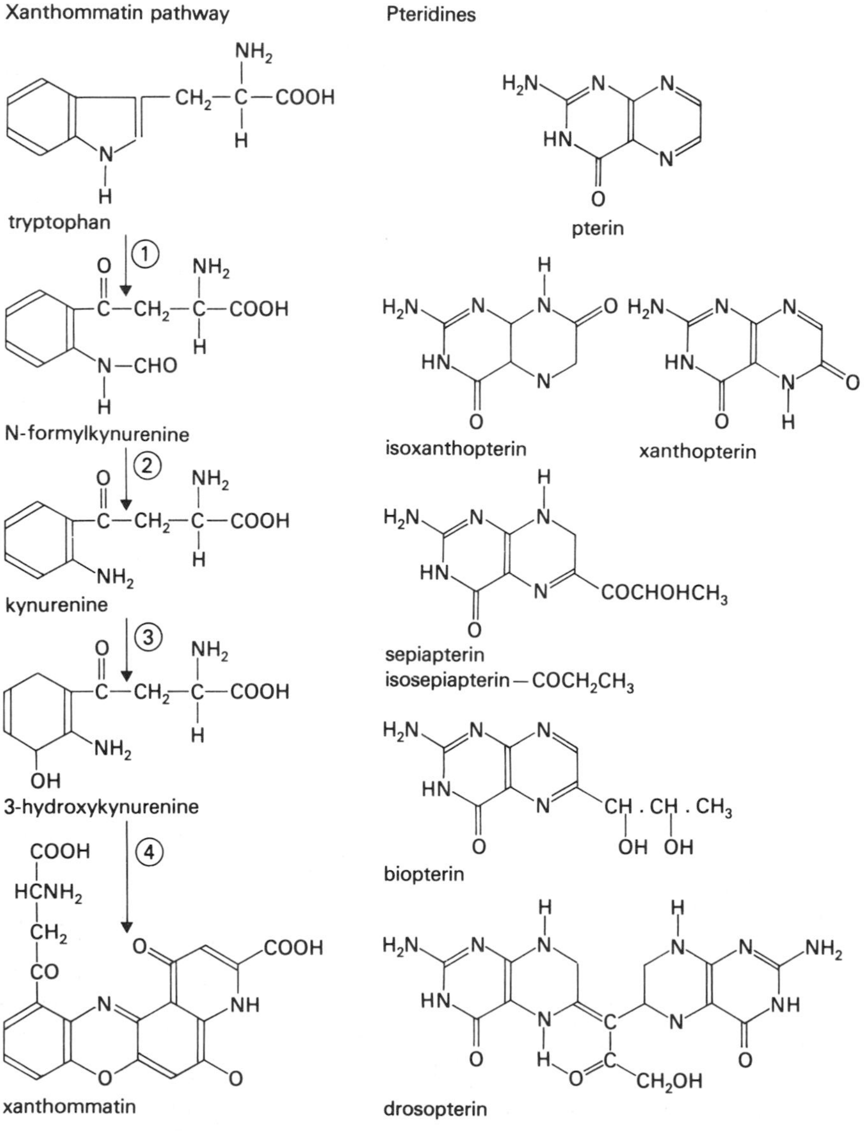

Xanthommatin pathway

tryptophan

① N-formylkynurenine

② kynurenine

③ 3-hydroxykynurenine

④ xanthommatin

Pteridines

pterin

isoxanthopterin

xanthopterin

sepiapterin
isosepiapterin—$COCH_2CH_3$

biopterin

drosopterin

mutant maps as the right-most allele of the *w* complex, adjacent to the $w^{prx}$ mutations already implicated as regulatory sequences. Further, mutants at one of these sites (e.g. $w^{ch}$) suppress *z* (see above). Being dominant, $w^{DZL}$ cannot be tested for complementation, but it is interesting that the doubly mutant male ($z\ w^{DZL}/Y$) has light-yellow eyes, not the red that either mutation alone would cause. This, and the interaction with $w^{ch}$, suggests that *z* and $w^{DZL}$ affect the same process, and may act synergistically. $w^{DZL}$ and *z* appear to be complementary.

The deficiency for *z* carries a lethal and the homozygote cannot be tested, but when the deficiency is combined with a putative *z* amorph ($Dfz/z^a$) the eye colour is wild-type. The $z^+$ function is therefore not necessary for pigmentation. It follows that the *z* allele must repress *w* locus activity (Jack and Judd, 1979), but it does this only when $w^+$ is in its usual location on the two X chromosomes, or if both $w^+$ loci are transferred to the third chromosomes, or in other situations where the two loci can synapse. If one $w^+$ is on the X and the other on the third chromosome ($z\ w^+/z\ Dfw$; $Dpw$) zeste is not expressed and the eyes are red. The $w^+$ loci must be somatically paired, or closely juxtaposed, for *z* to interfere with pigmentation, and this physical apposition suggests that *z* (or some *z*-dependent product) interferes with *w* transcription. Post-transcriptional interference would imply discrimination between the products of the asynapsed loci. $w^{DZL}$ is similarly dependent on synapsis, for a rearrangement which prevents somatic pairing between it and another $w^+$ allele renders it recessive to wild-type (Bingham, 1980). The dominance of $w^{DZL}$ therefore depends on the repression of the white locus transcript *within* the synapsed region. The simplest, but unproved assumption is that the repressor is an RNA.

The cloning of white (Fig. 5.5) illustrates the kinds of mutation we have considered, all within the confines of a single complex locus. Some mutants seem to be due to conventional sequence changes, others are deletions, and among the thirteen spontaneous mutations examined by Zachar and

---

**Fig. 5.6** The ommochromes and pteridines of *Drosophila melanogaster*. The xanthommatin pathway starts from tryptophan and the enzymes involved are (1) tryptophan pyrrolase, (2) formamidase, (3) kynurenine-3-hydroxylase, (4) phenoxazinone synthetase. The vermilion (*v*) mutation blocks (1) and the cinnabar (*cn*) mutation blocks (3). A number of mutants, scarlet (*st*), cardinal (*cd*), lightoid (*ltd*) and white (*w*), block the last, complex step of (4). In some cases (e.g. *v* and *cd*) the precursors accumulate, but in others they do not. The pteridine pathway has still to be worked out since many expected intermediates are ephemeral and the known chemical structures are difficult to relate to one another. The pteridines derive from GTP, and the *Drosophila* set from 'pterin', or 2-amino-3,4-dehydro-4-oxypteridine. The drosopterins are red, sepiapterin and isosepiapterin are yellow, and biopterin colourless. No pteridines are found in the mutants brown (*bw*) and white (*w*), although some are present in pupae. The sepia (*se*) mutant blocks drosopterin synthesis and causes the accumulation of sepiapterin. rosy (*ry*) blocks isoxanthopterin synthesis and accumulates sepiapterin, biopterin and pterin. Many, like maroon-like (*ma-l*), cause a reduction of drosopterin synthesis and accumulation of precursors, and this is so for the coloured white alleles, like white-apricot.

Bingham (1982) seven are due to insertions of various sizes, including copia. The $w^{DZL}$ mutant is particularly interesting because it carries a 13 kb transposon which has many homologies with the foldback elements already described (p. 66). The long, internally repetitious, inverted terminal repeats of this transposon generate instability and chromosomal rearrangements (Levis *et al.*, 1982b). The molecular heterogeneity of mutations needs no further general comment.

The white locus also shows that complex loci may be heterogeneous and polarized. The distal group seems to carry only structural sequences, but in view of its size one must question its apparent genetic complexity. The proximal group regulates, but only when the chromosomes are synapsed, and this is also true of the more distant $w^{DZL}$ locus, which represses $w^+$ in *cis*. What this means in terms of molecules has still to be found out, but it clearly implies that we are concerned with long-range effects of changes in the DNA.

## Modifying mutations

Not all mutations are directly expressed as we have just seen for zeste, and these modifying mutations are identified because they specifically alter the expression of other mutants, either by suppressing them (*su*(*x*)), or enhancing them (*en*(*x*)). They are usually named after the mutant they were first found to affect, e.g. suppressor of Hairy wing is *su*(*Hw*), although it suppresses mutations at a dozen other loci. Suzuki *et al.* (1976) give an up-to-date list of the *Drosophila* mutations, which are usually without phenotypic effect on their own account although a few cause sterility, or are lethal as homozygotes. Suppressors outnumber enhancers by two to one, but this may be because they are more obvious. They have not yet been used for the fine-structure analysis of the eukaryote genome in the powerful way that studies of prokaryotes and unicellular eukaryotes show to be possible (Hartman and Roth, 1973). Consequently, we shall consider only the major classes of modifier mechanisms.

The first class results from mutations affecting the protein synthesizing system, causing an error in this which corrects for the nonsense (or missense) mutation of the structural gene. For example, the suppressor mutation of sable (*s*) in *Drosophila*, *su*($s^2$), is said to suppress the mutations sable, purple, speck and vermilion because the enzyme processing $tRNA^{Tyr}$, $tRNA^{Asp}$ and $tRNA^{His}$ to their functional, secondary forms is ineffective. Consequently, the anticodons of these tRNAs remain largely unmodified, and associate the 'correct' amino acids with the 'mutant' mRNA site (White *et al.*, 1975). The mRNA is then read as if it were non-mutant, and the proper functional protein is formed. This conclusion has recently been questioned (Wosnick and White, 1977), but the example serves to illustrate the principle.

The second general class comprises mutants that produce metabolic suppression, where the second mutation so alters the metabolic flux that the product not made as a consequence of the first mutation is either formed by another route, or is preserved if it is not made in sufficient amount, or becomes effective as a consequence of an ionic change, etc. For example, the rudimentary (*r*) mutation blocks pyrimidine synthesis in *Drosophila* (p. 149). The suppressor of rudimentary (*su*(*r*)) greatly reduces the efficiency of the catabolic enzyme dihydrouracil dehydrogenase, thus sparing the degradation of pyrimidines obtained from the diet, and allowing normal development of *r* larvae (Bahn, 1973). Or again, melanotic tumours (*tu bw* strain) of *Drosophila* are suppressed by a mutation (*su* (*tu bw*)) which apparently interferes with energy metabolism and prevents the tumour-forming cells (haemocytes) from transforming into the necessary aggregating state (Sang, 1969).

Enhancer mutations can be assumed to act in the opposite fashion, but few cases have been studied. Rose and Hillman (1974) found an unexpected example while working with the *Drosophila* mutant Abnormal abdomen ($A^{53g}$). In this case, expression of the major mutations is virtually dependent on the presence of three enhancing genes which cause a more effective than normal aggregation of the tRNA aminoacylating enzyme system. This over-efficiency results in a developmental imbalance, and the abnormal mutant phenotype. Other *en* mutations may be associated with gene duplications, but there is no information as to their biochemical action.

## Conclusions

We have been concerned with what we mean by a mutated gene and what it tells us about normal gene activity. Our emphasis has been on non-vital lesions, setting aside the largest mutant class of lethal loci for later consideration (Ch. 7). This has allowed us to compare the actions of allelic series in the expectation that the grades of change then manifest might indicate the role of the gene. In fact, most of the mutants considered bear the earmarks of structural gene changes, but it is sometimes difficult to distinguish this type of change from associated *cis*-dominant regulatory mutants. Where the gene product is known, or where the allelic series runs from amorphs, through hypomorphs to isoalleles, we are likely to be dealing with structural gene mutations. Many such mutants are used as tools, either for marking (identifying) chromosomes and cells, or for perturbing development more precisely than can be done by the embryologist's scalpel.

We have also seen that dominance and recessivity tell us little about gene action because of the complexity of metabolic interactions, and our lack of understanding of them. And it is not surprising that other mutations can disturb this metabolic flux and so modify the effects of the primary mutation. However, it seems worth paying attention to some dominant mutations—antimorphs, neomorphs, overproducers—for they may involve the regulation of structural genes. More particularly, complex loci suggest a functional organization of the genome especially worth study. Before we consider these mutants in their proper context, we shall examine what can be exposed by the more refined genetic analysis of genes of known function.

The surprising finding that many spontaneous mutants result from the insertion of transposable elements into the cistron, provides a new tool for making mutants, as we saw for the white mutations produced by hybrid dysgenesis. In addition, since these elements are scattered through the genome, strains can be identified where the P element (or copia, as we saw earlier) is associated with a particular gene of interest. A gene bank from this strain can then be probed for P (or copia, etc.) and the clones then identified by annealing to the salivary gland chromosomes where the clone (or clones) matching the required salivary gland band can be expected to carry the desired gene. Also plasmids made from transposable elements can be engineered to transport known genes into flies and cells (p. 324). Transposable elements therefore give us a new handle for exploring genome organization, and it is likely that they will first be used with the best-analysed gens, which we shall now consider.

# 6 The genetic analysis of genome organization

Considerations such as those just outlined have led us to investigate the general problem of the genetic control of developmental and metabolic reactions by reversing the ordinary procedure and, instead of attempting to work out the chemical bases of known genetic characters, to set out to determine if and how genes control known biochemical reactions.

G. W. Beadle and E. L. Tatum, 1941

As Beadle and Tatum say, it is experimentally more profitable to work from a known enzyme to the phenotypic effect caused by a mutational change at its genetic locus, rather than conversely. No-one has progressed from a mutant phenotype to the identification of the relevant enzyme or protein without previous knowledge of the biochemistry of the cell or tissue affected, except by accident. Thus we can relate a disease syndrome to an abnormal globin gene (p. 58), or a defective muscle to a faulty myosin gene because we already know about blood and muscle cells. The analysis of the 'genetic control of development', also on Beadle and Tatum's agenda, is a different matter: we do not know what gene products to look for. The genetic study of known genes with known products, as opposed to the molecular analysis already considered, has therefore centred on examination of (1) the organization of the gene and its contiguous regions, which may tell us if there are regulatory elements associated with it, and (2) the identification of genes which alter the developmental profile of the structural gene concerned. Both approaches involve genetic fine structure analysis and our first example is one which has pursued this approach nearly to the limit of the technique.

## Xanthine dehydrogenase genes

### The rosy locus of 'Drosophila'

Xanthine dehydrogenase (XDH) is the most thoroughly studied gene–enzyme system in higher eukaryotes. The enzyme is a complex one, requiring $NAD^+$ as a cofactor, and containing molybdenum. It catalyses the oxidation of hypoxanthine to xanthine, and of the latter to uric acid. In *Drosophila*, it also catalyses the oxidation of 2-amino-4-hydroxypteridine to isoxanthopterine in the pteridine pathway (see Fig. 5.6), and its failure results in a deficiency, or absence, of the red drosopterin pigments in the fly's eye, which is then rosy-brown or maroon coloured. Both reactions can be used to assay the enzyme, and the fluorometric assay of pteridine oxidation is sufficiently sensitive to be used on a single fly. The first relevant eye colour mutant was named rosy (*ry*), and homozygotes contain no detectable isoxanthopterin, uric acid or XDH. We give evidence below that *ry* is the structural gene for XDH. The gene is on the third chromosome at map distance 52.0 (3–52.0), and there are no other eye colour genes immediately adjacent to it (Hilliker *et al.*, 1980).

Three other mutant genes produce similar eye colours, and the flies either lack, or have a depressed level of, functional XDH: they are maroon-like (*ma-l*, 1–64.8), low xanthine dehydrogenase (*lxd*, 3–33) and cinnamon (*cin*, 1–1.0). Unlike *ry*, they also depress the levels of two other molybdenum-containing enzymes, aldehyde oxidase and pyridoxal oxidase. This suggests that these genes may be involved in some post-translational modification of all three oxidases. And by using gel electrophoresis to characterize the charge and shape of the enzyme protein, Finnerty and Johnson (1979) have shown that the XDH protein is made in the presence of *ma-l* (and of *lxd*), but that its form is different when *ma-l* is wild-type or mutant, confirming this assumption (Wahl *et al.*, 1982). The same seems to be true for the action of *cin* (Bentley and Williamson, 1979). Since the *ry* product is made (but not modified to the active form) in the presence of these three mutations, they cannot be involved in *ry* regulation. They are complementary, pleiotropic genes with an interesting and unexpected mode of action.

The first problem is to decide if *ry* is a simple cistron and to define its cytogenetic limits by performing a fine structure analysis of the locus. This required more mutants than were available, and these were obtained by finding eyes of the right colour when third chromosomes from mutagenized males were placed over either the standard null mutant $ry^2$, or a deficiency of the region. The 'cleaned' mutants were then positioned in relation to each other and mapped following Whittinghill's (1950) scheme (Fig. 6.1). Constructing the series A and B stocks is laborious, and a purine-selective system was employed subsequently. This depended on Glasssman's (1965) observation that *ry* (or *ma-l*) homozygotes die when purine is added to the culture medium, whereas wild-type larvae survive. Thus, only wild-type

Female | Male | Recombinants

A *M34 Dfd* + + $ry^1$ + + + + / + + *cu kar* + $ry^{26}$ *l26 Sb Ubx* — *cu kar* $ry^+$ + + + + wild-type

X *In (3) MRS M34* + $ry^2$ *Sb* + / + + *Dfd* $ry^2$ + *Ubx*

B *M34 Dfd cu kar* + $ry^{26}$ *l26* + + / + + + + $ry^1$ + + *Sb Ubx* — + + $ry^1$ $ry^{26}$ *l26* + + mutant

**Fig. 6.1** Whittinghills's lethal selector system so arranges recessive lethal mutants around the gene being studied that only crossovers within the locus survive. In this example, one chromosome in the female carries a pair of recessive lethals on the left of *ry* (here Minute 34 (*M34*) and Deformed (*Dfd*)) and the other carries a lethal pair on the right (here Stubble (*Sb*) and Ultrabithorax (*Ubx*)). Non-selective flanking marker mutations (curled (*cu*), karmoisin (*kar*) and lethal 26 (*l26*)) are included to check the recombinants. The two mutants to be fine structure mapped are introduced into both chromosomes, and the mirror image series A and B females generated. The male parent carries one from each of the selective lethal pairs on either side of a standard *ry* mutant (there is no crossing-over in males). If $ry^1$ is to the left of $ry^{26}$ (as shown) then $r^+$ individuals will survive in series A only as a consequence of crossing-over between the *ry* mutants, whereas the recombinants in series B will be *ry*. The order of the mutants is thus determined and the frequency of $r^+$ recombinants measures the map distance between the mutations. The purine selector system (see text) eliminates the need for constructing these elaborate lethal chromosomes since only $r^+$ crossovers survive and need to be scored. Controls are run without purine to provide estimates of population size from the cultures. Both procedures were employed to make the *ma-l* genetic map of Fig. 5.3.

recombinants survive (the equivalents of series A in Fig. 6.1), and need to be scored; the selective lethal genes are dispensed with since the *ry* mutants act as lethals, and the outside markers are used to check the crossovers. Further, the non-reciprocal event of gene conversion (i.e. "the replacement of the genetic material at a particular site in one chromatid by the genetic material from a precisely corresponding site from a non-sister chromatid" (Fincham and Day, 1965) is also picked up using purine selection. This selection technique therefore permits the very large numbers required for fine structure mapping to be handled as with a microorganism and, by this test, shows that there is no difference in gene organization of the two classes, as we shall see (reviewed by Chovnick *et al.* (1977)).

The map of the non-complementing alleles is shown in Fig. 6.2*a*, and this gives us the conventional genetic map of the locus. The mutants that exhibit interallelic complementation are mapped in Fig. 6.2*b*, and these demon-

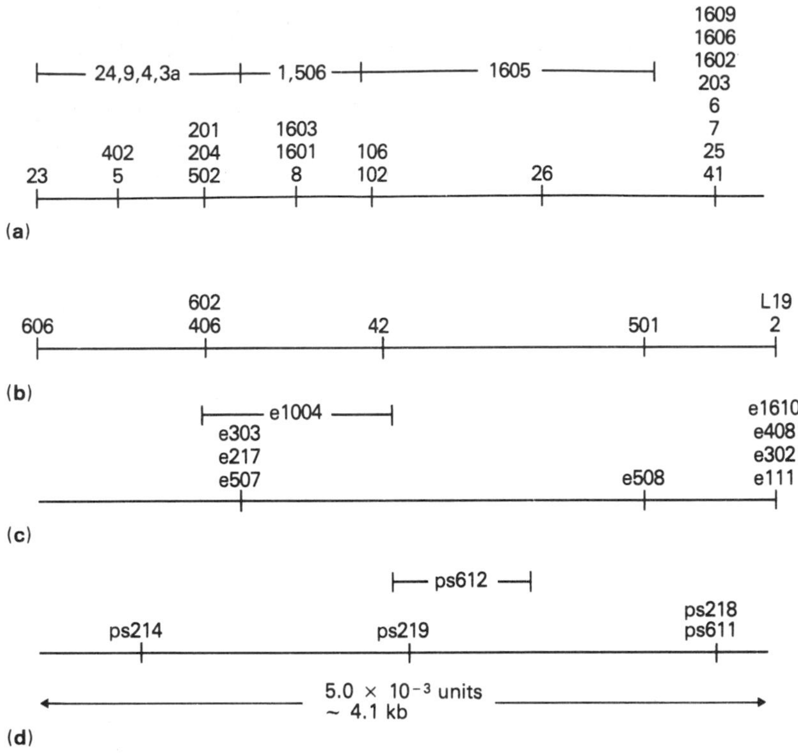


**Fig. 6.2** The fine structure maps of the rosy structural gene. (**a**), Single site inactivations of gene function; the XDH⁻ non-complementing mutants. (**b**), Mutants showing interallelic complementation. (**c**), Electrophoretic mobility sites. (**d**), Purine sensitive, or leaky mutants (see text). The map units are converted to kb of DNA on the assumption that a 0.001 map unit is equivalent to 0.82 kb (data of Chovnick *et al.*, 1977).

strate that the XDH structural protein functions as a dimer, but with a complex configuration since the complementation map is circular in form (Chovnick *et al.*, 1978). Electrophoretic variants of wild-type have also been isolated from laboratory stocks and from wild populations and these, like the preceding classes, are distributed at sites throughout the locus (Fig. 6.2*c*). Finally, a group of purine-sensitive mutants (Fig. 6.2*d*) was found which have low XDH levels and wild-type eyes, but are killed by purine; they are leaky *ry* mutants. These, too, are distributed across the locus. The last three classes are unambiguous structural gene mutant groups, and their members are not sited at one end of the locus as they might be if part of it was allocated to regulatory functions and the remainder to structural gene sequences. The three classes can be mapped in relation to one another, defining the left-most border as $ry^{606}$ and $ry^{ps214}$, and since there is no recombination between $ry^{606}$ and $ry^{23}$ the latter mutant must mark the left limit of the standard map. The right hand border is then defined by $ry^{2}$

and the cluster of electrophoretic mobility sites below it. It will be interesting to see if the amino and carboxy-termini of the XDH polypeptide correspond with these limits (Gelbart *et al.*, 1974).

From what we have already said (p. 133), it is not altogether surprising that all these many mutations are in the structural gene, and none in regulatory regions. However, among the electrophoretic variants of the $ry^+$ isoalleles some were found which made more XDH ($ry^{+4}$) and some less ($ry^{+10}$) than normal ($ry^{+0}$), and these provided the material which led to the identification of regulatory elements adjacent to the structural gene (McCarron *et al.*, 1979). These elements are active throughout development, do not alter the tissue distribution of XDH or affect its stability or other properties (Edwards *et al.*, 1977; McCarron *et al.*, *loc. cit.*). They therefore alter the level of $ry^+$ activity: $ry^{+4}$ makes four times as many XDH molecules as normal, and $ry^{+10}$ about half and does not survive on the purine-selective medium.

Since $ry^{+4}$ possesses a faster electrophoretic mobility and has a higher level of activity than normal, it is possible to see if these properties can be recombined. This led to the identification of a site (named *i409*) to the left of, and outside the limits of, the XDH structural gene. $ry^{+4}$ carried a high activity allele (*i409H*) and $ry^{+0}$ a normal activity allele (*i409N*). A similar analysis showed that $ry^{+10}$ carried a site (*i1005L*) different from the normal (*i1005N*) and that this, too, was located to the left of the structural gene.

The discovery of these oppositely acting elements raises the question: are *i409* and *i1005* variants of the same element, or are they also separable? This can be tested by mating + *i1005L* +/*kar i409H* $ry^{406}$ females to tester males and rearing the progeny in purine-selective medium which kills the *i1005L* and $ry^{406}$ progeny. If there is just one site, all surviving recombinants will show the high activity associated with *i409H*, and all conversions of *i1005L* will be *i409H*. In $1.3 \times 10^6$ zygotes sampled, nine crossovers were

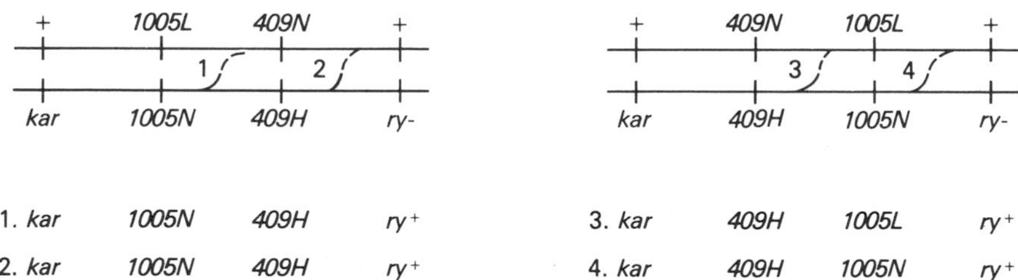


**Fig. 6.3** The relative positions of the two rosy control sites cannot be determined as long as the effect on XDH activity of the double variant *i1005L i409H* is unknown. Crossovers at both 2 and 4 give high activity recombinants (nine were found). The three normal activity recombinants also found can be accounted for by crossovers at 1. The data are compatible with *i1005* lying to the left of *i409*, as shown on the left. Unfortunately a crossover at 3 *might* also result in normal XDH activity which would validate the alternative arrangement. This random strand mapping test gives an ambiguous result (data of Chovnick *et al.*, 1980).

high but three were normal, and while three conversions were H; there was also a normal one. The elements are clearly separable. Unfortunately this does not allow the ordering of these elements because of the ambiguity made plain in Fig. 6.3.

Finally, one can ask if these elements are *cis-* or *trans*-acting? Since XDH is a dimer, a heterozygote for fast and slow forms of the enzyme will give three bands on electrophoresis; homodimers of the fast and of the slow form, and an intermediate heterodimer of both fast and slow. If *i409H* is present in association with one variant, the other being normal, then all three bands will be increased in amount if it is *trans*-active, but only its associated variant and the heterodimer will be increased if it is *cis*-active. The latter is the result found (Fig. 6.4). Levels of *ry* activity are *cis*-regulated by elements closely associated with, and adjacent to, the structural gene. What this means at the molecular level depends on current analysis of the cloned gene, for this fine detailed work on rosy makes it an obvious locus for relating DNA structure to the regulation of function.

### *The hxA locus of 'Aspergillus'*

Use of the purine selector system with rosy means that we have been concerned with purine catabolism, since it is the failure of the 'mutant XDH' to cope with purine breakdown that kills. This pathway is also being studied in *Aspergillus nidulans* where more has been done on its regulation. There are two purine hydroxylases in this mould, but we are concerned only with

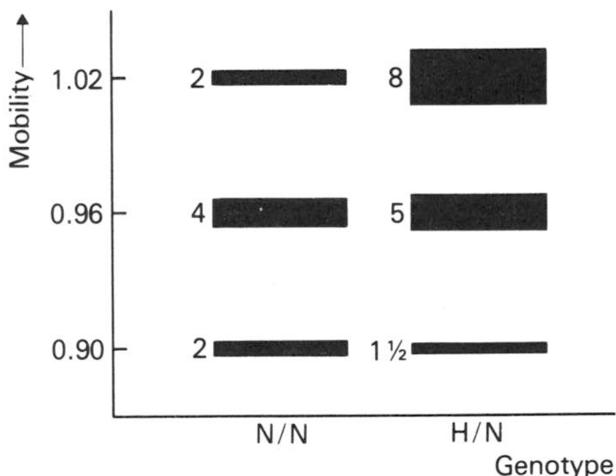


**Fig. 6.4** The diagram on the left shows the electrophoretic pattern of XDH from the heterozygote *i409Nry*$^{+1}$/*i409Nry*$^{+12}$, where that from *ry*$^{+1}$ has a mobility of 1.02 and that from *ry*$^{+12}$ a mobility of 0.90. Their hybrid molecule has the expected intermediate mobility of 0.96. The heterozgote on the right is *i409Hry*$^{+4}$/*i409Nry*$^{+12}$, and the XDH from *ry*$^{+4}$ has the same mobility as that from *ry*$^{+1}$. If the high producer, *i409H*, is *trans*-active we should expect the same pattern as before but with broader bands. Instead, only the fast variant and the intermediate are increased in amount. The *i409H* element regulates only the *ry* structural gene on the same chromosome: it is *cis*-active.

purine hydroxylase I which has its structural gene at the *hxA* locus where null, inactive, cold-sensitive and revertible mutants have been found (Scazzocchio and Sealy-Lewis, 1978). Purine hydroxylase I is equivalent to the XDH of *Drosophila*, of the chick or in milk and, as with the first, there is a group of loci (*cnx* loci) apparently involved in the synthesis and processing of the Mo-containing cofactor which, where mutant, inactivates both purine hydroxylases (and nitrate reductase). Nothing has yet been done to see if *hxA* has adjacent control elements.

A positive regulatory gene (*uaY*) has been identified which controls the expression of six (or more) genes of this catabolic pathway (Scazzocchio and Gorton, 1977). The protein encoded by this gene has been isolated and shown to bind to DNA. Uric acid combines with this protein as a specific inducer of the enzyme pathway. The uric acid xanthine permease gene *uapA* is under the control of *uaY*, and also has a *cis*-acting gene (*uap*-100) which regulates permease gene expression. A $uaY^+$ gene has to be present in the same nucleus for a *uap*-100 mutation to be expressed and then urate uptake is at constitutive levels (Scazzocchio and Arst, 1978). Although the detailed mechanisms of this system are complex, it demonstrates that the prokaryote type of gene regulation is found in single cell eukaryotes. It would therefore be very interesting to see if equivalent genes are found in multicellular eukaryotes, or if the segregation of function to specialized cells and tissues (*Drosophila* XDH is found only in larval fat body and Malpighian tubules, for instance) implies different control mechanisms.

## The Aldox 1 locus of *Drosophila*

*Drosophila* aldehyde oxidase is a dimer, like XDH, and also requires $ma\text{-}l^+$ etc. for its activity. It is named after its experimental substrates (acetaldehyde, benzaldehyde, farnesol) but its natural function is unknown. It is not a vital enzyme since null homozygotes survive. The enzyme increases (per mg protein) during larval development, and then shows a notable rise around pupation and again just before eclosion. There are strain differences in the ratio of pupal to adult levels, which is usually about 85 percent but may be as low as a 40 percent pupal/adult ratio if the strain shows little or no pupal increase. This difference is genetic, and heterozygotes are intermediate. Since electrophoretic variants exist (which allow the structural gene to be mapped to 3–56.7) this regulatory element can be combined with them. It is again found, from an electrophoretogram like Fig. 6.4., that the element is *cis*-active. Immunotitration and enzyme kinetic measures show that the mutant element reduces the amount of enzyme made (Dickinson, 1975). However, in this case the control element is active only during the late larval-early pupal stage, and not earlier or later. It must itself be regulated in a temporal fashion which is not yet understood.

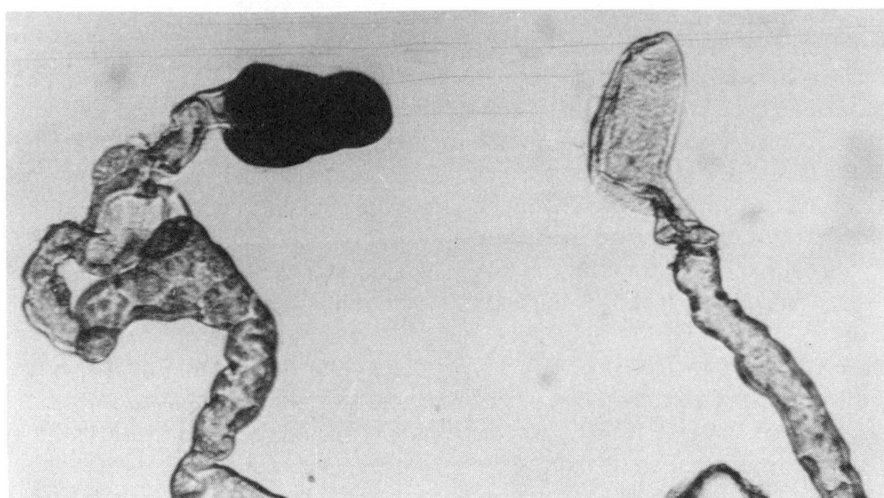

**Fig. 6.5** Aldox staining of tissues can be used to identify cells. In this case wild-type blastoderm cells were injected into maroon-like, and therefore aldox negative, blastoderm eggs. In the hatched adult one patch of chimeric tissue was found at the termination of one Malpighian tubule (left), whereas the other (right) was *ma-l*. The donor embryo survived, and was found to have one short Malpighian tubule. The transformed aldox positive cells had therefore integrated into the tubule of the host and could be identified as such by their staining properties. Photograph by Dr. A. Simcox.

The electrophoresis staining technique can be applied to gluteraldehyde-fixed larval and adult tissues, and then the enzyme is found to have a very specific distribution (Fig. 6.5). For instance, cells of the proximal region of the gastric cecae are heavily stained but the morphologically identical cells of the distal region do not stain at all, and two rings of staining cells in the hind-gut are separated by non-staining cells. As we shall see later this gene activity is cell autonomous: the aldehyde oxidase found in a cell is made there, and an unstained cell is presumably not transcribing the gene even if the cell is morphologically indistinguishable from its active neighbour. As might be expected, different strains and species show pattern differences. If the species hybrid is viable one can take advantage of the inevitable electrophoretic differences of the parental enzymes to ask two questions: is the tissue distribution the same, qualitatively and quantitatively, in the two species; and if there are differences, how are they expressed in the hybrid? In the example shown in Fig. 6.6, there are obvious tissue differences between species: a disparity of amounts in fat body, and a near equality in ovary and in head (not shown). However, differences between individuals may reflect disparities in organ size etc., so the data from a hybrid is more impressive because it is internally controlled. Here we find in all cases that the ratio between homodimers and heterodimer is as expected from the binomial distribution. (The random association into

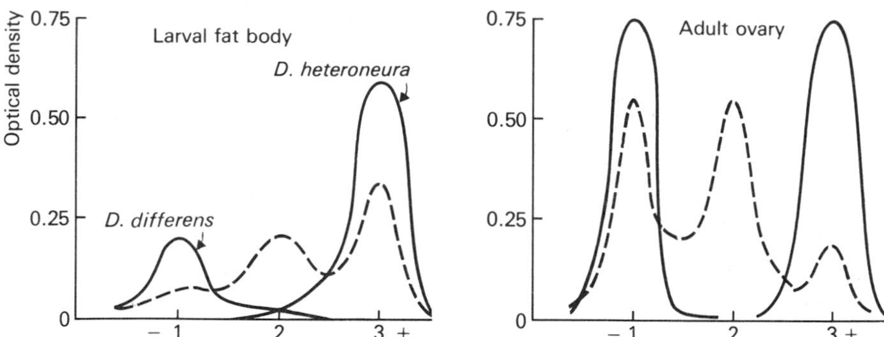


**Fig. 6.6** The expression of aldehyde oxidase in *D.differens* and *D.heteroneura* and in their hybrid (broken line). In the hybrid larval fat body the *heteroneura* gene is expressed at about twice the level of the *differens* gene, whereas the converse is found in the ovary. There is a tissue regulation of product accumulation, and possibly of gene transcription. The curves (Dickinson, 1980) are optical density scans of electrophoretic patterns like those in Fig. 6.4.

dimers should be in proportions corresponding to $p^2 + 2pq + q^2 = 1$, where $p$ and $q$ are the frequencies of the two kinds of subunits.) So, "each allele is expressed according to the developmental programme characteristic of the parent from which it was derived" (Dickinson, 1980). The *D. differens* allele is expressed in the ovary (Figure 6.6) in about twice the amounts of the *D. heteroneura* allele, whereas the converse is found for fat body. These regulatory differences segregate with the structural genes in back crosses, providing *a priori* evidence that differences in pattern expression are under the control of one or more *cis*-acting elements associated with the parental loci.

There is a long tradition of using species hybrids to explore development (see Davidson, 1974) and the novelty of the above example lies in exploiting hybrids to examine the expression of a particular gene. However, the method has limitations since we do not know whether the genes are on homologous chromosomes, and we lack the genetic marker stocks so successfully exploited for the analysis of rosy. Nevertheless, this example shows that *cis*-regulatory elements of one kind or another are involved in the developmental programme which results in differences of expression in separate tissues.

## α-Amylase genes

### *The Amy locus in 'Drosophila'*

The gene for the starch-digesting gut amylase of *Drosophila* (*Amy*, at about 2–78) has been located using electrophoretic enzyme variants. It is a single gene in *D. hydei*, but is tandemly repeated in *D. melanogaster*. The enzyme is inducible, and larvae reared on sugar diets have suppressed enzyme levels (Doane *et al.*, 1975). If the paired *D. melanogaster* genes code for different electrophoretic variants of the enzyme, they are demonstrably not equally co-induced when sugar-fed larvae are moved onto starch. There is a surprising lack of coordinate control of these tandem genes, which is not yet understood. The genetics of this control system clearly merit further study.

In larvae, the amylase secreting region extends from the anterior to the middle of the posterior mid-gut, as can be determined by placing the gut on a thin starch film and then staining the digested starch blue with iodine. There seems to be no variability in this larval pattern. Adult mid-gut derives from imaginal stem cells of the larval mid-gut, which it replaces during metamorphosis, and here there are differences. There are two subdivisions of secretory cells, each heritable. The whole posterior mid-gut may be secretory (PMG12), or only its anterior subdivision (PMG10), or none of it (PMG00). PMG12 is dominant over PMG00, and incompletely dominant over PMG10. These differences are due to alleles at a single locus, the mid-gut activity pattern (*map*) locus. This has been mapped at two units to the right of *Amy* (Abraham and Doane, 1978). Similarly, there is another locus adjacent to *map* which determines the enzyme distribution within the anterior mid-gut, but this has not been as fully studied. This, and *map*, are clear examples of genes which control the activity of other structural genes.

The separation of gene control of anterior and posterior mid-gut function is probably not surprising since the two rudiments have different embryonic origins. It is not so easy to understand how one member of an allelic series (*map*-PMG10) can affect one part of the posterior mid-gut, but not the other. Since *map*-PMG12 is dominant, *map* must be *trans*-active, and this is demonstrably so: an isoallele linked to *map*-PMG00 is expressed in the heterozygote (Doane, 1977). The *map* locus is the first clear example of a *trans*-regulatory gene specifically controlling the pattern of tissue expression of a structural gene.

*map* acts in an all-or-none fashion, as a switch. On the other hand, *Amy* expression in the anterior mid-gut is additive (i.e. is intermediate in strong/weak heterozygotes) and this suggests that there is also a *cis*-acting control associated with *Amy* which functions in the anterior mid-gut, but not in the posterior (Treat-Clemons and Doane, 1980). It is therefore interesting that in another organism there is a *trans*-regulatory element, some 37 units from another structural gene (*Cat2*), which affects the amount of catalase synthesized in the scutellum of developing maize, as can again be shown from crosses of strains producing different enzyme levels 7 days or so after the seed is caused to germinate (Scandalios *et al.*, 1980). There is therefore evidence of regulatory genes associated with, and distant from structural genes, and they may be quantitative or qualitative in their effects.

### *Amy1 in the mouse*

Amylase is a tissue specific enzyme in the mouse which is made in the salivary gland, liver and pancreas. A single gene (*Amy1*) encodes the salivary and liver enzymes which are identical proteins, and a second locus (*Amy2*) codes for the quite different pancreatic enzyme. There is a 100-fold difference in the amount of α-amylase mRNA in the cytoplasm of the salivary gland (2 percent) compared with liver (0.02 percent), and there are differences in the 5′ untranslated sequences of the two *Amy1* mRNAs which may be responsible for this quantitative dichotomy. However that may turn out, *Amy1* provides a quite unexpected picture of tissue specific transcription (Young *et al.*, 1981).

By identifying and sequencing DNA clones for the 5′ quarter or so of this locus, it was found that the gene occurs only once in the genome and has the same organization in all tissues, whether it is transcribed or not. The segments that specify the two different 5′ untranslated sequences lie 2.8 kb apart, and the first of the 11 exons common to the two mRNAs lies a further 4.5 kb downstream from the liver 5′-end leader sequence (Fig. 6.7). The leader sequences are of different lengths, 47 nucleotides in the salivary mRNA and 158 in the liver mRNA. The tissue specific mRNA synthesis therefore involves quite different splicing events to produce the dissimilar 5′-non-coding leader sequences, whereas the splicing of the coding sequence is the same. In addition to the tissue specific splicing there must also be tissue specific promoters, for there is no evidence of either a selective passage of RNAs to the two different cytoplasms or a selective processing of a single nuclear RNA. Probably this promoter difference accounts for the difference in product level in the two tissues, although this is still not proved. The situation is different in other mammals which have

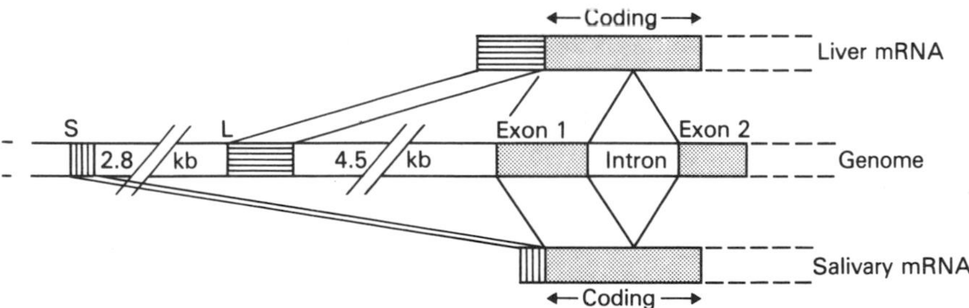


**Fig. 6.7** The organization of the 5′ end of the mouse Amylase 1 genetic locus and its relation to the different mRNAs in liver and salivary gland. Transcription of the salivary mRNA starts at S and the short, shaded, leader sequence is the only one found preceding the coding sequences. Transcription of liver mRNA starts at L, and again the short, shaded, leader sequence is the only one found preceding the coding sequences. The intervening sequences are therefore handled as if they were introns, but differently in the two tissues.

a number of non-allelic *Amy* genes, and the mouse fortuitously provides unique material for the study of this fascinating tissue regulation of transcription.

## The hormonally regulated dopa decarboxylase locus (Ddc)

Dopa decarboxylase (DDC) is a homodimeric enzyme of about 110,000 molecular weight, encoded by a single gene in *Drosophila* and expressed in the nervous system, proventriculi and epidermis. It is a key enzyme for the sclerotization of insect cuticle and for the synthesis from tyrosine of biogenic amines in the nervous system (Fig. 6.8). Inhibitors have been developed in connection with nerve activity studies.

In *Drosophila*, the enzyme is cyclically induced by the pupation hormone, 20-hydroxyecdysone, and increases in amount at the larval moults, pupariation and emergence when chitin is being laid down by the enzyme-containing epidermis (Fig. 1.13). Proof of this relationship comes from exploitation of the temperature-sensitive ecdysonless ($ecd^1$) mutant (Garen *et al.*, 1977), since suppression of ecdysone synthesis then results in disappearance of DDC mRNA, and feeding 20-hyroxyecdysone restores mRNA synthesis (Kraminsky *et al.*, 1980). There is one anomaly: the mutant does not block the mid-pupal hormone increase, and the relation of this increase to enzyme activity 60 h later is not yet understood (Marsh and Wright, 1980). Nothing is known either about the molecular mechanism of the hormone induction of gene activity, but the recent cloning of the gene may provide the tool for exploring this problem (Hirsh and Davidson, 1981).

The first attempt to locate the *Ddc* gene used the substrate analogue α-methyldopa (αMD) which inhibits the enzyme. The inhibitor causes larval death at moulting, but not between moults, as would be expected from interference with sclerotization, and a level was determined which allowed selection for resistant mutants and for dominant hypersensitive mutants (i.e. mutants where +/m die on αMD levels which allow +/+ to survive). Two of the three resistant mutants found involved pairs of mutations and the third did not have elevated levels of DDC. On the other hand, all seven hypersensitive mutants were found to be alleles, and were homozygous recessive lethals. They were designated *l(2)amd*, since they are on the

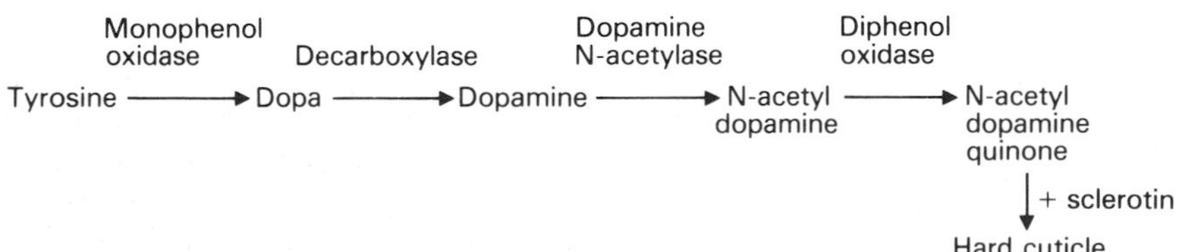


**Fig. 6.8** The enzyme-regulated reactions for the conversion of tyrosine, and leading to the hardening of insect cuticle.

second chromosome. They had no measurable effect on DDC activity, rather surprisingly, and their only phenotype is sensitivity to αMD. You get just what you select for!

A second, independent, approach to finding the *Ddc* locus exploited Lindsley *et al.*'s (1972) technique for creating segmental aneuploids across the entire genome (Fig. 6.9). The argument was that the *Ddc* structural gene could be identified where local hyperploidy caused an increase of enzyme activity and a reduction of activity where the structural gene was covered by a deficiency. Aneuploidy at only one chromosomal region (map location 36EF–37D on the left arm of chromosome 2) gave the expected proportional gene dosage effects (~150 percent activity when three copies of the region were present and ~50 percent with one copy, when compared to the normal diploid). It was taken that the *Ddc* gene was in this region, but it was quite unexpected when *l(2)amd* turned up there, too (Hodgetts, 1975).

Sixteen overlapping deficiencies were made covering the 36EF–37D region, and used to define the limits of the two genes genetically and cytologically, and one of them (*Df(130)*) defined the DDC dosage-sensitive region as 37B10–C6 (Wright *et al.*, 1976a, b). *Df(130)* was then used to select four lethal mutations (*Df(130)*/m larvae without DDC die through failure to form chitin) and examination of one group of 17 alleles showed that they had 25–75 percent normal activity when heterozygous with wild-type. Interallelic complementation and a ts mutation confirm this group as mutations of the DDC structural gene at 2–53.9. Very surprisingly, *l(2)amd* was only 0.002 map units to the right, implying that the lesions in the two structural genes are separated by little more than 1000 nucleotides. *Ddc* and *amd* are conventional adjacent loci (Wright *et al.*, 1982).

The *amd* mutants do not affect DDC levels in any way or modify the enzyme protein structure. *Ddc* mutants are not hypersensitive to αMD, even when mutants with very low (5–10 percent) levels of DDC are tested against the analogue. Double heterozygotes of the two loci are completely viable. So there is no evidence that the mutations have anything in common other than their interactions with the structurally related compounds, dopa and α-methyldopa. However, there is preliminary evidence that one of the mutant loci involved in αMD resistance also results in increased DDC production. This may imply that this element (lying between 2–53 and 2–54.5) regulates both *amd* and *Ddc*, but it will be difficult to prove this while the *amd* gene product remains unknown (Marsh and Wright, 1979).

We should note, in parenthesis, an interesting fact about the next enzyme in this pathway, dopamine acetyltransferase (DAT) which acetylates the dopamine formed by DDC. Unlike DDC it does not cycle with the larval instar changes, nor does it greatly increase before pupation. There is therefore no clear evidence that it is regulated by ecdysone, or of any coordinate regulation of two sequential enzymes of the sclerotization pathway (Marsh and Wright, 1980). DDC illustrates very clearly that our theoretical expectations (p. 105) will not always be fulfilled.

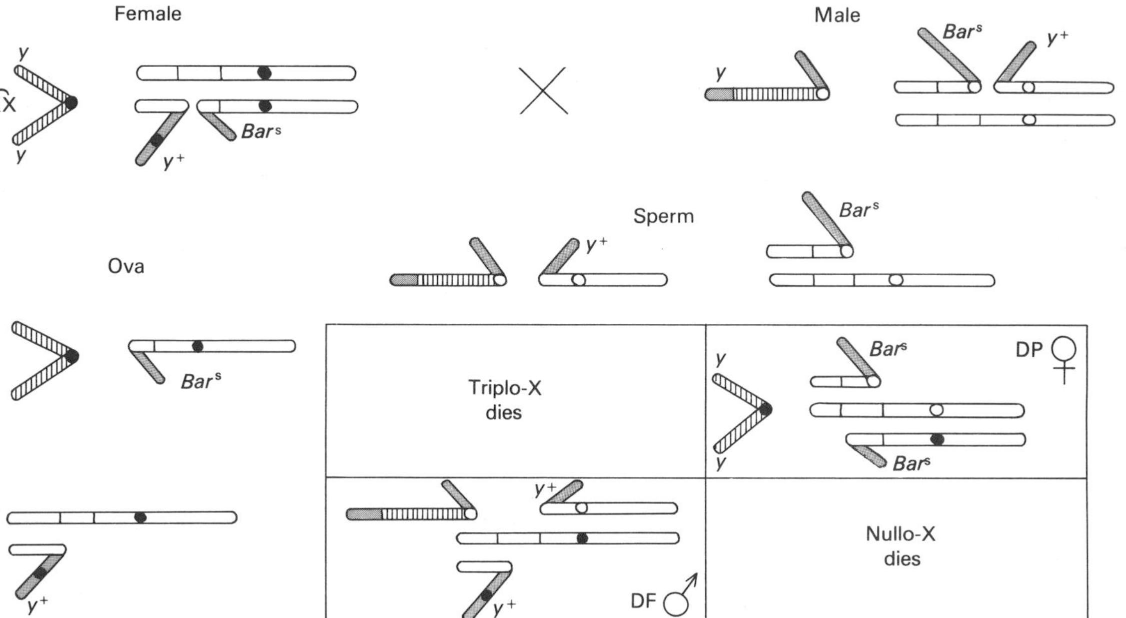


**Fig. 6.9** The use of Y-autosome translocations to generate segmental aneuploids, in this example for the central region of the left arm of the normal (unshaded) autosome. Different pairs of translocations are used to generate the required range of segments. The translocated Y chromosome is stippled; the Y short arm ($Y^s$) carries yellow $^+$ ($y^+$) and the long arm Bar (*Bar*$^s$). X chromosomes are striped. The attached X ($\widehat{XX}$) and the compound $\widehat{XY}$ are homo- and hemizygous for *y*. The right arm of the autosome can be marked with a dominant homozygous lethal (e.g. Curly (*Cy*) for the second or Ultrabithorax (*Ubx*) for the third chromosome) and the genotypes of the duplication female (DP♀) and the deficiency male (DF♂) can therefore be checked (after Lindsley *et al.*, 1972). Similarly, the X chromosome can be analysed by mating $\widehat{XX}$ females to males carrying reciprocal X-Y translocations which link a portion of the paternal X to the Y centromere, and the remainder to the X centromeres. Females will be either euploid or duplicated for one or other end of the X depending on which centromere is recovered. Dosage compensation (p. 125) complicates the results of these X chromosome manipulations.

**The rudimentary locus of *Drosophila* pyrimidine synthesis**

Only one complex locus has been analysed in its biochemical details, and it is the morphological mutant rudimentary (*r*, 1–54.5), a recessive causing size reduction and other abnormalities of the wing (rudimentary wings) and female sterility. This sterility is incomplete and is 'rescued' by the presence of a + gene in the female. Also, some mutant *r*/*r* or *r*/Y progeny of homozygous mutant mothers escape death in ordinary cultures: 1 percent of the lethal class are 'escapers', and there is no sex difference in this proportion. We therefore have:

1. *r*/+ × *r*/Y → *r*/+, +/Y(wild-types), *r*/*r*, *r*/Y (rescued survivors);
2. *r*/*r* × *r*/Y → *r*/*r* and *r*/Y (escapers only);
3. *r*/*r* × +/Y→ *r*/+ (wild-type), *r*/Y (escapers only).

Carlson (1971) ingeniously took advantage of these properties of this female sterility to analyse 45 *r* mutants.

Previous work using the wing phenotype had shown that some alleles complemented and others did not. Carlson rechecked this complementation pattern, using isogenic stocks, by making females heterozygous for all *r* combinations ($r^1/r^2$, $r^1/r^3$ etc.) and mating them to wild-type males (mating (3) above). For non-complementing mutants the surviving male progeny will be escapers and account for 0.5 percent of the eggs, whereas complementing mutants (responding like mating (1) above) will have more than this proportion of males in a culture. The complementation map (Fig. 6.10) shows 16 complementation groups, but as these overlap this can be reduced to 6 complementation 'units' (Jarry and Falk, 1974). The wing phenotype follows the same complementation pattern, which is consistent with earlier studies.

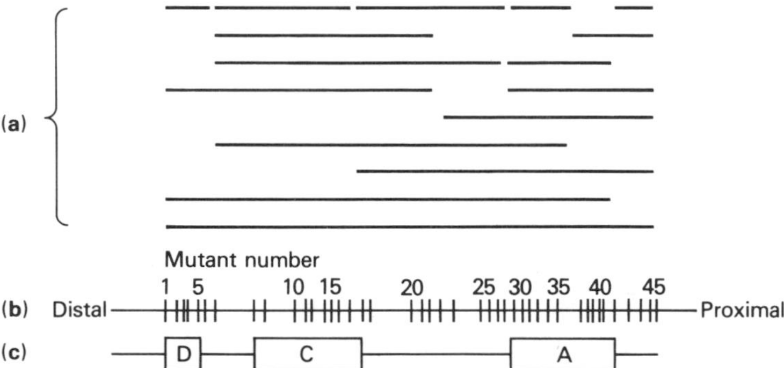


**Fig. 6.10** The complementation map of rudimentary alleles (**a**) placed above the crossover map (**b**), based on Carlson (1971). Mutations which affect the enzymes are indicated in (**c**) where C is carbamyl phosphate synthetase, A is aspartate carbamyl transferase and D is dihydroorotase (see text).

Knowing the complementation map simplified making the recombination map of the locus by mating $r^x/r^y \times r^x/Y$. The rare recombinants are $+/r^x$ and $+/Y$, and survive; non-recombinants are $r^x/r^x$, $r^x/r^y$, $r^x/r^y$, $r^x/Y$ and $r^y/Y$ and the 1 percent escapers among them provide an estimate of the total population. Two non-complementing mutants ($r^7$ and $r^{42}$) at the distal and proximal ends of the locus were used to order the mutants, and the precise position of each mutant was fixed by locating it within its complementation group. This fine structure map is shown in Fig. 6.10. The complementation map is co-linear with it. Since the alleles at each extreme of the map give 35 recombinants in $10^5$ progeny, the closest alleles 2.6 recombinants in $10^6$ progeny (and double recombinants are not recovered), the total map distance works out at 0.07 units, and the closest distance at 0.00052 units. Carlson (1971) concluded that he had defined a single cistron whose alleles showed a complex pattern of interallelic (intracistronic) complementation.

This conclusion did not stand for long. Norby (1970), trying to improve the yield of his cultures of rudimentary, found that supplementing the diet with RNA cured both phenotypic defects. And he was able to show that the pyrimidines of the RNA were the effective substances. By testing intermediates, he also demonstrated that *r* blocked pyrimidine synthesis at an early step (Fig. 6.11). Almost contemporaneously, Falk and Nash (1974) were selecting purine and pyrimidine auxotrophs of *Drosophila*, using axenic culture techniques (Sang, 1978), and they found that the 20 pyrimidine-requirers that they had isolated all mapped at the *r* locus (designated $r^{pyr}$), although only 11 displayed the *r* phenotype. Clearly *r* was the site of a structural gene (or genes) involved in pyrimidine synthesis. This was subsequently confirmed by gene dosage studies and the identification of temperature-sensitive mutants (Jarry, 1979).

Examination of enzyme production by representatives of the complementation groups (references in Rawls and Fristrom, 1975) showed that the first three enzymes (CPSase, ATCase, and DHOase) of the pathway were affected by *r* mutations, but that the fourth enzyme (DHOdehydrogenase) was not. It was then possible to relate the complementation map to the apparent cistrons for each enzyme (Fig. 6.10). While there is necessarily some ambiguity about the limits of these genes, DHOase is delineated by the five most distal mutations, CPSase by mutants 8 to 17, and ATCase by

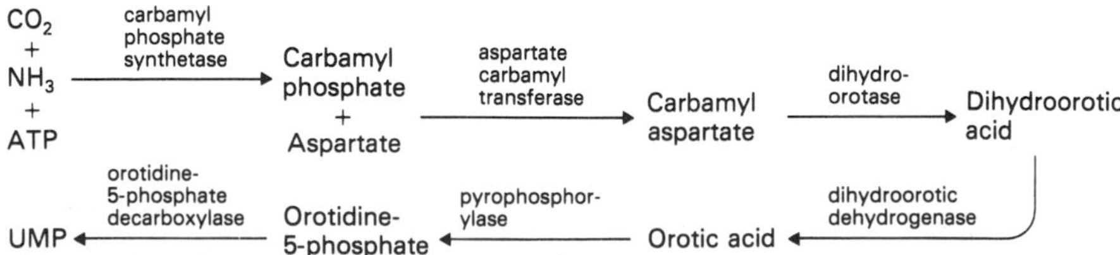


**Fig. 6.11** The pyrimidine synthesis pathway and the enzymes involved. The three enzymes coded in the rudimentary locus are in the upper line and they are not ordered in the genome in this functional sequence.

mutants 28 to 41. In the region proximal to 17, mutants may complement like CPSase or ATCase mutations, which suggested that the three enzymes might form a functional complex and thus give ambiguous complementation. Jarry (1976) was able to isolate this large complex of 800,000 molecular weight and to show that it was a single, multifunctional polypeptide (Brothers *et al.*, 1978). The complex is, in fact, formed from a primary translation product of 210,000 molecular weight, which may be broken down to its constituent enzyme components by proteolysis in the cell, as is the analogous complex found in a wide range of animals. There is a large discrepancy between the size of this translation product (equivalent to 6.6 kb) and the estimate of DNA in the locus (25 kb, according to Jarry and Falk (1974)), and it is not known if this is accounted for by introns.

It is now easy to see how the pyrimidine deficit in *r*/*r* eggs causes their non-viability, and Okada *et al.* (1974) have confirmed that this is overcome by injecting pyrimidine nucleosides into *r*/*r* eggs. Equally obviously, *r*/+ females would lay pyrimidine-containing eggs. The wing phenotype results from a disproportionate reduction in the number of cells in the posterior region, due to cell death. But why these particular cells are vulnerable, and the precise mechanism of their disruption, as it relates to pyrimidine metabolism, is still unknown (Fausto-Sterling, 1980). Nor is there much evidence of developmental regulation of enzyme synthesis (Mehl and Jarry, 1978), or that enzyme synthesis is stimulated by pyrimidine dietary deficiencies, although these are rate-limiting for growth and development (Falk *et al.*, 1980).

Before leaving *Drosophila*'s pyrimidine metabolism, we can note that a further complex locus containing the structural information for both orotate phosphoribosyltransferase and orotodylate decarboxylase has been identified at 3–70.4. This locus has been named rudimental, and its phenotypic effects are identical to those of *r* (Rawls, 1980; Lastowski and Falk, 1980). The remaining enzyme of the pathway, mitochondrial dihydroorotate dehydrogenase, has also been identified and mapped to about 3–48 (Rawls *et al.*, 1981). Thus the genes for all the enzymes of *de novo* pyrimidine synthesis have been identified and located in *Drosophila*, and that for the first time in any animal. The analysis of their presumed coordinate regulation awaits further work.

## The *intA* gene of *Aspergillus*

There seems to be only one claim for identification of a eukaryote integrator gene; a positive regulatory gene (Arst, 1976). This locus (*intA*) on linkage group II of *A. nidulans* probably controls three structural genes: acetamidase (*amdS*) on linkage group III, GABA transminase (*gatA*) on linkage group VII, and GABA specific permease (*gabA*) on linkage group

VI, all enzymes of the γ-aminobutyric acid (GABA) catabolic pathway. The biochemistry is complex, as is the genetics. However, *intA* mutants affect the expression of more than one of the genes listed, and this implies a redundancy of receptor sites, not of integrator genes, in the Britten and Davidson model. *amdS* may also be specifically controlled by another regulatory locus (*amdA*), suggesting the association of more than one receptor site with one particular structural gene, in agreement with the model. Although this system is not a developmental one, its further study is certain to tell us more about the functional organization of eukaryote genes.

## The ocelliless gene of *Drosophila*: Chorion proteins

The insect ovary is a useful organ for developmental studies: the stages of ovum growth and development are laid out sequentially along the ovariole and the egg cysts can be separated by size, or dissected out, so that one stage can be compared with another (Fig. 3.14). In particular, the proteins of the endo- and exochorion are synthesized in a precise and coordinated fashion during the short time-span of the last phase of egg formation, and the changes which then occur have been examined for the silk moth (Kafatos *et al.*, 1977) and for *Drosophila*. We shall look at the simpler *Drosophila* case as an example of the use of molecular technology for studying the steps between gene product and gene organization. It has the added interest of providing an explanation for one of the pleiotropic effects of the sex-linked recessive, ocelliless.

The chorion or shell (s) proteins are laid down by the follicle cells during stages 11–14 of egg formation (King, 1970). These proteins can be labelled by injecting [$^{14}$C] amino acids into actively laying females. The egg chambers of successive stages can then be dissected from the ovaries and their solubilized chorion proteins separated by two-dimensional gel electrophoresis. About 20 distinct proteins are synthesized during this 5-hour period, some for longer and others for shorter periods, as diagrammed in Fig. 6.12. They range from 15,000–150,000 molecular weight and are classified by their size in kilodaltons (s15, s16 etc.), although some size classes contain proteins with different charges (Waring and Mahowald, 1979).

Knowledge of the pattern of protein synthesis facilitates the search for their relevant mRNAs by the analogous procedure of separating mRNAs from the appropriate egg chamber stages. cDNA clones have also been prepared complementary to the poly(A)-containing RNAs of the follicle cells, and these and the mRNAs have been used for *in situ* hybridization to salivary gland chromosomes (Spradling *et al.*, 1980; Griffin-Shea *et al.*, 1980). Both approaches show the genes for chorion proteins to be clustered, and that there are clusters on the first, second and third chromosomes. The

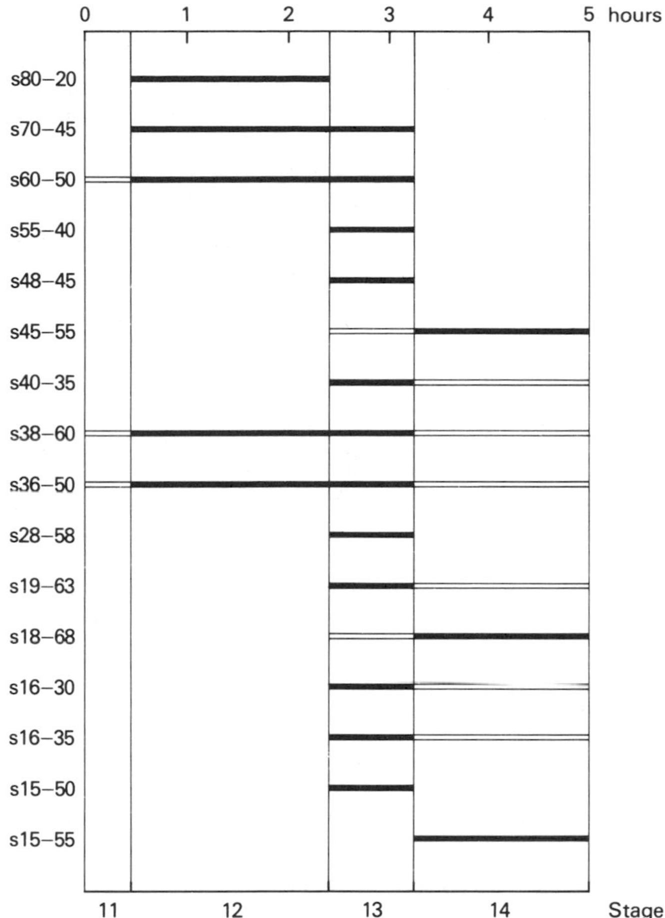


**Fig. 6.12** The pattern of chorion protein synthesis in *Drosophila* egg chambers during stages 11–14. The shell proteins are designated by their molecular size, in kilodaltons, and by their relative migrations from the acidic to the basic end in the first dimension of the gel. Thus s36–50 is a protein of 36,000 daltons which migrates half way up the gel. Maximal synthesis is shown by the solid lines, but detectable synthesis (open lines) may be found to precede this and often to follow it (data of Waring and Mahowald, 1979).

most interesting of these is the pair of genes active during stage 12 whose mRNAs cluster at the cytological location 7E11. This is a region already known to contain the female sterile mutation, ocelliless (1–23.1), one of whose pleiotropic effects is on chorion formation (Beatty, 1949).

In normal ovarian follicles, s36 and s38 are made in about equal amounts and during the same stages (Fig. 6.12). There is still an equality in follicles of the *oc* mutant but the amount of the proteins is reduced. By making heterozygotes for electrophoretic variants of the proteins, it can be shown that the amount of s36–s38 protein produced by an X chromosome is independent of the ocelliless mutant present on its homologue. The heterozygote makes the normal level of s36–s38 from its $oc^+$ chromosome and the reduced amount from its *oc* chromosome: the *oc* mutation is *cis*-acting (Spradling *et al.*, 1979). *oc* would appear to regulate two (or more?) closely linked structural genes.

Cloned cDNA covering the s36 protein gene was used to probe the mechanism of reduced production, and surprisingly showed that chorion protein genes are amplified during oogenesis. The Southern blot technique enables the quantitative assay of the number of copies present in the whole DNA digest which are complementary to the cloned genes, relative to a control which, in this case, were embryonic sequences not expressed in the ovary (Spradling and Mahowald, 1980). There were no differences in the relative amounts of DNA complementary to the two plasmids when tested against DNA from males, or from carcasses without ovaries, or against early egg chamber stages. When DNA from chambers with active follicle cells (stages 11–14) is tested, 12 times as much s36 DNA is found as control DNA. The comparable figure for the *oc* mutant is 4.2, suggesting that this decreased gene dosage may account for the chorion defects which the mutant produces. Since the s36 and s38 genes are separated, the amplified region may include a number of genes.

Follicle cells are polyploid by stage 10 and contain about 45C of DNA, but the chromosomes are not replicated subsequently. It is at about this stage, or a little earlier, that specific amplification of the chorion genes occurs and this affects some chorion clusters other than s36 and s38 (Spradling and Mahowald, 1980). Presumably this regimen reflects the need for high gene copy numbers to match the short burst of protein synthesis necessary for chorion formation. How *oc* interferes with this process, which is not unique to the chorion genes (p. 10), is still uncertain. Nor has any relationship yet been worked out between the mutant's effect on the chorion and its earlier impairment of ocelli differentiation, the associated disturbance of head bristle organization, and lack of accessory glands in the female reproductive system. The molecular tools for studying all these phenomena are now available.

## Conclusions

We are at the beginning of the analysis of genome organization, and the examples considered above were selected to illustrate the more interesting techniques as well as the progress being made. The last example, the chorion proteins, links the more purely genetic studies to the molecular (Ch. 2), and this marriage of the two technologies, sometimes accidental as in the case of the ocelliless mutation, is now the favoured future approach. The two methodologies are complementary, for while we can anticipate that examination of a cloned rosy region will tell us about its associated regulatory elements, we only know to look for these elements, and where we may find them, as a result of the careful genetic analysis just described. Identification of mutants is generally a prerequisite for effective molecular analysis.

The preponderance of structural gene mutations in our examples is striking. There is no reason to think that regulatory sequences should be unaffected by the mutagenic treatments, but the explanation of the apparent disparity may be trivial: regulatory sequences may be short and few. Unfortunately this cannot be a firm conclusion since some of the mutant selection systems may be biased towards the choice of null structural genes.

The identification of regulatory elements comes from looking at phenotypes in a different way, and has depended on exploiting the variants of these elements found in different strains. This procedure has its inherent limitations, and presumably is valid only for non-vital enzyme systems since suboptimally-regulated vital enzymes could be lethal. The first group of regulatory elements is *cis*-active and, if we can generalize from rosy, where they lie hundreds of base pairs from the structural gene, it is unlikely that they are promoter or operator mutants. The second class is *trans*-active and, if we can generalize from the *map* example, the element is even more distantly sited. But of course we cannot yet generalize from these few examples, and we can only consider it significant that both *cis* and *trans*-acting elements have been found. Nor do our data testify to *cis*-acting sequences being commoner; only that they are easier to recognize.

The data from unicellular eukaryotes, which we have only briefly outlined, is more conclusive since it shows that many structural genes are regulated together. We have still to confirm, though, that this prokaryote-type of organization regulates metabolism in higher eukaryotes where the pathway is, at least sometimes, tissue-limited. The temporal regulation of the luxury proteins of developmental systems may imply other forms of organization, so it may not be surprising that the dopa pathway (Fig. 6.8) is not simply coordinated by ecdysteroids. Certainly the unexpected organization of mouse *Amy* sequences, and the amplification of chorion genes, suggest that there may be more kinds of regulatory mechanisms than we have dreamed of (and see p. 262). What is also clear is that we have too few *developmental* systems under study to warrant any conclusions about coordinate regulation of gene activity. And in many ways we know too little about the characteristics of developmental systems, and we must next look at what genetic manipulation tells us about this.

# 7 Lethal factors

Lethal factors occupy a special position in biological research. Since a high proportion of the mutations in all organisms are lethal, they contribute a large body of material which needs to be incorporated into any general theory of the gene and its mutability. In addition, the Mendelian lethal factors are a striking illustration of the extensive role played by the genetic material of the chromosomes in the fundamental processes of development. Finally, each newly arisen lethal mutant sets up a highly specific experiment, shedding light on the functional relationships between individual mutational states and the processes leading to the formation of characters.

E. Hadorn, 1960

Lethals are the commonest mutations. An estimated 60–80 percent of loci on the *Drosophila* X chromosome can mutate to a lethal state (Abrahamson *et al.*, 1980), and the more detailed studies of Shannon *et al.* (1972) and of Hochman (1976), already referred to, put the proportion even higher at 80–90 percent. Thus recessive lethals (dominants eliminate themselves) provide a great variety of mutants which we should be able to exploit for understanding development, as Hadorn (1960) postulated in his classic text, *Developmental Genetics and Lethal Factors*. It is only recently that Hadorn's expectation has met with some success, for early work concentrated on detailing the morphological consequences of many lethals to the neglect of the mechanisms underlying their action.

There are three principal reasons for past failures to exploit lethals:

1. Many mutations will affect housekeeping genes and so block necessary cell processes. How this is expressed depends on how 'leaky' the mutation is, and how far particular cells or tissues are vulnerable to a deficit of the gene product. The rudimentary alleles illustrate this situation well enough, for some members of the series have no effects on wing or ovary, while others affect one or both, and null, or near null, mutants are lethal on pyrimidine-free diets.

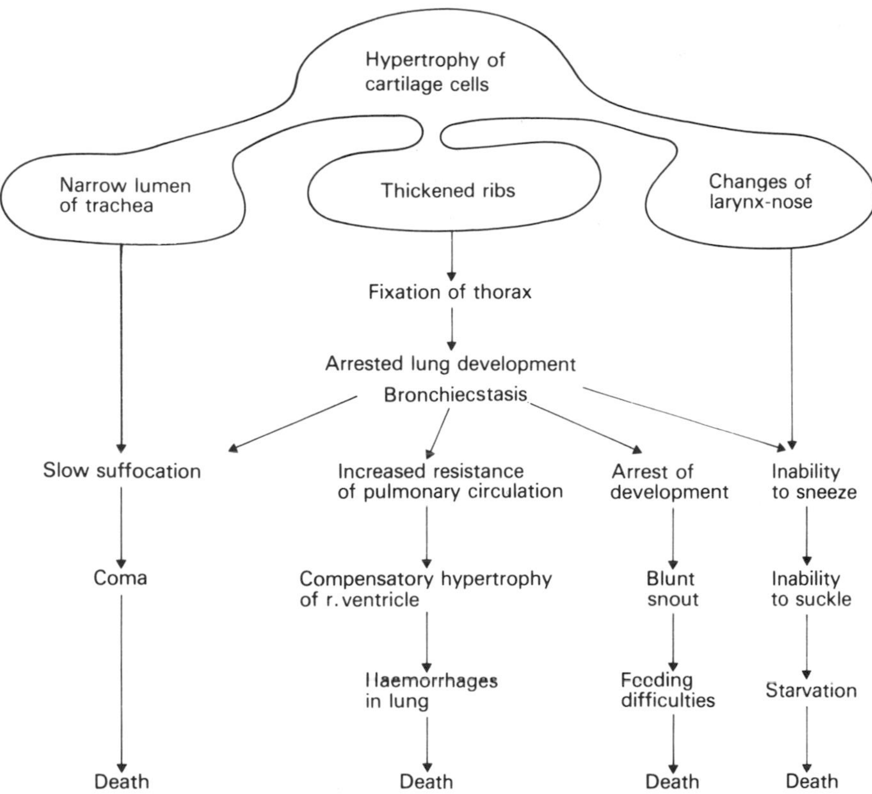


**Fig. 7.1** Consequences of a lethal mutation in the rat, simplified from Grüneberg (1938). The genetic lesion causes over proliferation of cartilage cells: such cells from lethal embryos hypertrophy when transplanted into normal rats, while normal cells transplanted into lethals grow normally. The gene is autonomous in its effect on cartilage development. In the lethal embryo it causes overgrowth of (some) cartilage and bone, and all subsequent anomalies flow from this and from the physiological compensations then induced. The circulatory changes also include increased haemoglobin and raised erythrocyte counts as well as hypertrophy of the heart, presumably in response to reduced lung development. There are interactions among these secondary, heteronomous, effects and the direct cause of death is variable. The apparent pleiotropic effects of the mutation are false.

2. Just as nutrition is important in this last example, so also is the passage of maternal gene products into the egg. Using cell-lethal deficiencies, Garcia-Bellido and del Prado (1979) were able to show that the lethality patterns in lethal homozygotes depend on the maternal genotype: homozygous lethal mothers make eggs which die earlier than eggs from heterozygous mothers. Uncertainty about the role of these maternally transmitted products (mRNA or protein), which may *perdure*, or remain available during many cell generations, may invalidate any conclusions about the time when the zygotic gene product is required for proper

development. Estimates of the phenocritical period, the stage when the effect of the mutant is manifest, are often of doubtful value for this reason.

3. The cause of death may also be ambiguous if failure of one developmental event has a sequence of indirect, secondary consequences. Figure 7.1 is an example of this kind of false pleiotropy, where the defect produced by the mutant gene initiates a series of physiological responses in no way directly due to the mutation. Only careful analysis can separate false from true pleiotropic gene action.

These difficulties are often compounded, and since we have no method of identifying mutants of luxury genes as such, the interpretation of the developmental effects of lethals is often uncertain. We shall illustrate these difficulties below.

Lethal mutations may cause death at any time during development but, as Hadorn (1960) generalized, many cause overt death just at the various crises of development (gastrulation, ecdysis, birth, etc.) suggesting that many genes may be under coordinate regulation at these 'critical phases'. This may be a proper generalization since we know that these phases are sometimes associated with hormone pulses, rapid growth and the like; but they are also periods of stress and some death may occur then for that reason. Further, some lethals have multiple phases of activity (p. 181) and, like rudimentary, may cause sterility. We shall look first at lethal factors causing sterility.

## Mutations affecting gametogenesis

Gametogenesis is to a greater or lesser degree isolated from the other processes of development. It follows a well-defined course of distinct, and often synchronized, morphogenetic changes, and it has been intensively studied in *Drosophila* whose sterility mutants are easily isolated. Figure 7.2 shows the mutants which affect each of the steps of the complex process of egg production.

### *Agamety*

The grandchildless (*gs*) mutant of *D. subobscura* causes vacuolation of the egg cytoplasm which, apparently mechanically, prevents migration of nuclei to the polar cytoplasm in time for pole cell formation (Fielding, 1967). Although it results in agametic adults (cf. Fig. 1.3) the lesion is exclusive to the egg, and it is not a polar granule mutant (Mahowald *et al.*, 1979), although these also exist and would similarly cause sterility of both sexes.

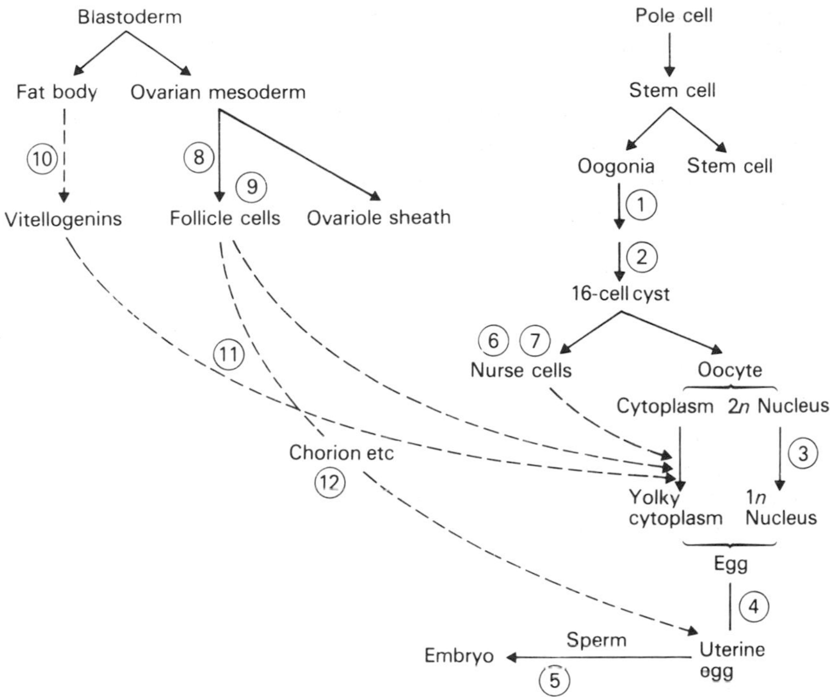


**Fig. 7.2** The development of the *Drosophila* ovum and the mutation classes known to affect this, simplified from King and Mohler (1975). Cell and tissue relations are indicated by solid lines, and their products by broken lines. The hormonal regulation of oogenesis is ignored. The mutation classes are: (1) suppression of cystocyte division, as with *Fs(2)D*; (2) initiation of tumorous divisions as by *fs(2)B* and others mentioned in the text; (3) meiotic mutants like abnormal oocyte (*abo*); (4) ovulation blocking mutations, such as lozenge ($lz^{34l}$); (5) female-sterile mutations (see text); (6) mutations resulting in abnormal ribosomal synthesis (bobbed) or (7) which modify nuclear differentiation, such as singed; (8) mutations affecting ovariolar differentiation; (9) mutations modifying follicle cell migration (tiny) or their differentiation; (10) mutations which block vitellogenin synthesis (*fs(3)A17*), or (11) vitellogenin uptake (*fs(3)A16*); (12) mutations affecting the egg membranes, such as tiny, singed, etc. A full listing of members of these classes is given by King and Mohler (1975) and the actions of some are discussed in the text.

Grandchildless is an instance of false pleiotropy causing a curious, tell-tale pattern of inheritance (see also Thiery-Mieg, 1982).

### *Female sterile (fs) mutants*

In the gonad, pole cells become stem cells which bud off oogonia (or spermatogonia) in succession. The oogonia divide four times to make a remarkable 16-cell cyst, where the most posterior member becomes the oocyte proper and the remaining 15 cells differentiate as nurse cells whose nuclei become polytene (see King, 1970). The division pattern is completely regular; each subsequent division being at right angles to its predecessor (Fig. 7.3). The cells also remain attached by ring canals at their points of separation (incomplete cytokinesis), and these canals are subsequently used for feeding the cytoplasmic products of the nurse cells into the oocyte. Only two cells will have four ring canals at the last division, and one of them becomes the ovum.

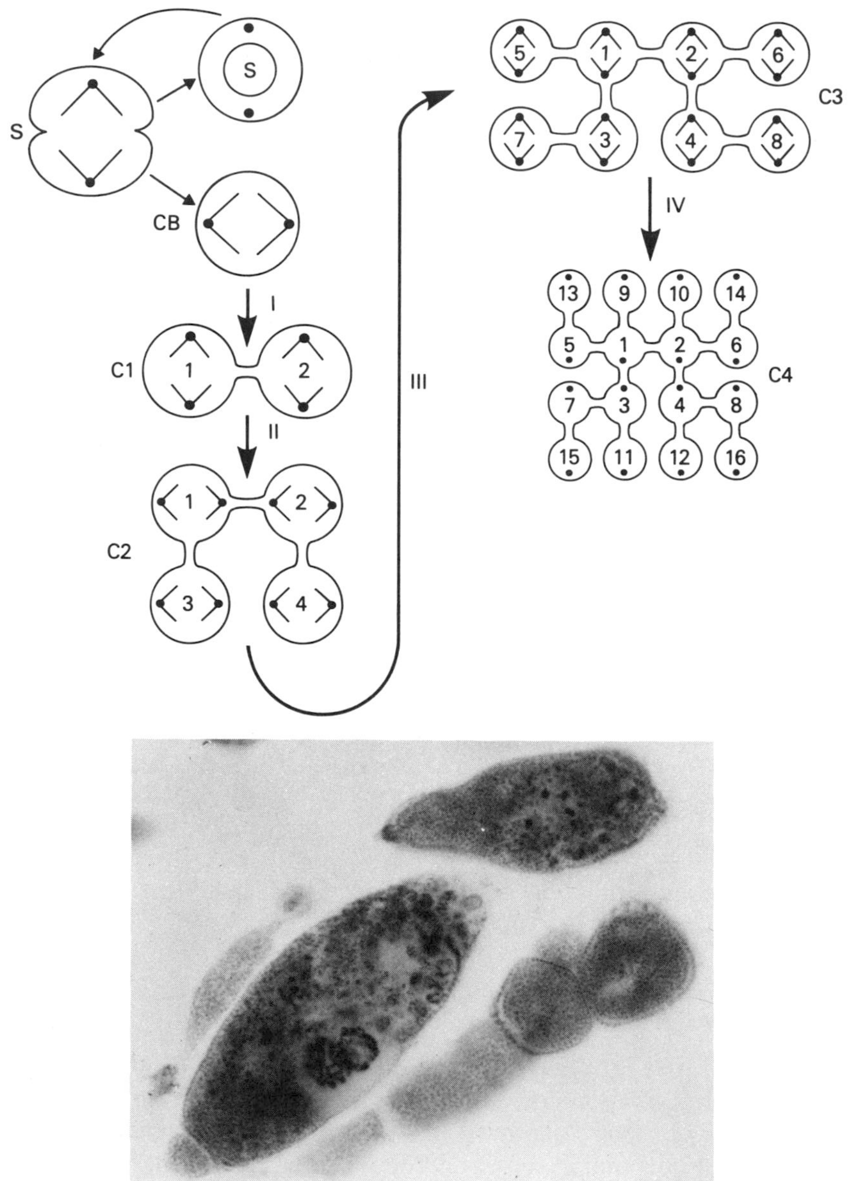

These regular divisions are blocked by the dominant mutation *Fs(2)D*, which is inactive in males, allowing the stock to be kept. *Fs(2)D* egg chambers have only two cells (cystocytes) on average, and no oocytes are formed. Contrariwise, a number of mutants (the wing mutants fused (*fu*) and narrow (*nw*), and *fs(2)B, fs231*, and others) permit divisions to continue so that the ovarial cyst becomes filled with an oogonial tumour of thousands of cells. On rare occasions some of these cells may have four ring canal connections, and they then differentiate as oocytes which never form eggs. Johnson and King (1972) therefore postulate that the four ring canal configuration is the signal for oocyte differentiation which, in turn, signals the cessation of cystocyte division. Since these tumorous cystocytes undergo complete cleavage (no ring canals) about 40 percent of the time, the mutants must affect ring canal formation specifically, by an as yet unknown mechanism. *fs231* has no effect in males or in the follicular epithelium which surrounds the egg cyst (King *et al.*, 1978).

About a dozen other genes disturb nurse cell differentiation and function in a variety of ways, including the mutant bobbed (*bb*), which has a reduced number of the repeated (130 times) cistrons for rRNA. However, *bb/bb* females lay eggs containing normal amounts of rRNA, but they lay them slowly. Mohan (1976) suggests that there is a regulatory feedback system which ensures that an egg is not laid without an adequate complement of rRNA. There are other mutants (e.g. abnormal oocyte (*abo*)), believed to affect rRNA synthesis (Mange and Sandler, 1973) which do not show this physiological regulation. And still other mutants (singed (*sn*), raspberry (*ra*) etc.) alter the normal course of nurse cell chromosome polytenization.

These female-sterile mutations well illustrate the point that a single locus may be active in a variety of tissues (*fu, nw, bb, sn, ra*), and that the obvious consequence of their mutation, after which they were named, gives no indication of their molecular role. Mutants affecting spermatogenesis may similarly cause anomalies of eye, wing or abdomen. Since we frequently know only one mutant of a locus its effect may be misleading: $sn^2$ only singes bristles, while $sn^1$ also causes female sterility. The 'strength' of the allele

---

**Fig. 7.3** Diagram of the development of the 16 interconnected cystocytes which form the oocyte and nurse cells of a single *Drosophila* egg chamber, after Koch *et al.*, (1967). The stem cell (S) makes two daughters, one of which repeats the stem cell cycle while the other becomes a cystoblast (CB). After four divisions (I–IV), a cluster of 16 interconnected cystocytes is formed. Each division is at right angles to its predecessor (indicated by the position of the centrioles and astra rays prior to the next division). Cytokinesis is incomplete, so the cystocytes remain connected by ring canals. When the divisions are terminated (C4) the cell cluster is symmetrical about the two cystocytes bearing four ring canals (1 and 2). One of these becomes the oocyte and the remaining 15 cells become nurse cells. The nurse cells become polytene and their products are fed, through the ring canals, into the growing oocyte. The mutation *fs(2)B* affects this remarkable and regular cystocyte differentiation to give tumorous egg chambers. Even in these some nuclei may become polytene, as seen in the photograph.

and the thresholds of the responding systems both affect the pleiotropic expression of the mutant locus. But even allowing for this, it seems likely that some *fs* mutants (*fs(2)B, fs(2)D* and *fs231*) specifically block oocyte differentiation in an interesting way, and what is surprising is that others (*fu* and *nw*) do the same. The mutant combinations which might tell us if these genes act sequentially (i.e. are phenotypically equivalent to the stronger mutant) or additively, do not seem to have been made, except for the combination *fu/fu*; *fs(2)B/+* which demonstrates additivity or perhaps synergism. King (1970) therefore deduces that the mutants affect the two separate polypeptides of a heterodimer, but other models are possible. Homozygous fused embryos from *fu/+* mothers, incidentally, develop normally while those from *fu/fu* mothers die, illustrating the importance of maternal gene product transmission (see below).

### Follicle cell mutants

Egg cysts are intimately surrounded by follicle cells which are of mesodermal origin, as well as by the ovariole sheath. Mutations active in this mesodermal and extraovarial tissue can be distinguished from those of the germline itself by transplating pole cells, or larval gonads, from the mutant into wild-type hosts. The mutants $sn^1$, *fu* and *fs(2)B* etc. then behave autonomously showing that they directly affect oogonial differentiation even in a wild-type somatic environment. The converse is true with *lz* and apterous (*ap*) which are 'cured' of their defects. So also are the more obvious examples of the malformed eggs of *fs(1)K10* which are normally shaped when surrounded by wild-type follicle cells (Wieschaus *et al.*, 1978), and *HW49c* eggs which are laid instead of remaining blocked in the oviduct (Holzworth *et al.*, 1974). Transplantation between genotypes enables location of the action of the mutant in the tissues of different origin which make up the gonad.

Follicle cells must also regulate the flow of materials from the haemolymph into the egg, including the vitellogenins formed by the adult fat body. The *ap* mutant fails to form vitellogenic oocytes and is female sterile for that reason. Juvenile hormone is required for vitellogenin uptake by oocytes, which suggested to Postlethwait and Weiser (1973) that the apparent uptake defect in the follicle cells might result from a hormone deficiency, and this proved to be correct. Vitellogenesis in $ap^4$ could be restored by applying a juvenile hormone mimic to adult females. Bioassay confirmed that $ap^4$ made less than the normal amount of juvenile hormone (Postlethwait and Handler, 1978). The defect is not in follicle cell function, although this might have been our deduction from the transplantation experiment of the preceding paragraph. Interestingly, too, there is no evidence that the wing and haltere defects which give the mutant its name, or the gut abnormalities and precocious adult death which are also part of the mutation syndrome, are directly related to the hormone deficiency (Wilson, 1980).

We have already considered the synthetic functions of follicle cells (p. 152) and their interesting gene activation programme. Here we need note only that there is one mutant which causes defects in both the vitelline membrane and in the chorion, and this is also female sterile (*fs(2)A9*) (Margaritas (1974) quoted in King and Mohler, 1975).

### *Meiotic mutants*

Since the discovery of the first meiotic mutant, *C(3)G* in *Drosophila*, by Gowen and Gowen (1922), the genetic control of meiosis has attracted ever-increasing attention (reviewed by Baker *et al.* 1976). We shall not be concerned with these many fascinating studies using most experimental organisms from *Neurospora* and yeasts to man, and where mutants are usually identified by their effects on fertility. Instead, we shall take a single example from spermiogenesis in *Drosophila*, which shows that abnormal meiosis does not necessarily greatly disturb differentiation.

The gonial cells of *D. melanogaster* undergo four divisions to form primary spermatocytes which then go through the first and second meiotic divisions to give 64 spermatids which differentiate into sperm. The Y chromosome, which is apparently inactive in all tissues other than the testis, forms lampbrush loops (p. 98) during the primary spermatocyte stage, and absence of the Y blocks spermatogenesis at the second meiotic division in *D. hydei*. On the other hand, XO males of *D. melanogaster* proceed to form spermatids. In both cases, the Y chromosome products are necessary for meiosis since even in *D. melanogaster* the second meiotic division shows a chaotic assortment of chromatin bodies in its absence (Lifschytz and Hareven, 1977). The two species are therefore differently organized with respect to cytodifferentiation, which proceeds in *D. melanogaster* despite the failure of meiosis. Similarly, the male-sterile mutation *ms(1)516*, which causes the two centrioles to go to one astral pole, allows differentiation to spermatids, and *ms(1)413* which causes premature mitochondrial aggregation (*nebenkern* formation) at the primary spermatocyte stage, blocking meiosis thereby, allows the differentiation of tetraploid spermatids, without the usual cell division and haploidization (Lifschytz and Hareven, *loc. cit.*).

These examples show that, even in the differentiation of a single cell type, the developmental programme is not necessarily tightly linked to the sequence of events which we might consider to be the essential steps in the whole process, as it may be in *D. hydei*. There are in a sense parallel programmes of differentiation, and a block in one need not impede the other.

### *Male rescue*

The cytoplasm of the egg usually, but not invariably, stores sufficient material to allow development up to gastrulation, after which translation of male genome products can be identified (e.g. by the appearance of a male-borne isozyme). It follows that many early events in embryogenesis are maternally determined since sperm do not carry this information store. Formally, then, such maternal effects are expressed through the homozygous mutant mother but not through the homozygous mutant father. However, if a wild-type allele transmitted by the father is activated in time to make up for the deficit in the egg from a homozygous mutant mother, heterozygous offspring will be viable (male rescue). If it is not activated in time, both heterozygous and homozygous offspring from such mothers will die: their fate depends on the mother's genotype alone.

The autonomous eye colour mutant deep orange (*dor*) of *Drosophila* is a type example of male rescue. Homozygous daughters of heterozygous mothers survive, while homozygous daughters from homozygous mothers die at gastrulation, thus demonstrating the maternal effect. On the other hand, about half the heterozygous daughters of homozygous mothers survive to the adult, demonstrating male rescue. Rescue can also be effected by injecting cytoplasm from unfertilized, normal eggs into syncytial stage *dor/dor* embryos, showing that the defect is truly cytoplasmic (Garen and Gehring, 1972). Just over a dozen *Drosophila* mutants are subject to male rescue (Gans *et al.*, 1975), as is the lemon (*lem*) mutant of *Bombyx mori*. Among vertebrates, the *f* mutant of the axolotl (*Ambystoma mexicanum*) can be rescued by normal sperm, which corrects the overproduction of blastocoel fluid found in homozygotes, but does not respond to cytoplasm or nuclear sap injections. Conversely, the autonomous *O* mutant of the axolotl is rescued by injection but not by wild-type sperm. In this last case, it seems that a protein found in the germinal vesicle may be involved and that this is concerned with the activation of RNA synthesis (Briggs (1973) discusses amphibian mutants). Male rescue and cytoplasmic rescue provide systems which may allow the bioassay of particular gene products defective in maternal-effect lethals. The rudimentary gene is the only one for which this possibility has been fully exploited.

### *Non-rescuable mutants*

A great variety of maternal-effect mutants have been isolated in *Drosophila*, some of which block fertilization or early cleavage, or make the blastoderm haploid or otherwise malformed (Zalokar *et al.*, 1975). Two are of particular interest for the light they throw on the functions of the egg cytoplasm, as

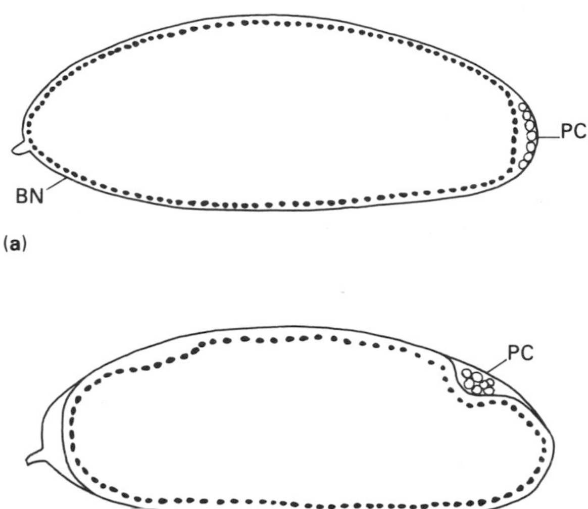


**Fig. 7.4** In the 3 hour *mat(3)1* embryo (**a**) more or less normal pole cells (PC) form, but no cell walls grow around the syncytial blastoderm nuclei (BN), as would normally be the case at this age. At about four hours (**b**) these pole cells migrate dorsally and there is an invagination of the acellular blastoderm. In the wild-type, the posterior mid-gut rudiment forms a pocket of cells which migrates dorsally carrying the pole cells with it. This result of Rice and Garen (1975) shows that the dorsal migration of the pole cells is independent of cellularization and is a property of the cytoplasm.

such, on the normal morphogenetic movements during early embryogenesis, and on determination.

The *mat(3)1* mutation blocks cellularization of the blastoderm, but normal pole cells are made (Rice and Garen, 1975). These pole cells are normal by the transplantation test (Regenass and Bernhard, 1978), and in the mutant they migrate dorsally in the normal fashion (Fig. 7.4). Since there is no cellular blastoderm there can be no germ band extension, although this pushing backwards and upwards by multiplying ventral cells has long been taken to be the mechanism which transports pole cells to the dorsal region (see Fullilove and Jacobson, 1978). The *mat(3)1* mutant therefore proves that this movement results from properties of the activated egg cytoplasm itself. The *mat(3)6* mutant confirms this conclusion. In this case, blastoderm cells form only at the anterior and posterior ends of the egg, leaving an acellular gap in the middle. Yet germ band extension still occurs as judged by movement of the posterior mid-gut rudiment which does differentiate (Rickoll and Counce, 1981). Further, culture *in vivo* (p. 221) of the anterior and posterior cells shows that they are determined to form anterior structures (such as eye, and antenna) or posterior structures (genital disc) in the usual way. Complete cellularization is not necessary for the establishment of region specific cell determination, and this must lie in some property of the cytoplasm (see below). The temperature-sensitive *mat(3)3* mutation can be used to make localized blastoderm defects by exposing mothers to the restrictive temperature. Hatching adults then have missing cuticular structures, such as wing, leg or an abdominal segment (Rice *et al.*, 1979), confirming the extreme mosaicism of these eggs. Interestingly, *Drosophila* embryos ligated 40–50 min after laying are also capable of forming anterior and posterior structures although intermediate segments are missing (Schubiger *et al.*, 1977; Vogel, 1977).

We might expect to find fewer examples of maternal effects in the mouse since the egg contains less and is regulative, and because there is evidence that the embryo's own genes are active by the 2 or 4-cell stage (Wudl and Chapman, 1976). There are two proven cases. The *T/t* allele, ($T^{Hp}$, Hairpin), is an embryonic lethal only when transmitted from the mother, and it is not rescued by wild-type sperm (Johnson, 1974). The ovum mutant (*om*) of the DDK strain shows a curious reciprocal of the male rescue situation: when the female is outcrossed to other strains the embryonic trophectoderm is abnormal and the blastocysts die, which does not happen when she is mated to DDK males. The data suggest that the *om* egg cytoplasm contains some substance incompatible with another substance in non-mutant sperm, and this combination is lethal to the embryo (Wakusagi, 1974).

### *Determination mutants*

The *mat(3)* mutants suggest that the cytoplasm of the *Drosophila* egg carries information which causes blastoderm cells (or more likely *some* blastoderm cells) to become determined as soon as they are formed, and cell transplants confirm this (p. 7). Indeed, there is evidence, again from transplants, that nuclei from the acellular blastoderm are determined when they reach the egg cortex, but perhaps not as stably as when they are subsequently enclosed by the cell membranes which grow around them (Kauffman, 1980). This agrees with the observation that transcriptional activation of informational sequences (poly(A) RNAs) occurs shortly after nuclei reach the cortex, and increases in the formed cells (Lamb and Laird, 1976; Zalokar, 1976). There must therefore be mutants which alter these possible determinative processes, and they will be maternally inherited.

The mutant bicaudal (*bic*) on the second chromosome of *Drosophila* is just such a maternal-effect mutant (Bull, 1966; Nüsslein-Volhard, 1977). When fully penetrant, it causes a longitudinal, mirror image duplication of the posterior abdominal segments, with consequent elimination of head and thorax (Fig. 7.5). The hemizygote constructed using a *vg* deficiency ($bic/Dfvg^B$) gives almost 100 percent of flies showing the phenotype. *bic/bic* females produce rare bicaudals, as do $+/Dfvg^B$ females, and *bic*/+ gives wild-type. The mutation is a hypomorph. Penetrance is notably age dependent, declining from about 90 percent of differentiated eggs on the

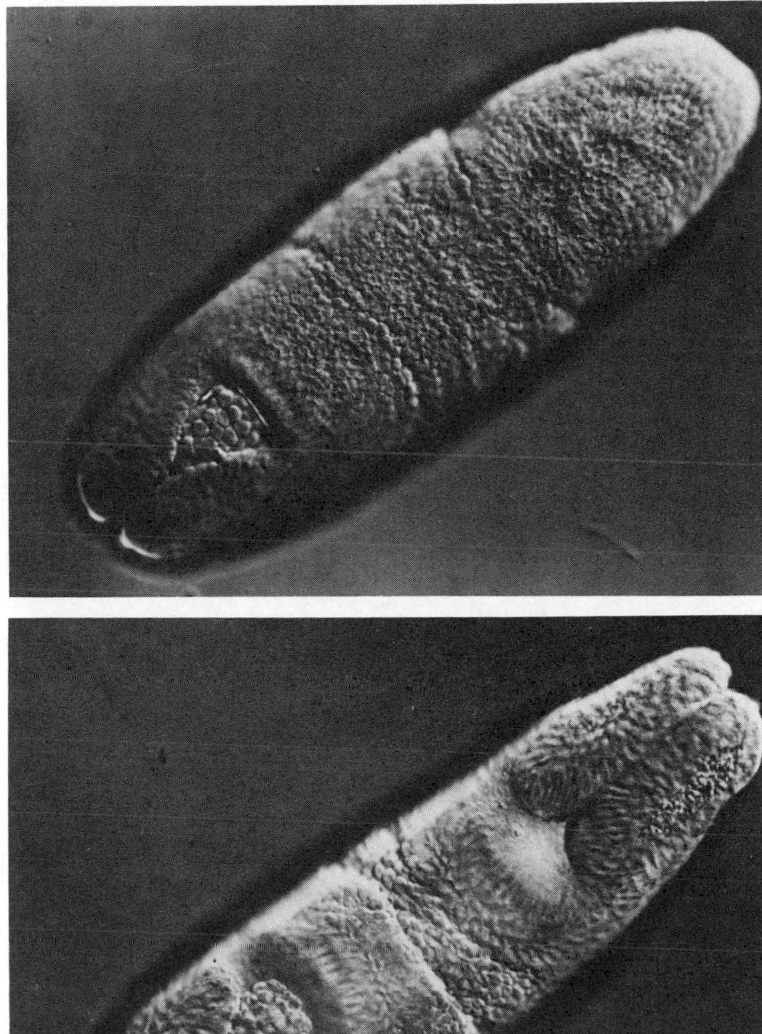


**Fig. 7.5** Normal (**a**) and bicaudal (**b**) embryos at about 4½ h old, as seen using Nomarski optics. In this dorsal view, with the head to the right, in (**a**) the cephalic furrow is easily seen running across the embryo about a third of the way from the anterior end. The cluster of relatively large pole cells has been displaced from its original polar position by the extension of the germ band and now lies within the posterior mid-gut invagination. In (**b**) we see the mirror image mid-gut invagination at the anterior, but without pole cells which form only at the posterior. Photographs kindly supplied by Dr. C. Nüsslein-Volhard.

first day to 10 percent or less by the third day after emergence, and subsequently. High temperatures (29 °C) raise penetrance, particularly in eggs entering vitellogenesis, but have no effect on the developing eggs. The *bic*$^+$ product is therefore produced during ovogenesis. Its absence is without effect on the morphology, viability or fecundity of the females. As we should expect from such a sensitive system, penetrance and expressivity are variable, ranging from reduced embryo size, through head defects and headlessness, to asymmetrical bicaudal through to the full phenotype illustrated. It has no influence on pole cell formation, or on the egg membranes; but absence of the embryonic/larval thorax results in loss of the thoracic imaginal discs. The *bic*$^+$ gene product determines the longitudinal organization of the embryo in some way.

Lohs-Schardin (1982) has described a semi-dominant third chromosome mutant, dicephalic (*dic*), which has the opposite effect to *bic*, and an entirely different mode of origin. In this case, the first evidence of mutant expression is the clustering of nurse cells at opposite poles of the oocyte, most commonly in groups of six and nine. Micropyles are usually formed at both poles of the egg instead of only at the anterior, and respiratory horns may develop, too, although most usually at only one pole. Eggs generally fail to develop, but those that do may show the cuticular patterns of two anterior parts of opposite polarity. Since penetrance and expressivity of this mutant are very variable, a wide range of abnormalities is associated with it, but they can all be related to the labile organization of the cystocyte stage of ovogenesis. We may speculate that a dicephalic mutant, acting through the establishment of stable spatial coordinates, or gradients, in the egg may still be found.

A third maternal-effect mutant, dorsal (*dl*), alters determination in the dorso–ventral axis in a similar fashion (Nüsslein-Volhard, 1979). Embryos from homozygous mothers make no ventral structures and they are but yolk-filled tubes of dorsal hypoderm (Fig. 7.6). The cuticle of the tube bears dorsal hypoderm hair patterns, and there are no internal organs since these are of ventral origin. Unlike the *mat(3)* mutants described above, the blastoderm is fully cellular and determination of the ventral cells must therefore be abnormal, but shortage of distinctive cuticular 'marker' structures makes it uncertain whether we again have a mirror image duplication, as with bicaudal. Since *dl/Df(2)137* hemizygotes have the same fully penetrant phenotype as *dl/dl*, the mutant is probably an amorph and, as this would imply, it is not temperature-sensitive. However, eggs from heterozygous females show a dominant phenotype at high temperatures, which is also modified by the genetic background. Presumably this reflects the shortage of the *dl*$^+$ gene product in what are effectively hemizygotes for the locus. The embryos are then defective in muscle, which arises from mesoderm which is the most ventral region of the embryo (Nüsslein-Volhard *et al.*, 1980). Some mesoderm is formed in less extreme phenotypes, giving the impression that

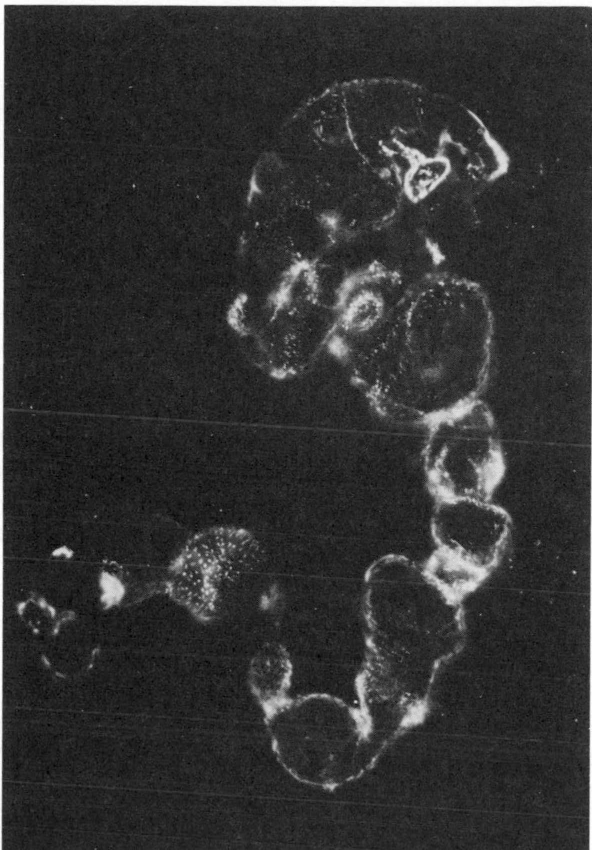

**Fig. 7.6** An embryo carrying the mutation dorsal released from the egg, showing that the embryo is a single cell-layer tube with many infoldings. There is no evidence of the ventral setal bands (Fig. 7.7) and sections show that cells do form in the ventral and lateral regions of these embryos. Photograph by Dr. C. Nüsslein-Volhard.

determination depends on a gradient system in which the amount of $dl^+$ product available to a cell settles its phenotype.

Two additional loci, gas and easter, also cause dorsalization of embryos when mutant, and dorsal and easter can be rescued by injection of wild-type egg cytoplasm. Even more interestingly, dorsal cytoplasm rescues easter, and vice versa, and gas cytoplasm rescues both. These substances may therefore be required for the normal ventralization of embryos. A dominant mutation, Toll, ventralizes embryos, and cytoplasm from Toll eggs rescues all three dorsal mutants, presumably because the necessary gene products accumulate there. Extracts of Toll cytoplasm are also effective, so there is a bioassay system which, for the first time, may permit identification of molecules involved in the events of dorso–ventral determination (Anderson and Nüsslein-Volhard, 1982). At present it is useless to speculate about these mechanisms.

Nüsslein-Volhard (1979) interprets both *bic* and *dl* as *coordinate mutants* (Sander, 1976) whose normal gene products establish stable spatial coordinates, or gradients, in the egg as the first process in cell determination. These coordinates are then interpreted at a second step by zygotic genes, as we shall see below. However, the maternally inherited coordinate system is possibly more complex than this. Just as the scarlet eye colour phenotype results from a mutation in the ommochrome, not the pteridine, eye colour pathway (Fig. 5.6), so must *bic* represent a mutation of the head determination process, and *dl* a mutation in ventral determination. We can therefore anticipate that the coordinate system must involve head–abdomen–ventral–dorsal elements.

A phenocopy, which anticipated *bic*, can be made by UV irradiation of the pre-blastoderm embryo of the chironomid midge, *Smittia* (Yajima, 1964; Ripley and Kalthoff, 1981). This 'double abdomen' embryo depends on the UV inactivation of cytoplasmic RNP particles localized near the anterior pole (Kalthoff, 1979). The UV effect is completely photoreversible by blue light; RNase treatment also makes 'double abdomens' when applied to the anterior pole, but has no effect when used elsewhere. These results therefore suggest that an anterior-determining factor (masked mRNA?) is laid down in the anterior of the egg. If these eggs are centrifuged anterior→posterior, a proportion of embryos develop as 'double cephalons' and the abdomen is not formed, showing a rather direct action of the RNP particles. Unfortunately these procedures do not work with *Drosophila*, and we can only note the parallel.

Particular proteins, 'indicator proteins' unique to abdomen or cephalon, can be identified using two-dimensional gel electrophoresis of the material from anterior and posterior fragments of the *Smittia* egg. Anterior fragments from late blastulae make a protein not found in the posterior, and its synthesis is renewed when UV-treated anterior fragments are photoreversed with light. Double abdomen embryos do not make the protein; instead, both their anterior and posterior fragments make another, large, protein during early blastoderm. This protein is missing from double cephalon embryos. Although different and complex patterns of protein synthesis follow in both cephalon and abdomen development, it is not known if these indicator proteins then perform any regulatory function (Jäckle and Kalthoff, 1981). Nevertheless, both phenocopies provide evidence for maternally inherited information, RNA or protein, associated with, or possibly directly involved in, the determination of the longitudinal structures of the embryo; the *Drosophila* mutants provide evidence of the same sort. There is, of course, the problem of how mirror image double abdomen or double cephalon embryos become organized, but we shall look later at mirror imaging in situations where we have less speculative information (p. 224).

Finally, when mutated to the null state, the second chromosome locus extra sex combs (*esc*) has a maternal effect not on segmental organization, but on the qualitative nature of the segments themselves (Struhl, 1981a). Homozygous embryos from homozygous mothers develop into segmented larvae, but these are serial repeats of the eighth abdominal segment. Homozygotes from heterozygous mothers develop into normal adults, and there is a degree of male rescue which is greater if the zygote carries two copies of the wild-type gene. This suggests that the gene product is required early in development, and that the amount stored in the egg or made by the zygote has to reach a critical level for normal determination at that time. Where it does not, differentiation is 'patchy' and cannot be corrected during subsequent development. *esc* demonstrates another class of determination mutants, of novel importance for differentiation of virtually the entire array of developed structures. We shall have to return to this aspect of its action (p. 258).

## Embryonic lethality

We expect mutant housekeeping genes to be lethal, but how that lethality is expressed will depend on the reserves stored in the egg, the sensitivity of the cell type, and so on, as we noted at the beginning of this chapter. Therefore, not all apparent developmental lethals are important to us, although some mutant housekeeping genes may provide tools which we can use to manipulate development (p. 213). The significant mutations are those that alter the normal pattern of development, and we shall consider non-lethal members of this class later.

### *Interpretation mutants*

The maternal-effect mutations just described involve genes that establish a molecular framework in the egg, or so it seems. Zygotic genes then interpret their position in that structural plan, respond to it, and become determined in an orderly fashion to give the regular structures of the embryo. We therefore expect, and find, lethal mutations that fail to carry out this interpretation process properly. To be more exact, we find mutations that fail to make particular structures and they express their inadequacy in a surprisingly regular way. If we have correctly understood these mutations they provide the strongest evidence that development is an epigenetic phenomenon even in this most mosaic of eggs.

The *Drosophila* embryo is segmentally organized and this segmentation is first visible as transverse infolding of the germ band about 1 h after gastrulation. Segmentation is seen in the developed first instar larva as denticle bands at the anterior of the segment in which the denticles point posteriorly. The posterior of the segment is naked, and its border coincides with the anterior of the next denticle band. Segment polarity in the abdomen is also shown by the shape of the band, as well as by denticle orientation, and these denticles are heavier and more pigmented than those in the fine thoracic bands. There is sufficient diversity between bands to allow segment identification, and for the examination of mutant effects (Fig. 7.7). Fifteen lethal loci, scattered through the genome, have been found to affect the larval segmental pattern (Nüsslein-Volhard and Wieschaus, 1980). None affects the overall polarity of the egg, and they can be grouped in three classes that have features in common.

1. *Gap mutants* are the most extreme class, where up to eight larval segments may be entirely deleted. Three mutants (Krüppel (*Kr*), knirps (*kni*) and hunchback (*hb*)) have this effect; *Kr* eliminates thorax and anterior abdomen, *kni* removes all except the first and last abdominal segments, and *hb* lacks meso- and metathoracic segments (Fig. 7.8). Although variable in expressivity, the mutations show a remarkable specificity of action. Only the development of *Kr* has been studied in any detail (Gloor, 1954), and this shows not just a failure to form the median segments but also disintegration of the ventral ganglia, lack of central tracheal elements, distortion of gut and no Malpighian tubules. However, anterior and posterior morphogenesis and histogenesis are normal, making proper mouth structures and salivary glands, and (distorted) posterior ganglia and stigmata (cf. *mat(3)1*). *kr*/+ genotypes survive, but the adults may have missing or defective thoracic structures. Clearly the amount of $kr^{+}$ gene product is important, and the thoracic segments have the highest relative requirement for it.
2. *Pair rule mutants* cause deletions in alternating segments, and each of six such mutants has its specific pattern of deletions (paired (*prd*), even-skipped (*eve*), odd-skipped (*odd*), barrel (*brr*), or hairy (*h*), runt (*run*) and engrailed (*en*)). Typical patterns are shown in Fig. 7.8. Again the embryogenesis of only one (*prd*) has been studied, and this shows that the repeated defect is not due to cell death. Instead, *prd* embryos form only half the normal number of transverse infoldings of the germ band, and these are spaced twice the normal distance apart. The double origin of these segments is clearly shown by the appearance of two paired tracheal pits instead of the usual single pair in each enlarged segment. Denticle morphogenesis is normal except that the belts are twice the usual size (Sander *et al.*, 1980). This mutation affects the initial establishment of segmental borders without disturbing the identity of the individual segments (see also Nüsslein-Volhard and Wieschaus, 1980). While we cannot be sure that the other pair rule mutants similarly alter segment borders, it is difficult to avoid the conclusion that the repeating units of the embryo are defined in terms of double segments at some early stage.

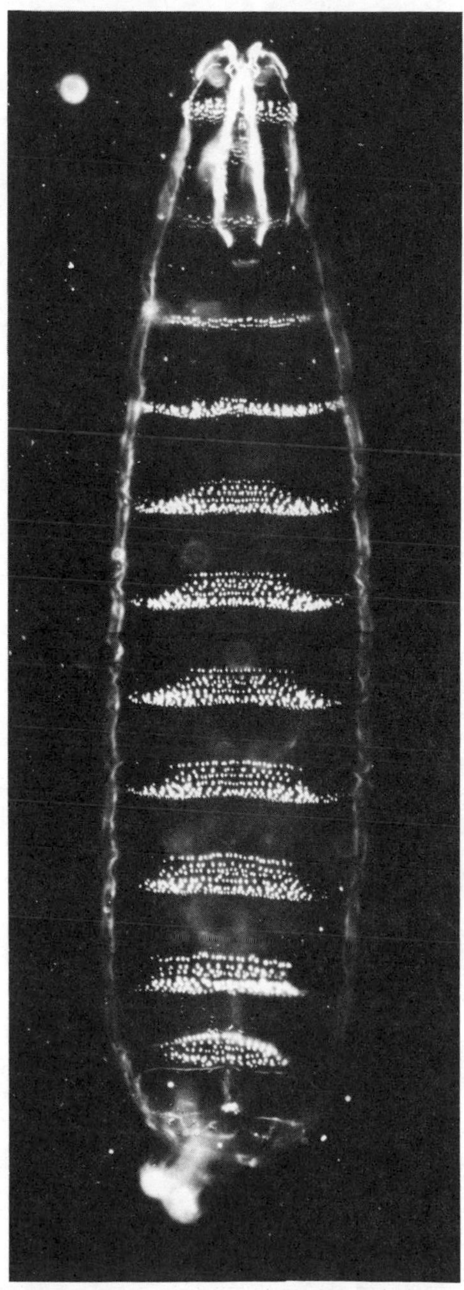

**Fig. 7.7** The ventral setal patterns of the 1st instar *Drosophila* larva. Each segment is defined anteriorly by a band of denticle hook rows, which are narrow and fine in the first three thoracic segments, and broader in the eight abdominal segments. The mouth hooks are seen centrally in the head (top) and the posterior spiracles protrude at the posterior end of the larvae. Dark field photograph of cleared larva kindly provided by Dr. C. Nüsslein-Volhard.

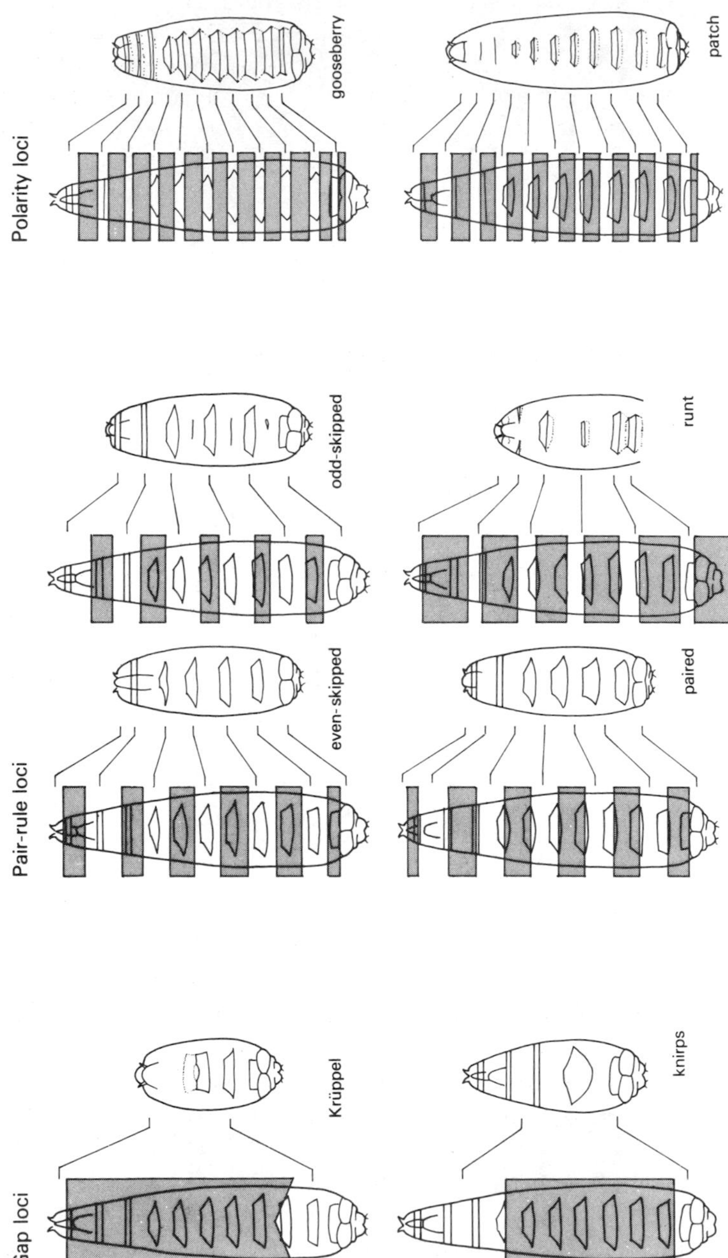


**Fig. 7.8** Mutants affecting segment number and segment polarity, after Nüsslein-Volhard and Wieschaus (1980). The shaded regions of the normal embryos are those missing in the mutants. As we should expect from the metameric segmentation of the embryo, these missing structures are segmentally repeated, except for the gap loci which are a different class (see text).

3. *Segment polarity mutants* are an equally large class: (cubitus interruptus$^D$ ($ci^D$), wingless (*wg*), gooseberry (*gsb*), hedgehog (*hh*), fused (*fu*), and patch (*pat*)). They do not alter the number of segments, but they delete a specific part of the normal pattern, and the remainder is duplicated posteriorly and in reversed polarity (Fig. 7.8). Rather surprisingly, they are all zygotic lethals, and the mutation is expressed only in homozygotes. Except as noted below, the naked posterior section of the segment is deleted and replaced by the anterior in mirror image, to give a segment almost entirely covered in denticles. In *wg* and *hh* the segment boundaries are apparently lost, and in *pat* the duplicated region includes some naked cuticle anterior to each denticle belt, so this mutant has twice the usual number of denticle bands. These genes must be involved in specifying the basic pattern of each segment, but unfortunately we have no information about their embryogenesis.

With the possible exception of the gap mutants, this interesting collection is involved in segment determination and in interpretation of the information laid down in the egg by genes like $bic^+$. That there should be so many zygotic genes involved in this primary process, and that they are as specific as they are, is perhaps surprising. This complexity makes it difficult to interpret how these genes might act (but see Nüsslein-Volhard and Wieschaus (1980) for a possible model) and we must await further studies before pursuing this point.

As with the maternal-effect mutants, there are zygotic genes which are independently concerned with the qualitative differentiation of segments. The gene Polycomb (*Pc*) acts like *esc* (and may also have a maternal effect (Denell, 1982)), and converts all segments to eighth abdominal. In combination with *prd* the segment array is just as with *prd* alone, again demonstrating the separation of the two kinds of determination. We shall look at other aspects of *Pc* later (p. 258).

### Metabolic mutants

The Notch (*N*) locus in *Drosophila* has long been studied (Poulson, 1940) for its significance in the regulation of tissue differentiation in the embryo, and even as a model integrator gene set (Britten and Davidson, 1969) since the great array of mutants isolated show many complex lethal and visible effects (see Wright, 1970). In practice, these mutations (Fig. 7.9) fall into three general classes.

1. About 10 *N* alleles behave like deficiencies of the region, and are non-complementing amorphs. They cause hypertrophy of embryonic nerve differentiation at the expense of hypoderm in homozygotes and in hemizygous males (Poulson, 1945). The heterozygotes have notches

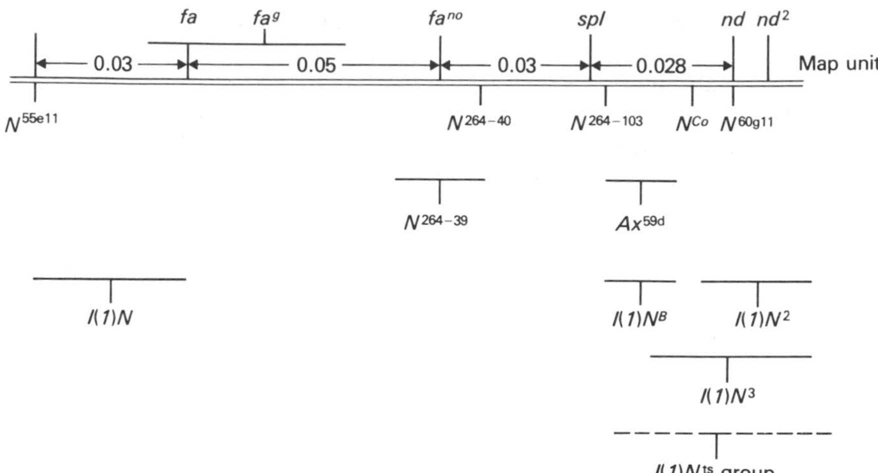


**Fig. 7.9** The genetic map of the Notch locus, after Welshons (1971) and Shellenbarger and Mohler (1978b). The recessive visibles are shown above the line, and recessive lethals below it. The mutants which show no wing notching in heterozygotes are symbolized *l(1)*, and those with notching as *N*, or with interrupted (Abruptex) wing veins as *Ax*. There are four mutants in the ts group, not yet critically positioned. *fa* is irregular eye facets; $fa^g$ is facet-glossy; $fa^{no}$ is facet-notched (apical wing nicking); *spl* is split which has small eyes and split bristles, and *nd* is notchoid which has notched wings and thick wing veins.

at the tips of their wings, and malformation of some wing veins and thoracic bristles. Deficiencies which abut and span the locus show that the mutable region is localized to chromomere 3C7 and its adjacent interband.

2. *l(1)N* mutants are also recessive lethals, but they kill the larvae not the embryos, and they do not express the dominant notched wing phenotype. They appear to be non-complementing hypomorphs.
3. Recessive visible alleles like facet (*fa*), facet-glossy ($fa^g$) and split (*spl*) affect the eyes, while notchoid (*nd* and $nd^2$) produces wing notching, exaggerated in combination with *N*. They are not lethal. facet-notchoid ($fa^{no}$), on the other hand, is lethal with *N*, but displays wing notching only in homo- and hemizygotes. The Notch mutants show a complex hierarchy of effects: from a very specific embryonic disturbance to more or less trivial upsets of adult differentiation, and their interpretation presents an interesting challenge.

The non-complementing *N* mutants are spread throughout the locus (Fig. 7.9) suggesting that *N* is a single cistron (with large introns?) and arguing for a unitary interpretation of its action. Wright (1970) suggests that "activity of the locus is necessary for a product which causes neurogenesis to be repressed in ectodermal cells", which explains the three-fold excess of neurones in *N*/*N* embryos. Subsequently, "certain groups of ectodermal cells

. . . by virtue of their developmental history achieve a particular dual state of developmental potential from which they are unable to differentiate into non-neural, ectodermal derivatives unless the *Notch* locus is sufficiently active . . .". These later developing cells will have different threshold requirements for the Notch$^+$ product (and some would have none at all), which might explain the differential action of the hypomorphic mutations. One test of this rather complex hypothesis is to see if *N* affects non-ectodermal cells. Culture of cells from *N* embryos confirms the abnormal production of nerve but it also shows that myotube formation is abnormal; and muscle is of mesodermal origin (Cross and Sang, 1978).

This problem has been re-examined using the temperature-sensitive allele $l(1)N^{ts1}$, which lies between *spl* and $N^{60g11}$, and is non-complementing with all *N* alleles at the non-permissive temperature of 29 °C. At 18 °C the mutants are normal, but for rare wing nicking and extra chaetae, which allows homozygotes to be bred (Shellenbarger and Mohler, 1978a). In practice, some of the mutant embryos survive at 29 °C, suggesting that the gene product is still partly active and that temperature is affecting the functional conformation of the relevant protein. The temperature-sensitive period (TSP) for lethality can be defined as starting with the earliest shift down to 18 °C which decreases viability, and ending with the earliest shift up to 29 °C which raises viability. The TSP can be further refined by applying short, high-temperature pulses within the period of temperature sensitivity. Non-lethal effects are scored by monitoring morphological defects similarly. All the temperature effects are prevented by an additional dose of $N^+$, indicating that they are *N* specific.

Pulses and temperature shifts show that the TSP for embryonic lethality is during the first half of embryogenesis (Fig. 7.10), which includes, but is longer than, the stage of neurogenesis. There are three additional TSPs, which would not be evident using lethal *N* embryos, of course. The second TSP is at, or near, the first larval moult, and the third during the final three quarters of the 3rd instar. The fourth lethal phase starts about mid-way through, and finishes about the end of, the pupal stage. The $N^+$ product is therefore required not just for embryogenesis, but also subsequently. Lethality gives only a crude assessment of the times when the gene product is needed, for its absence at other times might have non-lethal effects. So it is especially interesting that Shellenbarger and Mohler (1978b) also present evidence for morphological effects. Twelve-hour heat pulses during the 3rd larval instar cause scarring across the eye (identical to the effect of $shi^{ts}$ noted below) and pulses near the middle of this instar mimic the allele *spl*. These pulses also cause wing notches and other wing margin defects typical of *N*. Fusion of leg joints is also common. Pulses late in the 3rd instar cause bristle defects (extra macrochaetes), and pupal pulses mimic $fa^9$, and add or delete microchaetes. In short, the $N^+$ product is repeatedly required for normal fly differentiation, but at different times in different tissues, and this may explain the array of phenotypes found with different alleles. The

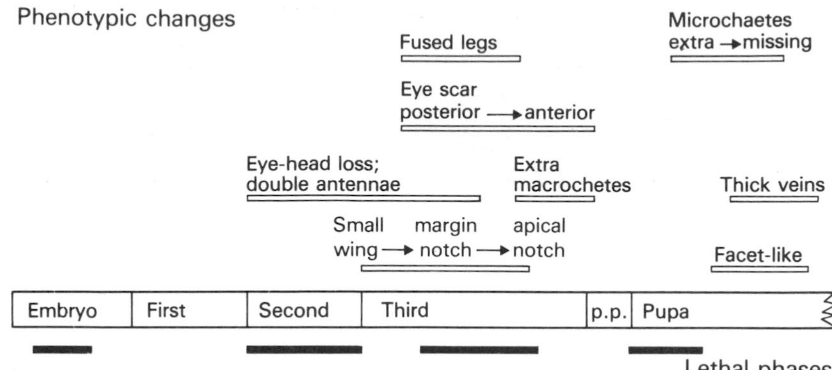


**Fig. 7.10** The temperature-sensitive periods of $l(1)N^{ts1}$, after Shellenbarger and Mohler (1978a). The TSPs for lethality are discrete, and are found in all developmental stages except the 1st instar. The TSPs for adult morphological defects are also found at all stages after the 1st instar, and progressive and multiple effects are induced during the 3rd instar. Missing bristles and wing vein gaps are the only Notch locus phenotypes not found when $l(1)N^{ts1}$ is temperature shocked.

abruptex mutations appear not to be primary cell lethals, which presumably reflects their antimorphic nature (Portin, 1981).

The paralysis mutation shibire (*shi*$^{ts}$) exposes very much the same array of phenotypes as $l(1)N^{ts1}$ when similarly heat shocked (Poodry *et al.*, 1973) and culture of embryonic *shi* cells again shows that more than ectodermal cells are affected by the mutation (Cross and Sang, 1978). This parallel suggests that both may code for housekeeping genes required by many cell types. There is now evidence that this is the case for *N* (Thörig *et al.*, 1981). $N^8$ affects the levels of NADH oxidase, NADH dehydrogenase, succinic dehydrogenase and α-glycerophosphate dehydrogenase, all mitochondrial enzymes, and so also does the putative point mutation $N^{264-40}$. Addition of a duplication of the locus to the genome gives the expected gene dosage-related effects. Since the dehydrogenases are flavoprotein enzymes (and there is evidence that other flavoprotein enzymes are reduced by $N^8$), it is argued that this multiple effect of a single mutation may be due to a defect in this common component. Certainly these results could account for *N* cell lethality, first reported by Demerec (1936), but it leaves the hypertrophy of embryonic nerve unexplained.

The Notch locus is not unique in mutating to a hierarchy of alleles with graded and different phenotypic effects, and the short tail (*T*/*t*) locus on chromosome 17 of the mouse is another classic example. The semi-dominant *T* allele (Brachyury) is homozygous lethal, and the heterozygous mouse has a short tail (Table 7.1). About one hundred recessives (*t*) have been identified (superscript w, from wild populations), and although two thirds of them are viable they all give tail-less mice when combined with *T*: the combination $T/t^x$ ($t^x$ haplotype) is the test for *t* alleles. The three dozen

**Table 7.1** Embryological defects and lethality in some *t* alleles of the house mouse

| Genotype | Stage of death | Overt defects |
|---|---|---|
| $t^{12}$, $t^{w32}$ | Preimplantation, 30-cell stage | Nucleoli defective, reduced RNA synthesis |
| $t^{w73}$ | Survives 8–9 days, blastocyst | Development stops at 2-layer stage |
| $t^{6}$, $t^{0}$, $t^{1}$, $t^{30}$ | Survives 7 days, egg cylinder fails to lengthen | Egg cylinder fails to form embryonic and extraembryonic parts |
| $t^{w5}$ (and other $t^{w}$) | Normal to about 6 days | Embryonic region pycnotic at 6.5 days |
| $t^{9}$, $t^{4}$, $t^{w18}$, $t^{w32}$, $t^{52}$ | Die at 8–9 days, often with axial duplication | Endoderm abnormal, over developed neural folds, mesoderm undifferentiated. |
| *T* | Death at 10.5 days | Abnormalities of neural folds, notochord and somites, reducing posterior of body |

Development is measured from time of coitus.

homozygous lethal *t* alleles can be arranged in six complementation groups by the test of lethality. Heterozygotes between groups live, but are male sterile. The complementation groups can be ranked by the time, or stage, of onset of lethality: $t^{12}$ causes death at the morula–blastula stage, $t^{w73}$ at the immediate post-blastula, $t^{6}$ ($t^{0}$, $t^{30}$) just after the primary endoderm has differentiated, $t^{w5}$ (and others) after the separation of embryonic from extraembryonic ectoderm, $t^{w18}$ (and others) through a disturbance (often doubling) of the neural tube and other defects, and so on up to $t^{w1}$ (and others) which upset only cephalogenesis and may sometimes allow birth. The pattern is typical of a hierarchy of alleles of decreasing effects, which suggests that lethality can be explained by a single cause; the quest for this cause has been pursued for over 50 years (Bennett, 1975).

The simplest hypothesis is that we are again dealing with cell lethality, possibly also due to abnormal mitochondrial function, and that different cells have different response thresholds (Hillman and Hillman, 1975; Wudl and Sherman, 1976). This has been explored by studying the survival of blastocyst cells cultured *in vitro* when we might expect that cells from early acting *t* alleles will die, whereas cells from late acting alleles will survive and differentiate. This is what is found in general: $t^{6}/t^{6}$ blastocysts die, as do cells from $t^{w5}$ homozygotes, but the late lethal $t^{w18}$ blastocysts give the same range of cells as wild-type and they survive well beyond term in culture. Unfortunately culture conditions do not allow all possible cell types to differentiate, and some do not differentiate properly, e.g. $t^{6}/t^{6}$ trophoblast cells fail to become properly polyploid even though they make $\Delta^5$, 3β-hydroxysteroid dehydrogenase, which is a biochemical marker of their normal differentiation (Wudl *et al.*, 1977). The simple lethality hypothesis is difficult to prove using current *in vitro* techniques, although these methods are instructive (see McLaren (1976a) for other examples).

The morula-blastula cells of $t^{12}$ embryos fall apart, suggesting some alteration to their surface which might prevent cell-cell interaction, or recognition, causing death through failure to become properly organized (Bennett, 1975; Sherman and Wudl, 1977). Silver *et al.*, (1979) have taken advantage of the fact that the testis has the same antigenic properties as the embryo to search for specific t-haplotype surface proteins. By detergent solubilizing these proteins from radioactively labelled testis and running two-dimensional electrophoretic gels, they have shown that one protein, p63/6.9 (molecular weight 63,000, and isoelectric pH 6.9) is uniquely different in all members of the six *T/t* complementation groups, compared with wild-type. p63/6.9a, the variant found with the *t*-alleles, is also present on the cells of *t*-lethal embryos. This is the first haplotype-specific protein to be identified, and its locus has been genetically located to the proximal part of the *T* region. Arguing from the high level (2–5 times normal) of galactosyltransferase found in all *t* mutants, Shurr and Bennett (1979) further deduce that p63/6.9b (the wild-type protein) is an inhibitor of this enzyme and wild-type sperm mixed with *t*-sperm do inhibit the transferase activity of the latter by some 80 percent. Finally, this protein analysis shows that $T^{Hp}$ (p. 166) is a null mutant, and probably a deletion, which eliminates that part of the locus responsible for p63/6.9 and thereby increases transferase activity some three-fold. These *t*-allele induced changes in glycoproteins and glycolipids might well account for the altered surface antigens of the *t*-haplotype embryos, but it has still to be proved that they account for the phenotypes.

The *T/t* locus is complex, and the mutations for taillessness and for embryonic mortality, for example, are separate, the latter being distal to the former. It is difficult to decide, therefore, if early acting *t*-mutants are lethal because they change different members of the complex from those modified by late acting, non-lethal, mutants. It could well be that the search for a unitary explanation of *T/t* locus action is misplaced; we may not need to chose between the 'cell lethal' and 'organizational failure' hypotheses outlined above!. What we now require is a more exact understanding of the number and nature of the genes in the complex, since the complex is very interesting from the point of view of genome organization, including as it does genes which distort segregation and suppress recombination ($t^{w5}$) as well as ones which produce three phenotypes as already described. The *T/t* complex illustrates the thorny problems often exposed when we attempt to proceed from apparently related phenotypes to their genetic basis.

### Plant embryonic lethals

Embryonic lethals have long been known in plants, and it has generally been assumed, without any formal proof, that morphogenesis involves the switching on of gene sets, on the animal model. Only recently have seeds or pollen been mutagenized and the segregants from heterozygous parents studied specifically for developmental defects, particularly in the crucifer, *Arabidopsis thaliana* and in *Zea mays*. Corn has the advantage of a great wealth of genetic information behind it, but *Arabidopsis* is easy to culture, has a 5–6 week life cycle, sets 30–60 seeds per fruit and has genetically marked chromosomes. In both cases, recessive mutants have been found in which development is blocked at specific stages. In the case of *Arabidopsis*, which has the same embryology as *Capsella* (Fig. 1.13), these blocks occur at the pre-globular, early globular, globular and globular-heart stages, while one mutant causes abnormal suspensor growth (Meinke and Sussex, 1979). Similarly, at least four maize mutants have been found in which development is blocked at different stages prior to the formation of the first leaf primordia (Sheridan and Neuffer, 1982). Since mutant embryos with normal endosperm can be bred, the effects of these mutations have been shown to be exclusive to the embryo, although in some cases mutant endosperm can impair growth. In other examples, the lethal embryos have nutritional defects, since they can be rescued by growing them in supplemented culture media.

We cannot draw conclusions from these few data, or others summarized by Meinke and Sussex (1979), although the stage specificity of gene action is what we should expect. However, the exploitation of embryo and cell culture techniques should soon provide more precise information about the action of these genes, and we may anticipate that study of plant lethals will prove as profitable as the work on animal embryos which we have already described.

## Later lethals

Not all lethals act on all cells, either directly or indirectly, and the lethal giant larva (*l(2)gl*) of *Drosophila* is an example of a mutant whose pleiotropic effects are expressed selectively. The larvae grow normally until the end of the 3rd instar, when they fail to pupate and become bloated or giant (Hadorn, 1960). Examination then shows a remarkable array of pleiotropic effects of the mutation (Fig. 7.11): the ring gland is abnormal, with a reduced prothoracic gland region deficient in smooth endoplasmic reticulum, which accounts for its failure to make pupation hormone; the salivary glands are small and underdeveloped; the foregut is abnormal; the testes are non-functional, and the ovaries behave much like *fs(2)B* when transplanted into wild-type larvae and metamorphosed; the lymph gland is greatly enlarged, as is the brain; and the imaginal discs are disorganized and incapable of metamorphosis on transplantation (Gateff and Schneiderman, 1974). If imaginal disc tissue from 16 h, or older, *l(2)gl* embryos is trans-

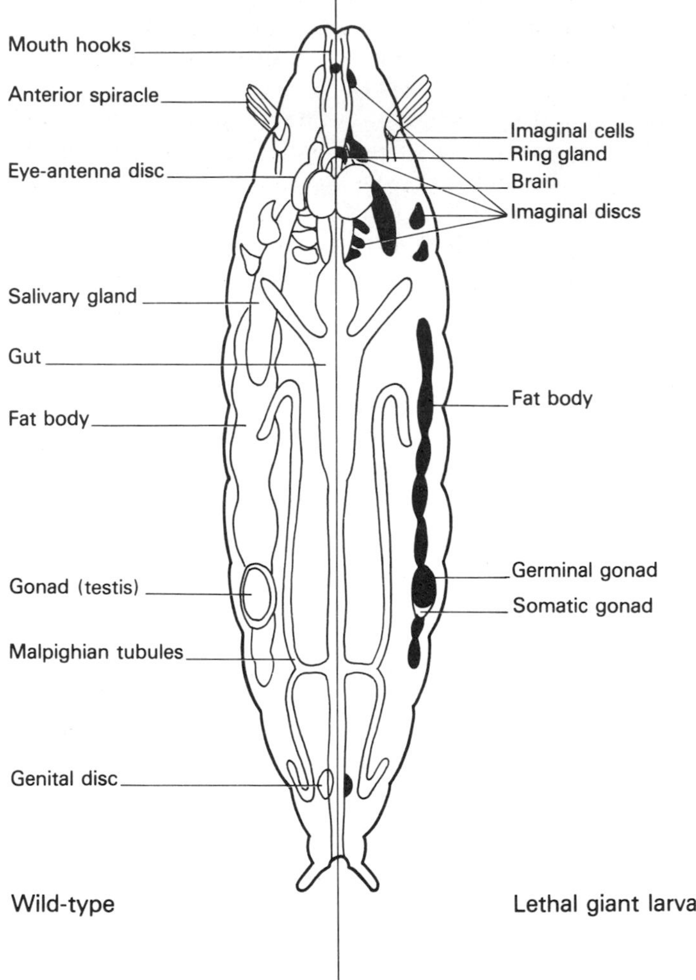


**Fig. 7.11** The pleiotropic effects of *l(2(gl)/l(2)gl* as seen in the 3rd instar larva, after Hadorn (1945). The structures of the normal larva are displayed on the left of the diagram, and the abnormalities of the mutant on the right. The structures which are degenerate or retarded in development in *l(2)gl* are shown in black, and are described in detail in the text.

planted into adult hosts it proliferates indefinitely as a non-invasive neoplasm. It ceases growth when exposed to ecdysone after transplantation into a metamorphosing larva, but secretes no cuticle and behaves like an atelotypic disc (p. 222). Transplanted brain, on the other hand, becomes an invasive tumour in adult hosts, and the cells which multiply are exclusively those which normally differentiate to form the optic centres of the adult fly. They are in this neoplastic state by the 10th hour of embryogenesis. Like the imaginal disc cells, they cease multiplication on exposure to ecdysone, but they lose this response after serial culture in adults (Gateff and Schneiderman, *loc. cit.*). Wild-type discs, or brain cells, transplanted into *l(2)gl* larvae behave normally: the *l(2)gl* larval environment is not the cause of the cancerous change.

Since most larval cells, including brain and ganglion cells, differentiate normally, the $l(2)gl^{+}$ gene product must be required exclusively for the regular functioning of the particular cell types just listed. It is therefore unlikely to be a housekeeping gene, and what is surprising is that the mutation results in degeneration of some cell types (prothoracic gland) and apparently blocks the differentiation of others (disc and brain) causing them to become neoplastic. All these changes are initiated well before larval death at the pupal stage, which results from this multiplicity of causes. Transplantation exposes the nature of some of these changes (e.g. the neoplastic growth) which would not otherwise be discerned due to larval death, and shows that the gene action is cell autonomous, at least in ovaries, discs and brain. Unfortunately we have no clear evidence that we are dealing here with a structural gene defect, or with failures of gene regulation; both are possible.

The distinction we have made between early and late lethality is not just one of convenience, although it may be so. For example, the *Drosophila* mutant lethal meander (*l(2)me*) results from a deficiency of gut proteolytic enzymes causing poor larval growth (and meandering tracheal stems which are relatively too large to pack into the ill-nourished, short body) and, ultimately, larval death (Chen and Hadorn, 1954). Similarly, muscle or nervous system mutants of *Drosophila* and of *Caenorhabditis* are known which interfere with movement and feeding, and thus cause late mortality. As with *l(2)gl*, the genetic lesion operates early in development in these cases, but death is delayed. But *l(2)gl* and these examples also show that late lethals may sometimes reflect a tissue specificity of gene action which it is valuable to know about.

## Temperature-sensitive lethals

We have already noted the use of temperature-sensitive mutants in the context of Notch, where temperature shocks (and presumed gene product inactivation) were used to explore when, and in what tissues, the $N^{+}$ gene product is required for normal development. Temperature-sensitive lethals can also be used for 'genetic surgery' (Girton and Bryant, 1980), and we shall look at refinements of this technique in Ch. 9. Here we shall consider the effects of switching organisms to the restrictive, inactivating, temperature at different developmental stages. This may cause limited and localized cell death which then exposes both normal and abnormal morphogenetic events.

In *Caenorhabditis elegans*, normal embryogenesis depends on a very precise pattern of cell division and of cleavage rates in the cell lineages, each of which has its own 'clock' (Fig. 1.7). Temperature-sensitive embryonic (*emb*) lethals upset both the pattern (Fig. 7.12), and the rates and sequences of divisions (Miwa *et al.*, 1980; Cassada *et al.*, 1981). Out of 54 *emb* mutants, maternal (unshocked) gene expression of about 43 is necessary and sufficient for normal embryogenesis. The egg is programmed by these, and presumably many more, maternal-effect genes, one of which (*emb3*) codes for germline determinants (Schierenberg *et al.*, 1980). At gastrulation, two cells of the E lineage (Fig. 1.7) migrate internally and then divide to form the precursor cells of the intestine. In the *emb5* mutant, division precedes migration, and migration is subsequently slowed. As a result, the intestinal precursors find themselves in an abnormal cell environment and the intestinal primordium is malformed. Yet the gut cells, which become fluorescent

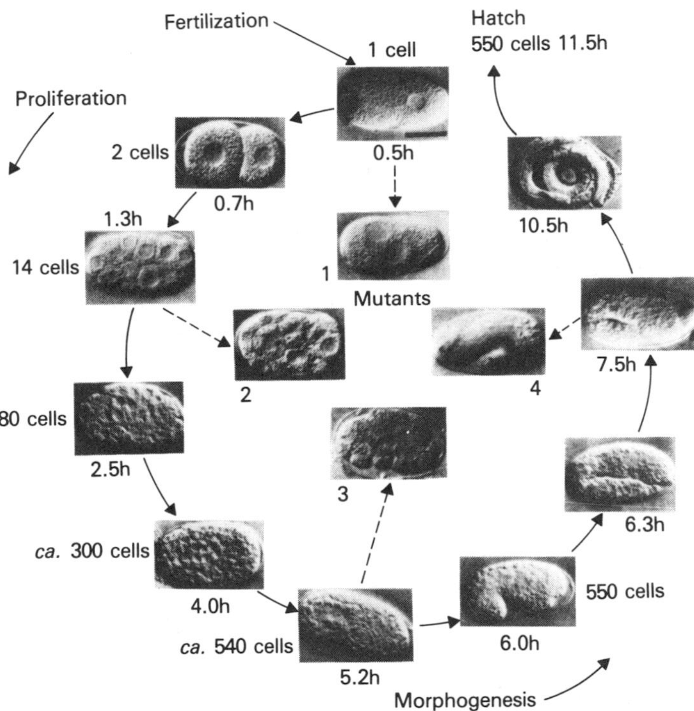


**Fig. 7.12** The normal development of wild-type *Caenorhabditis elegans* (outer circle), and the terminal phenotypes (centre) of four classes of temperature-sensitive *emb* mutants reared at 25°C, from Shierenberg *et al.* (1980). Note that proliferation is complete prior to morphogenesis in the normal (see Fig. 1.7), and that morphogenesis follows a distinct, and easily identified, sequence of forms. Mutant class 1 is arrested at the 1-cell stage, mutant class 2 at the 14–15-cell stage, mutant class 3 predominantly arrests at the 'lima bean' stage at the commencement of morphogenesis, and mutant class 4 at the early 'pretzel' stage. The gross abnormality of classes 2 and 3 will be obvious.

and so can be identified, differentiate independently of their position in the embryo in this, and in other mutants which upset embryonic cell order. Conversely, some mutants block differentiation without interfering with cell division. In all these examples exposure to the non-permissive temperature is effective prior to, or at the time of, the overt abnormality in embryogenesis, but not subsequently.

A great many *Drosophila* temperature-sensitive lethals have also been identified (Ripoll and Garcia-Bellido, 1973; Murphy, 1974; Russell, 1974; Arking, 1975), and some appear to affect all cells while others specifically damage particular tissues. As had been found in the early work of Goldschmidt (1938) and others when they exposed developing larvae to temperature shocks, particular structures are affected when different larval stages are briefly exposed to high temperatures. Typically, wing and leg abnormalities are common when 3rd instar *Drosophila* larvae are shocked, and shocked early pupae have defective abdomens. The imaginal cells of these structures are undergoing extensive division at these times, and the induced defects reflect their vulnerability then. When stocks carrying temperature-sensitive lethals are similarly treated the main difference is that the adult organs have regular missing parts. For example, using *l(1)ts726*, Russell (*loc. cit.*) found that transfer to 29 °C at 3 days for 48 h produced severe head and leg defects. In addition to the deficiencies then produced, a proportion of adults bore mirror image duplications associated with the defect. These duplications represent the tissue response to cell death (a phenomenon already known from mutants like eyeless and antennaless (Fristrom, 1969) which cause cell death during the development of these organs) and we shall look at this and other regenerative phenomena in Ch. 9. Temperature-sensitive lethals allow regular manipulation of cell death, and hence controlled exploration of regenerative events.

## Conclusions

The experiments we have outlined show that lethal mutations are powerful tools for exploring development. They may also be confusing, as the data on Notch and on the tailless alleles show. When null mutants are available, or when good ts alleles can be manipulated, some of these ambiguities can be overcome, and mutations of housekeeping and of 'developmental' genes may be distinguished from one another. If the biochemistry of Notch is confirmed, it may become a type example of the morphological and developmental effects of a lethal housekeeping gene mutation causing an array of structural defects, as already demonstrated to a lesser degree by rudimentary. Viable alleles of most lethal housekeeping genes will result in defective structures and organs, and we can also see that they will cause phenotypic variability since their effects will be strongly influenced by environment, genetic background, and cell and tissue sensitivities. These will be the earmarks of such mutant loci. Nevertheless, ts alleles can be effectively used for genetic surgery, and thus to expose the events which follow the killing of particular cell groups.

The more interesting lethals are those that disturb determination and differentiation. Among mosaic eggs we should expect to find many maternal-effect lethals which act by upsetting determination, and this is so for *Caenorhabditis*. Out of an estimated 2000 essential genes, between 200–550 are required for normal embryogenesis and, if the sample of identified mutants is not biased, about 80 percent of them are maternally sufficient (Cassada *et al.*, 1981). On the other hand, an equivalent mutant screen for maternal-effect mutants in *Drosophila* (Gans *et al.*, 1975) produced no single example of a recessive, maternally influenced morphological abnormality. This may argue, and bicaudal, dorsal, and extra sexcombs would support the argument, that determination in *Drosophila* and in the nematode have entirely different bases. But the argument is still open, for there is the old observation that the thoracic bristle pattern in *D. funebris* is maternally regulated by the polychaeta mutation (Timoféeff-Ressovsky, 1935), a predetermination of thoracic epidermal organization more like that found in *Caenorhabditis* than in the fly.

The interpretation mutants, complex as are the examples described, expose new developmental relationships and organization, and much further work will be required before their significance is properly understood. It is not clear how far the disturbed patterns are abnormalities, for example, or if they indicate an underlying set of segmental subpatterns brought to light by the mutations (but see p. 255). Later lethals, on the other hand, indicate when the gene products are involved in differentiation. Only *l(2)gl* suggests that regulation of a number of genes might then be concerned, but the data cannot yet bear such an assumption. The descriptions of the effects of mutations causing both male and female sterility suggest that these systems have advantages for exploring regularly-repeated defects, and these in easily available adults. In all these cases, however, the problem of identifying the molecular defect is difficult since we have few clues to the nature of the molecules involved.

The cell lineages of *Caenorhabditis* are strictly determined, and in only a few cases does ablation of a cell result in its replacement by a neighbour, demonstrating limited cell interaction even in this most mosaic of eggs (Kimble, 1981). *Drosophila* development is apparently more regulative since both ablation and induced lethality cause regeneration and duplication, at least, of ectodermal structures. However, we cannot conclude that cell lineage sets no restrictions on developmental potential for this, or for the patently regulative embryos of the mouse and other organisms. Our next question is therefore to ask about normal cell lineages and their variability, if any.

# IV Tissue determination and regeneration

Mutations can be employed as tools. They have been used for investigating the origins of tissues and organs, and for causing cell death and regeneration. Regeneration is the antithesis of determination, implying either the dedifferentiation of cells, or the presence of reserves of uncommitted cells, which subsequently multiply and differentiate. There must be rules for this accurate replacement of lost parts, and they may tell us, indirectly, about the processes of normal development. Most of the experiments probing the stability of the determined state and the curiosities of regeneration depend on surgery, and we shall describe them. But before we do this we shall look at what chimeric and mosaic organisms tell us about the normal sequences of development leading to the final, determined cell states.

The problems of tissue origins can be examined with some precision by inserting genetically marked (mutant) cells, or nuclei into early embryos. The marked cell clone can then be followed in the resulting chimeric individuals. After blastocoel transplants in the mouse, most tissues are found to be of mixed origin indicating considerable early cell movement; whereas in *Drosophila* they are not. Autonomous development of mouse tissues is not established until the primitive-streak stage, but most *Drosophila* cells are determined at blastoderm. Also, in the mouse, but not in *Drosophila*, the defect in a mutant clone may be corrected by association with normal cells, and teratocarcinoma cells may even give rise to normal progeny, suggesting that the tumour is a non-genetic defect. Natural mosaics (through X-inactivation in the mouse and position effect variegation in *Drosophila*) confirm the data obtained from chimeras.

In *Drosophila*, chromosome loss is also used to generate mosaics; and sex mosaics (gynandromorphs) have been exploited to make fate maps relating adult and larval structures to the initial egg-surface location of their primordia. Major structures derive from groups of cells in this detailed fate map, and are therefore polyclonal in origin. Sequential, timed X-ray induced mitotic recombination is used to monitor the orientation and growth of these polyclones. Unexpectedly, each clone is restricted either to the anterior or to the posterior of wings and limbs; the engrailed mutation allows clones to transgress this 'compartment' boundary. Other compartments are established later. Mitotic recombination is also used to identify the tissue(s) affected by a mutation and to follow its action, including its time of action.

Similar manipulations can be undertaken with some plants. These indicate that only a small number of clone progenitor cells is set aside in the meristem and that final determination (leaf, bud etc.) is location dependent.

Determined cells have a stable cell heredity which can be perpetuated indefinitely, as judged from *Drosophila* imaginal discs cultured *in vivo*. Some cells of the disc may transdetermine to another, stable state. By excision of parts, a detailed fate map of each disc may be constructed which shows a more or less invariant relationship between disc cells and final adult structures. However, if discs are divided and cultured, the blastema at the cut surface either regenerates the missing part or duplicates the existing part. The simple rules which explain the patterns of regeneration apply not only to *Drosophila* imaginal discs, but also to regeneration of cockroach and amphibian limbs. The basic assumption is that a gradient of some morphogen is established within the disc or limb, and that cells interpret this 'positional information' and become appropriately determined. However, recent evidence shows that isolated disc blastemas can regenerate whole limbs, and regenerating amphibian stumps do the same when exposed to vitamin A. The problem thus becomes inverted, and we have to ask how the already determined cells of the disc fragment or stump regulate the pattern of growth and differentiation of the undetermined blastema cells.

Cell-lethal mutants can be used to make clones equivalent to those produced by the microsurgery of discs, and they cause the same patterns of regeneration and duplication of structures. Apart, possibly, from bicaudal and dicephalic mutants, there is only sketchy evidence for mutations affecting positional information gradients, and similarly, there is a surprising deficit of interpretation mutations. The problems of the determined state, and of how it is achieved, have to be approached, it seems, using other gene systems.

# 8 Tissue origins

To these observers the cleavage of the ovum presented merely a series of problems in the mechanics of cell division, and its accurate study was almost wholly neglected as having no interest for the historical study of descent . . . [But] in many groups of animals (though apparently not in all) the origin of the adult organs may be determined cell by cell in the cleavage stages: . . . the *cell-lineage* thus determined is not the vague and variable process it was once supposed to be, but in many cases as definitely ordered a process as any other series of events in the ontogeny.

E. B. Wilson, 1898; quoted in Willier and Oppenheimer, 1964.

Cell lineage plays a paramount role in the embryonic development of *Caenorhabditis elegans*. The clones which derive from the initial half-dozen stem cells normally have an invariant fate: the egg is strictly mosaic (p. 17). In this case, the embryo is a set of clones which differentiate to form specific structures or parts of structures, epitomizing the ordered progression which Wilson (1898) was concerned to emphasize. It is not possible to chart cell divisions accurately in organisms with thousands of cells, though approximate patterns have been worked out from histological sections of successive stages (e.g. Fig. 8.1*a* for *Drosophila*) and, as we saw in Ch. 1, by following the distribution to progeny cells of pigments regionalized in eggs, or by marking areas of the egg with dye spots. These crude fate maps can be greatly refined by using mutations to label cells: mutations for either enzyme or immunological differences which can be seen using appropriate stains, or mutations affecting terminal characters like pigmentation or hairs, or mutants carrying chromosomal differences which can be identified, or whatever is appropriate to the objectives of the study. We can therefore ask, using this refined technique which will distinguish between a marked (mutant) and an unmarked cell, if cell lineage always does play a significant part in development. And we can also ask if there is an interaction between the two clones; mutant and wild-type.

There are two ways of making these experimental organisms. We can construct a compound organism, or *chimera*, by transplanting genetically

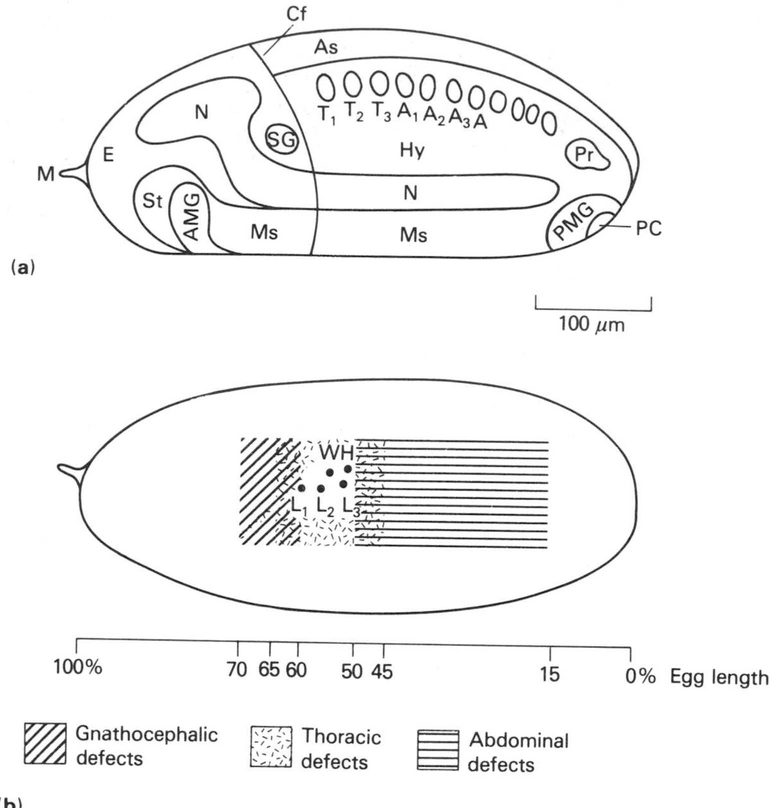


**Fig. 8.1** (**a**) Poulson's (1950) fate map of the *Drosophila* egg, with anterior to the left and dorsal above. As, amnioserosa; AMG, anterior mid-gut; E, ectoderm; HY, hypoderm; M, micropyle; Ms, mesoderm; N, nervous system; PC, pole cells; Pr, proctodeum; PMG, posterior mid-gut; SG, salivary gland; St, stomodeum; $T_1$, $T_2$ and $T_3$, thoracic tracheal pits and $A_1$ etc., abdominal tracheal pits. The cephalic furrow is Cf. (**b**), Adult thoracic primordia located by destroying blastoderm cells with a fine laser beam ($L_{1-3}$, legs 1–3; W, wing and H, haltere) based on Lohs-Schardin *et al.* (1979), and defects in larvae caused by removal of 50 blastoderm cells. After Underwood *et al.* (1980).

marked cells (or nuclei) from one organism into an embryo not so marked. Or we can make a *mosaic* organism by creating the genetic difference within the products of a single zygote, usually by irradiating early cleavage stages. This technique of induced somatic crossover has become a powerful tool for making cell mosaics in *Drosophila* where there is pairing of the somatic chromosomes (Fig. 8.2). In both classes, the useful marker mutations are cell autonomous, and the boundaries of clones (when present) are then clearly defined. However, since the chimera derives from two zygotes and the mosaic from one, we shall look at the two manipulations separately.

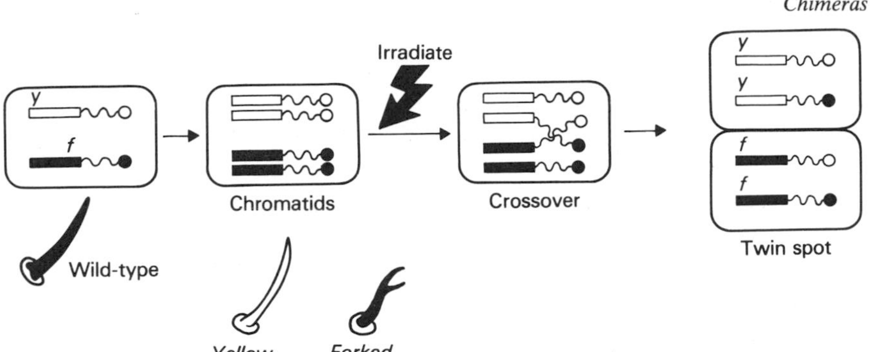


**Fig. 8.2** Diagramatic representation of induced mitotic recombination, Cells heterozygous for two recessives, here yellow (*y*) and forked (*f*), are irradiated to induce a crossover between the mutations and the centromeres (open and closed circles). The resulting cells are homozygous for yellow and forked, and the clones derived from them form twin spots of adult tissue identifiable by these markers, and surrounded by the wild-type, heterozygous tissue derived from cells which have not undergone crossing over. If a Minute (*M*) homozygous lethal, mutation is substituted for *f*, the heterozygous cells will grow slowly, the *M*/*M* crossover will die and the *y*/*y* cells will grow normally. There will be no twin spot, only a single clone of normally growing *y* cells in a background of slowly growing *M* +/+ *y* cells (see p. 208).

## Chimeras

The first experimental chimeras were made by aggregating 8-cell mouse embryos, culturing them *in vitro* for 1–2 days, and then transplanting the compound, single large morula so formed into the uterus of a pseudopregnant foster mother (Tarkowski, 1961). This technique can be used for embryos from the 2-cell to morula stages, but not for blastocysts which do not aggregate. Many embryos may be aggregated (p. 13), or isolated blastomeres or parts may be used, and such compounds regulate to give normal embryos within a few days, about the time when the preamniotic cavity forms. Alternatively, single cells or whole inner cell masses may be injected into blastocysts (Gardner, 1968), to provide more precise information. Both aggregation and injection chimeras, sometimes called allophenic or tetraparental mice, are visibly chimeric when different coat colour markers are used, showing that the products of both zygotes contribute to the embryo (Fig. 8.3). In no case is there evidence that a single organ or tissue is formed from only one component. Cells of these early stages are therefore not determined, and the progeny of more than one cell must follow the developmental pathway which leads to the formation of this or that tissue, or organ. This was clearly demonstrated by Mintz and Baker (1967) using

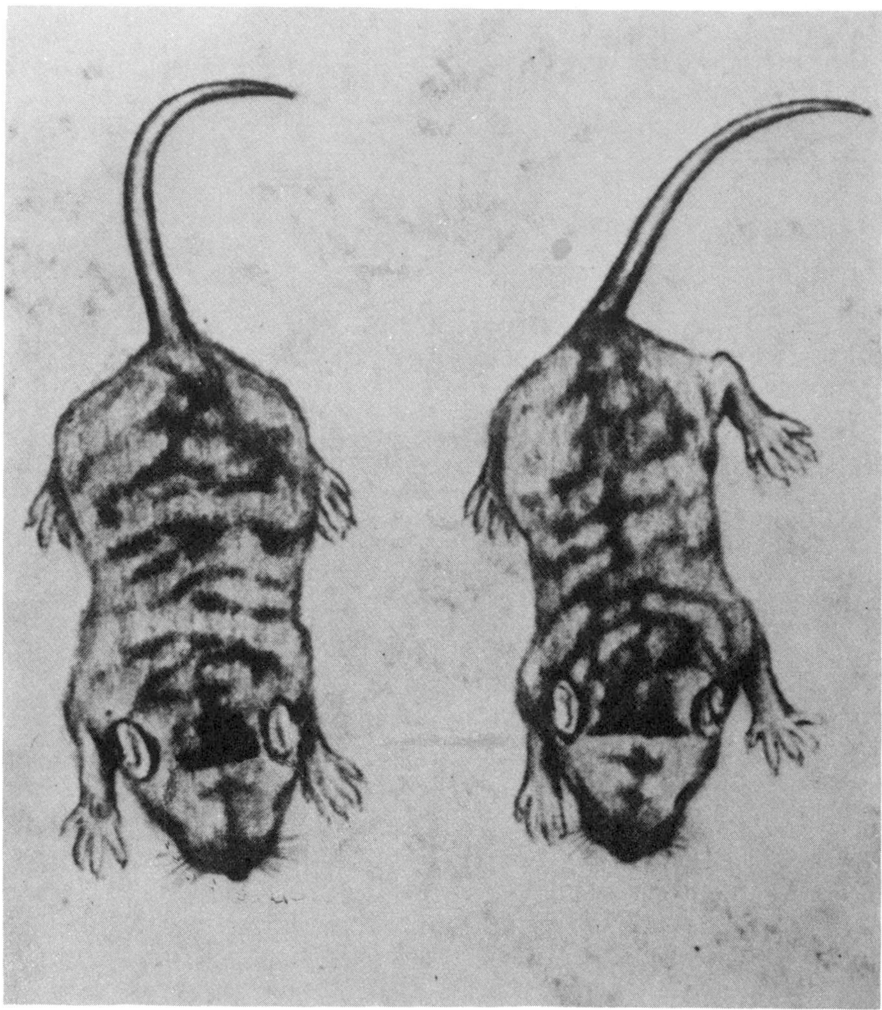

**Fig. 8.3** Two hair follicle chimeras showing black and agouti striping: photograph of $C_3H \leftrightarrow C57BL$ mice by permission of Dr. A. McLaren. The very distinct hair follicle bands suggest that they are clones of cells each deriving predominantly from one of the chimeric components. However, the agouti locus is expressed differently in the body regions—yellow on the belly, black down the spine and agouti on the flanks—and some of the apparent regularity may reflect differences in some systemic body gradient giving a 'wave' pattern to which the cells respond (see McLaren, 1976b). Which is the correct interpretation of this striking chimeric pattern has still to be resolved.

strains homozygous for electrophoretically distinct alleles of the dimeric enzyme, isocitrate dehydrogenase. Liver, kidney, lung, spleen and cardiac muscle each yielded enzyme of both variants (i.e. were chimeras). Skeletal muscle, in which the myotubes are formed by the fusion of separate myoblasts, yielded an intermediate, hybrid band in addition, showing that the enzyme subunits are assembled in the common cytoplasm containing nuclei of both types.

Cell lineage analysis is difficult in the mouse where there is a great deal of initial cell mixing. The progeny of a single marked cell injected into a blastocyst, for example, can contribute to every tissue, even one as small as a single lymph node (Ford *et al.*, 1975). This mixing seems to occur after blastocyst formation, and it appears to be a characteristic feature of mammalian development which applies even to the brain (Mullen, 1978). In another example, Gearhart and Mintz (1972) made aggregate chimeras from embryos marked by alleles for the enzyme glucose phosphate isomerase. Individual somites from 8–9-day embryos contained both enzyme variants in 30 out of 39 somites analysed, and the proportions of the two enzymes were similar in neighbouring somites. Each adult eye muscle, individually derived from a single somite, also contained both enzymes. Somites cannot therefore be derived from single cells. Nor can somites be derived from random groupings of cells since neighbouring somites would not then be alike. We can only conclude that there is both cell mingling and clonal descent involved in the development of these tissues (and similar evidence applies to others), but whether these are sequential, or concurrent, processes is not clear.

Chimeras sometimes show cell interactions where the recessive homozygote/wild-type is phenotypically like the recessive heterozygote. An interesting example was constructed by Peterson (1974) aggregating dystrophic embryos (also marked by an isoallele of malic enzyme) and wild-type. The chimeras were behaviourally indistinguishable from normal, and the dystrophic muscle was normal, too, even when the enzyme assay showed that 94 percent of it derived from the dystrophic strain. The chimeras show that the primary genetic defect is presumably in the nerve, not in the muscle, and behaves recessively. This dominance of wild-type in chimeras may imply that many mammalian genes are non-autonomous in their action, and that their product diffuses between cells.

The most exciting cell interaction of this sort was found when chimeras were made by injecting genetically marked, malignant teratocarcinoma cells, cultured *in vitro* for a long time, into blastocysts (reviewed by Mintz, 1978). These cells became incorporated in the inner mass, a few cells of which form the embryo. The mice born were normal, although all their tissues were genetic mixtures of the kinds already described. Some males produced sperm derived from this XY tumour, and their $F_1$ progeny were also normal (Illmensee and Mintz, 1976). The malignancy was cured. A few mice also carried tumours, probably because some of the injected cells did not become incorporated into the embryo and continued their growth just as if they had been injected into later embryos, thereby proving their continued malignancy. Since progeny of the injected cells were found in all tissues, the initial teratocarcinoma cells must have been totipotent, and their

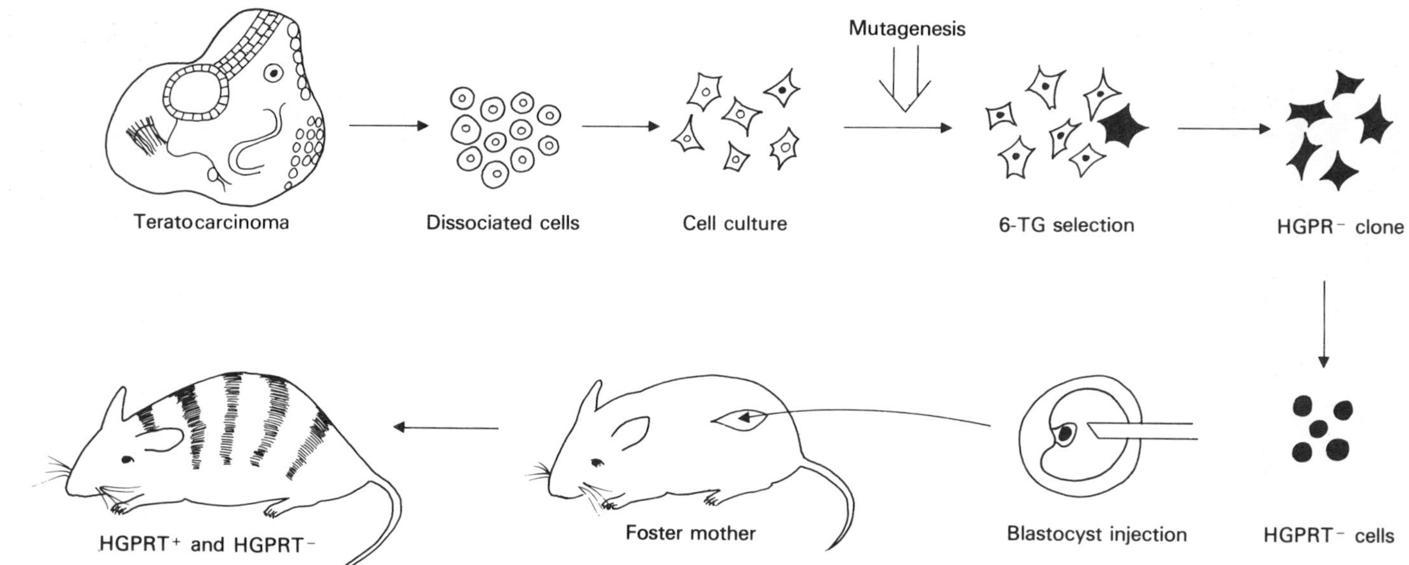


**Fig. 8.4** Cells from an induced teratocarcinoma are dissociated, grown in culture and mutagenized with a chemical mutagen. Progeny cells are then selected in medium containing 6-thioguanine (6-TG) where only the cells resistant (shaded) to this purine analogue survive. These cells lack the purine salvage enzyme, hypoxanthine: guanine phosphoribosyltransferase ($HGPRT^-$ mutants). Cells from the clone are injected into genetically marked blastocysts ($HGPRT^+$), and the blastocysts are then transferred to a mother previously made pseudopregnant by having mated with a vasectomized, sterile male. Among the progeny were mice with $HGPRT^-$ cells in their tissues and coats, along with $HGPRT^+$ cells derived from the blastocyst itself. After Mintz (1978).

malignancy is unlikely to be the result of an earlier mutation. It must be due, on this argument, to faulty development (gene regulation?), allowing proliferation at the expense of differentiation.

This technique further provides the very interesting possibility of making specific mutants of the teratocarcinoma cells in culture, and then incorporating the mutant into the gonads of chimeric mice so that they can be bred from (Fig. 8.4). This has been done by selecting cells for a deficiency of hypoxanthine phosphoribosyl transferase (HGPRT)—the known cause of Lesch-Nyhan disease in man—and making chimeras. As before, the mice are normal although their tissues are mixtures of the different cell types (Dewey *et al.*, 1977). Unfortunately, no male carrying the mutant cells has yet been born. The level of chimerism in these experiments employing teratocarcinoma cells is usually low (Papaioannou *et al.*, 1978), so completing the cycle from organism to cell to organism may take some time.

Mouse chimeras are also made in order to study the characteristics of the immune system, the genetic control of primary sex determination and so forth; but most of this work has given ambiguous results (McLaren, 1976b). One reason will become obvious when we look at mosaics: there is a great need for markers which will identify individual cells. The enzyme markers mentioned above have to be assayed in whole tissues (and indicate the proportions of the two kinds of cell in them), but this cannot tell us where the cells lie, which may be of great importance, for example, in sorting out the separate roles of the somatic and germinal parts of the gonad as they affect sex determination. In short, analysis of tissues containing many cell types is probably too imprecise to tell us about the possible clonal origins of individual cells. Specific mutants of teratocarcinoma cells may provide just the right sort of material for this necessary technical advance.

The youngest possible *Drosophila* chimeras are made by injecting blastoderm cells into blastocysts, and these behave quite differently. Cells taken from the posterior blastoderm and placed posteriorly in the blastocyst form posterior structures which frequently integrate into the developing embryo. If transplanted to the anterior they fail to integrate and form separate posterior structures such as abdominal or genital parts—and vice versa with anterior cells (Illmensee, 1976). This result confirms the Chan and Gehring experiment (p. 220) which showed that *Drosophila* blastoderm cells are determined. Unfortunately these manipulations are too imprecise to do more than prove regional determination, but the fact that the structures missing from the donor embryo correspond to those found in the host suggests a very precise local determination (Simcox and Sang, 1982).

Genetically marked nuclei can also be used to make *Drosophila* chimeras by injecting them into the acellular blastoderm, after which they become incorporated into the cellular blastoderm. The most striking result is that nuclei from long-cultured cell lines are then found to be multipotent, and capable of forming both larval and adult tissues (Table 8.1). How these determined nuclei come to be reprogrammed is not known (cf. the apparent

**Table 8.1** Tissues containing wild-type cells derived from single cultured cell nuclei injected into the posterior of *y* $sn^3$ *ma-l* cleavage embryos

| | Larvae | | | |
|---|---|---|---|---|
| **Nuclei source** | **Gut** | **Fat body** | **Malpighian** | **Hypoderm** |
| Dm1 | 3 | 1 | 2 | 2 |
| Dm2 | 1 | — | — | 2 |
| K | — | — | — | — |
| Blastoderm | 2 | 2 | 2 | 3 |

| | Adults | | | | | |
|---|---|---|---|---|---|---|
| | **Gut** | **Fat body** | **Malpighian** | **Thorax** | **Abdomen** | **Germline** |
| Dm1 | 2 | — | 1 | 1 | 1 | — |
| Dm2 | — | — | — | — | — | — |
| K | 1 | — | — | — | 1 | — |
| Blastoderm | 2 | 1 | 2 | — | 2 | 2 |

Blastoderm nuclei were used as controls, and the experimental nuclei came from Schneider's Dm1 and 2 cell lines and from Echalier's K line, each of which has been long cultured *in vitro*. Data from Illmensee (1976).

determination of the acellular blastoderm nuclei noted on p. 166). Nuclei from the inner cell mass of the mouse blastoderm, injected into a fertilized egg from which the zygote nucleus is subsequently removed, are also found to be totipotent. Both males and females so constructed produce normal progeny (Illmensee and Hoppe, 1981).

## Mosaics

The most important experimental mosaics are made by induced somatic crossing-over (Fig. 8.2) or through the use of strains constructed to generate chromosome loss. However, there are also two kinds of natural mosaics involving gene inactivation and we shall consider these first.

### *X-inactivation*

As a consequence of their dosage compensation mechanism (p. 125) all female mammals are mosaic for such heterozygous sex-linked genes as they may carry. The paternally-derived X is electively inactivated in marsupials, but in eutherian mammals this facultative heterochromatization of one or other X chromosome occurs at random in each embryonic cell, around the time of blastoderm formation (Monk, 1978). Apparently the cells of the trophectoderm and primary endoderm follow the marsupial pattern, but this does not concern us (Harper *et al.*, 1982). The descendants of the inactivated chromosome are then late replicating, permanently inactive and never transcribed. This inactivation can easily be shown by cloning fibroblasts from a female heterozygous for an X-linked gene such as glucose-6-phosphate dehydrogenase, when half the clones express one allele, the remainder express the other, and no clones express both. Unfortunately most of the X-linked genes of the mouse which can be followed through this inactivation process affect final pigmentation, like tabby. Even when they have a translocation added to them (e.g. Cattanach's translocation of part of chromosome 7, which carries albino and pink eye) they add little to the findings that we have already described for the equivalent chimeras. The most obvious differences are that mosaics have smaller, better defined pigment patches, and there is less variability between organisms when compared with chimeras. These differences may be accounted for by the later establishment of mosaicism, and the greater genetic homogeneity of the females (McLaren, 1976b). Little is known about the mechanism of X-inactivation in these difficult-to-handle blastoderm and preblastoderm embryos (Monk, 1978). However, the discovery that individual female teratocarcinoma cells show this change when they differentiate *in vitro* (Martin *et al.*, 1978), suggests that a system for studying the molecular genetics of this intriguing process may now be available.

### Position effect variegation

The second kind of mosaicism also involves gene inactivation in some, but not all, cells and similarly awaits the development of a system for its molecular analysis (see Spofford's (1976) review). The phenomenon was discovered by Muller (1930) when working on the Notch-white region of the *Drosophila* X chromosome. Heterozygotes of this strain with white ($w^+/w$) showed variegation of the pigment in their eyes, so that they were mottled with white instead of being the expected full red. The strain was found to have a fragment of the X translocated to the third chromosome, and it was this rearrangement ($R$), or position, of the X fragment which caused the variegation. The phenomenon is therefore called *variegation* (or *V-type*) *position effect* (Lewis, 1950), in constrast to the *stable* (or *S-type*) *position effect* already described for the *Bar* mutation (p. 123). Briefly, any chromosome break in the *constitutive* heterochromatin which repositions it adjacent to euchromatin, by inversion or translocation, inactivates the neighbouring euchromatic genes in some cells, producing a mosaic of clones which express, or fail to express, the gene (Fig. 8.5). The wild-type gene must be carried by the rearrangement, and its manifestation is usually monitored against a recessive mutant for the locus, in heterozygous combination. $R(w^+)/w$ and $R(w)$/Y show variegation, but $R(w)/w^+$ and $R(w^+)/w^+$ will not, in the above example. The inactivation is not permanent since reversion of the rearrangement restores the wild-type gene to its usual activity. It is the intimate *cis* association of the heterochromatin with the

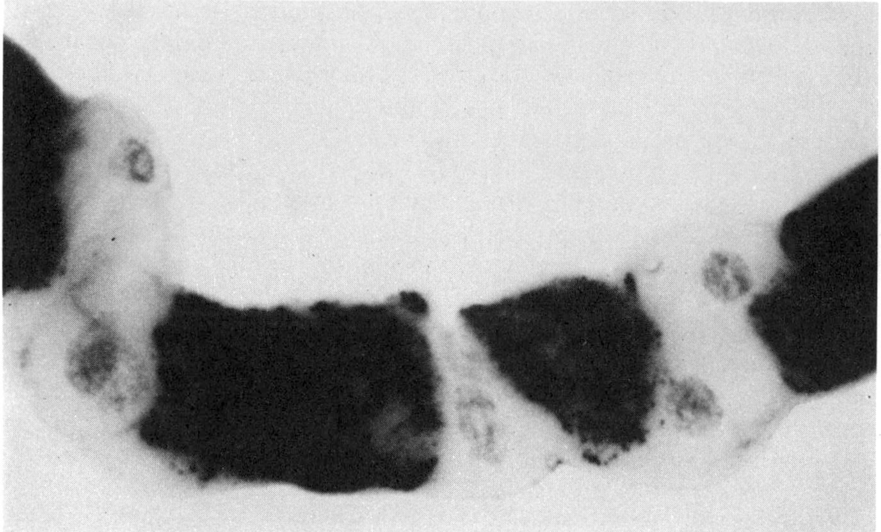

**Fig. 8.5** Position effect variegation occurs in all cells in which the gene is expressed. Although most work has been done on the eye (Fig. 8.6), the cell-by-cell basis of inactivation is very clearly seen in the Malpighian tubules. In this case variegation of *w* results in cells without pigment. There is sometimes an alternation of cells with ($w^+$) and without ($w^-$) pigment, as well as of 3–5-cell clusters. Photograph by courtesy of I. J. Hartmann-Goldstein.

euchromatic gene which inactivates the latter so that the mutant (*w*) is expressed, as in a haploid deficiency. *Cis*-dominance proves the 'position effect' principle of this variegation, and any euchromatic locus can be made to variegate by an appropriate rearrangement. Enzyme loci (amylase (Bahn, 1971), and 6-phosphogluconate dehydrogenase (Gvozdev *et al.*, 1973)) variegate in the same way as morphological characters, showing the effect to be through inhibition of structural gene transcription.

The many papers (Spofford, 1976) on V-type position effect show that inactivation involves the heterochromatization of the genes adjacent to the rearranged heterochromatin, and that this action spreads along the chromosome, sometimes for as far as 50 bands, usually giving a graded polarity effect so that the distal genes are less affected than those adjacent to the heterochromatin. Not all heterochromatin is equivalent for this activity, which can be suppressed by addition of extra Y chromosomes, or other heterochromatic fragments, to the genome. The cell therefore seems to have a limited capacity for making whatever is required for heterochromatization, and there is a maternal effect in the sense that transmission of the rearrangement through the egg is more effective than through sperm. Although heterochromatization usually occurs early, around blastoderm formation, there may also be other temperature-sensitive periods (high temperature usually suppresses the effect) and some are as late as the pupal state. We are therefore concerned with a very interesting and subtle physiological process, but we do not know what the mechanism is or why one cell shows this inactivation when its neighbour does not. V-type position effects have also been found with translocations in the mouse (e.g. Cattanach's translocation, which involves facultative heterochromatin) and for the plant, *Oenothera blandina*, so it is not exclusive to *Drosophila*. It seems unlikely that heterochromatization is involved in normal differentiation, nor is it related to determination which affects groups of cells (p. 204), whereas position effect inactivates genes in single cells (Fig. 8.5).

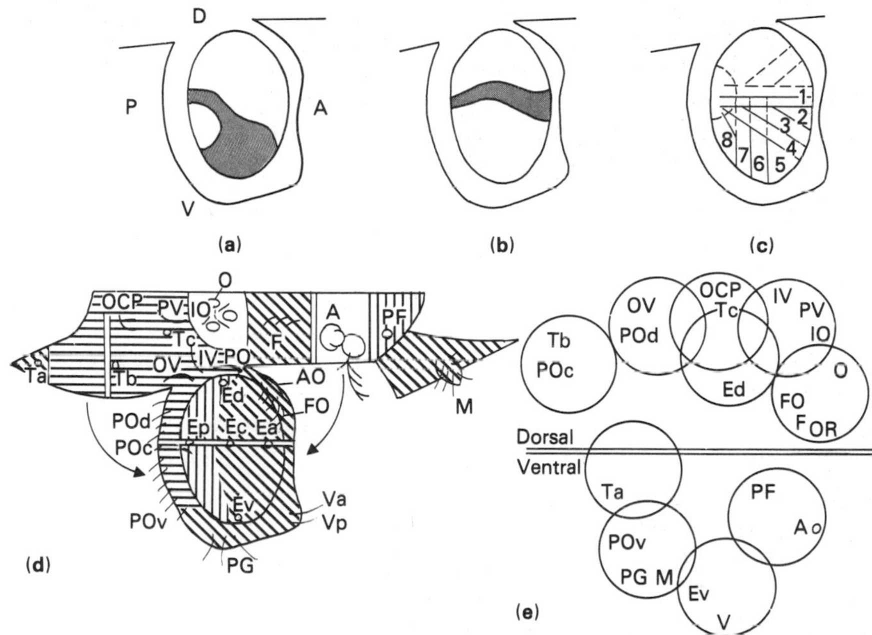


**Fig. 8.6** Eye and head mosaics. The eye–antenna disc gives rise to both these structures and to the associated parts of the head. The head forms by fusion and bending of the structures from both sides, and figure (**d**) shows their relevant parts after the head is cut and unfolded into a two-dimensional structure, as indicated by the arrows. (**a**) and (**b**) illustrate two clonal eye patterns resulting from position effect variegation of white: reciprocals (white on red) are also found. The general set of patterns disclosed is summarized in (**c**), which suggests the contributions of the eight eye cells of the late 1st instar to the ventral half of the eye. The entire head can be examined with mitotic recombinant clones (such as from *bw*; $M(3)i^{55}$*/mwh jv st*), and the relations of the parts are then as shown by the shading of figure (**d**). The only compartment divides the head into dorsal and ventral, indicated by the open double line. The gynandromorph fate map (**e**) locates the 10 blastoderm cells, each 2.5 sturts in diameter, which form the head. (Data from Baker, 1968, 1978b). A, antenna; AO and PO, anterior and posterior orbital bristles; E, virtual spots in the compound eye; F, frontal bristles; FO, frontoorbital bristles; IO interocellar bristles; IV, inner verticals; O, ocellar bristles; OCP, occipital bristles; OR, orbital bristles; OV, outer vertical bristles; PF, pre-frons; PG, postgena bristles; PV, post verticals; T, virtual spots at the rear of the head.

The eye mottling resulting from the V-type position effect has been carefully studied as a natural marker of the clonal descendants of affected cells (Baker, 1968). By combining the patterns from many eyes, it is possible to assign the origin of a clone to one or other of eight cells present in the eye anlage of the first instar larva (Fig. 8.6). We shall see that the same patterns are found using somatic crossing-over, confirming the validity of this technique. The two methods may be combined, when the crossover technique is found to be more precise and variegation to be more complex.

### Gynandromorphs

Sex mosaics, part male part female, were found early in *Drosophila* studies (Morgan, 1914), but they are rare, as they are in other species (e.g. the fowl). It was quickly established that these gynandromorphs arose through the loss of one X chromosome from XX *Drosophila* embryos, generating tissues which were either XX (female) or XO (male). But only when certain ring X chromosomes (*In(1)*$w^{vc}$, and, later, *R(1)5A*) were found, could sufficient XX/XO gynandromorphs be made for development studies, as nature clearly intended! These rings tend to be lost at the first nuclear division, but not during subsequent mitoses. Somewhat later, Sturtevant (1929) discovered that the claret (*ca*) mutation of *D. simulans* caused mitotic and meiotic anomalies in eggs from *ca/ca* mothers, resulting in the loss of the maternal X chromosome (and of the small fourth chromosome). This was the first mutant to be used for making gynanders. An analogous mutation, claret non-disjunctional ($ca^{nd}$), of *D. melanogaster* acts in the same fashion. Two further destabilizing mutant loci have since been identified: paternal loss (*pal*) and mitotic loss inducer (*mit*). The detailed technologies of all these mutants (Table 8.2) are discussed by Hall *et al.* (1976), and we need note only that gynandromorph data are mutually consistent whatever system is used to generate them.

Gynandromorphs are used for making fate maps, as first shown by Sturtevant (1929) and then elaborated by Garcia-Bellido and Merriam (1969). This procedure is possible with *Drosophila* because:

1. The boundaries between XX/XO patches can be identified by their sex differences of pigment and structure, and by exploiting recessive surface marker mutants carried on the male-expressed rod-X chromosome.
2. The orientation of the first zygote division is random (Parks, 1936) causing the first XX and XO nuclei to take up different spatial positions in each egg. Their subsequent mitotic products do not (or rarely) mix before they migrate to the blastoderm. Hence, assuming the two kinds of nuclei have the same division rates etc., as seems to be true, half the blastoderm will be populated with male, and half with female, nuclei. And there will be none of the intermingling which makes it difficult to follow cell clones in mouse chimeras. Also, as a conse-

quence of the random orientation of the first division, each blastoderm will have a different mosaic pattern (Fig. 8.7).

3. The mosaic patches will also be differently distributed on the surface of each adult. Their boundaries will follow the segmental organization of the adult, which results from the fusion of the imaginal discs

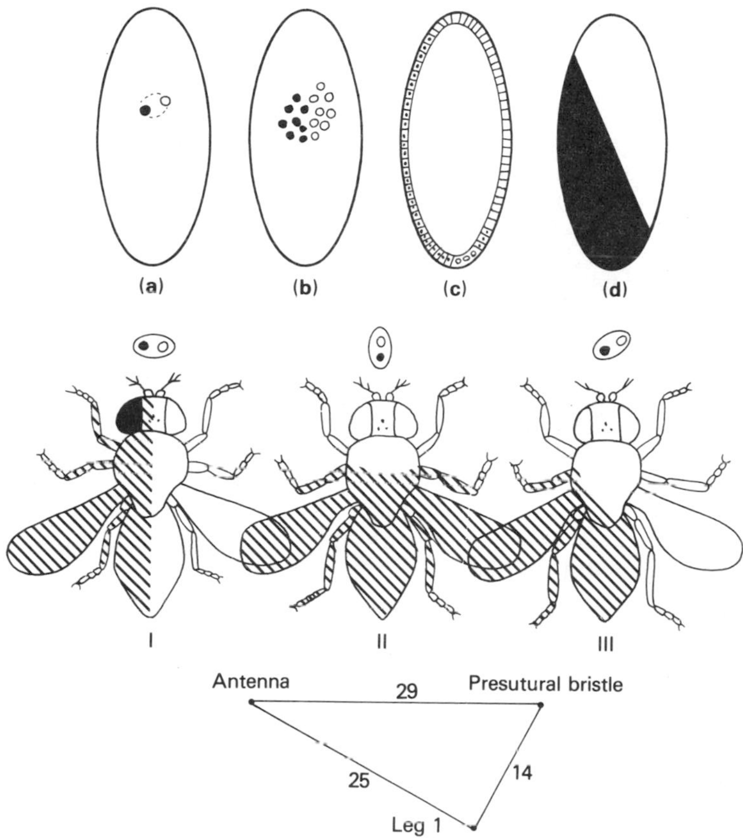


**Fig. 8.7** The development of gynandromorph mosaic flies. The loss of the ring-X chromosome generates a male (XO) nucleus and an XX (female) nucleus at the first division (**a**), indicated by the open and closed circles respectively. These nuclei multiply rapidly (**b**), move to the periphery of the egg which they populate, and where cell walls grow around them to form the cellular blastoderm (**c**). Examination shows that the male and female areas are usually distinct (**d**), although this is not always so (Zalokar *et al.*, 1980). The area of the egg populated by the male and female cells depends on the chance orientation of the first division, which is random. Three examples are illustrated: in I the adult fly is divided vertically by the two cell populations, in II the division is horizontal and in III much of the marked tissue is presumed to be larval so that less than half the adult is female. For sex differences to be apparent in this way, sex must be a cell autonomous phenotype (p. 281). The calculation of cell locations from gynanders is described in the text. The position of any three is determined by triangulation, which proceeds from structure to structure, as shown in Fig. 8.8.

**Table 8.2** Some properties of somatic chromosome loss-inducing systems

| | ring-X | $ca^{nd}$ | pal | mit |
|---|---|---|---|---|
| Transmission source | Either sex | Female | Male | Female |
| Chromosome lost | Ring | Maternal | Paternal | Either |
| Frequency of gynanders | 10–40% | 12–15% | 3–4% | 1–8% |
| Probability of XO structures | 0.5 | 0.45 | 0.3–0.4 | 0.1 |

The frequency of gynanders varies with strain, culture conditions, maternal age and other variables. The probabilities of finding male structures show a spread around the mean figures quoted and some of this variability between gynanders is due to assessment by adult structures, since it is easy to envisage a 50:50 gynander where most of the male tissue is larval, and therefore not identified, and vice versa. The mean probabilities therefore suggest that chromosome loss occurs at about the first division with the *ring-X* and $ca^{nd}$, perhaps involves the second division also with *pal*, and may extend as late as the third or fourth division with *mit*. We should also expect some variability around these stages. Data from Hall *et al.* (1976).

(p. 219). These discs in their turn derive from the single-cell surface of the blastoderm of the egg, and we have seen that there is a one-to-one relationship between these cells (or cell groups) and the imaginal discs (p. 30). It follows that the mosaic boundary separating two adult structures was earlier present as a boundary, unidentifiable as such, separating the blastoderm cells fated to become these adult structures. The problem is: how do we work out the blastoderm fate map of the cells which will form particular adult structures, from the distribution of these structures as adult male and female patches found in an array of gynandromorphs?

The logic is similar to that for constructing chromosome maps. If two structures on the same side are far apart they will be separated by many mosaic boundaries, but if they are close together the proportion falling in separate male and female patches will be small. Since the embryo is bilaterally symmetrical, data for the same structures (landmarks) can be combined from the two sides, and the proportion of the total falling in separate male and female patches calculated. This value, as a percentage, was named a *sturt* by Hotta and Benzer (1972), in memory of A. H. Sturtevant's original contribution. The sturt distance for two landmarks is therefore a measure of their separation and is calculated from the following matrix:

| | Landmark I | |
|---|---|---|
| Landmark II | ♀ (normal) | ♂ (mutant) |
| ♀ (normal) | *a* | *b* |
| ♂ (mutant) | *c* | *d* |

Where $a$ is the number of landmark I in female tissue with landmark II also in female tissue, and so on. The distance between landmarks is then:

$$D = \frac{b + c}{a + b + c + d} \times 100 \text{ (in sturt units)}$$

If the probability of generating ♂ and ♀ mosaic areas is not equal, as with *mit* or for mosaics made by mitotic recombination (see below), the formula has to be proportionally adjusted, as in Gelbart (1974). As with meiotic gene mapping, map distances are not additive over large ranges, and this disparity may be 'corrected' by using an exponential mapping function (Janning, 1978). Most maps are built up from adjacent landmarks, and this difficulty is then trivial.

Maps relate to the two-dimensional surface of the egg, of course, and not to a single dimension. A third landmark is therefore related to the first two by triangulation, by finding its difference in sturts from each of the first two (Fig. 8.7). This will always give two possible positions, and which is correct will depend on mapping a fourth landmark or on prior information about the structures involved (e.g. tergites would be expected to be dorsal to sternites, wings to legs, etc.).

Larval structures can also be fate mapped by exploiting staining for the cell autonomous enzyme aldehyde oxidase, specifically by using the recessive *ma-l* which inactivates the enzyme, and by using the pigment genes white and chocolate (Janning, 1974a, b). Similarly, other enzymes have been used to map internal organs (Kankel and Hall, 1976; Lawrence, 1981a). A map covering much of the available data is given in Fig. 8.8, which additionally marks some locations as rods or circles. These are the sites of structures which arise from more than one cell, and where mosaicism within the structure can be measured (e.g. between particular bristles or sensilla etc.). The frequency of this intrastructural mosaicism tells us about the size and shape of the primordium itself (Hotta and Benzer, 1972). The frequency is identical to the map distance for rod-shaped primordia, and half the circumference value for circular primordia (Wieschaus and Gehring, 1976a). The sizes of the circles and rods (Fig. 8.8) therefore indicate the area from which the structures are derived. Since we know the number of cells in the blastoderm to be 6000 (Zalokar and Erk, 1976), and that the surface area of an egg is about 18.7 ksturts$^2$, we can calculate the size of a sturt to average about two cell diameters or 7.4 μm (Janning, 1978), although there is some argument about this value (Wieschaus, 1978).

Figure 8.8 is a fate map which tells us only that a cell (or group of cells) formed at a particular location on the egg surface will duly become a particular structure *in the course of normal development*. This may imply some cortical organization which affects the likelihood that a cell will follow a particular course of multiplication and differentiation, but the map does not prove this. Nor does it imply that the cell is determined when it is formed. What the map provides is a more comprehensive picture of

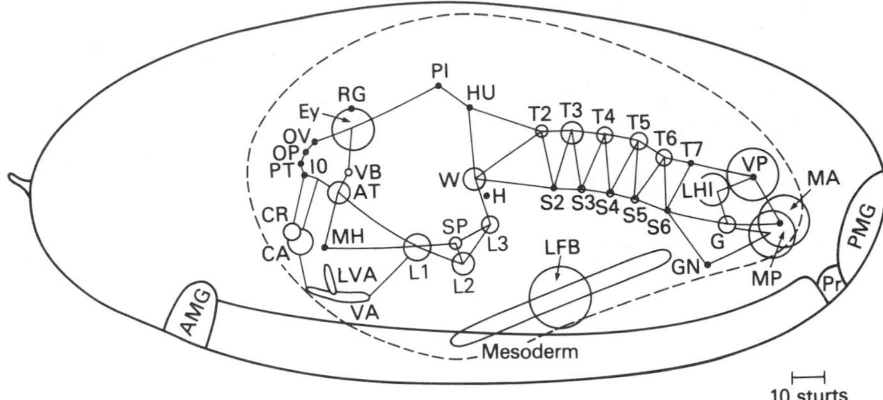


**Fig. 8.8** Some of the main features of the gynandromorph fate map of the *Drosophila* embryo, based on Janning (1978) and Underwood *et al.* (1980); cf. Fig. 8.1. Abbreviations: AMG, anterior mid-gut; AT, antenna; CA, cardia; CR, crop; EY, eye; G, inner and outer genitalia; GN, gonad; H, haltere; HU, humeral bristle; IO, interocellar bristle; L1–3, legs 1–3; LFB, larval fat body; LH1, larval hind gut; LVA, larval ventriculus; MA, MP, anterior and posterior Malpighian tubules; MH, mouth hooks; OP, occipital bristles; OV, outer vertical bristles; P1, proboscis; PMG, posterior mid-gut; Pr, proctodeum; PT, post vertical bristles; RG, larval ring gland; S2–6, sternites 2–6; SP, sternopleural bristles; T2–7, tergites 2–7; VA, anterior ventriculus; VB, vibrissae. The broken line shows the blastoderm border obtained from fate maps, as positioned in relation to the larval structures damaged by ablation (cf. Fig. 8.1**b**).

egg–blastoderm organization than earlier cytological studies (Fig. 8.1), and a better appreciation of this complexity.

Gynandromorphs (and the induced mosaics discussed under the next heading) have also been used to get estimates of the numbers of cells initially set aside to form a particular structure, say the wing. If this primary number is two, then the smallest possible mosaic area of the wing itself would be one half. If it is four, the minimum mosaic area would be one quarter. Thus, the reciprocal of the smallest fraction of the structure which is mosaic estimates the initial number of cells. Of course there are errors in such estimates: the sample of mosaics may miss the rare smallest one, growth may be unequal or cells may die, and identifiable landmarks may not properly cover the entire structure. Even so, analyses of this sort correspond reasonably well with direct cytological observations (Table 8.3), and fit the converse argument that the frequency with which a structure will be mosaic will be greater the larger the initial cell pool (Merriam, 1978). And all the data prove that the major structures of the *Drosophila* adult are *polyclonal* in origin: they derive from groups of cells, not single cells, except possibly for the genital disc.

**Table 8.3** Initial numbers of cells in *Drosophila* imaginal anlagen.

| Primordium | Counted cells | Estimated cells | |
|---|---|---|---|
| | | Gynanders | Mitotic clones |
| Antenna | 35 | 7 | 3–9 |
| Eye-head | 42 | | |
| Eye-antenna | 70 | 23 | 6–13 |
| Wing | 38 | 12–40 | 11 |
| Haltere | 20 | | |
| Leg | 36–45 | 7–20 | 18 |
| Genital | 64 | 2 | |
| Tergite | 19–21 | 1–8 | 8 |
| Sternite | 12–13 | | 5 |

The cell counts are from Madhavan and Schneiderman (1977), and the estimates based on clone sizes are from various sources. The discrepancies between the estimates and actual counts may be due to lack of landmarks, to the contributions of primordial cells to non-cuticular tissue (trachae, nerve, etc.) and to inadequacies in some of the experimental data. The counts provide the best estimates of primordium sizes.

### *Mitotic recombination*

Stern's (1936) thorough study proved that recombination occurs spontaneously in *Drosophila* somatic cells, although rarely. He also correctly deduced that the frequency of exchanges could be greatly increased by X-irradiation, as subsequently shown by Friesen (1936). Hence, X-irradiation of timed developmental stages, heterozygous for marker genes, became an additional, and powerful, tool for labelling cell clones. Mutations on all three major chromosomes can be used, and, by having different recessive markers on the two homologues, twin spots can be generated to prove the crossover (Fig. 8.2). The method has two limitations: the exchange occurs in a cell by chance and cannot be directed to the structure of special interest, and the crossovers themselves tend to be concentrated around the centromere. So chromosome manipulation is sometimes necessary, and many mosaics have to be made to find the relevant few. But they can be made at all developmental stages.

Taken overall, mitotic clones (counting both twin spots) tend to suggest smaller numbers of primordial cells for a given structure than does the minimum mosaic patch. In some cases, this may be explained by the discovery that early induced mosaics produce large clones which spread over both leg and wing (Steiner, 1976; Wieschaus and Gehring, 1976b; Lawrence and Morata, 1977). That is, the clone is established before leg and wing primordia have separated. And the same has been found for eye and antenna (Postlethwait and Schneiderman, 1971a; Baker, 1978a). The cell number estimate depends on the clone being made when the primordium has been established. We shall return to this point, noting here only that, in these examples, the mitotic clones show that leg–wing and eye–antenna originate each as single cell groups which only later become developmentally separated.

We should expect that clones made at later stages in development will be smaller since each cell then has fewer progeny. This is the case for disc cells, but not for histoblasts irradiated prior to pupariation. Disc cells therefore multiply during the larval period, and histoblasts do not. Direct counts (Fig. 8.9) have now replaced the estimates of division rates derived from clones (Madhavan and Schneiderman, 1977), but what clones show in addition is the orientation of cell divisions. For example, twin spots induced between 48–72 h are mainly transverse to the main axis of the leg, while later ones are longitudinal (Bryant and Schneiderman, 1969). Regional differences of mitotic activity in wing, leg and eye have been exposed by differences between clone sizes of twin spots (e.g. Becker, 1957). Twin spot clones expose the patterns of cell division during development, although not always during the last 8 h or so prior to puparium formation. Clones carrying the autonomous genes Hairy-wing (*Hw*), hairy (*h*) or achaete (*ac*) induced at this late period are not expressed in the wing, although marker genes are (e.g. singed (*sn*), yellow (*y*) and multiple wing hairs (*mwh*)); the phenomenon of *perdurance*, described previously (p. 157) applies (Garcia-Bellido and Merriam, 1971). In these cases, either sufficient wild-type gene

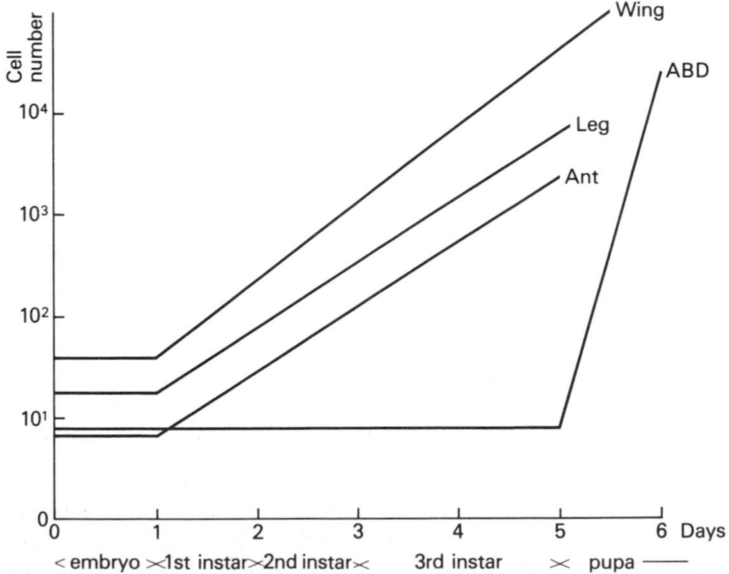


**Fig. 8.9** The growth patterns of some imaginal discs as estimated from genetic mosaics (from various sources). The cell number scale is logarithmic. ABD, abdominal histoblasts which start to multiply only at pupariation. Ant, the antennal disc which, like the others, starts to grow during the 1st larval instar.

product is accumulated and transmitted to the progeny cells, or the wild-type gene is not required at this late stage.

Mitotic clones can also be used to explore the action of mutant genes isolated in patches of normal tissue, and we shall look at the hypomorph engrailed (*en*) as an example. *en* is named after its scalloping effect on the scutellum, but it also alters the wing blade (see below) and introduces a secondary sex comb in mirror image position to the normal sex comb on the first tarsal joint of males. The question posed by Tokunaga (1961) was: can a small clone of *en* tissue produce secondary sex combs in an otherwise wild-type leg? One mating used to make such a clone is shown in Fig. 8.10. The result of the experiment was clear cut: even a small sector of *en* tissue falling in the secondary sex comb location produces a sex comb tooth, or teeth. The *en* gene is therefore autonomous, and it apparently responds to a prepattern, with an underlying singularity, present in the posterior half of

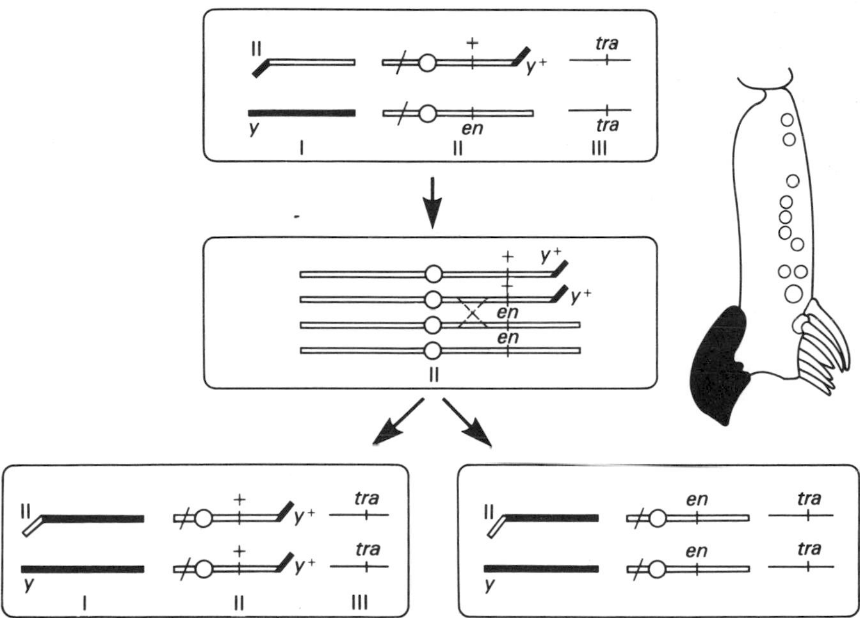


**Fig. 8.10** The generation of homozygous engrailed (*en*) clones in phenotypically male flies; after Tokunaga (1961). The heterozygous second chromosomes (II) carry *en* and a translocation of yellow+ ($y^+$), respectively. The first chromosome (I) pairs yellow and a translocation of the tip of the right arm of chromosome II. The third chromosome (III) is homozygous for transformer (*tra*) which converts females into males and thus allows expression of the sex combs. The X-ray induced cross-over is indicated by the dotted line in the middle diagram, and the daughter cells arising from mitotic recombination are below this. The left hand cell is non-*y* and non-*en*, and the right hand cell is *y* and *en*: both are phenotypically male. The 1st tarsal leg segment carries its normal sex comb on the left, and a *y en* clone on the right which allows expression of seven sex comb bristles. The clone runs the length of the segment.

the otherwise non-*en* leg. Wild-type cells cannot do so. This, and earlier work, suggested to Stern (1954a; 1968) that "development may be regarded as a sequence of prepatterns, each one being a realized pattern as compared to its predecessor, and a new prepattern as the basis for its successor". Stern and his colleagues therefore searched for mutants affecting prepatterns, but they found none (Tokunaga, 1978).

When the effect of *en* on wing morphology is examined it is found that the triple row of bristles characteristic of the anterior margin is duplicated on the posterior margin, replacing the usual double row of long hairs. Similarly, the alula of the posterior wing is replaced by a costa (anterior), or costa-like, structure. *en* transforms the posterior wing to anterior wing in mirror image arrangement (Fig. 8.11), although sometimes irregularly, since gene penetrance is variable (Garcia-Bellido and Santamaria, 1972). It is without effect on anterior wing tissue.

The further examination of *en* function depends on another discovery made using mitotic clones. The Minute (*M*) mutants are homozygous lethals which cause delayed larval development when heterozygous, due to slow cell multiplication (Brehme, 1939). So, induced heterozygous *M* clones in a wild-type background will be outgrown by the wild-type cells, which we should then expect to fill the major part of the wing area. This does not happen. Instead, the wild-type clones (marked by multiple wing hairs, *mwh*) fill most of the anterior, or most of the posterior half of the wing, but do not cross an otherwise invisible boundary line between the two, even when they run along half its length, or more (Fig. 8.11). Wings are therefore divided into 'compartments', anterior and posterior (Garcia-Bellido *et al.*, 1973). If the non-Minute chromosome also carries the *en* gene, this compartment restriction is lost: *en*-marked, rapidly growing clones cross the

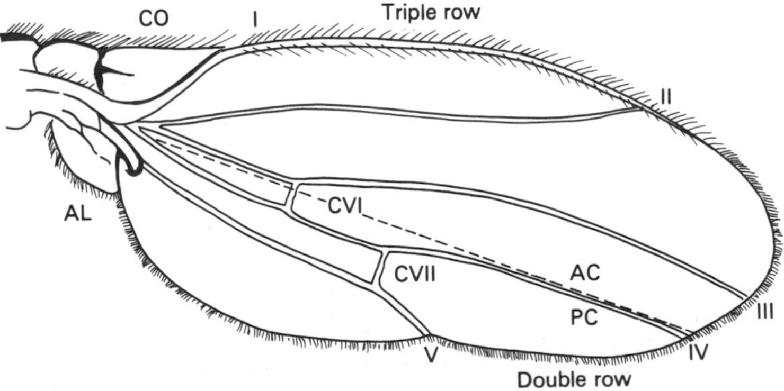


**Fig. 8.11** Diagram of the *Drosophila* wing, dorsal view. The wing veins are numbered I–V, and the anterior and posterior cross veins CVI and CVII. The anterior margin is distinguished by a triple row of bristles and the posterior and posterior-lateral by a double row of bristles. CO is the costa and AL is the alar lobe. The dashed line shows the dividing line between the anterior and posterior compartments (AC and PC), which is unrelated to any structural feature of the wing.

compartment boundary. $en^+$ is therefore the gene responsible for maintaining the straight boundary between cells of the anterior and posterior compartments, and Morata and Lawrence (1975) propose that $en^+$ acts by 'labelling' cells of the posterior compartment so that they do not mix with anterior cells during wing disc growth. It is not yet clear why the morphological polarity of these cells is reversed to produce a mirror image of the anterior wing, as is the case. Interestingly, staining mature prepupal wing discs for aldehyde oxidase shows that the anterior compartment expresses this enzyme, but the posterior does not. Also, the enzyme is found in both anterior and posterior compartments of discs from *en/en* prepupae (Kuhn and Cunningham, 1977). There is a true biochemical difference between compartments, as predicted, although it is unlikely that the oxidase staining indicates the direct biochemical cause of the clonal distinction.

$en^+$ is a special kind of gene (a *selector* gene in the model described on p. 268) responsible for maintaining the characteristics of the posterior wing compartment. *en* could have the same action on the foreleg in Tokunaga's experiment, since the posterior (compartment) sex comb there is a mirror image repeat of the anterior (compartment) sex comb, and the leg bristle pattern is similarly mirrored. There is therefore no prepattern.

These data strongly suggest that legs also have compartments, and Steiner (1976) demonstrated their existence using the Minute technique. In addition, he showed that early clones extended from wing to second leg and from haltere to third leg, and frequently between left and right forelegs, possibly because the last pair of primordia fuse during gastrulation. These clones were almost always restricted to one or other compartment. Compartment polyclones are therefore established before wing and leg primordia separate, as a primary event of differentiation. They have since been shown to arise similarly in the genital disc (Dübendorfer and Nöthiger, 1982), the head (Morata and Lawrence, 1978; Baker, 1978b) and the proboscis (Struhl, 1977). There is therefore an homologous arrangement in different discs. Further, in their pioneering work on the wing disc, Garcia-Bellido *et al.* (1973) showed that there were subsequent polyclones established which divided this disc into proximal–distal compartments, although this division is less rigorously maintained than the anterior–posterior compartment (Brower *et al.*, 1981). A mutant, Distal-into-Proximal (*Dipr*), affects this second compartmental restriction in the adult appendages (Kerridge, 1981).

We noted earlier that some *en* mutants are embryonic lethals. We might argue from this more extreme phenotype that they are null mutants, and that the results for the viable $en^1$ mutants, described above, do not necessarily define the role of this important locus. Lawrence and Struhl (1982) have examined this issue by making clones of $en^{lethal}$ genes (marked with straw and pawn bristle mutants) in a Minute background. Like $en^1$, clones of $en^{lethal}$ have no effects in anterior compartments, produce abnormalities in posterior compartments, and transgress the anterior–posterior compartment boundary. Also as expected, $en^{lethal}$ clones in the anterior compartment do not cross into the posterior (wild-type) compartment, supporting the assumption that $en^{+}$ 'labels' posterior cells so that their affinities become distinct from anterior cells. Not all the results are as clean as this would imply: there is a deficit of $en^{lethal}$ clones in the proboscis, and $en^{lethal}$ clones do not form secondary sex combs, for example. Thus, while details of $en^{+}$ function are still open to further study, this example illustrates a further use of mitotic recombination for the study of gene action.

The second, later round of compartmentalization (and there may be more) probably eliminates the possibility that compartments reflect only the fusion of primordia which were separate at some earlier stage of the evolution of the organism. Compartments are therefore important supracellular units of organization, polyclonal in origin (Crick and Lawrence, 1975), which ensure that the descendants of particular cells remain unmixed. This distribution mechanism segregates about twice as many cells to the anterior as to the posterior compartment of the wing, and about the same ratio to the wing–notum compartments. For the present, however, we can only speculate about the role of this basic, genetically controlled organization with respect to morphogenesis. Whether or not it occurs in other organisms has still to be explored.

We can bring together some of the issues raised by this clonal analysis by looking again at the *Drosophila* eye (Fig. 8.6). According to Baker's (1978a) gynandromorph analysis, only 10 blastoderm cells give rise to the head. Each progenitor cell may produce one or more terminal structures, but only a particular group of such structures. In this sense, a cell is fated in normal development to a limited repertoire of terminal states. By 27 h this head cluster of cells is divided into a dorsal (6-cell) compartment and a ventral compartment, the division running about midway across the eye proper but also including the postorbital chitin and its associated bristles. By 75 h a further compartment division is established at right angles to the first, separating a large anterior from a smaller posterior compartment (Baker, 1978b). Two other compartment divisions parallel to the first have also been identified in the anterior compartment (Campos-Ortega and Waitz, 1978). The first and second compartments appear not to respect clonal restrictions since progeny of the same blastoderm cell may end up in different compartments. These compartments bear some relation to the mitotic recombinant clones previously defined by Becker (1957) and to V-type position effect clones (Baker, 1967). However, this is true for only part of the eye, as Fig. 8.6 shows. So we find that cells of the same compartment give rise to different structures; ommatidia and head chitin. And cells of different compartments give rise to the same structures, the ommatidia. By using clones of white induced early in development, which eliminate pigment granules in both the pigment and receptor cells of the ommatidium, Ready *et al.* (1976) were able to show that mosaicism could be found in all cell types of the crystalline lattice-like structure of the ommatidium. The differ-

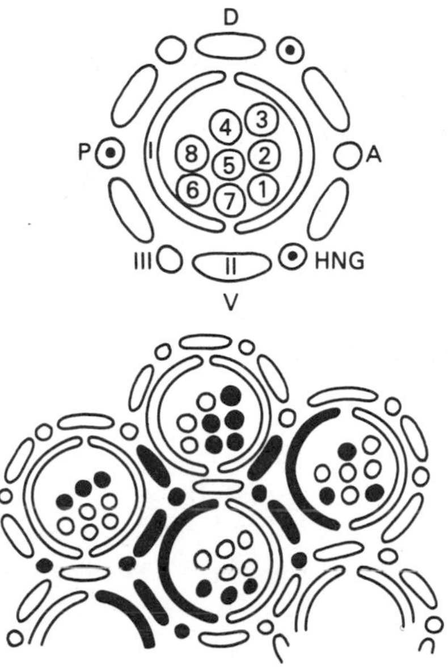


**Fig. 8.12** The cells of the *Drosophila* retina form a regular crystalline-like lattice, the details of which are shown above. A and P give the anterior-posterior orientation and D and V the dorsal and ventral orientation of this right eye. The retinula cells are numbered 1–8, the primary, secondary and tertiary pigment cell I–III, and the hair nerve group, HNG. Induced mitotic recombination to give pigmented cells in a white background permits coloured clones to be identified in sections of the retina. An example from the data of Lawrence and Green (1979) is shown which clearly illustrates the random combinations of the mixed cell types found. There is apparently no fixed cell lineage in ommatidial development.

entiation of the ommatidium is independent of the clonal lineage of the cells (Fig. 8.12), and which compartment they come from makes no difference either. Although there is early determination, as shown by the fate map, there must also be what we can loosely call inductive relationships during the development of the eye. It is not obvious what role compartments play in either of these processes, especially when the same differentiation spans a compartment boundary.

### Internal structures

The preceding studies with *Drosophila* show that the clonal history of cells plays an important part in the organization of adult cuticle. It has been more difficult to determine whether this is also true for internal structures which lack the obvious markers, like pigment differences or bristle shapes. Particular enzymes, like the autonomously expressed aldehyde oxidase, have been stained for (p. 142), but as we noted, such enzymes are not necessarily expressed in all tissues or even in all parts of an organ. Mutant non-staining clones may therefore be missed, or confused with regions of non-expression of the gene. It is only recently that a mutant of succinic dehydrogenase (*sdh*) has been found and developed as an apparently general cell marker which clearly distinguishes normal (stained) and mutant (unstained) cells in the soft tissues of mosaic larvae and flies (Lawrence, 1981a).

Using a variety of enzyme mutants, Ferrus and Kankel (1981) show that larval abdominal histoblasts (p. 30) give rise to clonally related epidermis, oenocytes and fat cells; that peripheral fat and epidermal cells have common precursors, at least in the notum; that patches of tissue lacking acetylcholine esterase are found in nerve tissue and that these differentiate abnormally, sometimes causing partial paralysis; and that adepithelial cells, which are found within the wing discs and give rise to its associated musculature, are not clonally related to the epidermal cells which form the discs. No evidence of compartments was found in any of these internal tissues.

The observation on adepithelial cells has been taken further by Lawrence and Brower (1982) who show that these cells are of mesodermal origin. By injecting fragments of wild-type wing discs (0–2 pieces of p. 227) into *sdh/sdh* mature larvae, they also find that the adepithelial cells contribute to adult muscle of abdomen and thorax, and are not restricted to the dorsal thorax as they would be during normal development. The cuticular wing structures, on the other hand, differentiate normally. Clearly there is a notable difference in the characteristics of the derivatives of ectoderm and mesoderm, and this corresponds to the embryological observation that damage to embryonic ectoderm affects mesoderm development whereas damage to mesoderm is without effect on ectoderm development (Sauer-Locher, 1954). In brief, development of mesoderm is dependent on its relationships to (at least) ectoderm, whereas the development of ectoderm is independent and quite differently organized, as we have seen. In view of this, it is perhaps surprising that Ferrus and Kankel (1981) find the mutant engrailed, and mutants of the bithorax complex (p. 255) which transform meta- to mesothorax, or cause a leg to form on the first abdominal segment, or are defective in dorsal appendages, are all without effect on internal tissue organization, or have trivial consequences unrelated to the mutant phenotypes. This may imply that these mutations act later than the determinative events which precede the differentiation of internal structures.

### Mosaic clones as tools

We have already seen that chimeras made by transplanting pole cells of female sterile *Drosophila* mutants can be used to check whether the defect is due to a malfunction in the pole cells themselves, or in the separately derived somatic tissue of the ovary (p. 159). Mitotic recombination can be used to make similarly instructive mosaics. Using *fs(k10)/mal* heterozygotes, and irradiating them at various stages of development, Wieschaus and

Szabad (1979) were able to follow the abnormal (malformed dorsal appendages) and aldehyde oxidase deficient eggs arising from the induced twin spots. Irradiation at 3 h after egg laying showed that only 7 of the 40 or so pole cells become incorporated into the ovary, confirming previous histological data and from later irradiations, that these cells increase in number during larval life, logarithmically in fact, to a maximum of 110 at 24 h after pupation. Mosaics induced in pupae and adults give a pattern of stem cell division (Fig. 7.3) which is quite different. The first *k10* eggs only appear 10 or so days after irradiation, implying that it takes this time for an irradiated stem cell to produce its first *k10* egg. Some females lay only a single *k10* egg and others only a few which are deposited irregularly and not in sequence. Apparently each ovariole possesses two or three stem cells which are variably active. Consequently, stained ($mal^+$) mosaic ovarioles are found to carry irregularly ordered cysts.

Clonal analysis can also be used to follow cell division in abnormal, mutant structures. The mutation costal (*cos*) causes large duplications (and sometimes triplications) of the anterior part of the wing blade in the region of the costa and triple row of bristles (Fig. 8.13). The induction of mitotic clones, marked by multiple wing hairs (*mwh*), permits measurement of clone sizes in both the normal portion of the wing and in the duplication,

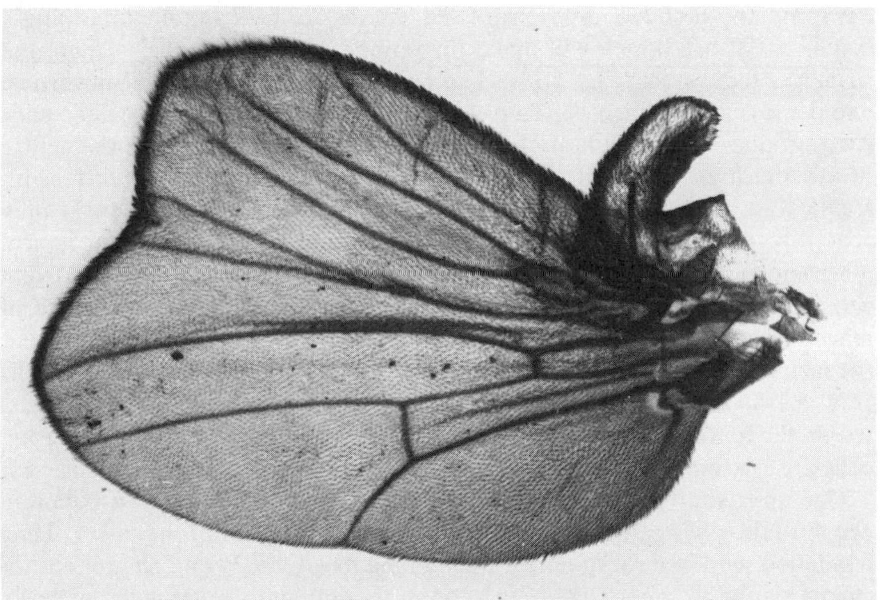

**Fig. 8.13** The mutation costal illustrates how a single gene change may cause a dramatic alteration of form (cf. Fig. 8.11). In this example we have a duplication in mirror symmetry around vein II, together with the typical outgrowth from the costal region, after which the mutation is named. The phenotype is very variable, and usually shows the costal outgrowth without wing duplication.

and of the initial number of cells contributing to the duplication. It is then found that the rate of cell division is similar in the two, but that division continues longer in the duplication. Variation in the sizes of the duplication depends on their initial cell numbers, not on growth (Whittle, 1976). Thus clonal analysis exposes the growth parameters within the duplication. Initiation of the duplication may be due to cell death, and we shall look at more certain examples of this in the next chapter where cell death is used as a tool.

Various other clone manipulations are also possible, and we have seen that mitotic clones can be used to place a group of cells of one genotype (*en*) in a normal environment to find how it responds (p. 209). However, making relevant clones depends on chance, and therefore on making large numbers of them, and this difficulty seems to have limited the exploitation of this non-surgical form of cell transplantation.

## Plant mosaics and chimeras

Somatic mosaics have been made with plants, particularly maize where we have the mutants which allow pigment patterns to be followed. In this monocot, the first two divisions of the zygote are horizontal, separating a distal, initial cell which will make the embryo, from the cells forming the suspensor below (cf. Fig. 1.13). The next division of the initial is vertical and the loss at this stage of a ring chromosome covering the white deficiency (*wd*) produces half-and-half plants where the separation runs up the centre of the midrib of each leaf to the midline of the terminal tassel (Steffenson, 1968). The plant is made up of two clones, but we cannot be sure if they are truly compartments. Subsequent initial divisions produce the apical meristem, and radiation-induced crossovers suggest that some meristem cells (see below) produce clones which run the length of the sequence of leaves formed during meristem growth. Each leaf is striped with pigment (or lack of it) as a result of the continued radial proliferation of single cells (Fig. 8.14). The established meristem produces a set of side-by-side clones, within the boundaries of the clones from the two initials. These proliferating cells are not the only derivatives of the initial pair, as we shall see below.

Coe and Neuffer (1978) have used clones to ask if there is a common origin to the physically separated sex cells of the ear and of the tassel. They irradiated seed heterozygous for the dominants (*A*, *B*, *Pl* and *Rr*) for anthocyanin synthesis, where loss of two, or more, dominant genes gives an easily identified green sector on a purple background. The embryo of the dormant seed already has 5 or 6 leaves, and a meristematic area (shoot apex) which will elaborate 15 more leaves, the ears and the tassel. Examination of 45 sectored tassels showed that most were 50 percent green, and that the average was about one third green. In other words, clone progenitors of

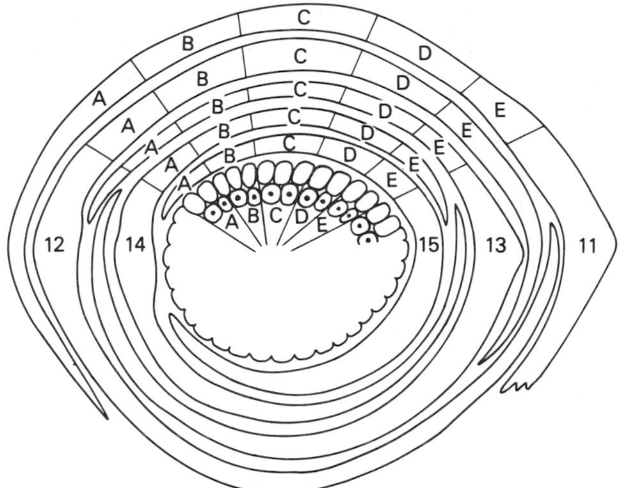


**Fig. 8.14** By following the progeny of radiation-induced gene-loss mutants, Steffensen (1968) was able to mark the radial proliferation of cells from the dormant shoot apex through to the generation of 15 leaves in maize. The above cross-section, reconstructed from a series, shows the sequence of clonally related leaf segments derived by the proliferation of apical cells A–E, as found in leaves 11–15. The leaves and stem (culm) are striped by the files of cells derived from the original apical mutants.

2–4 cells are set aside in the meristem, and are there determined to form the tassel and male sex cells.

The situation is quite different for the ears, which are modified branches appearing at the 11th–16th nodes of the lower half of the plant. Sector sizes suggest that an average of 32 cells contribute to an ear-bearing node, although there is the complication that some of these cells also contribute progeny to inferior and superior nodes. Within this set, 2–4 cells give rise to an ear and to the female sex cells, for their sectors again constitute half to one quarter. Since the side of the plant on which ears are formed alternates from node to node, it is argued that these ear-forming cells must enter from a body-forming meristem, from one side at one level and from the other side at the next level. Alternatively, there could be two sets of ear-forming cells, only one of which is expressed at each node level. Both possibilities imply late determination of these cells (unlike the male cells) and a flexibility of final differentiation which is typical of plant cell fates.

Plant chimeras can be made by grafting together mature, growing parts of many kinds. Their study adds two points to the work with corn (and many further details described by Stewart (1978)). If the meristem is a composite of two genotypes, as when a scion of one genotype is grafted onto the rootstock of another, the new growth may be sectored. These sectors often make up one half or one third of the plant, and they may run through many nodes of the stem (Stewart and Dermen, 1970). It follows that only two or three of the meristem cells are involved in new growth from the apex. It also seems likely that such cells must remain in a distal position in the meristematic layers, otherwise they would be left behind, and they must behave as a stem cell (in the cytological sense), budding off daughter cells while remaining in the meristematic tissue (Esau, 1965). It is not clear how these cells become determined, but the fact that a shoulder appears on the apex prior to the formation of a leaf and/or a lateral bud, and that leaves and buds are then formed from among the cells of the shoulder (i.e. by their division), suggests that determination is location dependent. This second point is supported by the observed regularity of leaf form and pattern even when single clones may contribute few, or very many cells to the leaf blade. Cell position is of great importance for the pattern of plant development, and it must ultimately decide how a plant cell differentiates within a clone covering a variety of the structures of stem and leaf.

## Conclusions

These studies show that mosaics are more powerful tools than chimeras for following the clonal history of cells. In particular, they allow the delineation of an exact fate map of the *Drosophila* blastoderm almost cell by cell, and this exposes a new problem. Unlike the clonal determination of *Caenorhabditis* structures from single blastomeres, the *Drosophila* adult ectodermal anlagen start as groups of cells (meso- and metathoracic legs from 42 and 24 cells, wing and haltere from 38 and 20 cells, respectively) and these numbers are remarkably constant between individuals (Madhavan and Schneiderman, 1977). It is difficult to imagine a mechanism of this precision, which would separate and identify these various cell groups within the general sheet of blastoderm cells. Further, these groups are subdivided into (at least) anterior and posterior compartment polyclones, and the evidence of the engrailed mutation shows that this more general subdivision is also regular. We cannot, as yet, relate this cell group determination to the actions of the maternal-effect genes described in the preceding chapter.

The same rules do not seem to apply to the precursors of internal structures of *Drosophila* which apparently depend on local signals for their differentiation. The data are still too few for us to be certain, but they suggest that mesoderm derivatives at least, regulate according to their position in the embryo. This local regulation is paramount in plant development where, at least in angiosperms, the cells of different organs retain their totipotence (p. 7). Position also plays a primary role in the early differentiation of the mouse (p. 13). Although the data from mouse chimeras and mosaics still require refinement, they agree with the expectation that clonal patterns of determination would not be found. But here there may be complications.

Monk (1981) suggests that mouse development depends on a stem cell system where only part of a pluripotent stem line changes state in response to extrinsic signals, the remainder continuing as a stem line which then repeats the process to form the next determined set of cells (Fig. 8.15). The primary germ layers are thus subclones branching from the continuing stem line. The data from mouse chimeras are insufficient to confirm or refute this hypothesis (or others which attempt to account for mouse cell lineage, as Gardner (1981) notes), but this model serves to emphasize that there may be situations where a clonal determination is important in development even when cell fates are subsequently modulated by extrinsic signals (cf. the data on plants). It seems clear (Snow, 1981) that the mouse embryo becomes determined by the primitive-streak stage since isolated parts then develop autonomously. Chimeras do not allow us to follow this process in the way that we can in *Drosophila* mosaics, and we cannot answer Snow's question: is development clonal?

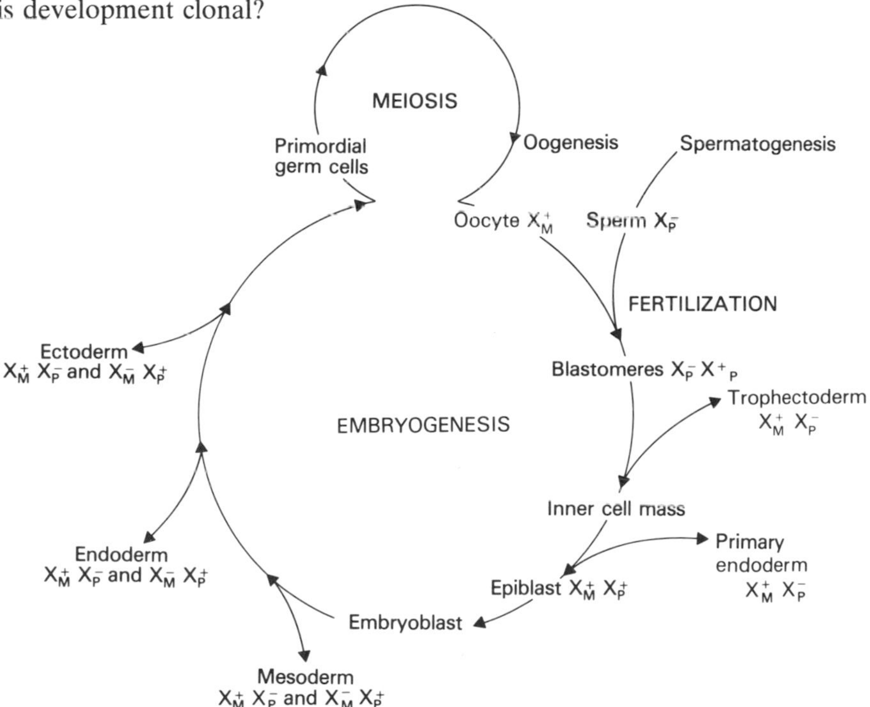


**Fig. 8.15** The stem line model of mammalian cell lineage is like a catherine wheel, throwing off polyclones (trophectoderm, primary endoderm, etc.) and itself becoming modified in the process. This restriction of potency is reversed during meiosis and returns the genome to totipotence. Part of the evidence for such a system comes from the pattern of X-inactivation in female embryos, as summarized in the diagram. $X_M$ and $X_P$ are the maternally and paternally derived X chromosomes, and their activity is indicated by + or −. Note that the paternal X is heterochromatic in sperm, is active in the blastomeres, is inactive in trophectoderm and primary endoderm, and that either X is randomly inactivated in the embryoblast, and progeny cells retain this state subsequently. Modified from Monk (1981).

# 9 Determination and regeneration

Apparently in multicellular organisms there are two types of determination, different in principle. One should distinguish between a *determination resulting in immediate differentiation* and a *determination resulting in a reproducible cell state propagated in the non-differentiating phase by cell heredity*. The first type of determination is characteristic of the cells of a *Drosophila* embryo that differentiate into the functional larval organs, whereas the cells of the imaginal discs are characteristic examples of cells belonging to the second type.

These two types of determination also occur in the vertebrate body.

E. Hadorn, 1965

## Imaginal discs

When Hadorn first exploited Bodenstein's (1943) discovery that an imaginal disc taken from a *Drosophila* larva could be grown in the abdomen of a fertile female, as in a culture vessel, he provided a new approach to the examination of the stability of the determined state. His thesis was that these cultured discs could be used as a model system, open to surgical, genetic and other manipulations, which should throw light on many 'unexpected and hard to explain events', particularly the regularly reproduced, determined cell state propagated during the growth of the disc prior to metamorphosis and terminal differentiation. This is the thesis we shall examine, so we must first look more closely at imaginal discs.

As we have seen from the fate map (Fig. 8.8) the cells which will make adult structures are segregated in the blastoderm, and are distinct from the cells for larval tissues which differentiate during embryogenesis. The cells which will make the adult soft tissues are set aside in the corresponding larval organs (gut, salivary gland etc.) and they will not concern us. Nor shall we consider the larval tissues which survive metamorphosis, like heart, Malpighian tubules and reorganized nervous system. We are interested only in the cells which form adult ectoderm, the exoskeleton of the fly (Fig. 1.14). The cells which will form the adult abdomen are segregated as histoblast nests in the wall of the equivalent larval segments, and the remainder are separated as well-organized structures, the imaginal discs. They can first be identified during late embryogenesis (Auerbach, 1936), and towards the end of that stage each forms a small ampulla by invagination from the embryonic epidermis. This single-cell layer sac, with the outside now inside, remains attached to the epidermis by a stalk. It is this structure that is called an imaginal disc. There are three pairs of discs in the head region (labial, clypeo-labrum and eye-antenna), paired dorsal and ventral discs for each thoracic segment (humerus region of prothorax, wing and haltere, and the three pairs of legs) and a single median, terminal disc (genital), as shown in Fig. 1.14. Discless mutant larvae develop normally to pupation (Shearn and Garen, 1974) so the discs have no larval role.

Unlike the surrounding larval cells which become polyploid (Bautz, 1971), disc and histoblast cells remain diploid and multiply during the larval period, each according to its own disc pattern (Fig. 8.9). Histoblasts only multiply at, and after, pupariation (Roseland and Schneiderman, 1979). Accompanying this growth is a distinctive folding of the tissue, and the resulting different disc shapes allow them to be identified. At pupariation, the discs are everted through the stalk so that the inside again becomes the outside, and changes of cell shape result in elongation of legs and antennae, as if the sections of a telescope were being pushed out. Hydrostatic pressure causes expansion of the wings. These changes, which are accompanied by differentiation of the cells, are induced by the large hormone pulse which occurs at the end of the 3rd larval instar (Fig. 1.13), and its effects can be copied by adding 20-hydroxyecdysone to 3rd instar discs cultivated *in vitro* (Mandaron, 1973; Milner and Sang, 1974). 20-hydroxyecdysone is therefore a general differentiation signal for disc cells which, by this stage, must have become determined to form the individual parts of each organ. The adult exoskeleton is therefore a mosaic of fused, differentiated discs which originate from the small cell clusters of the embryonic ectoderm which we located in the blastoderm fate map (see Bryant (1978) for details).

### *Imaginal cell determination*

We have already examined some of the evidence that imaginal cells are determined at blastoderm, at least with respect to their general fate (leg, wing etc.). The classic experiment which first suggested this used the technique of culturing cells in females. Genetically marked cells were taken from anterior or posterior halves of blastoderm (3 h) embryos and mixed with cells from whole embryos, differently marked (Fig. 9.1). The two groups

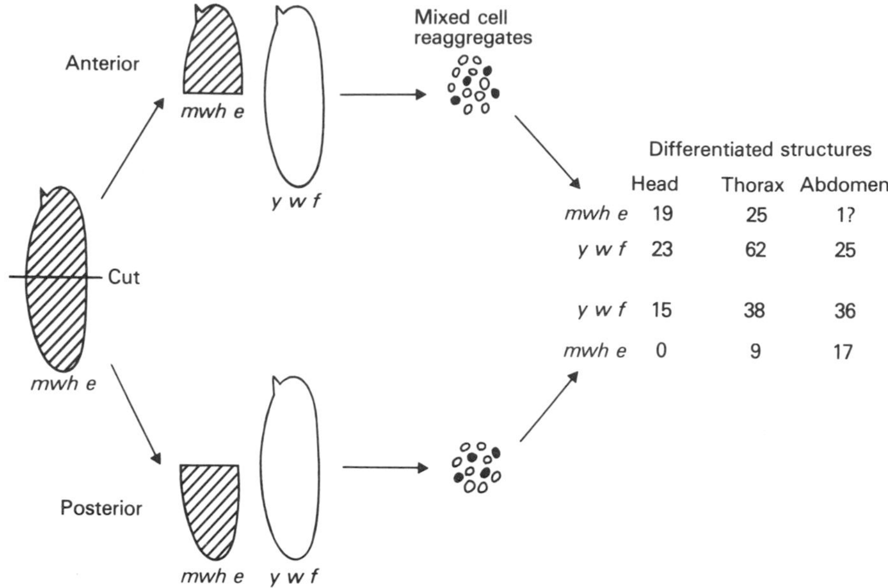


**Fig. 9.1** Blastoderm embryos, genetically marked with multiple wing hairs (*mwh*) and ebony (*e*) are cut in half and anterior and posterior halves separately intermixed with whole blastoderm embryos marked yellow, white and forked (*y w f*). The mixed embryos are dissociated into single cells and reaggregated into pellets which are then grown in the abdomen of females prior to metamorphosis (see Fig. 9.2). The differentiated structures are classified by genotype for the body parts made, when it is found that anterior halves make anterior structures, posterior halves make posterior structures, and the complete embryos make the entire range. Data summarized from Chan and Gehring (1971).

of cells were each intermixed in single cell suspensions, and then reaggregated by centrifugation. The pellets of mixed genotypes were implanted into females and allowed to grow for 2 weeks, after which they were removed and metamorphosed by transplantation into 3rd instar larvae. The differentiated implants removed from the adults which hatched produced structures from all segments of the whole embryo genotype, but only head and thorax structures from the marked anterior half embryo cells. Similarly the marked posterior half embryo cells produced only thoracic and abdominal structures (Chan and Gehring, 1971). At the very least, this shows that blastoderm cells are determined to produce anterior or posterior adult structures. Since the cells were randomly mixed, it seems likely that this determination is very stable and not influenced by the nature of contiguous cells, though we cannot be certain of this due to the design of the experiment where only identifiable adult structures were scored.

The determined state is, indeed, a very stable *cell heredity*. This was first demonstrated by Hadorn (1963; reviewed 1978) who used adult females in which to culture 3rd instar (i.e. fully grown) imaginal discs. The procedure is outlined in Fig. 9.2. A disc can be grown, subdivided and subcultured

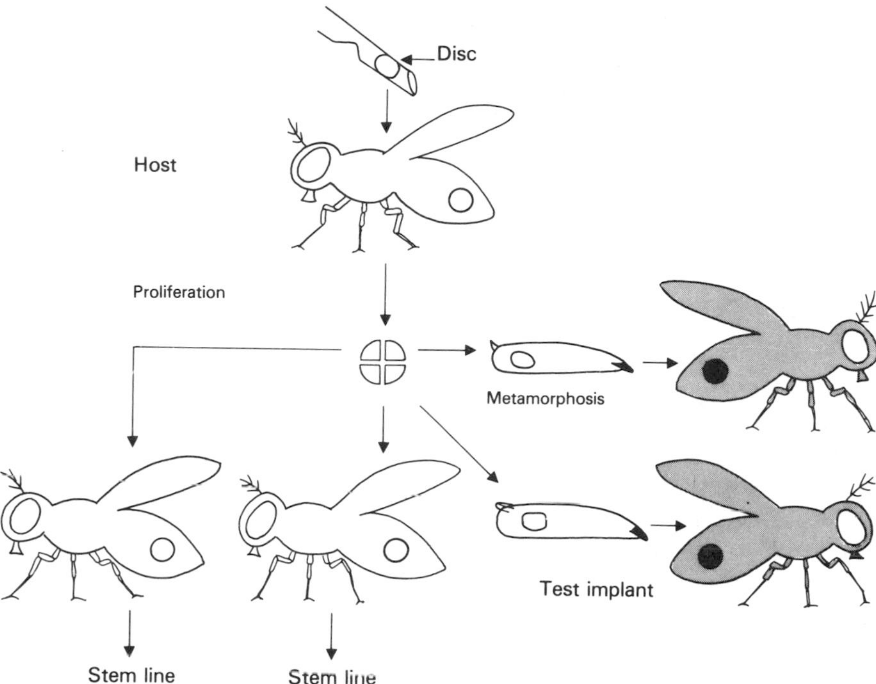


**Fig. 9.2** The cut imaginal disc is transferred by micropipette into the abdomen of a fertile female where it is allowed to grow for 2 weeks or so. The implant is recovered and divided, and one or two fragments (test implants) are transferred to larvae with which they undergo metamorphosis. The uneverted, differentiated implant is then checked for its characteristic identifiable structures. At the same time, the remaining implant fragments are again cultured in females, and the cycle repeated. These stem line cultures can be perpetuated indefinitely. As a precaution, disc and hosts are genetically marked to ensure that they are not confused (modified from Hadorn (1963)).

virtually indefinitely—Hadorn (1978) quotes 1500 cell cycles up to 1976. When metamorphosed, the disc is still capable of forming the original determined structures, such as male genitalia. The genome in each disc is locked into a particular pattern of genes accessible to transcription when metamorphosed. If the culture is made from only one area of the disc during the subdivision steps, say that part of the genital disc fated to make anal plate, then subsequent subcultures perpetuate this, and do not elaborate the other areas of the disc (but see the discussion of disc fate maps later (p. 224)). Since the disc grows during culture *in vivo*, it produces not one but a number of anal plates on metamorphosis. Determination is *arealized* in the unmetamorphosed, growing mature disc.

As was implicit in Chan and Gehring's experiment, the 20–40 blastoderm disc-precursor cells must elaborate a great variety of determined states during the growth of the disc to its final, prepupal state. However, leg disc precursors make only leg-type cells, wing precursors elaborate only wing cells, and so on. But disc cells are homogeneous in appearance, and one cannot identify, say, wing cells and leg cells until they differentiate. Further, as we have already seen for the eye–antenna disc (Fig. 8.6), the blastoderm cells are fated to form particular parts of the adult organ during the normal course of development. A disc therefore derives from a polyclone whose members undergo further, but covert, differentiation as the disc grows, and these changes are expressed when the hormone triggers differentiation.

### *Transdetermination*

Determination breaks down from time to time, and instead of the regular *autotype* being perpetuated a new *allotype* appears: the genital disc blastema produces leg, for instance. This change is called *transdetermination*, for we then have an extraordinary switch to the full determined state appropriate to another disc. These changes are not random but directional, and follow the regular pattern shown in Fig. 9.3. Some discs can transdetermine back and forth, like antenna and wing, while others transdetermine mostly in one direction. Once transdetermined, a blastema may suffer a second transdetermination, as just implied, with the result that they may then, or subsequently, become mesothorax, which seems to be the basic state from which they never, or rarely, return. Sometimes the transdetermination is from a *normotypic* to an *anormotypic* state when an abnormal structure is formed, such as a monster anal plate instead of the usual subdivided group of anal plate areas; sometimes the blastema is found to be *atelotypic* and incapable of responding to hormone as the normal *telotype* does. This last class behaves like a malignant neoplasm because differentiation cannot set a limit to its growth (Gateff, 1978a, b).

Much was made of the frequencies of the various transdetermination steps (Kauffman, 1973), but these data are probably invalidated by Strub's (1977) conclusion that only a small group of cells in the upper half of the first leg

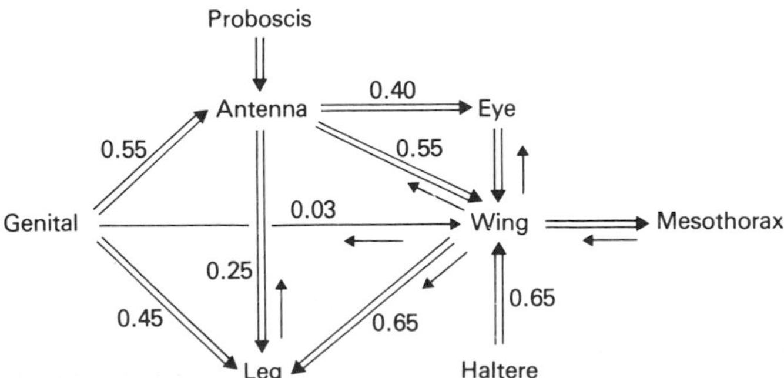


**Fig. 9.3** Transdeterminations are not random, but follow preferred sequences. The major directions, and approximate frequencies, are shown by the double lines, and reverse and infrequent transdeterminations by the single, short lines. Modified from Kauffman (1973). The significance of these data is discussed in the text.

disc can transdetermine; the other cells in the disc cannot. Chance could then decide the distribution of these cells, assuming they exist in all discs, when the blastema is fragmented prior to transplantation into new hosts. These special cells have to be at a cut surface or in a dissociated fragment, and must undergo division before they can transdetermine, which almost 100 percent do during the first culture passage. Apparently, more than one cell transdetermines at a time, since clones genetically marked with multiple wing hairs, and induced by somatic crossover to give *mwh* and *mwh*$^+$ cells, show contiguous patches in the allotypic structure (Gehring, 1967); the mosaic shows that at least two cell types therefore contribute to the transdetermined patch (Fig. 9.4). Transdetermination has also been shown to occur with the usual frequencies when discs are cultured *in vitro*, ruling out specific effects of adult haemolymph in this process (Shearn *et al.*, 1978).

Anormotypic and atelotypic changes occur at a sufficiently low frequency to be due to mutation, but transdetermination is too common for mutation to be a cause. For the same reason it is probably not due to the activation of transposable elements. It must represent some change in the state of the cells directly affecting the pattern of genes being transcribed. How this comes about is not yet known, but early discs (68–72 h larvae) and older discs (7–8 h prepupae) are not competent to transdetermine (Lee and Gerhart, 1973). Discs have to be of the right age (96–115 h) to be competent, which may reflect the state of the disc cells or their environment at that time. Strub's (1977) results question whether or not discs contain undifferentiated stem cells which are stimulated to divide when the tissue is cut, and then

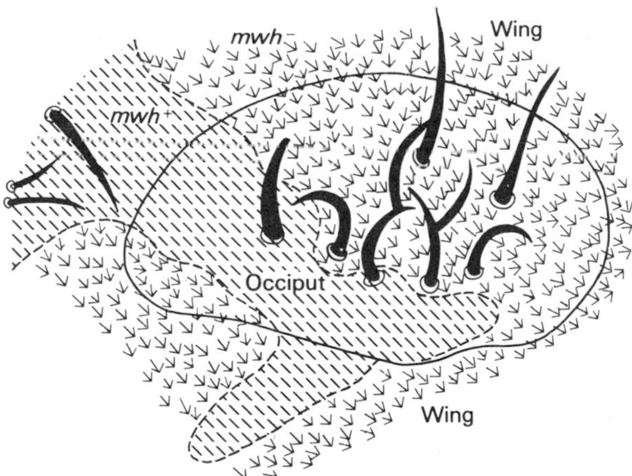


**Fig. 9.4** An induced clone of genetically marked multiple wing hairs (*mwh*) cells extends over the border (solid line) between transdetermined occipital—wing tissue, in this preparation of Gehring (1967). The *mwh* cells carry numbers of small trichomes variously arranged and the *mwh*$^+$ cells have large single trichomes orientated in one direction. Since the transdetermined patch contains cells of both types it must have originated from a synchronous change in at least two neighbouring cells (see text).

become determined. Apparently this is not so, for the *mwh* marked clone in Fig. 9.4 is continuous from head structures into transdetermined wing, showing that cells in the clone expressed their head determination before becoming transdetermined.

Although the mechanism of transdetermination is unknown, the phenomena summarized in Fig. 9.3 can be formally explained if we assume that the various determined states are regulated by a series of bi-stable circuits (Kauffman 1973; Kauffman *et al.*, 1978). A simplified version of Kauffman's model is shown in Fig. 9.5, to suggest that only a small number of genes may be necessary to establish the states of 'wingness' or 'legness', as is surely implied by the high frequencies of transdetermination.

### Disc fate maps: regeneration and duplication

If a 3rd instar imaginal disc is directly transplanted into another 3rd instar larva, it metamorphoses with the host according to its prospective fate; implanted leg disc becomes leg. Conversely, extirpation of a disc results in an adult lacking the corresponding structure (Zalokar, 1947), as we should expect from the blastoderm damage experiments. As arealization implies, the parts of a disc are determined at this stage, and a transplanted fragment differentiates into part of an adult structure while the remainder lacks that part when similarly metamorphosed. In this respect, a 3rd instar disc is a composite of determined parts. Although there is a minimum size of fragment which will survive, it is possible to cut discs into defined parts and, by metamorphosing them, to map the disc for the various identifiable structures such as bristles, sensilla etc. of each particular adult organ (Fig.9.6). All the discs have now been mapped in this way (Bryant, 1978). Younger discs do

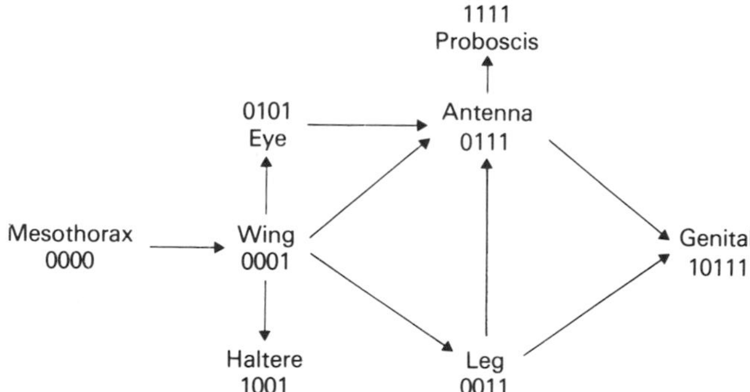


**Fig. 9.5** If we assume that transdetermination is due to the activation (1) or inactivation (0) of single 'switch' genes, and that mesothorax is the inactive state with respect to these genes, we can construct a pattern of gene activation like the above. At least four (1111) bi-stable switches, or circuits, are required to cover the eight transdetermination states, although Kauffman (1973) uses five to give a better approximation to the probabilities of Fig. 9.3. The significant point is that we need postulate only a small number of such genes.

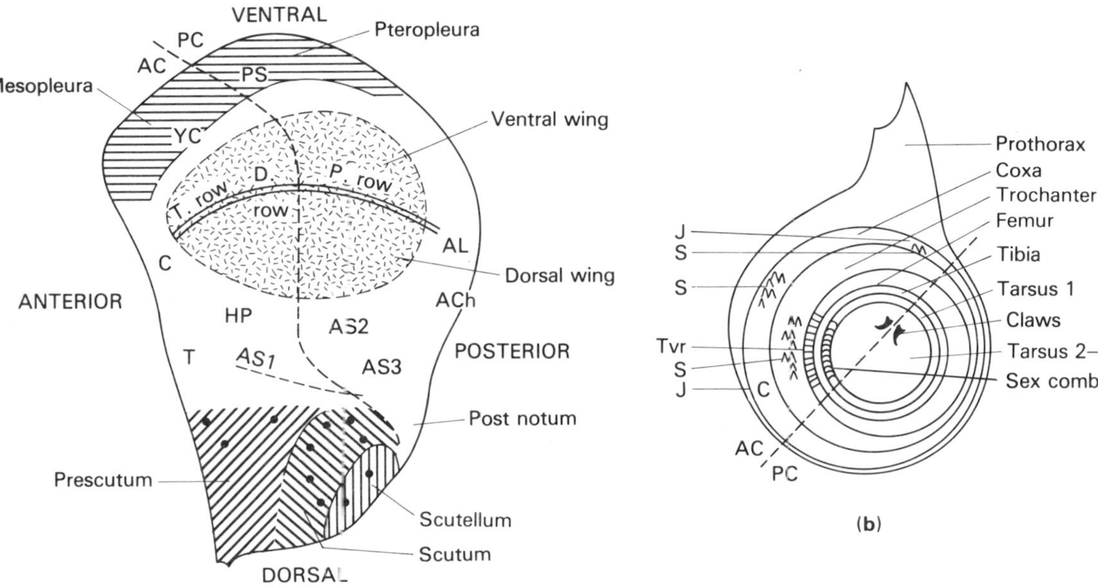


**Fig. 9.6** Fate maps of wing (**a**) and male first leg (**b**) discs. The presumptive wing blade is stippled and folds along the double line to form the dorsal and ventral wing surfaces. The anterior (AC) and posterior (PC) compartments of the disc are separated by the broken line. AL and ACh are the alar lobe and cord; AS1–3 are axillary sclerites; C is the costa; HP is the humeral plate; PS is the pleural sclerite; T is tegula; T, D and P rows are triple row and double row of bristles, and posterior row of hairs; YC is the yellow club. Major bristles are shown as closed circles (after Bryant, 1975). The leg disc is telescoped segment by segment, and the anterior and posterior compartments divided as shown by the broken line. Particular joints (J), sensilla (S) and bristles (Tvr, transverse rows of the tibia) can be identified, as described by Schubiger (1968).

not differentiate completely. Second instar discs differentiate only partially; e.g. proximal and distal regions of the leg differentiate first, and the gap between is filled in after subsequent growth (Schubiger, 1974). Other discs show different patterns for this early competence to metamorphose, wing hinge and blade first according to Bownes and Roberts (1979), and the change is always progressive within a disc as it ages. Since transplanted 1st instar discs do not differentiate, we cannot trace back the course of differentiation to the initial primordia.

If discs are transplanted into younger larvae (2nd or early 3rd instar) they have time to grow before the onset of metamorphosis (Vogt, 1946b). Except for labial discs which duplicate, undamaged discs do not grow. On the other hand, disc fragments, or the larva's own discs divided *in situ* (Bryant, 1971), regenerate the missing part, or duplicate themselves during this growth period. The same result is got when discs are allowed to grow in adults, usually for 4–8 days, prior to metamorphosis; and there is no growth beyond completion of the disc (or duplication of the part) during extended growth periods. Figure 9.7 shows that most bisections result in a half which regenerates, and a half which duplicates itself in mirror-image symmetry. By making different cuts it is possible to identify the disc region regularly associated with regeneration; the upper medial quarter in the case of the leg disc (Schubiger, 1971). However, data for the wing disc (Fig. 9.7) are difficult to fit into such a scheme (Bryant, 1975), since a fragment lacking the central 'regeneration' region can regenerate! The situation is therefore more complex than might be implied by the presence of a simple source of regenerating cells in one part of each disc. And we have also to explain how duplication comes about.

Since there is no regeneration or duplication without growth, the new structure cannot be due to a reorganization of existing cells (*morphalaxis*): it must involve the organization of new cells (*epimorphosis*). Autoradiography of discs labelled with [$^3$H]thymidine immediately after transplantation shows that the new growth is largely unlabelled, implying that the label is diluted out during the rapid cell proliferation (Wildermuth, 1968). When labelled later, most of the tritium is found in the new cells, again indicating that growth proceeds from the cut surface (Dale and Bownes, 1980). Clonal analysis using mitotic recombination similarly suggests the same rapid proliferation, and since there is no mirror-image symmetry of marked clones in the disc fragment duplicate, we can be sure that the daughters of a marked cell remain in the original fragment: the new replica does not derive by cell migration from the old original (Postlethwait *et al.*, 1971). The evidence, then, is that a blastema forms at each of the cut edges of the disc (Abbott *et al.*, 1981).

The new structures found when a fragment is grown for a short period, and metamorphosed, are those adjacent to the cut, as indicated by the disc fate map. Somewhat longer culture results in the appearance of structures more distant from the cut. Regeneration proceeds sequentially (Schubiger

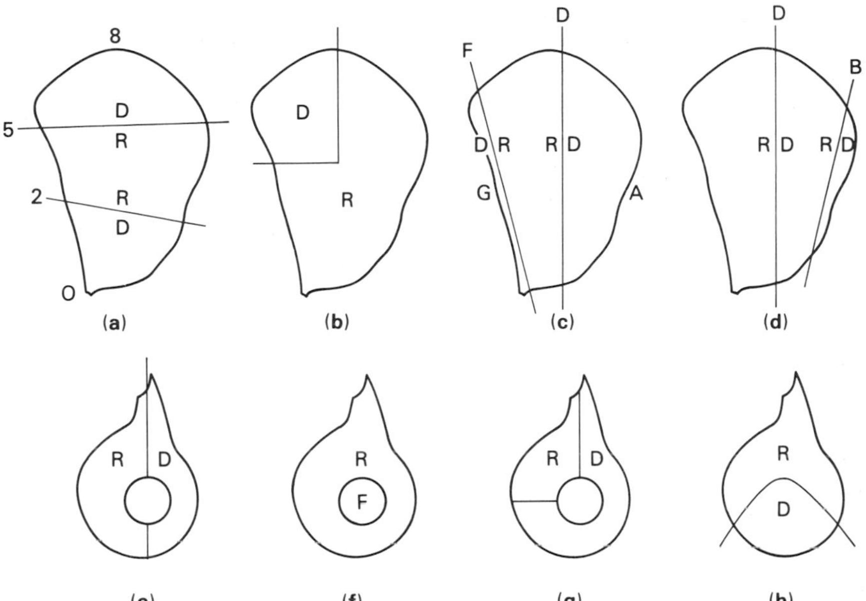


**Fig. 9.7** The competence of wing (**a–d**) and leg (**e–h**) disc fragments to regenerate (R), duplicate (D) or fail (F) to do more than make structures expected from the fate map, after culture *in vivo*. Bryant (1975) tested fragments cut horizontally (margins of disc numbered 0–8) and vertically (margins lettered A–D) through the wing disc, with the results shown. The 0–2 fragment is the notum, which regularly duplicates. The leg fragments are from the range tested by Schubiger (1971). In the first and third, the 'end knob' or tarsal fragment is excluded. It fails to regenerate or duplicate on its own. The fragments show that the regenerative potential centres on the upper medial quarter of the leg disc. The more complex wing data are considered in the text. They show that the upper medial quarter does not regenerate, nor does any other quarter (not shown) although the remaining three quarters always regenerates. Only a complex system could apparently account for the results shown.

and Alpert, 1975). While this morphological evidence is not entirely satisfactory due to the difficulty of identifying marker structures, a definitive description of disc fragment growth should shortly be available from studies of fragments grown *in vitro* (Davis and Shearn, 1977). The evidence suggests, for both wing and leg discs, that compartment boundaries are at first transgressed during duplication and regeneration. However, mitotic clones established during later growth respect compartment boundaries (Szabad *et al.*, 1979). Clonal restrictions become established during regeneration, and in the proper morphological position, clearly showing that anterior and posterior compartments at least are a necessary component of pattern regulation.

A description of growth does not tell us how differentiation is effected, and the obvious test for this is the embryological manipulation of grafting (p. 23). This is technically impossible with discs, but instead fragments can be mixed together, using fine tungsten needles, to give an equivalent operation, prior to culture *in vivo*. If an 02 wing fragment (notum) is mixed with a 68 fragment (ventral wing blade, hinge and pleura in Fig. 9.6), both of which would normally duplicate on their own, intercalary regeneration occurs and the genetically marked cells of the 68 fragment produce notum tissue. Similarly, anterior and posterior wing fragments intercalate intermediate structures. This conclusion explains the puzzling result of Fig. 9.7. The two cut edges of this fragment are found to heal together, and to allow intercalary regeneration (Haynie and Bryant, 1976). The cells of one of the 02 (or 68) fragments can be irradiated prior to mixing, and so prevented from multiplying. Nevertheless, such lethally irradiated tissue stimulates the production of structures which neither fragment would have made when cultured alone. This property of regulative interaction is also found when the 02 wing fragment is mixed with fragments from leg, antenna or genital discs, irradiated or not (Wilcox and Smith, 1977; Bryant *et al.*, 1978). Different disc parts interact in a fashion analogous to the inductive situations described in Ch. 1. But the experiments are not strictly comparable since we do not yet know if we are concerned with diffusible substances, cell contact relationships, or one of the many other possible interactions which might occur when disc parts are intermingled by folding them together with tungsten needles. It is particularly interesting that the reactions are not disc specific, and that the factors involved are common to most, possibly all, discs. Even embryonic tissue aged 16 h will serve when mixed with the 02 wing disc fragment (Fain and Schneiderman, 1979).

Proliferation from each cut edge of a divided disc produces the *same* structures, which we define as a regenerate or a duplicate, according to the fragment giving rise to it. These new structures are always distal parts, for example in the leg disc (Fig. 9.7), and in this respect they follow the 'distal transformation rule' previously found for regenerating amphibian limbs (Rose, 1962). Formally, this can be described by a simple model (Fig. 9.8) which implies a gradient of developmental capacity with a proximal high point and a distal low point, where growth from any one level can proceed only to lower levels. Such a model requires a gradient of some description extending the length of the organ (disc or limb), and also that cells are in some way capable of recognizing their position in that gradient and, hence, of differentiating according to their position; the positional information hypothesis of Wolpert (1969; 1971). The proliferating cells of a bisected leg disc generate only distal positional information on their own. The model can also be adapted to explain the results from wing disc fragmentation if we assume that there is a high point in the centre of the disc with multiple gradients radiating from it (Bryant, 1975), although Fig. 9.7 shows an exception to the distal transformation rule for wing (but see below). The model suggests that proliferating cells should follow a sequential, downhill pattern of differentiation, and this is what we see when regenerates are metamorphosed at successive times.

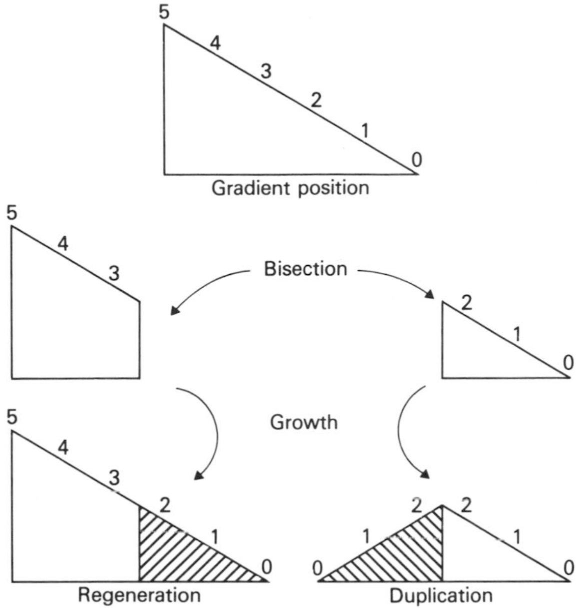


**Fig. 9.8** The interpretation of regeneration and duplication in terms of a gradient of developmental capacity. 5 is the source of the gradient substance, and 0 is the 'sink' where it is destroyed, giving a regular slope of gradient positions. It is assumed that only lower, not higher, gradient levels can be established from the bisected disc edge, and regenerating and duplicating fragments will therefore be produced (and triplicates if the fragments should adhere). This assumption also means that the new growth is identical in the two fragments.

### *The polar coordinate model of pattern regeneration*

These experiments with cut discs identify two processes involved in regeneration: intercalary growth and distal transformation. They also suggest that discs, at least at this stage, are organized as gradient systems, or fields in the embryological sense (p. 24). French *et al.* (1976) have used these facts to present a model which can account for the results just described, and also for the many different, and puzzling, regeneration patterns observed with cockroach and amphibian limb grafts. A model which can explain so much is a very powerful and exciting one, but it is too new for its inherent genetic assumptions to have been comprehensively tested.

The French, Bryant and Bryant model replaces the simple gradient of Fig. 9.8 with a set of polar coordinates, arbitrarily defined (Fig. 9.9). There is a set of circular (numbered) values as on a clock-face, the outer circle corresponding to the disc boundary. And there is a set of radial (lettered) values, such that the centre is the presumptive distal tip of the appendage (or the point on the dorsal wing surface already identified). Each cell has

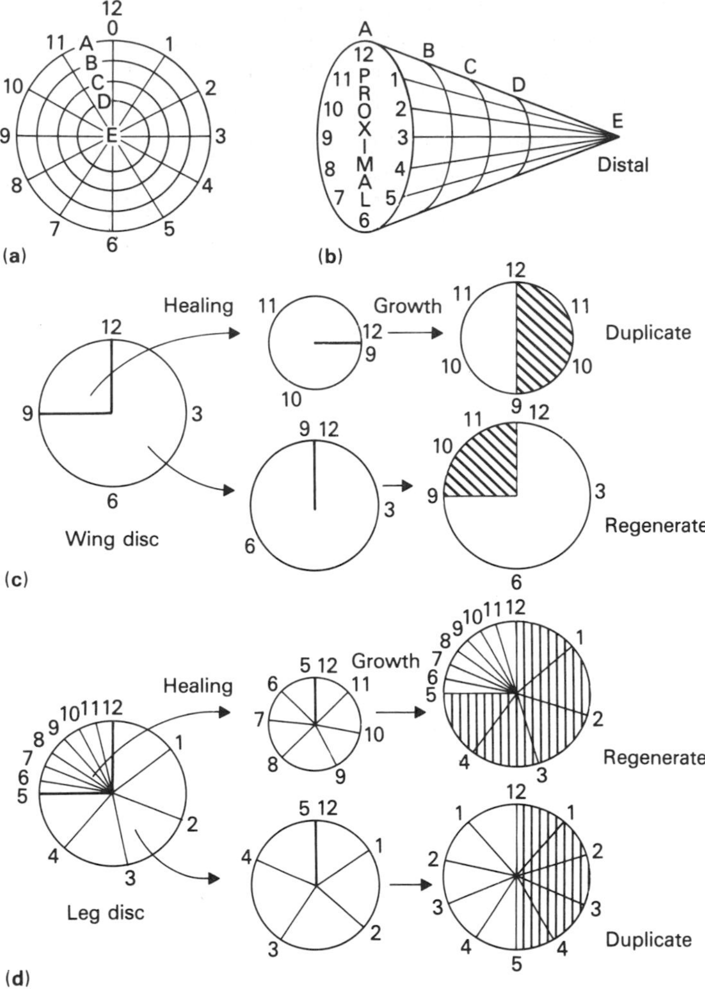


**Fig. 9.9** The polar coordinates of positional information in an epimorphic field are circumferential (0–12, where 0 and 12 are taken as identical), and radial (A–E). Each cell then has precise information as to its position in surface space. The field of an imaginal disc would be flat, with the proximal part at the edge and the distal part in the centre, as in (**a**). The surface of an appendage would be as in (**b**) with the distal point at the top of the limb. The results of culturing 90° and 270° sectors are shown for wing (**c**) and leg (**d**) discs. In the first, the values 9 and 12 heal together, and then by intercalary growth (shaded) result in duplication of the smaller fragment and regeneration of the larger. This explains the result of Fig. 9.7(**b**) (and of the other wing disc fragments, too). The equivalent experiment with leg discs (Fig. 9.7**g**) can be similarly explained if we assume that the circular values are not uniformly distributed around the disc, but clustered in the upper lateral quarter. The small fragment, having more than half the positional values, is then the one which regenerates; and the larger fragment duplicates (based on French *et al.* (1976)).

a locus which can be uniquely defined by a circumferential and a radial value. Of course, no disc is perfectly circular, but this departure from the ideal can be accommodated by altering the, in any event arbitrary, spacing between positional values, without invalidating the model.

Regeneration follows two rules:

1. *The shortest intercalation rule.* As we have seen, when two non-adjacent positional values come together as a result of a cut (or graft), growth occurs until all intermediate positional values have been intercalated; then growth ceases. The rule says that this is always done by the shortest route. That is, if the values 2 and 6 on the clock-face are juxtaposed, then 3, 4 and 5 will be intercalated, not 1, 0/12, 10, 9, 8, 7. This is a direct consequence of the coming together of the cut edge(s). And the same 'shortest route rule' applies to radial values.
2. *The complete circle rule for distal transformation.* If a cut exposes a complete set of radial values, or if such a set can be generated by intercalation, then growth will produce the more central (distal) values. This explains how distal transformation forms complete terminal structures. Fig. 9.9 applies these rules to wing and leg discs and shows that the model accounts for what appeared as anomalous results in Fig. 9.7. We shall see below that the rules also explain the more complex experiments with true grafts. Both rules suggest that growth is stimulated by the apposition of normally non-adjacent positional values, and as growth takes time it is not surprising that such transplants delay pupariation of the carrier larvae, by some unknown mechanism.

## Limb regeneration

Many invertebrates, especially arthropods, can regenerate broken limbs. In such species, limbs have a fracture plane which allows the organism to snap it off (*autotomy*) when caught by a predator, and a blastema forms at the stump. Adults cannot generally regenerate, although an adult limb transplanted to a younger stage can do so (Bodenstein, 1955); this capability of the younger stages allows a new limb to be made between moults, although if the break occurs late in the moult cycle of an insect or crustacean it may take a second moult to complete. In the cockroach (*Periplaneta americana, Blaberus coanifer* and other species), the limbs have this property of complete distal regeneration. Also, if a longitudinal strip of tissue (cuticle and epidermis) is removed from, say, the femur of a larval cockroach, the cut edges heal together, and by the next moult the leg is normal with all missing structures replaced. Intercalary regeneration has occurred, and all circumferential 'values' have been re-established, as the model predicts. Cockroach limbs are therefore ideal subjects for testing the model, and we shall look at two grafting experiments (from among the many that have been done) since proper grafts are not possible with imaginal discs.

Grafting a longitudinal strip of femur integument into a different circumferential position on the host femur confronts epidermal cells with different positions at the inner and outer margins of the graft, without affecting proximal–distal orientations. A variety of such grafts shows that the 'shortest intercalation rule' is always followed, and reversing the proximal–distal polarity does not upset this, since the circumferential position is independent of position along the length of the femur (French, 1978). When this intercalary regeneration forms a complete set of circular values in conjunction with the adjacent host tissue, a supernumerary distal regenerate is also formed, in accordance with the 'complete circle rule' (Fig. 9.10). The complete circle rule therefore accounts for the puzzling appearance of a supernumerary limb. Such grafts can also be used to create situations where the apposed positional values are 180° apart (9/3, 10/4 etc.) and the shortest intercalation may proceed in either direction (10, 11, 12/0, 1, 2 or 4, 5, 6, 7, 8). Such 'flip-flop' positions are indeed found where intercalation sometimes proceeds in one direction, sometimes in the other. This property can be used to determine which positional values lie opposite one another, are flip-flops, and hence how the positional values are themselves distributed for they are unlikely to be evenly spread around the limb. Positional values seem to be more or less equally spaced around the cockroach leg, as indicated in Fig. 9.10. And this arrangement seems to apply to other leg segments since grafts from them (e.g. tibia to femur) behave as we have described for the femur–femur grafts: the circular organization is sequentially repeated in the segments.

More complex confrontations of positional values can be made by grafting, say, a left leg to a right stump, and reversing the anterior–posterior polarity of the leg relative to that of the stump (Fig. 9.10). This creates two points of maximum incongruity (9–3) where a complete circle will be generated, and these circles will duly form two supernumerary limbs after

**Fig. 9.10** Intercalary regeneration in the cockroach limb (**a**) and in the newt limb (**b**). The graft between two levels of the cockroach tibia corresponds to distal levels A and E of Fig. 9.9. After two moults the missing values (B′, C′, D′) are filled in. Similarly, a graft of an early digit stage of the newt limb to the proximal stump of the upper arm results in regenerative intercalation. However, if contralateral transplants are made (**c** and **d**) so that anterior and posterior axes are opposed, supernumerary limbs develop after growth. The clock-face model (**e**) shows the positional values which are then confronted, and examples (4–8, 5–7) show the shortest route intercalations. There are two points of maximum incongruity where 3 confronts 9, and complete circles are generated there. Distal transformation then results in the generation of two supernumerary limbs, after growth. Both supernumerary limbs are of the same handedness as the stump, and since (**c**) and (**d**) show a left-handed graft on a right-handed stump, both supernumerary limbs are right handed. Correspondingly in (**e**), the complete supernumerary circles have the same orientation as the stump. Adapted from French *et al.* (1976).

distal transformation. The handedness of these extra limbs depends, as the model predicts, on the direction of the intercalary growth at the graft–host junction. Other, similar, manipulations give the expected results (French *et al.*, 1976). By using colour mutants, or parts of leg with a distinctive pigmentation, it can be shown that the new tissues generally derive from both stump and graft, although in some instances (Bohn, 1976) there is good evidence that growth derives from the distal regenerate. The active process of regeneration must involve continual interactions between the cells and surviving tissues.

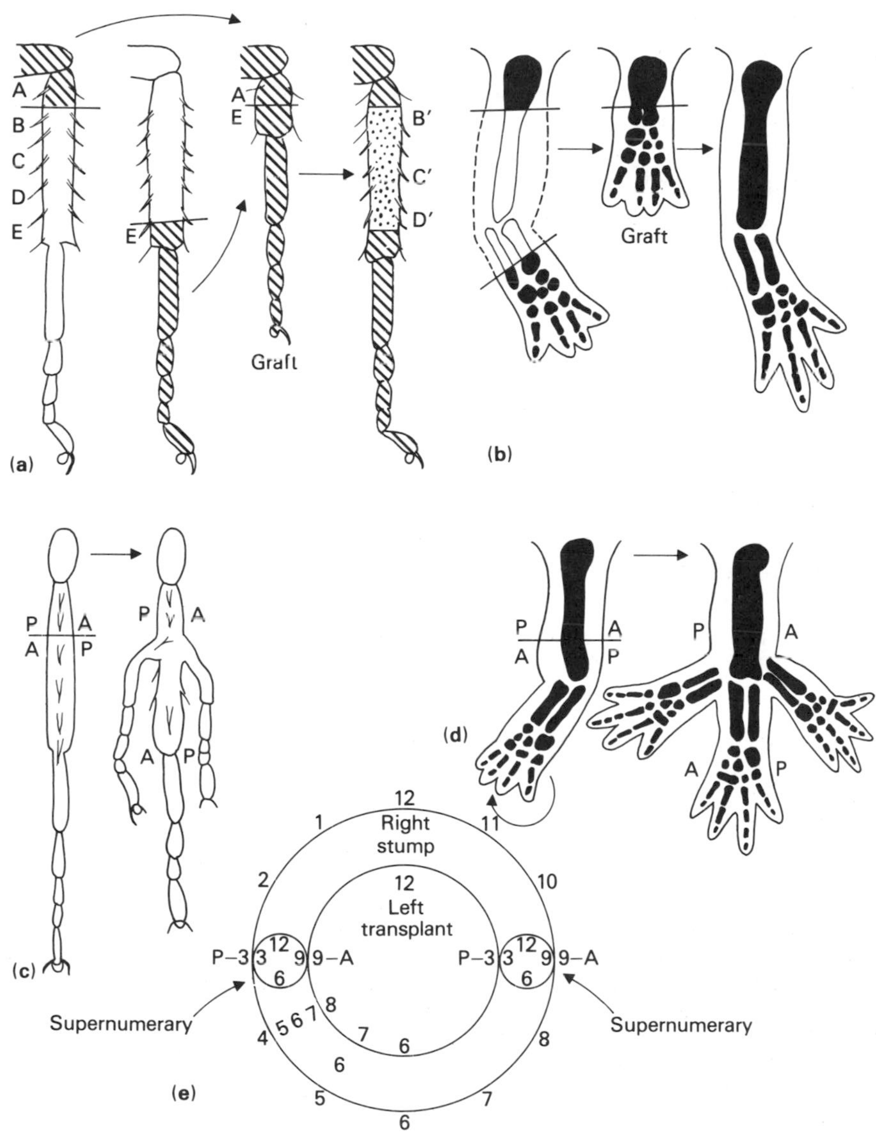

When looking at cockroach limb regeneration, it is fair to assume that we are dealing with a two-dimensional tissue (cuticle and epidermis), just as with imaginal discs. It is more difficult to make this assumption when we come to regeneration of the amphibian limb (newts, *Axolotl* etc.) which is obviously, and solidly, three-dimensional (bone, muscle, nerve, dermis, epidermis). Further, the blastema which forms under the healed epidermis following limb bisection arises through dedifferentiation of the cells of the adjacent, damaged tissues. It is apparently these cells which multiply and then differentiate in orderly continuity with the tissues of the stump to give distal (complete circle rule) regeneration (p. 231). French *et al.* (1976) argue that the positional information must be present initially in the patch of embryonic mesoderm, which we have seen (p. 24) is capable of inducing limb development when transplanted below 'non-limb' ectoderm, and they suggest that there is a hollow tube deriving from this mesoderm within the mature limb, which provides the required matrix for the theory! This is more likely than it might seem: if a limb is halved proximo-distally (or the cells killed by X rays), there is no intercalary regeneration unless the skin and mesoderm around the stump portion is left complete (Goss, 1957). The shortest intercalation rule is also implicit in limb grafting experiments similar to those already described for the cockroach (Fig. 9.10). In this, and other manipulations, the model correctly predicts the development of two supernumerary limbs, and their handedness. Bryant and Iten (1976) have also shown that most of the old embryological experiments on transplantation of vertebrate limb buds in abnormal orientations similarly fit the predictions of the model, particularly with respect to supernumerary limb differentiation. Thus both rules can also explain much, if not all, of the simpler regeneration data coming from studies of amphibian limbs.

Duplications and triplications are also found when imaginal discs are damaged. Early physical damage to the disc primordia (p. 226) may produce duplications, showing that positional values of some sort are established shortly after the disc cells are segregated in the embryonic ectoderm. (Regenerates, being normal, would not be identified). Mutants, particularly temperature-sensitive, autonomous cell lethals, can be exploited for a kind of 'genetic surgery', and they again result in the duplication of the final adult structures. For example, if larvae carrying the ts lethal *726* are exposed to the restrictive temperature during the first 48 h of the 3rd larval instar, pattern triplications are found. These follow Bateson's (1894) rule, with one complete leg of normal symmetry, and two incomplete legs, one with reversed symmetry and the other with normal symmetry. Bateson's observations came from examining anomalous organisms in nature, and this experiment suggests an explanation of their characteristic feature.

Programmed cell death is a regular event in normal differentiation, and Sprey (1971) gives examples of the part this plays in shaping arthropod appendages. We should also expect cell lethality to account for some mutant phenotypes, and we have already suggested (p. 213) that the costal mutant is a cell death-induced wing duplication. Other mutants like scalloped or vestigial wing, eyeless, antennaless and podoptera result from localized cell death during disc growth, and it is interesting that this sometimes also results in duplication. However, not all mutants affecting size and shape (e.g. wingless or tetraltera according to James and Bryant (1981)), show cell death, so we cannot generalize from the first examples. As we should expect from the disc surgery experiments (p. 227) it is cell death only in certain regions which results in duplications of the growing discs.

None of the models which have been invented to explain particular regeneration phenomena has proved adequate to explain all of them (reviewed by Tank and Holder (1981)) and the polar coordinate model is no exception. In particular, many exceptions to the complete circle rule have been found in which distal regeneration occurs when only incomplete circles are expected (Bryant *et al.*, 1981). These results may be explained if we assume that wounds heal irregularly, bringing together different circumferential values when they fuse. Intercalation would then make a complete circle if, it is assumed, the new cell is forced to adopt a positional value more distal than that of the pre-existing cell (Fig. 9.11): there is a distal outgrowth and distalization. This 'distalization rule' replaces the 'complete circle rule': it gives the same expectation when complete circles are generated, but it also explains the various different results of regeneration from symmetrical limb stumps and the results of a range of limb grafts (for details see Bryant *et al.* (1981)) which do not concern us. The important operative assumption is that the distalization rule depends on strictly local cell interactions, the state of the nearest neighbour, and is directional, i.e. distal.

We cannot comment on cell–cell interactions of this specific kind, but we can ask if there are mutants which affect the gradient system or the responses of cells to gradient signals. The data, unfortunately, are few.

## Gradients

Nothing is known about the molecular basis of any postulated gradient system. Nonetheless, the morphogen is taken to be non-specific and the experiments with grafts between parts of different discs would support this (Wolpert, 1971). Specificity lies with the responding cells and not with gradient singularities of the kind postulated as the basis for prepatterns (p. 208). There is no strong argument for double gradients in appendages (Crick, 1971b), and we must assume a 'source' which makes the morphogen and a 'sink' opposite to it where the morphogen is destroyed, and both must function to maintain a regular gradient which we can think of as a slope (Fig. 9.8). Its precise form will depend on the parameters of the appendage or disc, as they affect diffusion. Mutations affecting source, sink or diffusion pattern might all affect the array of positional information, and mutations of the first two would affect all structures utilizing the common morphogen.

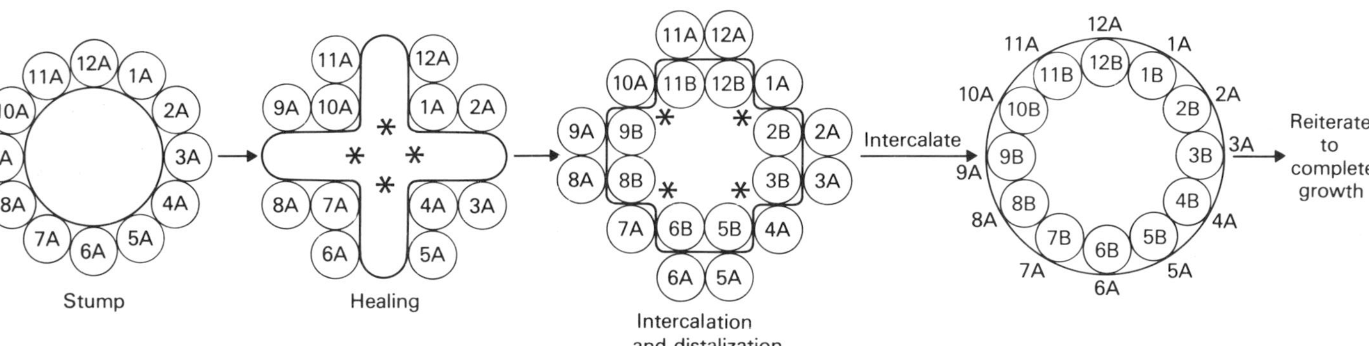


**Fig. 9.11** When a stump is cut (here at the A level) healing may be irregular bringing non-adjacent circumferential values together. The positions where intercalation will then occur are shown (✻). These intercalations will have values identical to those of pre-existing cells adjacent to them, and it is therefore assumed that they will be forced to the next most distal level (B). This is the distalization rule. Further intercalation will then complete the circumferential values at the B level, and this process will be repeated to give complete distal growth. Based on Bryant *et al*. (1981).

Since they affect thresholds, they would be dominant, non-autonomous mutations. There is no certain evidence for such mutations, although bicaudal and dicephalic are candidates, as we have seen.

Three *Drosophila* mutations have been claimed to alter the gradient system indirectly, eyeless Dominant ($ey^D$) and shibire$^{ts}$ ($shi^{ts}$) (Poodry and Schneiderman, 1976), and decapentaplegic (*dpp*) (Spencer *et al.*, 1982). The first two mutations cause breaks in the intersegmental membranes of the leg joints; $shi^{ts}$ when the gene product is inactivated by warming during the middle 12 h of the 3rd larval instar. Apparently as a direct consequence of the lesion, there is a local reversal of bristle polarity, and duplication of male sex combs and of the transverse bristle rows. Since the gradient system is taken to be segmentally reiterated, the polarity effect is compatible with a gradient reversal where the segment boundary is disrupted. Such a response had previously been described when the intersegmental membranes between the abdominal tergites of *Oncopeltus* are damaged (Lawrence, 1966). It can also be argued that the flow of morphogen through the membrane gap could cause some morphogen levels to be doubled and others reduced, which might explain the structure duplications. Since $ey^D$ causes cell death and $shi^{ts}$ blocks cell division, we can equally well argue that these consequences are the causes of duplication. And although both affect other discs somewhat irregularly, we cannot conclude that they provide good evidence for morphogen gradients.

Decapentaplegic (*dpp*) is a gene complex covering three salivary gland bands (22Fl–3), and > 100 kb of DNA. Its interesting feature is that four functional units (perhaps more) can be identified which increase in severity of action from proximal to distal members. In the most extreme situation, 15 of the 19 imaginal discs are affected (hence the name of the complex which was previously known from a single mutation heldout wing). Distal structures are abnormal or absent in mutant discs, and damage from more extreme alleles extends progressively to include more proximal structures. Members of the complex therefore have position specific effects on disc development, and this suggests that they may affect positional information. In conformity with this, the preliminary evidence from induced clones is that the mutations are not cell autonomous, although they are obviously disc autonomous. That is, the pleiotropic effects of the gene depend on interference with a process common to all the affected cells, just as the segmental mutants (p. 172) demonstrate by their disturbances of segmentation patterns, and this would also be expected from the assumed commonality of positional information signals.

Although there is a good *a priori* case for considering *dpp* the first positional information mutation, the case is not proven. In some instances duplication of legs or antennae occurs, which would be compatible with cell death, and it will be necessary to show that we are not dealing with a lethal mutation complex which specifically kills cells in the central core of imaginal discs, although this would be interesting in itself. This is a particularly important alternative because one of the functional units of the complex (EL) is an embryonic lethal. Since embryos survive without discs (p. 219), it is necessary to assume that *dpp* is a general epidermal signal mutant acting on positional information, but that it is a lethal if it does not.

### Interpretation of positional information

If a cell is to locate its position in a gradient, it must receive the morphogen, measure its amount and then respond appropriately. We have no information about these processes, and it is idle to speculate about them. What we can say is that mutations affecting the measuring/response system might leave the cell (or cells) incapable of differentiating (they would become tumorous) or cause maldifferentiation. Available data (Gateff, 1978b) do not allow us to relate known tumour genes to gradients, and only one mutant seems to cause maldifferentiation. This is Malformed (*Mal*), which causes either a bristle-bearing pit to appear in the middle of the *Drosophila* eye, or the displacement of eye facets by fronto-orbital-like tissue (Fig. 9.12). Clonal analysis shows that this orbital tissue arises from the ommatidium, not the orbit-forming, cells of the disc, (Baker and Tsai, 1977). The recessive, erupt (Glass, 1957), behaves similarly. Both mutations are therefore candidate gradient interpretation mutants, but analysis of their action is too incomplete for any firm conclusion.

Since few attempts have been made to relate the effects of mutations to the models derived from amputations and grafting, it not surprising that we cannot use genetic evidence to witness their validity. But just because the information is scanty, we must look more closely at regeneration before we leave the subject.

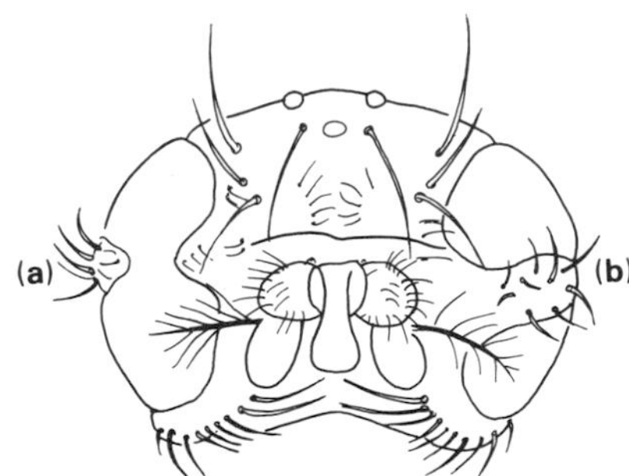


**Fig, 9.12** The erupt mutation has the same phenotype as Malformed, and shows the appearance of bristle eruptions in the centre of the eye facet area (**a**) as well as regions of orbital duplication within the eye (**b**). In Malformed, clonal analysis shows that the duplication often originates from a cell population different from that giving rise to the normal orbit. The argument is that these phenotypes result through misinterpretation of positional information.

## The nature of the blastema

All regeneration proceeds from a blastema, or zone of proliferating cells associated with the wound. In amphibia, the epidermis migrates to close the limb wound and forms a central apical cap, below which mesodermal cells dedifferentiate, accumulate and divide. The blastema cone grows, and the proximal cells next to the stump redifferentiate so that new parts are laid down in a proximal to distal order. Much the same sequence is followed in the cockroach and, as we have seen, in *Drosophila*; although there the sources of the undifferentiated blastema cells are not known with certainty. This pattern explains why regeneration is always distal (or its equivalent in discs). Further, they are, after a fashion, repeating the course of normal development from undifferentiated, but determined, cells.

As we saw, the cut surfaces of *Drosophila* discs come together and fuse in a day or so, permitting intercalary regeneration. However, if the blastema is removed from the cut surface prior to fusion, we can ask what is the potential of these dedifferentiated cells? Karpen and Schubiger (1981) have done this, selecting leg disc fragments which would normally not regenerate (i.e. have few positional values) as the source of the blastema. After culture in an adult for 6 days, two such blastemas made complete legs, and others regenerated variously. Note that these blastemas did not transdetermine since they did not originate from the area of the disc with competent cells. The limb-determined cells thus generate structures corresponding to *proximal* values in the polar coordinate model.

The same competence has been found during experimental manipulation of the *Xenopus* limb (Fig. 9.13). If the hand is removed, a new hand is regenerated. However, if the amphibian is kept in water containing the right amount of vitamin A, a whole new limb will regenerate from the stump, to give two limbs in tandem (Maden, 1982). Vitamin A has long been known to transform cultured chick epidermal cells into ciliated mucosa (Fell and Mellanby, 1953) but how this morphogenetic effect relates to proximal regeneration is unknown. This result, like the *Drosophila* experiment, suggests that blastema cells are capable of making all limb structures, but do not usually do so because they heed signals emanating from the differentiated cells of the stump. Any firm conclusion must, however, await further experiments.

In this context, it is relevant to note that disc cells multiply when dissociated early *Drosophila* embryos are cultured *in vitro* (Dübendorfer *et al.*, 1955). They form spherical vesicles of thousands of cells, each ball presumably originating from a single cell in most cases. The vesicles respond to 20-hydroxyecdysone by laying down chitin, bristles and trichomes somewhat irregularly. They metamorphose but form no recognizable patterns. However, if the vesicles are cultured *in vivo* for 9–10 days, the cells become columnar and the tissue folded, as in a disc. These transplants then form complete structures, like labium, as well as incomplete limbs and wings,

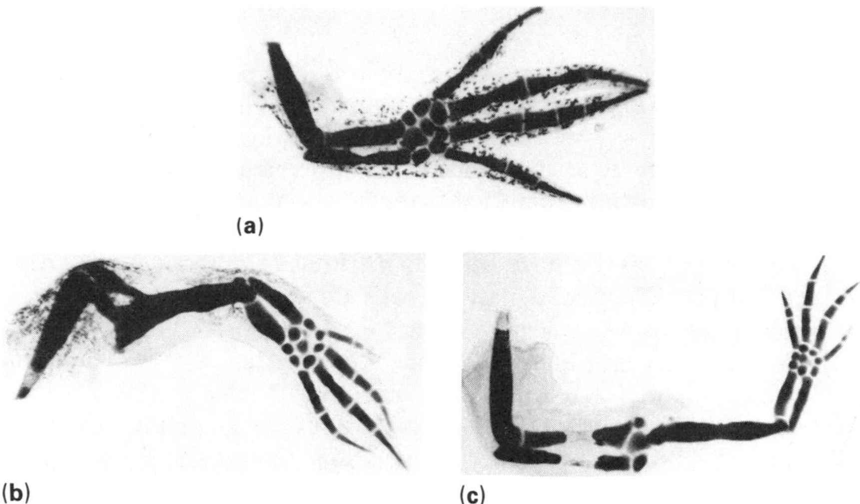


**Fig. 9.13** Regeneration of *Xenopus* limbs after amputation: in normal water (**a**), and after exposure to 300 mg/litre of retinol palmitate for 15 days (**b, c**) (**a**) is the control, where the amputation plane was through the radius and ulna, and the regenerate recreates the normal limb. (**b**) has the same amputation plane, but the retinol palmitate treatment causes a complete new limb to regenerate from the stump, i.e., we have the sequence: humerus, radius–ulna, humerus, radius–ulna carpals and digits. This is even more striking in (**c**) where the amputation was through the carpals, and a whole new limb now sprouts from the severed hand. In some cases, part of the limb girdle is also regenerated after longer exposure. The blastema therefore reverts to the properties of the entire embryonic limb field, which is expressed when the animal is returned to fresh water. Photographs kindly supplied by Dr. M. Maden.

after metamorphosis (Fig. 9.14). The progeny of a single determined cell are competent to form a complete appendage, and they do this by establishing some hierarchy of normal contact relationships with others when cultured *in vivo*. Nearest neighbours are clearly important, but whether they function as postulated in the Bryant *et al.* (1981) model remains to be seen. Presumably a cultured blastema becomes reorganized in a similar way.

## Conclusions

In completing the circle from the determined imaginal primordia of the embryonic blastoderm, through their normal development and responses to surgical and genetic insults, and back to the development of primordia in culture, we have shown that discs have most of the characteristic properties of developing systems in general. Indeed, they are even more interesting than Hadorn (1965) imagined. Our survey does not explain how they become determined in the first place, or the basis of the stability of that determination, and the phenomenon of transdetermination which so inter-

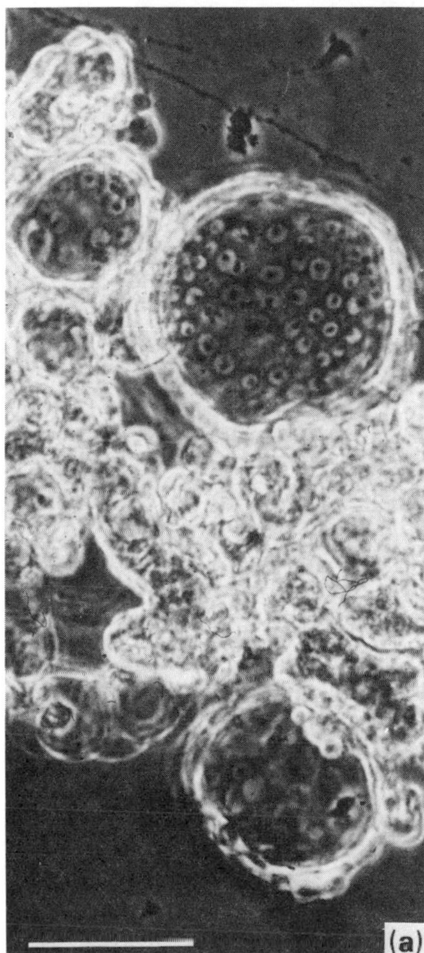


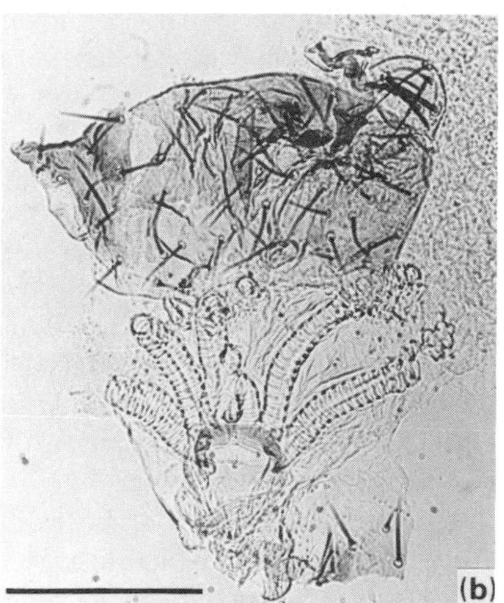


**Fig. 9.14** Imaginal disc vesicles arise when pre-gastrula dissociated embryos are cultured for 2–3 weeks, and they grow to a large size (**a**). When they are transplanted into adults and cultured prior to being metamorphosed, they become competent to form imaginal structures (in (**b**) a labium) which they cannot do if metamorphosed *in vitro*. The complete disc structure must arise from one cell, or very few.

ested Hadorn remains unexplained. Perhaps the technique of culturing discs *in vitro* will provide the additional clues required for the solution of this mystery. But we can note that duplicating disc fragments cannot transdetermine (and hence they arealize), implying that special cells have this property which is different from the characteristics of the determined blastema cells.

Transdetermination sometimes apparently occurs during limb regeneration: when broken off, a crustacean antenna may regenerate a leg (Bateson, 1894). But usually the regenerate follows the rules which are conveniently summarized by the polar coordinate model, and it seems particularly significant that the same model can be applied to such different systems as the *Drosophila* disc and regenerating cockroach and amphibian limbs. It does not follow that the systems are identical: e.g. amphibian regeneration depends on an adequate nerve supply to the blastema region since without it there is no growth, and there is no evidence of this need in the arthropods. The crucial part of the regenerating system is the blastema—the dedifferentiated but still determined cells which during their renewed multiplication apparently repeat the processes of normal ontogeny. These steps seem to involve a sequence of secondary and tertiary etc. determinations, which in the case of imaginal discs are expressed only on metamorphosis. The parallel with the delayed differentiation of the *Caenorhabditis* embryo (Fig. 7.12) is striking, but we know nothing of the lineages in discs.

The many models of regenerating events have stimulated a plethora of graft manipulations which test their validity. What they have not done is suggest the kind of molecular-genetic mechanisms that might be involved. If we are concerned, as the data now suggest (Mittenthal, 1981), with intimate interactions between neighbouring cells, we are back with the problems which have frustrated a generation of embryologists (p. 23). These are the problems of the role of cell surfaces, cell connections and the passage of small numbers of effector molecules within very small populations of initial cells. We must now ask, therefore, if we can find genetic clues to this general problem; if there are mutations which affect the development and differentiation of determined cells? Such genes may throw light on the processes which remain concealed during regeneration, and genetically engineered clones of these genes may allow us to identify the molecular basis of these processes.

# V Determination mutants

There are two genetic systems which demonstrate dramatic, gene dependent, changes of phenotype. The *homoeotic mutants* of *Drosophila* cause one segment (or some segments) to differentiate in the form normal to another segment; and mutations of sex genes cause one sex to differentiate in the form of the other sex. These are systems which match our requirement for genes which regulate other genes in a proper, orderly, fashion. We may anticipate, too, that they will give us some insight into the issue raised in the Introduction, namely, how DNA is organized to give tissue specific gene activation. While we have models which account for the facts obtained by genetic dissection of both systems, these remain only instructive models of very complex genetic relationships. Molecular analyses which test the models are only just beginning.

The characteristic feature of homoeotic mutants is the exact replacement of one part by another: an antenna by a leg, or vice versa. When replacements are incomplete, or somatic mosaic clones are made, this precision applies to individual elements of the transformed structures suggesting some general pattern of appendage specification which is interpreted differently by normal and mutant cells, even when they are contiguous. There are two major gene complexes in which mutants cause homoeosis; the Antennapedia complex (ANT-C) which covers structures anterior to, and including, the mesothorax, and the bithorax complex (BX-C) which regulates structures posterior to, and including, the mesothorax. The relationship between the two has still to be worked out, in detail.

We shall examine the action of the individual loci of each complex, but a general point of importance is that, where known, ANT-C genes function relatively late during the development of imaginal discs—halfway through the larval period—and, with the possible exception of the dominant *Antp* mutants which may be constitutives, we seem to be concerned with structural gene mutations. The situation may be different for the better understood BX-C where we now have a detailed model of gene regulation, segment by segment. The model assumes a gradient of a repressor (the Polycomb gene product) and of an inducer (the Regulator of bithorax gene product) along the length of the embryo which activate *cis*-regulatory dominant genes adjacent to structural genes which are then caused to function sequentially, segment by segment. The surprising genetic aspect of BX-C is that the more distal a locus is on the chromosome the more posterior is the segment it affects; and this organization has been confirmed at the DNA level.

The picture of how these genes function to establish determined states is not yet clear. The dominant loci are active early in development, and the recessives (or some of them) function later and continuously (or at critical times) during development. However, the molecular data have not yet confirmed that all the 'genes' of the BX-C are simply transcribed, and we have the possibility that some may function through their role in DNA organization *per se*. But the greatest obvious deficit is in our understanding of how a particular determined state results in the subsequent expression of a phenotypically distinctive gene array. And we have the same problem to resolve when we turn to sex differences.

Sex differentiation depends on the primary step of sex determination. There are two broad classes of sex determining mechanisms, one in which a Y(or equivalent) chromosome determines the heterogametic sex and the other where sex is determined by the balance of X chromosomes and autosomes; but the species-by-species organization is much more complex than this simplification implies. There is, indeed, a great variety of sex determining mechanisms, but the two broad classes of sex differentiation depend either on hormones, or are cell autonomous; plants being different, as usual. The germ cells play no role in this differentiation, except for the somatic gonad in *Drosophila* which develops only if populated by the right sex cells.

It is still not clear if the primary step in vertebrate sex differentiation depends on the production of a Y-borne histocompatibility antigen in males, or of the appropriate sex hormone. In either event, the hormones regulate subsequent development, and our understanding of sex differentiation will depend on the pursuit of the problems of hormone action.

The analysis of sex differentiation in *Drosophila* has started from examination of sex specific lethal mutations and from mutations which transform sex. Unexpectedly, some members of the first class also seem to be involved in dosage compensation. The maternal-effect daughterless$^+$ gene product seems to be essential for female development through its regulation of a dominant sex-lethal locus. Similarly, the transforming genes appear to regulate activity at the double-sex locus. Precisely how the two systems interrelate is now under active study. Since these loci (and others in other species) are cell autonomous in their action and are required throughout development, we clearly have a great deal to learn about how the same, and possibly single, molecules regulate gene repertoires in different tissues. The development control systems may not be so different in the two sex determination systems, since both seem to depend on the continued active presence of one, or a small number, of molecules.

# 10 Homoeotic mutations

One of the most interesting chapters of genetic interaction relates to the genetically controlled changes in primary determination processes. The best information is derived from the so-called homoeotic mutants in *Drosophila*, in which one type of segmental organ is changed into another: an antenna into a tarsus or whole leg, a wing into a leg-like structure or a haltere-like organ, a haltere into a wing, and so on.

R. B. Goldschmidt, 1955

The phenomenon of homoeosis was defined by Bateson (1894) as a morphological transformation by which "something has been changed into the likeness of something else"; or, more precisely, as a transformation which "consists in the assumption by one member of a meristic series of the form or character proper to other members of the series". Homoeosis is the development of a 'right' structure in the wrong place, so to speak, and it is easily seen in arthropods where the morphological specializations of the segments make any such anomaly obvious, as Goldschmidt points out. Homoeotic changes have also been described in other animals and in plants (Ouweneel, 1976) but, except for some work with the silkworm, homoeotic mutants have only been analysed using *Drosophila*. What, then, is the nature of this special class of mutational change?

Insects evolved from primitve arthropods with a body form of repeated segments each bearing similar appendages. Evolutionary specialization led to the loss of structures from some segments, and to their modification and elaboration in other segments. In *Drosophila*, the antenna and mouth palps are just such modified limbs: the three pairs of thoracic legs are each specialized (and identifiable); and the abdominal limbs have been lost, except for the modified, specialized genitalia. The origin of the insect wing is disputed, but there is no doubt that the haltere is a modified wing and we know nothing of the origin of the compound eye. If we look at the patterns of homoeotic genetic changes (Table 10.1) we see that the alteration is often

**Table 10.1** The main interdisc homoeotic mutants of *Drosophila melanogaster*

| Locus | Mutation | Homoeotic change | Reference |
|---|---|---|---|
| 1–1.4 | sparse arista (*sa*) | Leg II, III → Leg I | Rayle (1968) |
| 1–5 | ophthalmoptera (*opht*) | Eye → wing | Ouweneel (1970b) |
| 1–14.5 | tumorous head (*tuh-1*) | Arista → tarsus (see text) | Newby (1949) |
| 1–00 | reduplicated sex combs (*rsc*) | Leg II, III → Leg I | Yanders (1957) |
| 2–45 | ophthalmopedia (*oph*) | Eye → leg (palp) | Gordon (1936) |
| 2–54.9 | extra sex combs (*esc*) | Leg II, III → Leg I | Slifer (1942) |
| 2–67 | vestigial$^{W}$ ($vg^{W}$) | Haltere → wing | Shukla (1980) |
| 2–68 | Ophthalmoptera ($Opt^{G}$) | Eye → wing | Goldschmidt and Lederman-Klein (1958) |
| 2–00 | Hexaptera (*Hx*) | Humerus → wing/leg | Herskowitz (1949) |
| 2 or 3 | Haltere mimic (*Hm*) | Wing → haltere | see Lindsley and Grell (1968) |
| 3–47 | Polycomb (*Pc*) | Leg II, III → leg I | Lewis (1947) |
| 3–47.48 | Antennapedia complex (ANT-C) | See Table 10.2 | |
| 3–48.5 | tetraltera (*tet*) | Wing → haltere/leg | Goldschmidt (1940) |
| 3–51.3 | tetraptera (*ttr*) | Anterior haltere → anterior wing | Astauroff (1929) |
| 3–54.2 | Regulator-bithorax (*Rg-bx*) trithorax (*trx*) | Pro- and metathorax → mesothorax | Garcia-Bellido (1977) Ingham and Whittle (1980) |
| 3–58.5 | tumorous head (*tuh-3*) | Rostral hypoderm → genital | Postlethwait *et al.*, (1972) |
| 3–58.5 | spineless-aristapedia ($ss^{a}$) | Arista → tarsus | Balkaschina (1929) |
| 3–58.8 | bithorax complex (BX-C) | See Fig. 10.6 | |
| 3–94.1 | Pointed wing (*Pw*) | Antenna → wing | Bridges and Morgan (1923) |
| 3–103 | eyes-reduced (*eyr*) | eye → wing | Edwards and Gardner (1962) |
| 3–00 | aristatarsia (*art*) | Arista → tarsus | Ouweneel (1970a) |
| 4–00 | lethal (4) 29 (*l(4)29*) | Leg II, III → leg I | Gehring (1970) |

back to the primitive state, and Garcia-Bellido (1977) has called homoeotics *atavic mutations* in order to emphasize this. The argument is that wild-type homoeotic genes are responsible for the new organization of old structures. The similarity of this switching of states to transdetermination has often been remarked upon, and we shall have to consider if the two are different aspects of the same phenomenon or not. What they have in common, at least superficially, is the evocation of a change from one complete, organized set of structures to another; from haltere to wing, from antenna to leg, and so on (Table 10.1). This immediately implies that the genetic loci which regularly produce these changes when mutant must be 'master' genes whose normal products somehow control the activity of the batteries of genes peculiar to the specialized appendages. We have to see if this is so, for that would be a most exciting class of gene.

Not all homoeotics are limited in their effects to single appendages (single imaginal discs). We have seen that engrailed, by converting posterior to anterior compartment, comes within Bateson's definition. Kornberg (1981) has shown that this gene acts in the abdominal as well as in the thoracic, segments. It is therefore the locus responsible for all segment compartments, since the lethal null *en* alleles result in segment fusion (p. 209), Kornberg suggests that compartments, by establishing boundaries that are not crossed (just as in imaginal discs), are a necessary part of the segment-forming mechanism. If this is correct, and we are far from understanding segmentation, the homoeotic character of engrailed is general and secondary, though the direct function of the gene is not itself trivial, of course: it is the one proper intrasegmental homoeotic described so far.

Two further mutants with general homoeotic effects will also concern us: extra sex combs and Polycomb. In both cases, weak hypomorphs or the heterozygote result in the development of sex combs on the second and third legs of males (hence their names), but null mutants of both convert all larval segmentation towards the qualities of the last abdominal segment. These genes are therefore in some way, and unexpectedly, involved in the regulation of many other genes. We shall consider them in that context below.

## Antennal homoeotics and the general characteristics of transformations

Four loci have been identified which transform antenna to mesothoracic leg, in whole or in part (Fig. 10.1). They are: Brista (*Ba*) on chromosome 2, Antennapedia (*Antp*) and spineless aristapedia ($ss^a$) on chromosome 3, and lethal(4)29 (*l(4)29*) on chromosome 4. Where they have been tested, the transformed antennae no longer interact with mutations which affect antennal morphology, like thread which makes the arista a single filament, but they do respond to mutations like dachs or four-jointed which reduce the tarsal segment number (Waddington, 1939). By this test, and many others, the transformed antennae are proper legs. Each mutation has characteristic effects. *Antp* may result in the formation of a complete leg, $ss^a$ changes only the arista and its base to a complete tarsus, whereas *1(4)29* transforms the proximal antenna to coxa and trochanter. *Ba* may also make a complete leg, but frequently causes the loss of the distal parts. None of the mutants shows maternal inheritance, and all are variable in penetrance and expressivity. The last three have pleiotropic effects on legs, causing segment fusion etc.

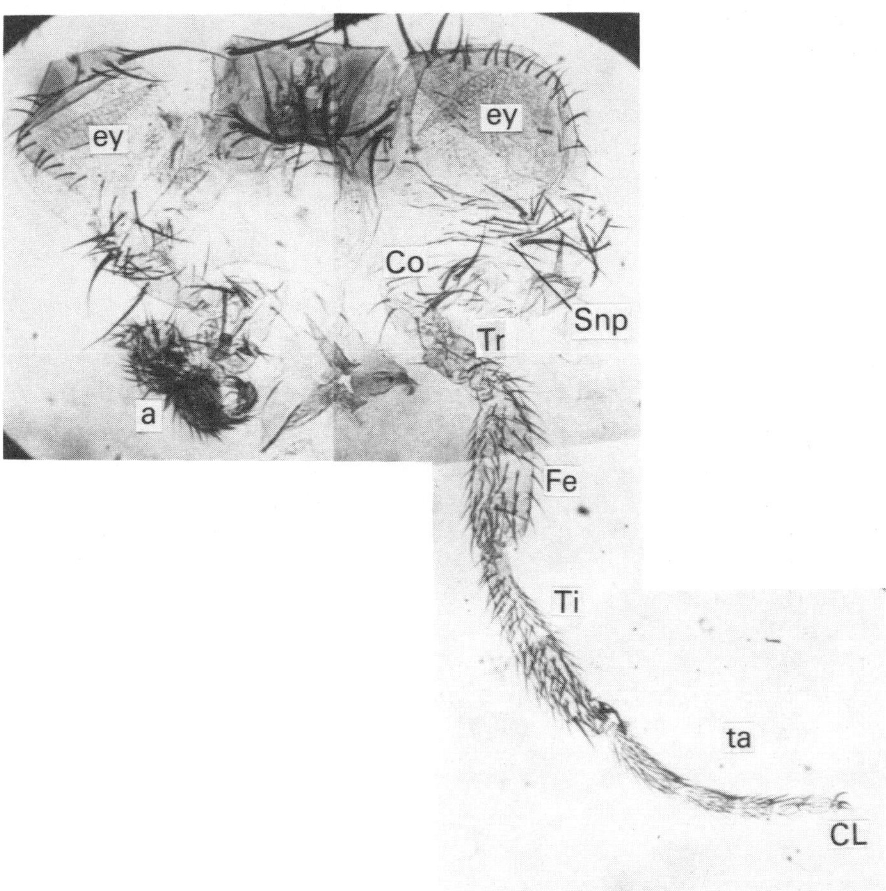


**Fig. 10.1** The homoeotic transformation of antenna to mesothoracic leg in *Antp*$^{Ns}$ homozygotes. As with most mutants of this complex, expressivity is very variable and most flies show incomplete transformations, some of which involve patches of transformed tissue surrounded by normal tissue. ey, eye; Snp, sterno-pleural bristles; Co, coxa; Tr, trochanter; Fe, femur; Ti, tibia; ta, tarsus and CL, claws. a is the opposite, partially transformed antenna. Photograph by J. P. Williams.

One general characteristic, which seems to hold for all homoeotics affecting appendages, is that there is a one-to-one-relationship between the normal and transformed structures (Fig. 10.2). This was clearly shown by Postlethwait and Schneiderman (1971) who took advantage of somatic mosaics to plot which leg structures replaced antennal structures throughout the range of 'phenotypic mosaics' which they could find. That is, a particular coxal patch always replaces a certain antennal sensillum, and so on. There is no ambiguity among cells at the edge of a transformed region; they are either antenna or leg, although sometimes physically distorted. Since a leg contains 3–4 times as many cells as an antenna, cell division as well as differentiation must be affected by the transformation. Put another way,

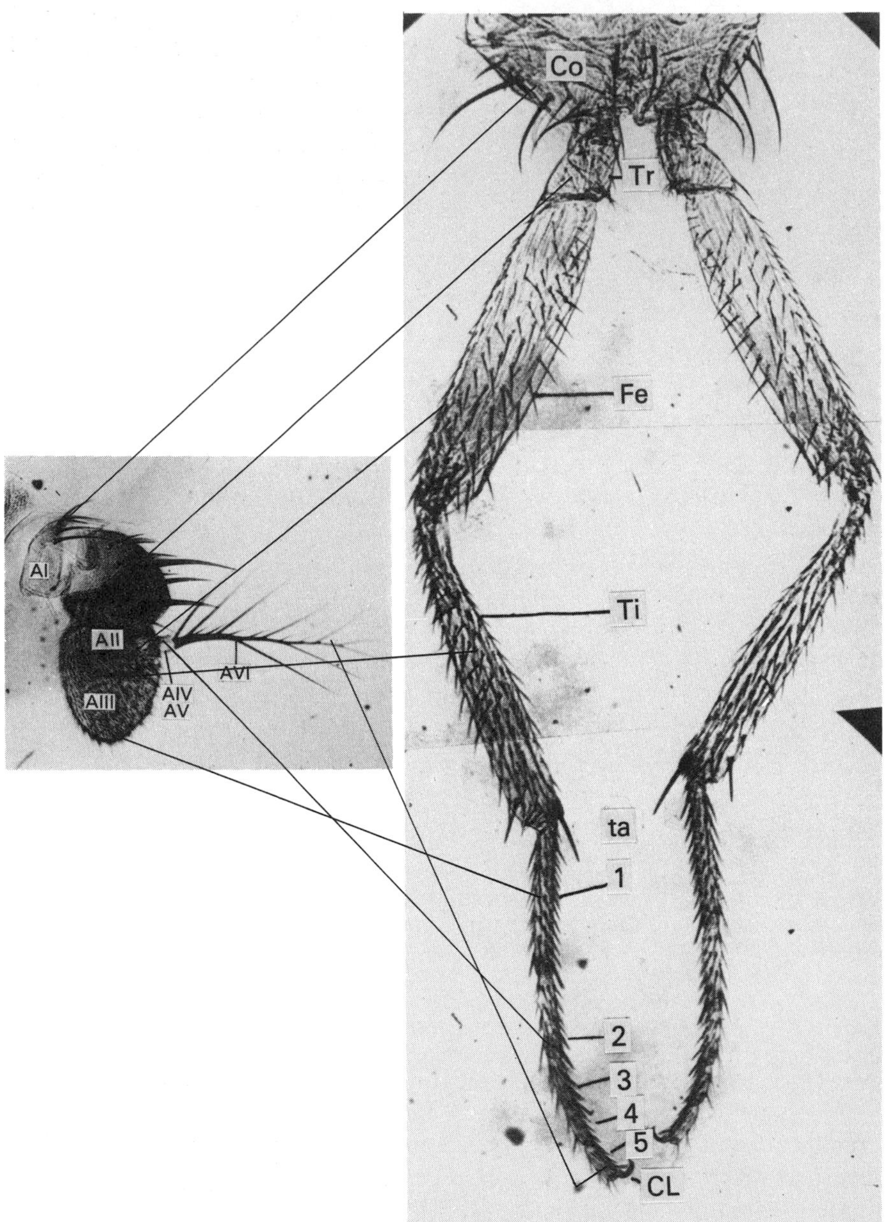


**Fig. 10.2** The homology between antenna and transformed leg, after Postlethwait and Schneiderman (1971b). AI, II and III, antennal segments I–III; AIV, V and VI are the arista. Leg segments as in Fig. 10.1.

there must be some kind of general 'information' in a disc which is interpreted in one way by an antennal cell and in another way by a leg cell. The basis of this 'specification' is unknown, but it allows an either/or, binary, type of switching like that already described (Fig. 9.5). An intriguing aspect of homoeotic loci is that many are clustered on the third chromosome, and in the case of bithorax this grouping together has a functional significance (p. 259). There is a similar clustering around the *Antp* locus, and this has been called the Antennapedia Complex (ANT-C) in the hopeful expectation that its organization will also reflect function (Kaufman *et al.*, 1980). We shall look at ANT-C first, since we have to refer to it when we come to the bithorax complex (BX-C).

## The Antennapedia Complex

The genetic map of ANT-C is summarized in Fig. 10.3, and the homoeotic transformations caused by the mutants are listed in Table 10.2. This summary of effects shows that ANT-C genes are important in determining the development of the anterior segments of the embryo. As we should then expect, deletion of the *Antp* plus *Scr* sites transforms all three thoracic segments to something like prothorax, as displayed by the cuticular patterns of the lethal larvae (Wakimoto and Kaufman, 1981). ANT-C has no effects on dorsal meso- and metathorax. Unfortunately we are still short of infor-

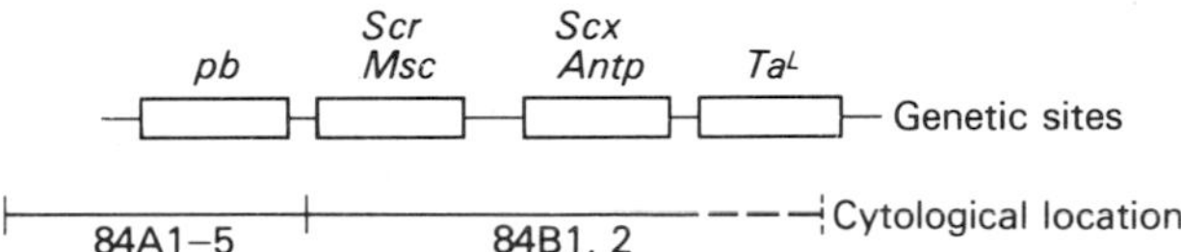


**Fig. 10.3** The ordering of sites within the Antennapedia gene complex as found by translocation-generated aneuploids (Kaufman *et al.*, 1980). The $Ta^L$ site is located from the lethal allele of Thickened arista, which transforms arista to tarsal segments. See Table 10.2 for a description of the mutants.

**Table 10.2** Complementation groups of the Antennapedia Complex

| Complementation group | Larval setal pattern | Adult transformation |
|---|---|---|
| proboscipedia (*pb*) | Normal | Labial palp → Antenna → Leg I |
| Sex combs reduced (*Scr*) | Ventral pro → ventral mesothorax, head abnormal | Leg I → Leg II |
| Multiple sex comb (*Msc*) | Meso-and meta- → prothorax | Leg II and III → Leg I |
| Antennapedia (*Antp*) | See text | Antenna → Leg II |
| *Antp* Extra sex comb ($Antp^{Scx}$) | Ventral meso- and meta- → prothorax | Leg II and III → Leg I |

Summarized from Lewis *et al.* (1980) and Wakimoto and Kaufman (1981).

mation which allows us to allocate an unambiguous function to each of the mutational sites.

### The ANT-C loci

The first discovered proboscipedia (*pb*$^{1}$) mutant is particularly interesting because the transformed labial palps become arista at 18 °C but prothoracic leg at 28 °C (Bridges and Dobzhansky, 1933). The fly may have sex combs on its proboscis! New alleles, which are not temperature-sensitive, give either the 18 °C phenotype (*pb*$^{4}$) or the 28 °C phenotype (*pb*$^{2,\ 3\ \text{and}\ 5}$). The first seems to be a hypomorph (it is dominant over the others) and the latter group are amorphs since they give the same phenotype when hemizygous (Kaufman, 1978). Thus *pb*$^{1}$ is a hypomorph which becomes, or approximates to, an amorph at 28 °C. The temperature-sensitive period for *pb*$^{1}$ is surprisingly late, during the 3rd instar (Vogt, 1946a). At this time, the presence of a non-functional putative structural gene product causes the labial palp to become prothoracic leg, a partly functional product makes antenna (arista), and a fully functional product gives the normal labial palp. It is difficult to envisage how a single gene product can function like this, and the only correlation is with growth rate (Fig. 8.9). As before, there is a regular and exact replacement of labial parts by antennal or by leg structures, but the mutation has no effects on other discs. We can only conclude from these data that the *pb*$^{+}$ product is exclusive to the proboscis, and while we may explain away the leg transformation as an atavism we can offer no explanation of the antenna transformation using that argument.

Less is known about the Sex combs reduced (*Scr*) and Multiple sex comb (*Msc*) complementation group. The effect in heterozygotes of the first is as its name defines, while the second transforms second and third to first leg (and sometimes reduces the sex combs on male first legs). Both are recessive lethals, and the setal patterns of dead *Msc* larvae show conversion of mesothorax and metathorax to prothorax (Wakimoto and Kaufman, 1981), as anticipated by Kaufman *et al.* (1980) on the basis of the described adult phenotypes. They also upset the formation of head and mouthparts. If mesothorax is the primitive thoracic state, this is a non-atavistic transformation which implies that *Msc* mutants activate (or stop the inactivation of) first leg genes in meso- and metathorax.

The Antennapedia extra sex comb mutant (*Antp*$^{Scx}$) similarly converts ventral meso- and metathorax to prothorax, and lethal embryos again confirm this conversion. In this case, dosage studies (Denell *et al.*, 1981) suggest that this dominant is an antimorph or a neomorph (i.e. an active, but abnormal, gene product is made). This again might imply that *Antp*$^{Scx}$ stops the inactivation of first leg genes in second and third legs. But we then have the problem of genes with like functions in two adjacent complementation groups, and this has not been resolved.

The problems of the genetics of the dominant Antennapedia complementation group are also great, but for another reason: most of the mutants are associated with rearrangements and all but two are recessive lethals. Lethal revertants (i.e. presumed null mutants which do not express the leg transformation) display the familiar conversion of meso- and metathorax to prothorax, as judged by the setal patterns of dead larvae. This implies that the $Antp^{+}$ product is necessary for the normal differentiation of meso- and metathorax (Denell *et al.*, 1981), but it does not explain the antenna–leg transformation. Since the mutant Nasobemia ($Antp^{Ns}$) shows this transformation but is not a homozygous lethal, it is clear that the two properties of the locus can be separated. Denell *et al.* (*loc. cit.*) account for this dominant phenotype by postulating a lesion in a "*cis*-acting regulatory function responsible for repression of $Antp^{+}$ in the antennal anlagen."

The situation has recently become even more surprising than these complications imply. Struhl (1981a) made a number of revertants of $Antp^{Ns}$, three of which were putative null alleles showing the terminal lethal phenotypes just described. Genetically marked homozygous clones of these reversions (e.g. $Antp^{Ns+RC3}$) were generated in blastoderm eggs and at later stages. Clones in the eye–antenna, head, proboscis, first leg and first six abdominal segments were normal wild-type. But clones in the second leg were expressed as antenna, and so also to a lesser degree were clones in the third leg (Table 10.3). This is the reverse of the obvious action of the dominant mutations of the locus (antenna to second leg). The conclusion is that the $Antp^{+}$ gene product is normally active in the mesothorax, promoting mesothoracic development, and is absent or inactive in the antenna. The revertant null mutant would then be inactive in mesothorax resulting in these limbs developing as antennae.

The corollary to this is that the mutations producing the dominant *Antp* phenotype must be active in the antenna if mesothoracic leg is to be made there (i.e. *Antp* would then be constitutive). $Antp^{Ns}$ clones generated by somatic crossing-over certainly form clones of leg tissue in the antenna, and the reciprocal (+/+ clones in $Antp^{Ns}/+$) produces antennal clones in the transformed 'antennal leg' (Kauffman and Ling, 1980). This latter genotype

**Table 10.3** The phenotypes of non-revertant and revertant $Antp^{Ns}$ clones in eye-antenna and mesothorax.

| **Compartment** | **Phenotypes** | |
|---|---|---|
| | $Antp^{Ns}$ **non-revertant** | $Antp^{Ns}$ **revertant** |
| Antenna A + P | Leg II | Antenna |
| Leg II A | Leg II | Antenna |
| Leg II P | Leg II | Antenna |
| Wingnnotum A | Normal | Abnormal |

A and P are the anterior and posterior compartments. Data from Struhl (1981).

can be used to find when $Antp^{Ns}$ acts by generating +/+ clones at different developmental times. It is then found that the gene is transiently active between 48 and 65 h of larval life, and produces a clonally stable state whose maintenance does not require further $Antp^{Ns}$ action. Clearly, $Antp^{Ns}$ must be active in the antenna and once having acted it establishes a cell-inherited state. This mutant is temperature-sensitive, and exactly the same times have been found for its temperature-sensitive period (Stepshin and Ginter, 1974).

Although $Antp^{Ns}$ has been taken to be a cell autonomous mutant, both the somatic crossover studies just described note that not all marked cells express the mutant phenotype. There must be a passage of primary or secondary gene products between adjacent mutant and wild-type cells: the effective products are diffusible. $Antp^{Ns}$ is also affected by its nutritional environment, for flies hatching early or late from cultures show different levels of gene expression. Jowett and Sang (1979) therefore checked the effects of dietary deficiencies on gene expression, using axenic culture techniques. Most deficiencies are without effect although they slow larval growth. However, deficiencies of thiamine and of pantothenic acid cure the $Antp^{Ns}$ lesion, contrary to expectation, causing the flies to have more or less normal antennae. Both thiamine and pantothenate deficiencies would result in a shortage of acetyl-CoA, and exploration of this area of metabolism suggested that restriction of fatty acid synthesis was the important consequence. Addition of oleic acid to the diet increases the expression of the mutant phenotype. Nutritional switches (Fig. 10.4) further show that acetyl

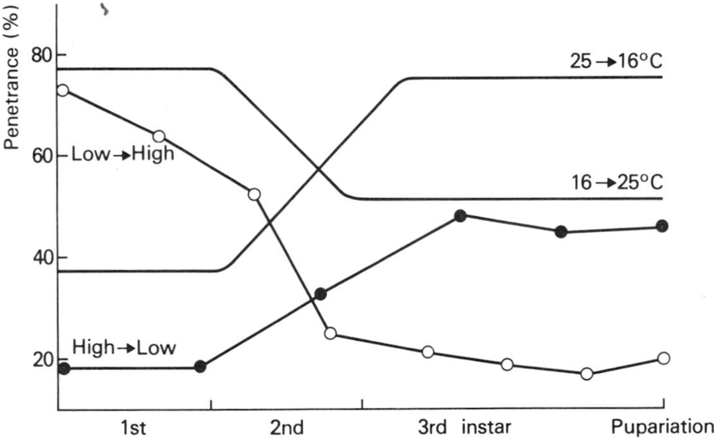


**Fig. 10.4** The temperature-sensitive and nutrition-sensitive periods of $Antp^{Ns}$ homozygotes, from the data of Jowett and Sang (1979). Larvae were shifted from normal to low temperatures (which raises penetrance), and conversely, giving the upper pair of curves which show that the temperature-sensitive period is between the first quarter of the 2nd larval instar and the beginning of the 3rd larval instar. The nutritional switches were between media with adequate pantothenate and media low in pantothenate (+ inhibitors of pantothenate usage) which raises penetrance. The nutrition-sensitive period is also during the 2nd larval instar.

CoA supplies affect gene expression at exactly the same time as the gene is expected to be active during the late 2nd and early 3rd instars. These results do not prove that *Antp*$^+$ is concerned with the synthesis of oleic acid (or some other related fatty acid) when 'making' mesothoracic leg in the antenna. But if it were, a deficiency of acetyl-CoA would prevent this, as is found.

Little is known about the last member of the complex, Thickened arista (*Ta*) which causes partial transformation of arista to tarsus-like structures. The quite separate locus, spineless aristapedia ($ss^a$), on the other hand, transforms the arista and its base to a complete tarsus. The temperature-sensitive period of this locus is during the mid-3rd instar, and follows the action of *Antp*. There is again evidence that a dietary deficit of acetyl-CoA permits normal antennal development. This suggests that *Antp* and $ss^a$ share a common metabolic requirement and act sequentially. The genetic compound *Antp* $ss^a/+\ ss^a$ generally shows a greater than additive transformation (Bulyzhenkov and Ivanov, 1978), as if there were an interaction among the sequential processes they affect. $ss^{a+}$ must specify arista development positively, as opposed to leg (Struhl, 1982b).

Two further issues have to be resolved before we can conclude that ANT-C is organized as a functional complex. The actions of properly defined null mutants have to be elucidated, since hypomorphs (like $pb^1$) may give misleading, if interesting, phenotypes. And we must have firmer evidence that transformation of larval segmentation patterns means that imaginal discs suffer the same change of fate, particularly since the mutant gene products, as we have seen for *Antp* and $ss^a$, function relatively late in development. We have also to explain the difference of response between prothoracic and mesothoracic (and possibly metathoracic) legs, for if *Ns* (and the other dominant *Antp* mutants) are constitutives we might anticipate that first legs, too, would be converted to antenna. This brings us to the nature and role of the bithorax complex whose mutants change mesothorax and later segments, and so shares in some way the control of body segment differentiation with ANT-C. We shall look at this relationship later.

## The bithorax complex (BX-C)

The first of the bithorax (*bx*) mutants was identified by its effect in transforming the anterior part (compartment) of the third thoracic segment into the corresponding part of the second thoracic segment: part of the haltere was made into wing (Bridges and Morgan, 1923). Subsequently, a number of non-allelic mutations were found affecting these two thoracic segments, and their genetic analysis showed that they were members of a pseudoallelic series (Lewis, 1955), or as we should now define them, members of a gene complex (BX-C). Two examples of the phenotypic effects of these mutants

(a)

**Fig. 10.5** (**a**) The thorax of the first six-winged dipteran, bred and photographed by Dr. P. Ingham. The genotype of the fly is $trx^1$ *y f/mDf(3R)red*$^{P93}$ (see text). Since the fly was unable to emerge, the wings are not inflated, but they can be identified as such by their typical sensilla and bristles. The *trx* homoeosis has converted first and third to second thoracic segments. (**b**) This fly is *bxd/Df(3R)bxd*$^{100}$, and has its first abdominal segment partly transformed into metathorax. It therefore has four pairs of legs, not three. These extra legs are often malformed, and in the example the fourth leg on the left has folded under the abdomen. Specimen prepared by Dr. S. Kerridge.

are shown in Fig. 10.5. At present, nine genes are known which when mutant result in partial or complete loss of a function necessary for the control of the level of segment development. Except for Ultrabithorax (*Ubx*) these are all recessives, and presumably structural gene mutations. And there are four dominant, gain of function, mutations which appear to be *cis*-regulatory. The map of the locus is given in Fig. 10.6, together with details of the genes (see Lewis 1981; 1982). The more distal a locus is on the chromosome the more posterior is the segmentation it affects (except for *bxd*), and the *cis*-regulatory dominants are adjacent to the genes they appear to regulate.

Study of the effects of mutation of members of the complex has been greatly advanced by following the changes they also make in the larval segmental patterns that we have already described (p. 173). Deletion of BX-C (*Df(3R)P9* homozygotes) changes larval segmental morphology from the third thoracic through to the seventh abdominal segment, so that they all develop as second thoracic segments. The eighth abdominal segment, rather curiously, develops characteristics of head and first thoracic segments as well

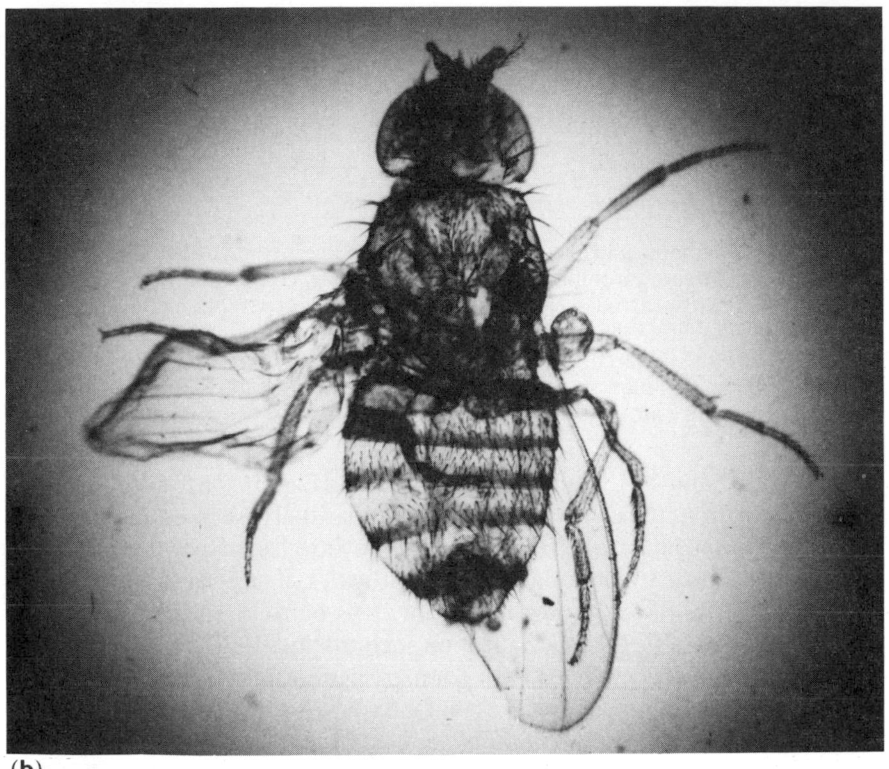

(**b**)

as of second thoracic, suggesting interactions with ANT-C (Duncan and Lewis, 1982); but in general *Df(3R)P9* defines the separate domains of action of the two complexes and shows that BX-C regulates all segments posterior to the second thoracic.

As before, it is assumed that the transformations of larval segmental patterns result in concordant changes in the segmental imaginal discs. This is certainly true for the interesting mutant bithoraxoid (*bxd*), which transforms the first abdominal segment into a metathoracic segment bearing halteres and legs (Fig. 10.5). The anterior histoblast nests are missing in the mutant, and are replaced by one or two leg discs ventrally. Clonal analysis shows that compartments are laid down early, but may subsequently be transgressed to allow adjustments of pattern formation, and that disc growth follows the leg pattern and is not delayed, as is division of the histoblasts (Kerridge and Sang, 1981). $bxd^+$ must be concerned with the proper organization and determination of histoblasts in the first abdominal segment where it functions early, possibly at blastoderm, and also subsequently (Morata and Garcia-Bellido, 1976). It is not yet known if the $iab^+$ genes act in the same way in later abdominal segments.

Using the deficiencies shown in Fig. 10.6 and bithoraxoid, Lewis (1978; 1981) was able to construct a model of the likely gene activities in the normal

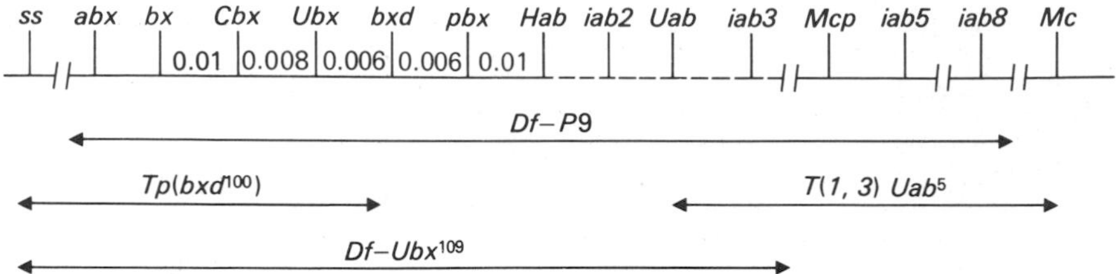


**Fig. 10.6** The genetic map of the bithorax complex, according to Lewis (1981). The mutations are: *abx*, anterobithorax; *bx*, bithorax; *Cbx*, contrabithorax; *Ubx*, Ultrabithorax; *bxd*, bithoraxoid; *pbx*, postbithorax; *Hab*, Hyperabdominal; *iab*, infra-abdominal; *Uab*, Ultraabdominal; *Mcp* Midcadastral pigmentation. The map distances are shown for the older members of the complex, and the deficiencies used in manipulating the complex are shown below. The flanking markers are: spineless (*ss*) and Microcephalus (*Mc*).

third thoracic and in the abdominal segments (Fig. 10.7). This model has two features: first, the genes are activated sequentially, segment by segment, in their chromosomal order; and, second, they are 'selector genes' (Garcia-Bellido, 1975) which determine specific pathways of segment development, as we previously saw that engrailed does. It is particularly interesting that *bx* and *pbx* show a further specialization such that *bx* transforms the anterior compartment and *pbx* the posterior compartment of third thoracic to second thoracic segments. Note, however, that *Ubx* inactivates $bx^+$ and $pbx^+$ in the *cis* configuration, and that *bx* and *pbx* are therefore covered by the model, if only formally.

### *Polycomb and Regulator of bithorax*

We have already seen that BX-C$^+$ expression depends on the prior, maternal activity of the extra sex combs locus, and that subsequently Polycomb (*Pc*) takes over a similar function (without either, only eighth abdominal-type segments are made). Mutations of both loci disrupt segmental determination by inhibiting the expression of BX-C genes. *Pc* is required throughout development (or at least into the 2nd instar) since mutant clones then induced form analia-like structures in head and thorax, but have no effect on the analia (Struhl, 1981b; Duncan and Lewis, 1982). Extra sex combs apparently initiates, and *Pc* maintains, the conditions for the proper determined state of each segment. *Pc* is likely to be the gene whose product represses the BX-C genes. Lewis (1978) therefore postulates a body axis (longitudinal) gradient of $Pc^+$ activity and a proximo–distal chromosomal gradient of increasing affinities of BX-C genes for a $Pc^+$ repressor. Thus, as one moves posteriorly along the embryo, $Pc^+$ activity would *decrease* in a graded fashion allowing genes with *increasing* affinities for the repressor to be sequentially activated. In the third thoracic segment, for example, only genes with low repressor affinity would be activated by the high $Pc^+$ activity

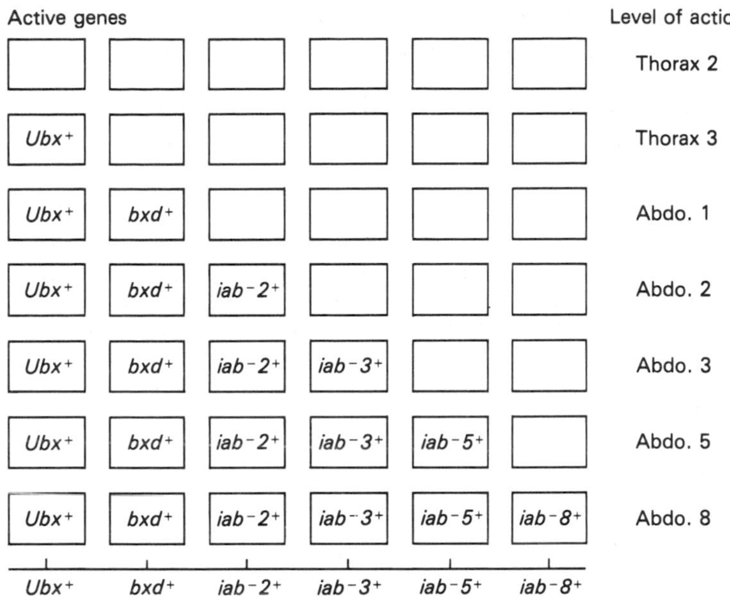


**Fig. 10.7** The segmental activity of BX-C genes, as proposed by Lewis (1981). The genes are shown below, and their active products which are necessary for the morphogenetic change are shown in the boxes. Thus the $bxd^+$ product is required in addition to the $Ubx^+$ product for the development of the first abdominal segment. $abx^+$, and *bx* and *pbx* (not shown) are also assumed necessary to give thorax 3. It is assumed that the substances act additively to give the sequence of segmental morphogeneses. *Df-P9* (Fig. 10.6) which elimates all the BX-C causes all segments to be thorax 2. $Tp(3)bxd^{100}$ transposes all the loci to the left of *Ubx* to the left arm of chromosome 3, and with *Df-P9* generates embryos which lack the genes *abx* through to *Ubx*, inclusive. The flies which hatch resemble *Ubx* homozygotes, with extreme *bxd*, *pbx* phenotypes. The changes in larval setal patterns are discussed by Lewis (1981).

there, and in the last abdominal segment all the complex would be activated by the near zero $Pc^+$ activity. $Pc^+$ acts as a negative regulator of BX-C. Since it also has other, more complex effects (Duncan and Lewis, *loc. cit.*) the case for this model is not yet watertight.

*Pc* is on the third chromosome at 78E, and at 88B there is another locus, Regulator of bithorax (*Rg-bx*) which causes anteriorly directed transformations: to mesonotum or wing on the third segment, or the seventh to the sixth abdominal segment, etc. in heterozygotes (Lewis, 1968). That is, *Rg-bx* acts in the opposite direction to *Pc*, and it is postulated that the locus makes inducer molecules responsible for the differential activation of the BX-C genes in the thoracic and abdominal primordia (Garcia-Bellido and Capdevila, 1978). One implication of this assumption is that *Rg-bx* should function early, and it is indeed found to have a maternal effect not completely zygotically rescuable (Ingham and Whittle, 1980). Another implication is that there should be a dosage interaction between $Rg\text{-}bx^+$ and

*Pc*$^+$. When the ratio of *Rg-bx*$^+$/*Pc*$^+$ is less than 3/2 the flies are wild-type, but higher ratios (2/1 and 3/1) give progressively more extreme Polycomb phenotypes, consistent with the assumption that the *Rg-bx*$^+$ inducer binds to the *Pc*$^+$ repressor (Capdevila and Garcia-Bellido, 1981).

As with *Pc*, exploration of *Rg-bx* functions has brought complications. An apparent allele has been named trithorax (*trx*) because it transforms both pro- and metathoracic structures to their mesothoracic equivalents, as well as anteriorizing abdominal segments, predominantly as first abdominal segment (Ingham and Whittle, 1980). Further, clones induced in the 1st larval instar show transformation of first and third leg structures to second leg, and of haltere to wing. Detailed examination shows, too, that *trx*/*trx* clones in the wing also transform posterior compartment structures to anterior, mimicking engrailed (Ingham, 1981). The *trx* mutations show that the locus functions later than would be expected, after embryogenesis, and that it may play some part in maintaining compartment differences as well as determining segmental differences. It is difficult to fit these observations into the inducer/repressor model just described.

### Phenocopies and BX-C dominants

A characteristic of the *Rg-bx* phenotypes is that they are patchy; the transformations are incomplete but show no cell overlap with wild-type. Similar *bx* phenocopies can be induced by exposing pre-blastoderm and blastoderm embryos to ether vapour (Gloor, 1947). It is particularly interesting, then, that the proportion of phenocopies is reduced in *Pc* heterozygotes but increased three-fold in *Rg-bx*$^+$ hemizygotes, and halved in genomes with three copies of *Rg-bx* (Capdevila and Garcia-Bellido, 1981). While these observations fit the theory in a general sense, there is still an unexplained peculiarity of ether phenocopies: the haltere to wing and leg III to leg II transformations are always confined to the anterior compartments of the segments (Bownes and Seiler, 1977).

Similar phenocopies can be induced by exposing blastoderm embryos to a heat shock of 37 °C for 15 min (Santamaria, 1979). Presumably this shock switches off all genes other than those making heat-shock proteins. In this case the phenocopies show transformations corresponding exclusively to the *dominant* mutations of the BX-C (Fig. 10.3). These mutations are taken to affect *cis*-regulatory elements for adjacent, distal structural genes, acting as their constitutive derepressors (Lewis, 1978). Correspondingly, the phenocopy of Contrabithorax (*Cbx*) by transforming meso- to metathorax mimics the action of *Ubx*; the phenocopy of Hyperabdominal (*Hab*) by transforming metathorax and first abdominal to second abdominal segments mimics *iab-2*; the phenocopy of Ultraabdominal (*Uab*) by converting first abdominal to second and third abdominal segments mimics *iab-3*; and the phenocopy of Midcadestral pigmentation (*Mcp*) by transforming fourth abdominal to the distinctive pattern of the fifth abdominal segment mimics *iab-5*. The dominant mutations and the phenocopies activate presumed structural genes a segment ahead of normal (Fig. 10.7) but in the same way.

On Lewis's (1978) assumption, the dominant mutations block the synthesis of *cis*-active repressors, and heat shock could do the same. However, we should also expect heat shock to copy more than one dominant mutation at a time, and this is found: individuals may carry phenocopies of not less than three *cis*-regulatory mutants. These genes must normally act at the same time, so it is interesting that the effective phenocopying stage is limited to the short period (2 h 20 min–3 h 20 min) during which the blastoderm cell walls are being laid down and are finally completed. Thereafter their homoeotic determined state is clonally inherited. Of course, these data do not exclude the possibility that heat shock upsets the establishment or stability of a general repressor gradient, e.g. of *esc*. And we must also note that the mutation Ultrabithorax-like (*Ubl*), which also copies *Hab*, turns out to be an RNA polymerase II mutation, showing that disturbance of transcription can also have BX-C-like effects (Mortin and LeFevre, 1981; Mortin and Kaufman, 1982).

Not all the genes of the BX-C have yet been definitely identified, as the following example shows.

Ultrabithorax (*Ubx*) in *trans* with *bx*, or *pbx* or *bxd* produces the phenotype of the recessive allele. Over the deficiency for BX-C, (*Df*(*3R*)*P9*), *Ubx* expresses all three recessive phenotypes. Morata and Kerridge (1981) induced *Ubx*/*Df*(*3R*)*P9* clones in blastoderm embryos and found that dorsal clones were as expected for the triple deficiency of *bx*, *pbx* and *bxd*, but clones in the posterior compartments of both second and third legs were unexpectedly transformed into posterior first leg. (The anterior compartments of first and second legs were normal, but in the third leg they were transformed to second as in *bx* mutations). This was the first evidence that BX-C mutations transform prothorax. Clones induced 7 h or later after egg laying behaved as originally expected: second leg posterior clones were untransformed and third leg clones were transformed into second leg (i.e. expressed the *pbx* phenotype). This suggests that there is a wild-type gene different from $pbx^+$ which is required early in metathoracic leg development to prevent its determination as prothoracic leg, and some 4 h or so later $pbx^+$ is required to prevent its differentiation as mesothoracic leg. By using appropriate deficiencies and transpositions it was shown that there was indeed a gene to the left of *pbx* which, if mutant, transforms the posterior compartments of both meso- and metathoracic legs into prothorax. Although not yet precisely located, it has been named postprothorax (*ppx*) and it is inactivated by *Ubx* alleles just like *bx*, *pbx* and *bxd*. Unlike these loci, however, it is required early in development and is then dispensible, whereas *bx* and *pbx* are first required later and continue to be transcribed until the end of the larval period.

### *BX-C DNA*

The identification and isolation of the DNA sequences of BX-C by Bender, Spierer and Hogness (reviewed by Akam (1983)) presents us with an exciting and instructive picture of the molecular organization of the complex. In the first place, more than 200 kb, and possibly as much as 300 kb, of DNA is committed to the complex, which is more than we might have anticipated from its known cytology (Fig. 10.8). Second, the majority of mutants, many previously classed as point mutations and assumed to result from base changes, turn out to be due to insertions of 5–10 kb, including copia-like, 'gipsy' and other sequences repeated throughout the genome. Third, some mutants are due to breaks, deletions and transpositions. In the last class, the *pbx*$^1$ mutation is a 17 kb deletion which moved about 43 kb proximally, in reverse orientation, to give the *Cbx*$^1$ mutation, which arose at the same time (Lewis, 1964). Nevertheless, these mutations are in their proper order according to the genetic map (Fig. 10.6).

Perhaps even more surprising are the *Ubx* mutations just described. A dozen mutants are found to result from breaks or rearrangements lying randomly within about –30 kb to –100 kb, suggesting that this entire 70 kb region is a single transcription unit. If this unit is broken, or the ends separated, it no longer functions. A probe made from the right (distal) side of the region (*Ubx*$^1$) picked up three small mRNAs (3.7, 4.3 and 4.7 kb) and these also matched a probe from the left-most end of the region, but not any of the DNA in between. It therefore seems likely that the *Ubx* transcription unit is made up of small terminal exons, and a large intron which is spliced out to give the functional RNAs, which are found in post-blastoderm embryos and in larvae.

There is a difficulty with this interpretation of *Ubx* organization: the postulated intron carries the *abx* and *bx* genes! To justify our previous statements about them we must assume some segment-specific splicing mechanism which permits the genes to be transcribed as required. There is no

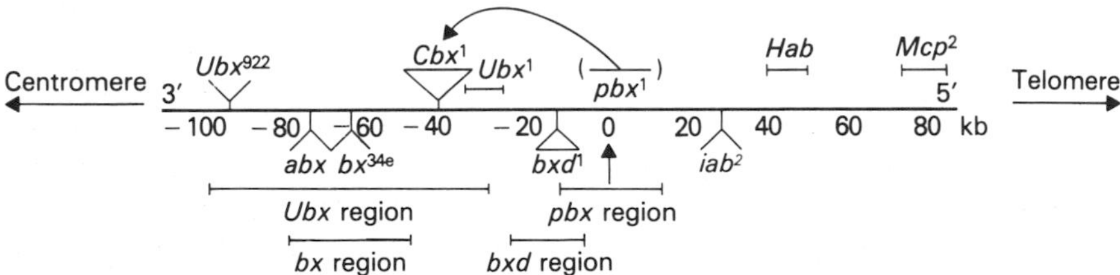


**Fig. 10.8** The molecular organization of BX-C, based on Marx (1981). The chromosome 'walk' started by finding a cloned sequence which mapped to an inversion, giving a starting point in the *pbx* region taken as 0. More than 100 kb on either side of this has been mapped for the known mutations. These mutations are either inversions (⊢⊣) or deletions ( ), like *pbx* which is transposed and inverted to give the *Cbx* insertion. Other mutant sites are due to insertions (⅄) of a variety of transposable elements, like copia and gipsy. None of the mutations seems to result from simple base changes. See text for further details.

evidence for *abx* or *bx* mRNAs at present, whereas *bxd*, which is distal to the *Ubx* domain, makes a 1.4 kb mRNA after splicing out a 19 kb intron. The same difficulty applies to *Cbx*. The *pbx* deletion shows that this 17 kb fragment encodes the instruction to make posterior haltere, not wing. The insertion of this instruction in the wrong place may cause it to be 'read' earlier, as implied in Lewis's model, and so to convert wing to haltere, i.e. to function in mesothorax instead of metathorax. In fact, both genes can be shown to function in $Cbx/Ubx^{130}$ heterozygotes where flies grown at 29 °C express a strong *Cbx* phenotype (wing → haltere) but flies grown at 17 °C show strong *Ubx* phenotypes (haltere → wing). If the weak *Ubx* allele ($Ubx^{61D}$) replaces $Ubx^{130}$ the flies have a *Cbx* phenotype at both temperatures (Kauffman, 1981). The transposition may have made *pbx* a constitutive mutation, as Lewis (1968) suggested, which we now call *Cbx*.

These data suggest that we cannot ignore specific splicing systems as mechanisms for gene regulation, but we shall only be able to think more precisely about this when the relevant mRNAs have been isolated and their products identified. Preliminary data suggest that these mRNAs appear early in embryogenesis and are required throughout development. This is not what we should expect of determinative switch mechanisms, followed by a fixed cell heredity; but it is what we might expect if the DNA 'programme' had to be re-established every cell division. The cloned genes of BX-C should allow us to resolve this issue, and others.

Finally, we must note the range of other mutants with bithorax-like effects (Hexaptera, tetraltera, metaplasia, tetraptera, etc.). While we know little about their actions, or of their interactions with BX-C, they caution us against assuming that the secrets of segmental differentiation lie exclusively within the two major complexes.

### *Specification of the thoracic segments*

We have seen that three dominant loci are involved in the specification of the thoracic segments, namely, *Scr*, *Antp* and *Ubx*; and we noted the apparent anomaly that *Antp* clones made antenna only in the mesothoracic legs. Struhl (1982b) has examined the consequences of combinations of these loci, using induced mitotic clones in a Minute background. His results identify the role of each wild-type gene in segment specification, and the dependence of one on the activity of the others. The genes illustrate an unexpected combinatorial relationship which suggests that they might have regulatory functions, and this also accounts for the unique activity of *Antp* in the mesothoracic leg when it is the only mutant locus.

*Scr* clones, Struhl (*loc. cit.*) confirmed, convert first leg structures to second leg structures, and also labial to maxillary palps in the proboscis. They have no effects in the eye–antenna or in meso- and metathorax. $Scr^{+}$ is required to specify, at least, ventral prothorax, and labial development. As we have seen (p. 253), *Antp* clones transform mesothoracic leg to antenna. They have no effect elsewhere, except rarely in the other legs. *Antp*$^+$ specifies the mesothoracic state of the second leg. *Ubx* clones transform metathorax (haltere and third leg) to mesothorax, but mutant clones induced early transform posterior compartments of meso- and metathoracic legs to prothoracic leg compartments. *Ubx*$^+$ is therefore required in the metathorax to determine meta- as opposed to mesothoracic development, and has an early function in posterior leg compartments which blocks prothoracic development in the second and third legs.

Two of the gene combinations are particularly interesting. Clones homozygous for both *Scr* and *Ubx* transform pro- and metathorax to mesothorax, i.e. behave additively and have no effect on mesothorax. There is no effect of their combined early loss on posterior leg compartments, unlike the effect of *Ubx*$^+$ loss just described. The difference may be a consequence of early activity of *Scr*$^+$ when *Ubx*$^+$ is removed at blastoderm. *Scr*$^+$ may therefore be active first in meso- and metathorax and then become inactive. As we might expect from this result, triply mutant clones are also additive, and antennal clones are then formed in all three pairs of legs, the remainder of the leg being mesothoracic in type. This finding suggests that *Antp*$^+$ is required in all three segments, but that its activity is overridden or modified by *Scr*$^+$ and *Ubx*$^+$ in the pro- and metathorax, respectively. In sum, *Antp*$^+$ is active in all three segments, *Scr*$^+$ only in the prothorax and *Ubx*$^+$ in the metathorax, with the qualification that there are the additional early activities which we have noted.

The argument we made above that mesothorax is the segmental 'ground state' (see also Lewis, 1978) is supported by the actions of *Scr*$^+$ and *Ubx*$^+$ in changing that state, but controverted by the assumed activity of *Antp*$^+$ in all three segments, and by the apparent different early role of *Ubx*$^+$ in posterior leg compartments. This latter function implies that one locus has more than one kind of activity, so it is interesting to ask if *Ubx*$^+$ has an early function in the transformed antenna of an *Antp* mutant which is replaced by a morphologically normal mesothoracic leg (Fig. 10.1). In the same experiment, *Ubx*$^-$ clones established at blastoderm form prothoracic structures in the posterior compartment of the mesothoracic leg, as before, but they fail to show this transformation in the 'cephalic leg' which replaces the antenna (Morata and Kerridge, 1982). It is also interesting that mesothoracic leg compartments are established at blastoderm (3 h) but the anterior and posterior compartments of the antenna arise later, at about 72 h of development. The 'cephalic leg' follows the antennal pattern of late compartment formation, but whether or not this accounts for the difference between cephalic and mesothoracic legs, or if this implies that *Scr*$^+$ is inactive in the eye–antenna disc, remains to be sorted out. It is, in fact, idle to speculate about these processes until we know more about the gene products involved. What is surprising is that this difference between thoracic and cephalic legs questions the significance of compartments for leg morphology (see also Karlsson, 1983).

## Homoeosis in the eye disc

Most homoeotics show variability of expression, but it is not always clear how far this is due to the erratic working of 'weak' hypomorphs, or to the direct effects of nutritional deficits etc., as already described, or to both. There is also a homoeotic class which requires more than one gene mutation for its expression, and transformations of the eye disc exemplify this group. As might be expected, these transformations are particularly prone to environmental effects and to modification by the rest of the genotype.

When combined with eyeless, which reduces eye size, the mutation Ophthalmoptera ($Opt^G$) has no effect on eyes when larvae are reared at 17 °C, but results in the replacement of eye tissue by wing if the larvae are reared at 29 °C. There is no ambiguity about the wing then formed, and there is a regular replacement of region by region: anterior eye transforms to the triple row of bristles, the fronto-orbital region makes costa and so on. It is surprising that there seems to be a gradient in the eye disc common to wing, but that is one interpretation of the transformation. The temperature-sensitive period for this response is during the 2nd larval instar, and raising larvae to 29 °C during the 1st or during the 3rd instar is without effect (Postlethwait, 1974).

The separate mutation ophthalmoptera (*opht*), along with the eye-reducing mutation loboid, behaves similarly. The temperature-sensitive period is again during the 2nd larval instar, but in this case penetrance is high at 17 °C and lower at 25 °C (Ouweneel, 1969). It is not clear if we are concerned in these two eye-wing transformations with heat and cold-sensitive mutants, respectively, or if the two mutations affect quite separate processes. It is certainly difficult to argue, as Ouweneel (*loc. cit*) does, that the low-temperature effect on *opht* is a consequence of increasing the relative growth rate of the eye disc, which would not apply to the converse response of $Opt^G$, and that "enhanced proliferation is a prerequisite for many homoeotic phenomena". What the two situations have in common, and as a necessary prerequisite for homoeotic expression, is the presence of mutations affecting eye growth and the survival of eye cells.

More detailed work has been done on the tumorous head mutants (*tuh-1* and *tuh-3*) which somewhat erratically transform antenna to leg, but may also replace part of the head with abdominal tergites and the 'rostalhaut' by genital elements (Postlethwait *et al.*, 1972). The mutations also cause cell death (and embryonic mortality) and the expected deficiencies and duplications of head structures often result from this. The *tuh-1* mutation has a temperature-sensitive maternal effect, and increased temperatures throughout the entire period of ovogenesis raise the penetrance of the transformation, as do extra doses of the mutant gene. The *tuh-3* gene is similarly temperature-sensitive, but penetrance is raised only during the 8th–12th hour of embryogenesis when the developing eggs are exposed to 29 °C (Bournias-Vardiabasis and Bownes, 1978). It is argued that some substance is laid down in the egg (*tuh-1*$^+$ gene product) which is then metabolized in some way by the *tuh-3*$^+$ gene product to give normal eye development.

The eye region of the eye–antenna disc does not stain for aldehyde oxidase, but a central staining region appears in *tuh-1* and *tuh-3* transformed discs. If this region is dissected out, and metamorphosed by transplantation into a late larva, it differentiates structures of the abdominal tergites and genitalia. The aldehyde oxidase staining reflects a metabolic change and signals the transformation which becomes overt only after differentiation. Both histoblasts and genital disc stain for aldehyde oxidase, but the former cannot be cultured *in vivo*, so it is surprising that central parts of the *tuh* eye discs from 3rd instar larvae can be continuously cultured and will then form abdominal and genital tissue on metamorphosis (Kuhn *et al.*, 1979). This stability of the homoeotically transformed state during culture *in vivo* is also found with loboid-ophthalmoptera discs (Ouweneel, 1970c).

The *tuh-3* locus lies adjacent to, or in, the BX-C region, at 89E4 or 5 (Kuhn *et al.*, 1981a). It may therefore be the last gene of this ordered complex, and responsible for the determination of posterior and genital structures. However, there is also one lethal mutant (Shearn *et al.*, 1971) which affects genital disc determination. When the mutant genital disc is cultured *in vivo* and metamorphosed, tarsal and antennal tissues are found adjacent to anal plate. The relationship of these mutants is not known, but merits study.

The eye disc transformations bring a new dimension to the phenomenology of homoeosis. The ophthalmoptera pair depend on some developmental instability caused by the eye-reducing genes, which facilitates the transformation to wing, and the specificity of this response involves not one gene but two. Unfortunately the interaction of *Opt*$^G$ and *opht* does not seem to have been studied, and we have no indication of their possible sequential action, or if they code for the two components of a single effector protein, or whatever. Nor is there any reported homoeotic change of wing to eye, although such a transdetermination is recorded (Gehring, 1972).

So far there is no ambiguity about the need for two genes in the *tuh* transformation. This raises a further problem since the gene pair has at least three actions: antenna to leg, as well as eye to abdominal or to genital tissue. This range of action suggests some more general role for *tuh* gene products which contrasts with the 'master switch' activity attributed to homoeotic genes generally. However, we have much still to learn about the homoeotic potential of the eye disc since for instance, the mutant ophthalmopedia, now lost, converts the same central area of the disc to a leg-like, or palp-like structure (Waddington and Pilkington, 1943). Many genes must be involved in the normal development of the eye, just as many are concerned with leg development (Table 10.1), but in this case the surprising consequence is that mutations of separate loci cause quite unrelated homoeotic changes. It is hard to believe that eye tissue, so anciently evolved, goes through a sequence of determinative steps such that blocking one or another results in the coherent development of the structures described above.

## Conclusions

We shall ignore the remaining mutants in Table 10.1 and merely note that, overall, homoeotic changes encompass the range of currently known transdeterminations (Fig. 9.3), and more. Since transdetermined discs originate from particular, and *apparently* undetermined, cells and, like a blastema, make a whole disc structure, we cannot say that the parallel between transdetermination and homoeosis is *not* due to the same mechanisms. The transdetermination, for example, from haltere to wing might result from inactivation of post bithorax$^+$ during growth, and until we can identify the mechanism involved there is little point in trying to answer the question. There is an exception: we should not expect any transdetermination to match the homoeoses due to inactivation of maternal-effect genes like *esc* or *tuh-1*. But maternal-effect genes are a minority, and the great majority of homoeotic loci are required to function throughout larval development. This raises an interesting point, since most of the homoeoses we have described are intersegmental (or intercompartmental) and there is no evidence that there are segmental maternal-effect genes responsible for the primary steps of segment determination. It is for this reason that we have assumed that segmental determination depends on the recognition of, and specific response to, gradients of *Rg-bx*$^+$, *esc*$^+$ and *Pc*$^+$ gene products, and possibly others. When we remember that a segment spans only three cell diameters in the fate map, it is obvious that we still have much to learn about this primary step of segment determination (Lawrence, 1981b).

Lewis's (1978) model of the orderly derepression of the recessive loci, in response to segmentally interpreted gradients, explains the sequence of primary determinations from thorax to posterior abdomen (Fig. 10.7). The predictive power of this model is well illustrated by the identification, *post hoc*, of the *iab-2* locus (Kuhn *et al.*, 1981b) which, when mutant, converts the second to first abdominal segment. Further, it lies to the right of bithoraxoid, confirming the left–right gene order on the chromosome, possibly implying a functional significance for the arrangement of genes in the BX-C. Unfortunately we do not yet have comparable data for the ANT-C, and therefore have little understanding of the developmental regulation of the anterior segments of the fly. This is a major gap in our knowledge.

The argument that the recessive loci function as 'activator' genes at blastoderm, or just subsequently, is supported by the data on phenocopies of BX-C mutations (but not ANT-C mutations) and we should expect a hierarchy of other loci functioning as 'selector' genes transmitting the determined state, and 'realizator' genes allowing the expression of these states prior to overt differentiation (Garcia-Bellido, 1977). However, not all mutants seem to follow this formula (Table 10.4). If haltere discs transformed by *bx* or by *pbx* are appropriately cut and allowed to regenerate by culture *in vivo* (Adler, 1978), anterior haltere disc fragments carrying *bx* regenerate posterior haltere (untransformed tissue), whereas posterior fragments carrying *pbx* regenerate anterior wing (transformed tissue). The two recessives behave differently; the latter, like an activator or selector, establishing a cell inherited state, and the former being required throughout development. Similarly, we have seen that $Antp^{Ns}$ clones remain determined after a certain stage, whereas $ss^{a40a}$ action can be reversed by temperature switches during development (Schubiger and Alpert, 1975). Thus, we still require more information about the order and sequence of gene activities, the failure of any of which might be expressed as a homoeotic transformation (Shearn, 1979).

**Table 10.4** The stepwise regulatory control of developmental pathways according to Garcia Bellido (1977).

| **Gene class** | **Process regulated** |
|---|---|
| 1. Maternal cortical inducer (*esc*, etc?) | Placement of maternal information in egg |
| 2. Early embryonic activators (*Rg-bx*, *Pc*) | Repressor products inactivated by combination with inducer. |
| 3. Selector genes (*Antp* and *bx* complexes) | Establishment of segmental determined states during early embryogenesis. |
| 4. Realization genes (?) | Expression of the determined state. |

According to Garcia-Bellido (1977). As in the Britten-Davidson model, the selector genes activate different realizator combinations, but here necessarily on a segmental basis.

Nor can we be sure that the dominant homoeotic genes are *cis*-regulatory as Lewis (1978) postulates, for we have seen that *Antp* and *Cbx* may be operator-constitutive mutations. What is certain, of course, is that the extensive information which has been accumulated about these mutants provides a firm base for the exploration of the molecular mechanisms which underlie the determination and development of *Drosophila* segments and their associated organs. And the cloned gene sequences, no doubt soon to be available for ANT-C also, provide the tools for this analysis.

# 11 Sex differences

Where species consist of two distinct types, males and females, the analogy with Mendelian genotypes (one sex being the homozygote and the other a heterozygote) was very early apparent and was emphasized by Correns (1907) among others. But where the species consists of only one type which acts both as a female and as male a different mode of determination had to be considered. Only one genotype was present in, for example, maize, with the female functions localized in the ear shoot and the male functions in the terminal tassel. Both the chromosomes and the genes are supposedly the same in these two sexual regions. Hence the difference in sex must be put on the same basis as that of any other *organ differentiation* in plant or animal, for example, the difference in legs and wings in birds—both modifications of ancestral limbs.

C. B. Bridges, 1939

The evolution of a great variety of sex mechanisms presents the geneticist with an equally diverse array of developmental problems, most as yet untouched. They range from the situation in monoecious plants and other hermaphrodites where the problem is the development of the two sex organs in one individual, as Bridges notes, through monomorphic species where the sex difference is essentially in the gonad and its associated apparatus, to dimorphic species where, in the extreme case, every structure of the male differs from that of the female. These sex differences are usually invariable and, like the homoeoses which they resemble, apparently result from the activity of a small number of genes. Mutations are generally picked up through the appearance of intersexes which can be analysed genetically, and this explains why we shall be concerned mainly with sex in dimorphic species, where the differences are most obvious. In fact, there are two interrelated problems: sex determination and sex differentiation. The first has attracted most attention in the past, and it is only recently that mutants involved in the more general aspects of sex differentiation have been studied. Since the sex determining mechanism may predicate differences in the machinery involved in subsequent developmental processes, we shall consider this first.

## Genotypic sex determination

There are two broad classes of sex determining mechanisms: environmental and genetic. In the former, sex is determined by the environment to which the immature form is exposed, and there is usually no evidence of the heterogamety characteristic of the genetic systems. For example, the larva of the marine worm, *Bonellia*, becomes female if it develops in isolation, but is a parasitic male if it settles on the proboscis of a female. Or, again, among turtles, some lizards and crocodiles, incubation of the eggs at a high temperature (32 °C) makes the clutch all of one sex, while a lower temperature (below 28 °C) gives a hatch of the other sex. The temperature-effective period is fairly specific, and short (Bull, 1980; Ferguson and Joanen, 1982). Sex determination is tied to particular environmental cues in these cases, whereas in the primitive annelid, *Dinophilus*, the cue is cytoplasmic. Females produce large and small eggs. On fertilization, the former give rise to females and the latter to males. These instances remind us that all development depends on genotype–environment interactions, whether of the cellular environment or the external environment. The genetic sex determining mechanisms have evolved to minimize the effects of possible environmental insults, and to limit phenotypic variability.

We have heterogamety when part of the genome segregates according to sex. Whole chromosomes, chromosome parts, or single loci may segregate, and the unit may not always be cytologically identifiable. Where the sex chromosomes are differentiated, and designated as XX-XY or ZW-ZZ systems, the Y is usually heterochromatic and a YY or YO genotype is lethal, indicating that an X is essential for development (but see p. 280). In some cases (e.g. Hemiptera) there is no Y chromosome and the male is XO. In half a dozen groups of insects (e.g. bees), females are diploid, and males

**Table 11.1** Chromosome complement and sex determination

| Genotype | Sex phenotypes | | | |
|---|---|---|---|---|
| | **Drosophila** | **Bird** | **Mammal** | **Silene** |
| 1. 2A:XX (ZZ) | ♀ | ♂ | ♀ | ♀ |
| 2. 2A:XY (ZW) | ♂ | ♀ | ♂ | ♂ |
| 3. 2A:XO (ZO) | ♂ | — | ♀ | — |
| 4. 2A:XXY (ZZW) | ♀ | ♂ | ♂ | ♂ |
| 5. 2A:XXX (ZZZ) | ♀ | ♂ | ♀ | ♀ |
| 6. 3A:XXX (ZZZ) | ♀ | ♂ | ♀ | ♀ |
| 7. 3A:XXY (ZZW) | ⚥ | ⚥ | ♂ | ♂ |
| 8. 3A:XYY (ZWW) | — | ♂ | — | — |

A is the haploid autosomal set, and X and Y the sex chromosomes. In birds, some reptiles, amphibia and insects the female is the heterogametic sex, and to avoid confusion the sex chromosomes are designated ZW (♀) and ZZ (♂). In the literature *Silene dioica* often goes under its older name, *Melandrium*.

are haploid and arise by parthenogenesis. Unfortunately, we have comprehensive data on chromosomal sex determination only for samples from a few phyla. Table 11.1 summarizes some relevant information.

Birds (and some other groups) differ from *Drosophila* and mammals since the female, not the male, is the heterogametic sex (Table 11.1, rows 1 and 2). *Drosophila* and the mammal are unlike since the Y chromosome is unnecessary for male fly development (rows 2 and 3), whereas it is essential for male mammalian development, irrespective of the number of X chromosomes or of autosomes (rows 4 and 7). The Y chromosome of mammals therefore carries male determining elements, and this would seem to be true also for the caryophylaceous plant, *Silene* (*Melandrium*). In both cases recorded XXXY genotypes are male. Contrariwise, the Y chromosome of *Drosophila* has no sex determining function; and the datum of row 4, which is from a single example, suggests this may also be true of the Z chromosome of birds (see McCarrey and Abbott, 1979). Thus we have two general chromosomal sex determining systems: the first being the male-inducing Y.

In *Drosophila*, Bridges (1921) proposed that genetic sex depends on a balance between female factors on the X and male factors on the autosomes. Thus, 2A:1X (rows 2 and 3) is male; 2A:2X, 3A:3X and 2A:3X combinations are female (rows 1, 5 and 6) and 3A:2X (row 7) is an intersex. And the same for the fowl but with the sexes reversed. This 'genic balance' theory predicts that changes in the amount of X material, by deletion and duplication, should modify the sex phenotype. Using *Drosophila* intersexes as sensitive indicators of sex change, Dobzhansky and Schultz (1934) showed that addition of X fragments variously increased the expression of the female phenotype while deletions acted in the male direction. The entire Y, and the proximal third of the X chromosome, were found to be without effect on the sex phenotype. As also would be predicted for an autosome, male factors have been located on the third chromosome (Bedicheck-Pipkin, 1959). So the evidence favours an X:A balance system of sex determination in *Drosophila*, and birds may also operate some of the features of this balance. There are many other sex determining mechanisms which have not been studied in detail (Lewis and John, 1968), but we shall confine ourselves mainly to the further examination only of the groups in Table 11.1. Of these, the Y male-inducing mechanism seems the simpler: a single 'switch' gene might be enough.

## Mammalian sex differentiation

Early embryonic development is sexually indifferent in mammals; in particular, the reproductive apparatus of both sexes is identical, up to about the seventh week of gestation in humans, for example. The primitive gonad is made up of an outer germinal epithelium, or cortex, which proliferates

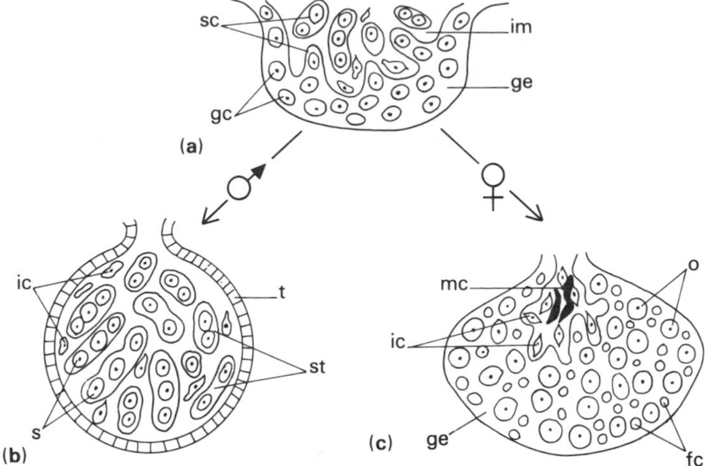


**Fig. 11.1** Early development of the vertebrate gonad. The indifferent gonad is already differentiated into a distinguishable cortex (outer) and medulla (inner). As shown in (**a**), it is made up of a germinal epithelium (ge), primitive sex cords (sc) and an interstitial medulla (im). Both the epithelium and the sex cords are populated with sex cells (gc). This differentiates further, and in the male (**b**) the sex cords form the seminiferous tubules (st) in which the spermatogonia (s) develop, and the cortex becomes the tunica albuginea (t). The interstitial cells (ic) become the Leydig cells. In the female (**c**) the germ cells become the oogonia (o) and future follicle cells (fc), and the medullary cords (mc) regress.

primitive sex cords into the loose mesenchymal cells of the medulla. The sex cells migrate into both these major components, cortex and medulla, and sexual differentiation results from elaboration of one component and regression of its alternative. The cortex forms the ovary, and the medulla forms the testis (Fig. 11.1). Interstitial Leydig cells, of uncertain origin, are the male hormone (androgen) secreting cells of the testis, and cells from the primitive sex cords are estrogen secreting in the ovary. Testis differentiation starts earlier than ovary differentiation, which suggested to Mittwoch (1973) that differences in cell division rates might be a primary sex determinant.

Similarly, the duct systems of the internal genitalia are all present in the indifferent embryo (Fig. 11.2). The Müllerian duct degenerates in males and the Wolffian duct differentiates as the epididymis, vas deferens and seminal vesicle. The converse occurs in the female: the Müllerian duct becomes the fallopian tube, uterus and upper vagina, and the Wolffian duct degenerates. Differentiation of the reproductive organs therefore involves the selective elimination of pre-existing structures, usually according to nuclear sex, and the growth of the surviving structures. Differentiation of the external genitalia from common anlagen follows from this by hormone action (see Ham and Veomett (1980) for illustrations of these changes in man), and subsequent secondary sexual differentiation also depends on the hormone

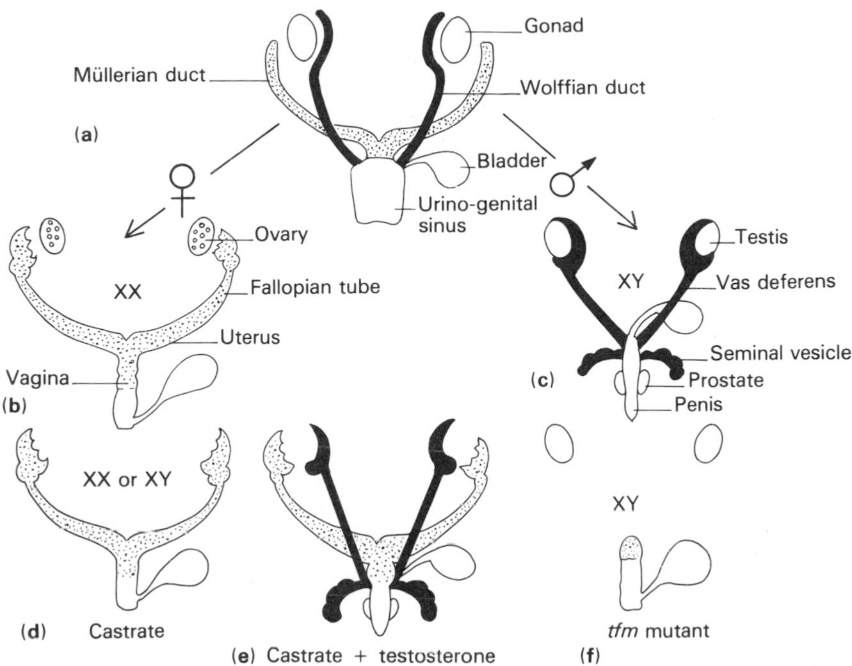


**Fig. 11.2** The mammalian sex ducts start from an indifferent system (**a**) which consists of paired Müllerian ducts, which will form the fallopian tubes; paired Wolffian ducts, which will form the vas deferens, and a central urino-genital sinus. These transformations are diagrammed in (**b**) and (**c**). If the embryo is castrated early enough, only the female ducts differentiate (**d**), irrespective of genetic sex. If the castrate is exposed to high levels of testosterone, the male ducts also differentiate (**e**). In the androgen-insensitive *tfm* mutant (**f**) all ducts are missing and the external genitalia are female, but blind. See text for details.

secretions of the gonads. Gonadal development is therefore the primary event.

The role of the gonad has been shown by its removal at the indifferent stage from rabbits or from mice (Jost, 1965), when both sexes then develop as females, i.e. they retain the Müllerian duct and lose the Wolffian duct (Fig. 11.2). The female is therefore the neutral, or null, sex. Dominance of the male state was first noted in heterosexual cattle twins (Tandler and Keller, 1911) when there is a fusion or anastomosis of the vascular systems resulting in the passage of XY cells and of male hormone into the female. The female is masculinized as a result and becomes a freemartin, with ovarian reduction and sterility, and regression of its Müllerian ducts. Similarly, by chance association, allophenic mice provide three sex combinations (XX/XX, XX/XY, XY/XY) yet the progeny show a great excess of males, and a low proportion of hermaphrodites (McLaren, 1976b). These data suggest that male sex is dominant over female sex, and is imposed on the null state of the latter. The sex states are not antagonistic, as once supposed. This dominance relationship is nevertheless difficult to reconcile with hermaphroditism (i.e. the coexistence of both kinds of gonad) which remains unexplained. Further, male dominance consigns female development to a 'black box', and although there are chromosomal defects which affect ovary differentiation, like the XO condition of Turner's syndrome, we have no mutants which allow us to explore ovary development (cf. *Drosophila* p. 281). Turner's syndrome, incidentally, also presents a problem since dosage compensation involves inactivation of one of the pair of X chromosomes to give the equivalent of the XO state. This anomaly is usually explained on the assumption that there is no X-inactivation in the germ cells, and the ovary degeneration typical of the syndrome results from the need for two functional X chromosomes in these cells. But this does not explain other Turner characteristics, like small size and webbing of the neck, etc.

Male dominance is established through secretion of three hormones by the testis: a Müllerian regression factor, testosterone which causes the virilization of the Wolffian ducts, and dihydrotestosterone which induces the masculine pathways of development of the external genitalia. Since the last two hormones are steroids, suitable receptor proteins must also be synthesized in the responding tissues. There is at present no evidence for a Y-borne gene (or genes) regulating these syntheses (but see below), which would be the simplest hypothesis, although they have been postulated (Ford, 1970). The most recent theory of mammalian sex determination suggests a more indirect mechanism.

Skin can be grafted between members of an inbred strain of mice without rejection (as they are histocompatible) except when male skin is transplanted onto female hosts. There must therefore be a histocompatibility antigen in male tissue which is absent from female tissue, and the relevant structural gene (or its regulatory locus) must be carried by the Y chromosome which is the only genetic difference between the sexes. This antigen has been called the H-Y antigen (histocompatibility of the Y chromosome), and it has been found in all mammalian species so far examined. It was therefore proposed that this evolutionarily conserved antigen plays a fundamental role in sex determination (Silvers *et al.*, 1968; Ohno, 1976; Wachtel, 1977) since there is a high correlation between testis formation and the presence of the H-Y antigen.

There are data which support this hypothesis, like the fact that 2A:XYY males have more H-Y antigen than normal, which implies that the Y carries the structural gene (Wachtel, 1977); or, the evidence that the short arm of the human Y chromosome carries a locus involved in H-Y expression (Koo *et al.*, 1977); or the finding that 2A:XY wood lemmings which develop as females are H-Y negative (gene loss?) (Wachtel *et al.*, 1976); or the apparent specificity of gonadal cell-membrane antigen receptors for H-Y antigen (Ohno, 1979). On the other hand, there are exceptions which challenge the theory (Silvers *et al.*, 1982). For example, a strain of mice has been found where one third of the XY progeny develop as females but have the H-Y antigen (Silvers *et al.*, 1982) and mutant male mice have been described which are H-Y negative (Melvold, 1977). The case for the male determining activity of the H-Y antigen locus is by no means proven, therefore, and identification and cloning of this locus must now have high priority so that its role can be directly tested by gene transfer into gonads.

A few mouse mutants tell us about sex determination, gonad development and hormones, though the story is still incomplete. The first mutant is Sex reversed (*Sxr*), originally identified as an autosomal dominant by Cattanach *et al.* (1971). This causes sex reversal in females together with H-Y antigen expression, and the gonads are agametic. Sex cells play no part in gonad differentiation, as has been confirmed by eliminating them both surgically and chemically.

The location of *Sxr* has been explored by Singh and Jones (1982) using the minor satellite DNA from the W chromosome of the banded krait (*Bkm* DNA) as a probe. This shows, on hybridization *in situ*, that there are equivalent sequences near the centromere of the normal Y chromosome, and that they are repeated as an additional element in *Sxr* carriers. Further, during meiosis this distal element, recognized by *Bkm* sequences, is transferred to the terminus of the X chromosome by crossing-over. This exchange has been confirmed cytologically by following the transfer of a terminal dark-staining body from one Y chromatid to one X chromatid during meiosis in *Sxr* males (Evans *et al.*, 1982). *Sxr* males apparently carry a translocation (duplication) of the *Sxr* locus, and so produce sperm with a normal X, an X carrying *Bkm* sequences, a normal Y, and a Y with extra *Bkm* sequences. This would give the apparently autosomal inheritance ratios of XX ♀♀:XX *Sxr* ♂♂:XY ♂♂ and XY *Sxr* carrier males, in equal numbers.

Sex-reversed females are XX-*Sxr*, and by using Searle's translocation *T(X:6)H*, it is possible to organize the inactivation of the X-*Sxr* chromosome. Such females will have a normal active X, and should be normal and fertile, although they carry *Sxr*. This is what is found (together with intersexes and sex-reversed females) in the progeny of the appropriate mating (Cattanach *et al.*, 1982; McLaren and Monk, 1982). These results strongly suggest that the normal, centromere-associated, *Bkm* sequences on the Y chromosome are, or are linked to, a male-determining (or female-suppressing) gene. It may prove to be so, but there are complications. Two dominant, testis-determining genes have been identified, one on the Y near the *Bkm* sequences (*Tdy*) and the other autosomal (*Tda*), but their relation to *Sxr* and to antigen production has still to be worked out.

A second mouse mutant, X-linked testicular feminization (*tfm*), renders androgen-receptive tissues unresponsive through lack of properly functional receptor proteins (Fig. 11.2). It is the regulatory or structural locus for these proteins (Ohno, 1977). Testes develop and secrete normal amounts of androgen, but Wolffian ducts and secondary male sexual characteristics fail to differentiate. The protein anti-Müllerian duct hormone is made by the foetal testis and the females lack oviducts, but they do synthesize sufficient estrogen for more or less normal secondary female differentiation, giving a male pseudohermaphrodite: genetic males with undescended testes, and totally female external development. As the receptor deficiency has no effect on testis differentiation, testosterone activity cannot be involved in the first steps in differentiation of that organ. This again is supported by the, somewhat ambiguous, results of administration of exogenous hormones to embryos, which certainly affect secondary sex characters but fail to switch the ovary permanently to testis, or vice versa (see McCarrey and Abbott, 1979). There are human inherited conditions identical to these two mutants.

Mammalian sexual differentiation must involve many complex and precisely integrated mechanisms, and our picture of their genetic control is still very incomplete and one-sided. The assumption that there is a basic female programme of development, and that the Y chromosome switches regulation to testis development, leaves the female programme unexplained. Thereafter, the testicular hormones initiate a cascade of secondary changes from female to male pathways of differentiation, and precisely how these many genetic programmes are selected is unknown. Each affected tissue or cell must have regulation of the alternative programmes built into the functional organization of its DNA, but we have no clue as to what this involves. This, of course, is the problem of gene regulation by morphogenetic hormones which we have already considered (p. 28). The primary issue is the first step; control of the change from ovary to testis development and its dependence on the Y chromosome. The attraction of the H-Y antigen hypothesis is that it provides a simple genetic and molecular explanation of the initiation of testis development. We must therefore return to this point after looking at sex in birds where we can ask: does the H-Y antigen induce differentiation of the heterogametic ovary?

## Sex differentiation in the ZW-ZZ system

The ZW sex chromosome system of birds inverts the sex determination rules just described (Table 11.1), and there is no evidence that the ZZ male is dosage compensated (Cock, 1964). This raises an interesting, but unexplained, problem of how males adjust to double the dose of sex chromosome gene products found in females. Gonad development in birds follows the same pattern as in mammals, except that only the left ovary differentiates in females. The cortex of the right gonad is sparse and without the potential for further development (Mittwoch, 1971), whereas the medulla of this indifferent stage is normal and differentiates as a testis in males. The heterogametic (female) gonad differentiates first, but its growth rate (as judged by DNA and protein synthesis) is lower than that of the testis at 6 days of incubation (Gasc, 1978), which argues against Mittwoch's theory of different gonadal growth rates as a cause of sex differentiation, due to an advantage in the heterogametic sex.

Many experiments on removing the sex cells prior to their migration, which is a simpler and more conclusive operation in the fowl since they are located at the periphery of the blastodisc, show that both male and female indifferent gonads can differentiate without these cells (McCarrey and Abbott, 1978). Again, although the gonads become endocrine glands early in their development, the addition of hormones to the indifferent gonad cultured *in vitro* gives only partial sex reversal, although there is some evidence that any sex steroid causes ovary differentiation and that testis development results from the absence of hormone (Carlon and Erickson, 1978). The evidence for a role for hormones in primary sex determination of fowls is therefore still ambiguous. There is no doubt, on the other hand, about the female's being the dominant sex since opposite sex twins with a common circulation in double yolked eggs hatch as females and feminized males. And early castration demonstrates that the male is the neutral or null state (Wolff and Wolff, 1951). The heterogametic sex is apparently the dominant sex in vertebrates. Correspondingly, a cross-reacting antigen to the mouse H-Y antigen is found in female chicks (Bacon, 1970), suggesting an active H-Y system of primary sex determination. This has not been analysed in the same detail as the mammalian system, and mutants like *Sxr* and *tfm* have not yet been identified (except for hen feathering (*Hf*) an autosomal dominant which causes males to have female plumage, perhaps implying loss of a tissue-specific androgen receptor).

Secondary sexual differentiation is hormone dependent and regulated in birds. As implied in the last paragraph, estrogen production must then dominate over androgen production to fit the contrariety of sex dominance and heterogamety. Some fowls change sex during, or after development, and this is commonly from females to the null state of males, as the above assumption would predict. However, our shortage of information on the sex genes of birds remains frustrating, and the argument by analogy with the mammalian system is necessarily speculative.

Both types of heterogamety are found among amphibia: *Rana pipiens* is XX/XY whereas *Xenopus laevis* is ZW/ZZ, and the H-Y (H-W) antigen is present in the heterogametic sex. Indeed, this is a general rule for all vertebrate species studied, indicating the evolutionary conservation (with modifications) of the gene, and suggesting its general importance. However, gonadal sex is not rigidly determined by nuclear sex in these lower species, as we have already seen (p. 270), and developing *Xenopus* tadpoles can all be made functionally female by adding estradiol to the water in which they are living. Androgens, on the other hand, do not induce sex reversal in ZW female tadpoles, but transplantation of the testis from a newly metamorphosed froglet into a ZW tadpole does. This manipulation can be used to generate WW *Xenopus* females by mating these sex-reversed ZW males to normal

**Table 11.2** Presence of H-W antigen in somatic and gonadal tissue of *Xenopus*

| Phenotype | Karyotype | Brain | Liver | Gonad |
|---|---|---|---|---|
| Male | ZZ | — | — | — |
| Female | ZW | + | + | + |
| Female* | ZZ | — | — | + |
| Female† | WW | + | + | + |

* Estrogen sex-reversed male.
† Selected WW female (see text).
Antigen levels of ZW and WW females are indistinguishable. (After Wachtel *et al.* (1980)).

ZW females (Mikamo and Witschi, 1964). We can therefore ask: what is the status of the H-W (H-Y) antigen in three different kinds of females? All three express the antigen (Table 11.2).

Since H-W synthesis occurs in sex-reversed ZZ females (and in ZZ embryos of chick and quail, according to Muller *et al.* (1980)) the structural gene for the H-W antigen cannot be on the W chromosome. Nor, by the same argument, is it likely that the W chromosome carries a regulatory gene. Since the tissue distribution, and amounts, of the antigen are normal in WW females, it seems certain that the H-W structural gene is not on the W chromosome. The only common feature of the three types of female, therefore, is that they are female and that their sex hormone is estrogen, and this suggests that the appearance of the antigen is estrogen induced and is a consequence, not a cause, of their sexual differentiation. It is unwise to assume that the ZW and XY sex determining systems are more than formally similar, but both systems are subject to the same uncertainties concerning the primary step of gonadal differentiation and its genetics. The genetic role of the Y (W) chromosome is still mysterious.

## Sex expression in plants

Flower organization in angiosperms falls into three broad categories which can be arranged in an evolutionary sequence. Primitive flowers are hermaphrodite, with both anthers and ovules. Or the flowers may be unisexual, with staminate and pistillate flowers being borne on the same, monoecious, plant, one appearing before the other during growth. Advanced species are dioecious with staminate or pistillate flowers on separate plants. An XY mechanism is commonly found in this last class, as exemplified by *Silene* in Table 11.1, although there are also rarer cases of female heterogamety. We must therefore look at how dioecious plants have evolved from monoecious forms for what this tells us about their sex determination genes.

The major problem, of course, is how the floral primordia differentiate in the apical meristem thus initiating flower development, which is the major expression of the sex phenotype in plants. A great deal of work has been done on this, but without any positive conclusions. Flower formation can be affected by temperature, photoperiod, nitrogen and mineral supplies, plant size and age, and in some cases by exposure to hormones (gibberellins and auxins). But ". . . the floral stimulus remains a physiological concept rather than a chemical reality" (Zeevaart, 1976). Not surprisingly, the genetics of flower induction reflect only this lability; strains can be selected for early or late flowering, and the balance of male and female flowers on monoecious plants can similarly be selected in favour of one or the other. These quantitative responses may result only from indirect genetic effects on the physiological processes in developing apex.

In monoecious plants with unisexual flowers, both sets of primordia may be present initially in each apex, as they are in hermaphrodite flowers. One or other set is then suppressed during differentiation to give either male or female flowers in which the germ cells form. Suppression may occur so early in dioecious plants that the alternative primordia cannot be certainly identified, and only the chromosomally determined set differentiates. Suppression, or lack of initiation, therefore determines the sex of the flower, and at least two mutations will be necessary to transform a monoecious plant to a dioecious one: a mutation to suppress pollen production and one to suppress ovule production. Charlesworth and Charlesworth (1978) have developed a model of the evolution of this transformation which tells us about these mutations, and about XY sex determination.

The simplest evolutionary path is when the first mutation causes recessive male sterility, giving gynodioecy (i.e. a population of hermaphrodites and females). If this is followed by a dominant mutation causing complete female sterility, full dioecy will evolve if the mutants are tightly linked, and the male will be the heterogametic sex. We should therefore expect to find a 'Y' chromosome carrying a dominant female suppressor, and an 'X' chromosome with a recessive male suppressor. If, conversely, the male sterility mutant is dominant, the second mutation must be recessive for dioecy to evolve, and the female would then be the heterogametic sex. Since dominant male sterility mutants are rare, this conclusion also explains why male heterogamety is the commoner state.

Westergaard (1958) was able to show that the Y and X chromosomes of *Silene* (*Melandrium*) did, in fact, contain genes, or gene complexes, of the sort just described. The Y deficiencies illustrated in Fig. 11.3 show that the first segment of the Y contains a dominant female suppressor gene ($Su^F$), since flowers are bisexual and plant growth normal in its absence. Segment 3 carries genes for the late stages of anther development, and from the comparison of XX and all these XY arrangements, one can conclude that segment 2 has genes (*M*) for the initiation of anther development. The same

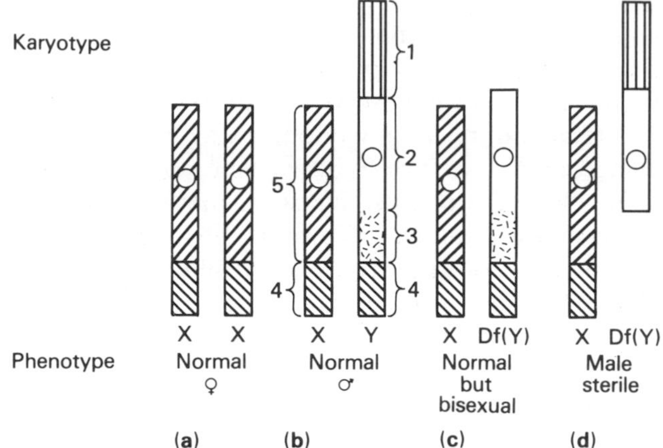


**Fig. 11.3** Westergaard's (1958) analysis of the effects of Y-chromosome deficiencies on sex in *Silene alba* (see also Lewis and John, 1963). (**a**) and (**b**) are the normal male and female sex chromosomes. The Y chromosome in (**c**) is deficient for segment 1, which must therefore carry a dominant female suppressor gene(s)—*SuF*. The Y chromosome in (**d**) is deficient for segment 3 and the phenotype is male sterile due to late anther degeneration. Segment 3 must therefore carry genes for anther development (*M*) and segment 2 must have the genes for anther initiation. Correspondingly, segment 5 must carry the genes for ovary initiation. Segment 4 is homologous in the X and Y, suggesting a common bisexual ancestor. The deficiencies have no effect on vegetative growth.

argument leads one to conclude that the X chromosome must have recessive alleles (*m*) of these male loci. Further, segment 5 must carry genes for ovary initiation. Segment 4 represents the homologous sequences of the X and Y which are capable of recombining. The X chromosome is therefore $su^F\ m$, and the Y is $Su^F\ M$. The Y chromosome determines sex because it carries genes which suppress female development and genes which initiate and complete male development; both dominant and tightly linked. There are other species which show similar arrangements (e.g. *Ecballium*) and in some of these 2A:YY plants survive, presumably because the X and Y differ at only one or two loci. Crosses producing aneuploids show that autosomes are also involved in *Silene* sex determination, and true bisexual strains can be made by progressive selection. The Y chromosome effect is therefore primary, but not exclusive.

Secondary sex differences, other than in the androecia and gynoecia, are uncommon in plants. Females may be larger (*Silene*, hemp etc.) or males may be larger (asparagus), and the often reported differences of corolla size are not consistent; females have larger flowers in *Silene alba*, but not in *S. dioica*. There is no evidence that any secondary sex differences are directly caused by differences of the X and Y chromosomes (Lloyd and Webb, 1977). Analyses of the kind done for *Drosophila*, which we shall look at now, are therefore not possible with flowering plants.

## Sex differentiation in Drosophila

The organization of sex differentiation in *Drosophila* is dramatically different from the cases already reviewed. We have seen from gynandromorphs (p. 200) that sex differentiation in *Drosophila* is cell autonomous. *Drosophila* has therefore no sex hormones, and each cell is competent to form either male or female structures according to the sex of its nuclear genome. Thus, when the gynandromorph boundary goes through the sex comb region of the first tarsal segment, only the XO section expresses this male secondary sexual character, and there is no ambiguity at the line between XO and XX cells. The organization of the genital disc is an important exception to this general rule.

The embryonic organization of the genital disc has been analysed using mosaics (through chromosome loss and mitotic recombination) and shown to contain about 13 cells at blastoderm, like the foreleg. The fate map also suggests that these genital disc precursors lie in a definite anterior–posterior sequence of: female genital structures, male genital structures and analia (Schüpbach *et al.*, 1978). The analia behave according to their nuclear sex like the foreleg sex comb region, and form either male or female anal plates (and hind-gut internally), as diagrammed in Fig. 11.4*a* and *b*. The situation is different for genital structures (Nöthiger *et al.*, 1977): where the XX area of a gynander contains the male genital and anal precursors and the XO area the female precursors, only female analia differentiate (Fig. 11.4*c*). Conversely, if the boundary apportions the proper nuclei to the primordia, both sets are made (Fig. 11.4*d*); or, if the gynander boundary divides the terminalia vertically (Fig. 11.4*e*), female structures are formed on one side and male structures on the other, and so on. Female genitalia differentiate only if their primordia are populated by female nuclei, otherwise they remain 'silent', presumably because the cells do not undergo the normal pre-differentiative changes. Correspondingly, male primordia are 'silent' unless they are populated by male nuclei. Partial structures may also differentiate when the gynandromorph line divides the precursor group of cells (Fig. 11.4*f*). In summary, we have the main system of sexual dimorphism where cells take the form dictated by their nuclei, and a second system of potential male and female cells which differentiate only if populated by the appropriate nuclei, otherwise they apparently come to nothing (Epper and Nöthiger, 1982). We do not yet know about the mechanism involved. The important point, however, is that each somatic cell expresses its sex genotype independently.

This cell-by-cell switching apparently according to the X:A balance, raises a problem when we remember that X chromosomes are dosage compensated, one and two X chromosomes making about the same level of product (p. 125). From Dobzhansky's and Schultz's (1934) work, and subsequent similar studies, it is usually assumed that there are many female-determining genes on the X (i.e. we are dealing with a polygenic character) and that

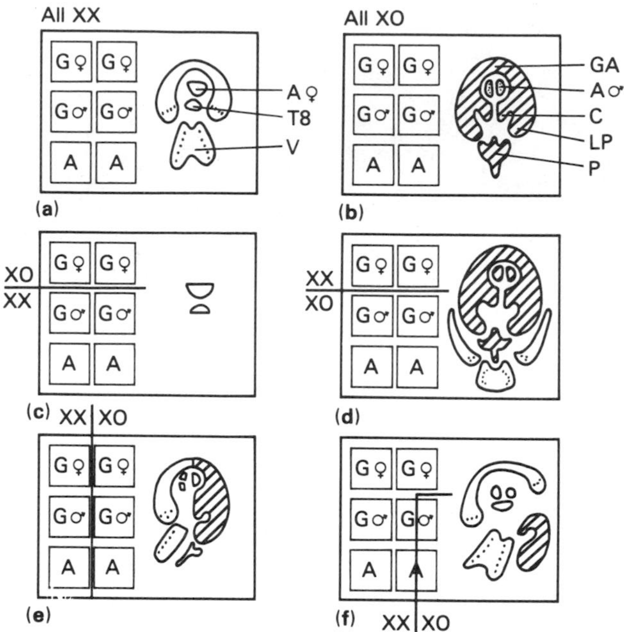


**Fig. 11.4** The effects of gynandromorph boundaries on the differentiation of the *Drosophila* genital disc, after Nöthiger *et al.* (1977). The paired genital primordia are shown on the left of each diagram (G♀ and G♂ are the genital primordia, and A the primordia for the analia). The normal structures formed are shown in (**a**) for the female and in (**b**) for the male. (Photographs of the structures shown here in diagrammatic form are given in Epper (1981).) A♀ and A♂ are the analia, V the vaginal plates, T8 the eighth tergite, GA is the genital arch, C the claspers, LP the lateral plate and P the penis. In (**c**) the XO and XX nuclei populate the 'wrong' genital primordia and only female anal plates differentiate. In (**d**) the converse occurs since the gynander boundary allocates the required sex to both sets of genital primordia and the fly bears both male and female genitalia (and male analia). When the XX/XO boundary runs down the midline of the embryo (**e**) the two sides are of opposite sexes. And when the boundary divides primordia (**f**) partial structures are formed according to the nuclear sex.

these genes are exempt from dosage compensation. This resolves the problem in theory, but that is far from proving that a difference in levels of some substance (or substances) results in cells being phenotypically male or female. So it is interesting that triploid genetic females (AAA:XX) are a mosaic of male and female cells, such that one forelimb cell may make a male sex comb bristle while its (female) neighbour fails to do so. This implies that the balance is then so critically set that minor environmental differences can tip it one way or the other in each cell (Hannah-Alava and Stern, 1957). We have therefore to ask what evidence we have for male- and female-determining genes. As before, we shall first consider early acting genes.

### *Sex specific lethals*

Females homozygous for the mutation daughterless (*da*) produce defective eggs which cannot support XX development, but allow XO males to survive. It is a sex specific lethal (see Marshall and Whittle (1978); and Cline (1978) for others), and the maternally inherited defect can be cured by injection of wild-type egg cytoplasm into the egg (Bownes *et al.*, 1976). Temperature shifts show that the maternal effect acts prior to blastoderm formation, and mosaic analysis proves that female cells in all tissues die even when surrounded by male cells. The *da* gene product is required early for female cell survival, but this does not mean it is necessary for the establishment of phenotypic sex. Its possible function in this latter respect depends on its relation to another sex-specific lethal locus.

A dominant sex-lethal (*Sxl*) locus on the X chromosome has the curious characteristic that mutants which lie only 0.007 map units apart have completely opposite effects, specifically killing males or females ($Sxl^{M1}$ and $Sxl^{F1}$). A single dose of $Sxl^{M1}$ enables daughters to survive the *da* maternal effect, but has no action on normal females. $Sxl^{F1}$, and an *Sxl* deficiency, enhances the *da* maternal effect, and kills normal females. The two loci are clearly involved with each other, and Cline (1978) has proposed a model which explains their relationship (Fig. 11.5). The assumption is that the $Sxl^{+}$ product is essential for female development but lethal to males. Thus, the two sex specific effects would occur if $Sxl^{M1}$ is a constitutive mutation of the control region, and if $Sxl^{F1}$ made a defective structural gene product, lethal to female cells. Regulation of the locus is delegated to the *da* maternal factor, modulated in some fashion by the X:A balance, as an activator of *Sxl*. The *da* maternal effect would then be equivalent to $Sxl^{F1}$ (no induction of *Sxl* product), and *da* rescue would result from the constitutive activity of

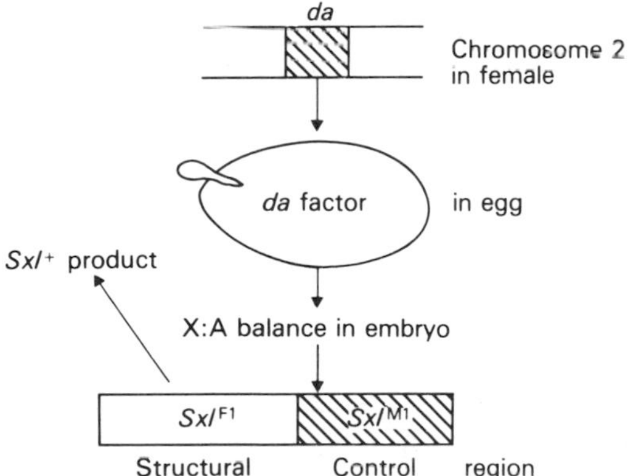


**Fig. 11.5** The daughterless gene product accumulates in the egg. In the developing embryo, this product, possibly under the influence of the X:A balance, activates the control region of the $Sxl^{+}$ gene, whose structural product is necessary for 'female expression' and which suppresses male expression. Mutation of the control region to cause constitutive transcription of the structual gene is therefore lethal to males.

$Sxl^{M1}$. Further, *da/da* females carrying $Sxl^{M1}$ produce defective eggs that do not produce daughters at higher temperatures, showing that $Sxl^{M1}$ acts on the embryo cells according to their genotypic sex.

$Sxl^{M1}$ has another interesting property (Cline, 1979): it causes haplo-X (male) tissue to differentiate as if it were female. In gynanders carrying $Sxl^{M1}$, for example, genetically marked regions of the first leg which should have carried sex combs, are either female in form or have incomplete sex combs (cf. the triploid intersexes above). The $Sxl^{+}$ product (assuming $Sxl^{M1}$ is a constitutive mutation) is therefore also involved in tissue differentiation, and specifically with the sex transformation.

Cline (1978) also suggests that *Sxl* (and *da*) might be involved with the dosage compensation mechanism, and this turns out to be true. Relative transcriptional activity can be measured by comparing [$^3$H]uridine incorporation into X-chromosome and autosome RNAs in salivary gland chromosomes. It is then found that females homozygous for $Sxl^{F1}$, and the heteroallelic pair $Sxl^{F1}/Sxl^{Fhu1}$ which permits better larval survival, show hyperactivity in the X chromosomes. Unfortunately $Sxl^{M1}$, which slows larval growth badly, causes under-replication of the salivary gland chromosomes which vitiates any attempts to assess transcriptional activity properly (Lucchesi and Skripsky, 1981). The evidence is that the $Sxl^{+}$ gene product induces the lower female rate of X chromosome transcription and the higher male rate would apply in its absence. Although this has not yet been shown, *da* should have the same effect as $Sxl^{F}$. Dosage compensation and sex determination have a common genetic control at the earliest stage of development (Fig. 11.6).

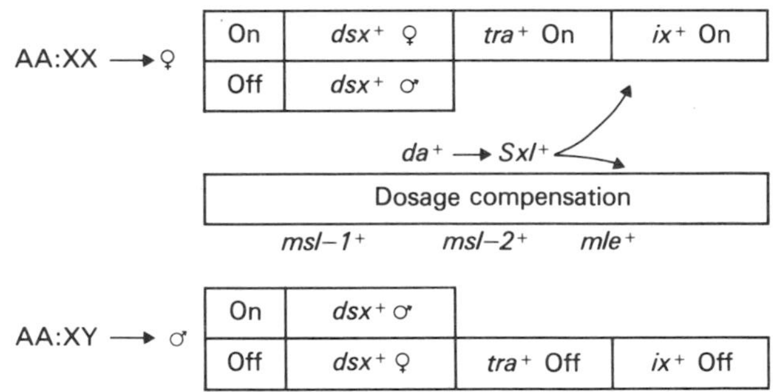


**Fig. 11.6** The interrelations of the known sex transforming genes (transformer and intersex) to the two sex phenotypes is suggested as depending on their switching double-sex to either a female or a male state. What role the X:A balance might play in this is unclear. Similarly, the effects of dosage compensation, particularly on the sex specific lethals (male sex lethal and male lethal), are unknown. The results described in the text do suggest that Sex lethal functions both in dosage compensation and in sex differentiation, thus linking the two phenomena. Note that the loci are separate and interactions must be due to their product, as is the case with daughterless.

Other male-specific lethals have also been found to regulate X chromosome transcription (Belote and Lucchesi, 1980). All three are on the second chromosome (male-specific lethal-1 (*msl-1*), male-specific lethal-2 (*msl-2*) and maleless (*mle*); and homozygous males, as judged by the activity of four X-linked loci coding for measurable enzymes in 3rd instar larvae, make only 60 percent of the same products found in wild-type controls. There is no difference between the mutant and control levels when assessed by enzymes encoded by four autosomal loci. The three genes therefore each block the dosage compensation mechanism. Except for a temperature-sensitive allele of *mle*, none of the mutants forms usable salivary gland chromosomes, but homozygous males of this allele have narrow, dark-staining X chromosomes (female-like) and they incorporate only 65 percent of [$^3$H]uridine into RNA compared with the chromosomes of their female sibs. Dosage compensation is blocked, as judged by this test, too. Precisely how these genes function in relation to *Sxl* is not known, but it is interesting that $Sxl^{F1}/+$; *mle*/*mle* individuals often develop as females bearing male sex combs (Lucchesi and Skripsky, 1981), again suggesting a role for these genes in sex differentiation (Uenoyama *et al.*, 1982; Skripsky and Lucchesi, 1982).

The sex specific lethality of these genes may be related to their functions in dosage compensation, for an early imbalance of products coded for by X-borne genes might easily be deadly. We cannot be sure of this until we know more about what they do, but exploration of this class of mutation opens up the prospect of apportioning sex determination in *Drosophila* to the activities of particular genes, and replacing the somewhat mystical concept of X:A balance as the agent of sex.

## Sex transforming mutations

A number of autosomal, sex transforming mutations have been identified, and most invert the expression of the female somatic phenotype making them phenotypic males or intersexes (Table 11.3). Generally, they are genes regulating sex differentiation, and their activity continues during larval development. They have no effect on the sex cells: pole cells make sperm or ova according to their XX or XY complement and are not affected by the sex transforming mutations, as can be shown by transplanting mutant pole cells into normal embryos (Schüpbach, 1982). We are therefore now concerned with genes regulating the sex differences of somatic cells. As before, the mutations are cell autonomous in their action, and mosaics can be generated by mitotic recombination to assess when they function (Baker and Ridge, 1980).

The sex transforming mutants are listed in Table 11.3, and the precise effects of the various genes on the sex phenotype of male and female forelegs, abdominal tergites, analia and genitalia are summarized by Baker and Ridge (1980). The two transformer loci, *tra-2* and *tra-3*, have the most

**Table 11.3** Mutations affecting somatic sex phenotypes of *Drosophila*

| Gene | Location | Characteristic effects |
|---|---|---|
| Sex-lethal (*Sxl*) | 1–19.2 | Dominant male and female lethals |
| daughterless (*da*) | 2–41.5 | Maternal effect female lethal |
| intersex (*ix*) | 2–60.5 | Females to intersexes, males normal |
| transformer-2 (*tra-2*) | 2–70 | Females to males, males sterile |
| transformer-3 (*tra-3*) | 3–45 | Females to males, males normal |
| double-sex (*dsx*) | 3–48.1 | Males and females into intersexes; and mutations affecting only one sex |

The *tra-2, tra-3* and *dsx* mutations are null alleles; *da* is a temperature-sensitive. The complexity of *Sxl* and *dsx* mutations is described in the text.

striking effects, converting females into males, except for their size and their gonads which differentiate abnormal oocytes. Except that *tra-2* sterilizes males, their consequences are indistinguishable, and it is tempting to think that they are the loci for a dimeric protein. This protein, or these proteins, are required up to the beginning of the 3rd instar for female differentiation of genital structures, since *tra/tra* clones induced earlier make male genitalia. And the $tra^+$ gene is required for even longer (120 h after egg laying) to ensure female differentiation of the analia (Wieschaus and Nöthiger, 1982). Since the sex phenotype can be switched from one state to the other during this period, it follows that the $tra^+$ product is necessary to promote female differentiation and to suppress male differentiation. The same conclusion comes from temperature switching a *tra-2* ts allele (Belote and Baker, 1982). Thus *tra* behaves like a selector gene of the homoeotic group (p. 268), but since the genital disc contains separate ♂ and ♀ primordia it is particularly instructive that the transformation inactivates the female primordium and activates the male primordium. That is, in this disc, the gene affects the differentiation of two structures, but in the foreleg and analia it determines the alternative states of single structures. At intermediate temperatures the $tra\text{-}2^{ts}$ allele gives intersex differentiation of cells (Epper and Nöthiger, 1982).

The intersex (*ix*) mutation results in females with male-like genitalia and incompletely differentiated ovaries, sometimes testis-like. It has no effect on males. The double-sex (*dsx*) mutation acts similarly, but it converts both males and females into morphologically identical intersexes, with intermediate bristles on the forelegs, poorly developed male and female genitalia (both primordia are activated (Epper, 1981)), and sometimes with ovary and testis. The mutation pattern at this locus is complex, and also involves dominants. Double-sex dominant ($dsx^D$) and double-sex Masculinizer ($dsx^{Mas}$) both transform only females into intersexes, and when they are heterozygous with a deficiency the females are made male. There is also a mutant, $dsx^{136}$, which transforms only males into intersexes. The mutant *dsx* locus must exist in different states, but it is not clear if this duality is in some way like that of *Sxl*, or if it involves a more complex mechanism like the one which we shall describe for yeast (p. 299). However this may be, mitotic recombination shows that $dsx^+$ is required throughout larval development, probably for longer than $tra^+$, to give normal females.

When these mutations are looked at in pairs (Baker and Ridge, 1980), flies homozygous for *dsx* and for any one of the three other mutations are intersexes, and identical to *dsx*. *tra-2* and *tra-3* give the *tra* phenotype, and *tra-2* or *tra-3* with *ix* give the *tra* phenotype. There is a clear epistatic sequence *dsx*>*tra-2*, *tra-3*>*ix*, which is consistent with these loci controlling the single steps of a pathway to the sex phenotype. There is also a most interesting interaction between $dsx^{136}$ and *tra* transformed females. X/X;$dsx^{136}$/*dsx* are normal, fertile females, but X/X;$dsx^{136}$/*dsx*–*tra-2*/*tra-2* flies are intersexes: that is, $dsx^{136}$ now fails to complement *dsx* as it would in normal females, and $dsx^{136}$/*dsx* is then epistatic to *tra-2*/*tra-2* and acts in the male-transformed cells.

It is hazardous to arrange these genes in a probable functional sequence when we know so little about what they do. We shall therefore order the data as a flow diagram (Fig. 11.6) according to the following logic (Baker and Ridge, 1980). First, since the genes are required throughout most of the growth period of the anlagen, later in abdomen than in leg, they are not switch genes but loci whose wild-type products sustain a particular sex state. Second, null alleles of *dsx* repress neither the male nor the female states, and both states are expressed, if poorly (Epper, 1981). The wild-type allele of *dsx* must be active in both sexes. This conclusion is supported by the combination of alleles which separately make males and females into intersexes: $dsx^D$ or $dsx^{Mas}$ over $dsx^{136}$ complement in males, and in females $dsx^{136}$ behaves as $dsx^+$. As we have noted, the roles of the dominant mutations require further study to confirm or refute this conclusion, which allocates a dual function to *dsx*. Third, the *tra* mutants, which we shall treat as one, make good males, and $tra^+$ is therefore necessary for female expression in the presence of $dsx^+$. But *dsx* is epistatic to *tra*, and the intersex level of female expression therefore occurs in the absence of $tra^+$. This argues against $dsx^+$ being the male-determining gene and against $tra^+$ being the female-determining gene; and suggests, instead, that $tra^+$ functions by setting $dsx^+$ in one or other of its alternative states, according to the X:A balance. Fourth, *ix* does not affect males or *tra*/*tra* transformed females; it makes females into intersexes. Thus *ix* prevents the expression of male functions in genetic females. Finally, we have too few data to fit the morphogenetic role of *Sxl* into this scheme, which cautions us against taking it as more than an interim summary of current knowledge.

The *tra* and *dsx* mutants have been used to test the possibility that dosage compensation is regulated by physiological sex. A number of studies show

**Table 11.4** The levels of X-linked glucose-6-phosphate dehydrogenase (G6PD) and of 6-phosphogluconate dehydrogenase (6PGD) in transformed flies

| Genotype | Phenotype | G6PD | 6PGD |
|---|---|---|---|
| +/*tra*; Y-attached X | Female control | 1.3 ± 0.1 | 1.6 ± 0.2 |
| *tra*/*tra*; Y-attached X | Transformed male | 1.8 ± 0.1 | 1.6 ± 0.1 |
| +/*tra*; $y^+$ | Male control | 2.0 ± 0.1 | 2.0 ± 0.2 |
| *dsx*/*dsx*; Y-attached X | XXY intersex | 3.0 ± 0.2 | 1.2 ± 0.1 |
| *dsx*/*dsx*; $y^+$ | XY intersex | 2.2 ± 0.2 | 1.8 ± 0.2 |

The female control value for G6PD may be misleadingly low. Data from Komma (1966).

that this is not the case (e.g. Table 11.4). In addition, when [$^3$H]uridine incorporation into salivary gland chromosomes is measured in gynandromorphs, the uridine incorporation into XO cells equals that of the paired XX chromosomes (Lakhotia and Mukerjee, 1969). Dosage compensation is therefore cell autonomous, and independent of physiological sex, *per se*. This is also taken account of in Fig. 11.6.

Sex determination in *Caenorhabditis elegans* follows the *Drosophila* pattern. Primary determination depends on the X:A balance: 2A:XX are hermaphrodites and 2A:XO are males. And again there is evidence that extra X fragments favour the female phenotype. Similarly, a number of sex transforming mutants have been identified, but they usually differ from their *Drosophila* equivalents by changing both soma and germline. Members of three transformer (*tra*) autosomal complementation groups make hermaphrodites into males, but have no effect on males (Hodgkin and Brenner, 1977). And a group of hermaphrodization (*her*) mutants convert males into hermaphrodites. One of these, *her-2*, maps to the *tra-1* site, and may be thought of as causing the constitutive expression of *tra-1*, for it feminizes both sexes and is a dominant. There is also an intersex (*isx*) mutant which feminizes only the gonads of both sexes. The parallels between these and the *Drosophila* mutants are obvious, but whether or not they can legitimately be fitted into a similar gene action sequence remains to be determined. Both sets of data leave the role of the X:A balance to further study, for it is hard to understand when our best feminizing genes (*tra*$^+$) are on autosomes.

## Other sex determination systems

There is evidence for a variety of other systems for the epigenetic control of sex (Lewis and John, 1968), but the genetics of most have still to be worked out. We shall therefore consider only two, which contrast with those already described.

### Male factor systems

Even among Diptera with heterogametic sex determination, the *Drosophila* X:A balance system does not always apply. For example, Rothfels and Nambiar (1981) working on hybrids between *Prosimulium multidentatum* and *P. magnum*, find that the backcross of *P. multidentatum* to the hybrid gives progeny that are male if they carry a Y chromosome (Y?, XY, XXY) and female if they do not (X, XX, XXX). This argues for a male, testis-determining factor on the Y (cf. *Silene*). Such a factor, or something like it, is not uncommon, and has been reported for other dipteran species. Populations may be polymorphic for the factor which maps to one linkage group in other strains, as shown by Beerman (1955) with *Chironomus*, and by Baker and Sakai (1976) for the mosquito where the male factor (M) is on linkage groups I or III in different strains. Such rearrangements may be due to translocations, but as Green (1980) points out, it would not be surprising if they were sometimes due to transposable elements. Either way, there must be one or a small number of male-determining genes.

This is the usual situation in the housefly (*Musca domestica*) which has a very similar pattern of development and morphogenesis to *Drosophila*. Generally females are XX and males XY, but aneuploids for the X are either male or female depending on whether or not they carry a Y. This, and cytologically recognizable Y autosome translocations which confer male control to the autosome, suggest a male factor (M), as before (Milani, 1975). Natural populations are found which are XX and with an autosomal M factor, and strains from them may sometimes also be MM in both sexes (Franco *et al.*, 1982). Such strains carry a feminizing factor (F), which behaves as a good mendelian gene on the first autosome. Sex is then determined as *Ff* (female) and *ff* (male). When F is introduced into the normal XY system, fertile XY females are generated (Rubini, 1967): F is epistatic to M. There is also a maternal-effect dominant which causes females to lay eggs predetermined to develop as males. Half the progeny are therefore transformed XX females; but as they are physiologically functional males they can be mated to XX females to produce all female progeny since they transmit no Y chromosome.

### Chromosome imprinting in Sciara and in coccids

It would be difficult to imagine greater differences than those between the sex determining mechanisms of *Musca* and *Drosophila*, but some Diptera go further and adopt the stratagem of chromosome inactivation. In these cases, though, the situation is much more complex than in the X dosage compensation of mammals, and involves identification, or 'imprinting' of the paternal chromosomes (reviewed by Brown and Chandra (1977)).

The zygote of the dipteran, *Sciara coprophilia*, has two sets of autosomes, but three X chromosomes, two being paternal in origin ($A^mA^pX^mX^pX^p$).

One paternal X is eliminated in the germ cells of both sexes, which are therefore identical in males and in females ($A^mA^pX^mX^p$). The somatic cells of the female also have this composition, but male somatic cells lose both paternal X chromosomes ($A^mA^pX^m$). Phenotypic sex is therefore determined by an XX-XO system, achieved by somatic segregation, and it is this somatic genotype which determines whether the genetically identical sex cells develop as ovary, or as testis. Meiosis in the testis is atypical. At the first division only one spindle forms, and the maternal chromosomes move to the pole while the paternal chromosomes are eliminated. At the second division, the two X chromosomes formed move together to one pole along with one of each of the autosomes and just this product ($A^mX^mX^m$) goes on to form sperm; the product containing only autosomes is lost. The paternal chromosomes of the zygote are thus exclusively maternal chromosomes of the previous generation (Metz, 1938): males transmit only maternal chromosomes, but these become 'imprinted' paternal chromosomes in their progeny and are eliminated as just described. Since *S. coprophilia* produces unisexual broods, there must be two kinds of females, XX′ producing daughters and XX producing sons. The X′ chromosome contains a large paracentric inversion, which could account for the maintenance of the differences between the X chromosomes (Crouse, 1960). XX′ causes the elimination of one X, whereas XX eliminates two!

These studies show three important points. The paternal chromosomes are 'imprinted' in some way which allows their selective elimination. Male somatic and gonadal cells exercise this elimination differently. Somatic cells determine the character of sex cell differentiation. It is difficult to envisage genetic mechanisms which could account for these ill-understood processes, but the XX-XO difference between males and females may imply that the nature of gonad differentiation depends on the amount of some substance produced by the X chromosomes of the somatic gonadal tissue.

Imprinting is found in a somewhat different form in coccids. In *Pseudococcus citri* (the mealybug) half the chromosomes of the male are heterochromatic and half euchromatic, whereas all the chromosomes of the female are euchromatic. There is no sex chromosome mechanism, and males are effective haploids as are some Hymenoptera. The origin of the heterochromatic male chromosomes can be traced with marker genes or by irradiating males and females. Irradiated females show damaged euchromatic chromosomes in their sons, whereas irradiated males show only fragmented heterochromatic chromosomes in their sons, even down to tiny fragments, suggesting there is no heterochromatizing centre. Once again, the paternal chromosomes are imprinted (Brown and Nelson-Rees, 1961). In this case, females produce progeny of both sexes, but environmental factors and age have marked effects on the sex ratio, which is largely determined by the physiology of the mother. Meiosis is normal in the female, but in the male both euchromatic and heterochromatic chromosomes divide equationally (mitotically), and the heterochromatic and euchromatic groups move to separate poles at the second division. The heterochromatic nuclei degenerate leaving the two euchromatic nuclei to form sperm, and the imprinted paternal chromosomes are again lost. Imprinted chromosomes undergo heterologous regulation during development. In some male tissues (gut, Malpighian tubule, etc.) their heterochromatization is reversed, and they apparently become active (Nur, 1967).

These two example^ show that heterochromatization may be facultative, that it may occur in autosomes as well as sex chromosomes, and that it is reversible. The molecular mechanism of imprinting is not understood, but it is obviously a highly regulated one, influenced by the gene products of surrounding cells in a highly specific fashion. Both systems provide a challenge for molecular genetics.

## Conclusions

These examples illustrate the somewhat daunting variety of sex mechanisms found in eukaryotes. Except for plants, these mechanisms cannot be ordered in an instructive evolutionary progression, and only a few features seem to be general. The sex cells, where they differentiate early, play no part in determining the development of the somatic gonad, although in *Drosophila* (but not in *Musca*) they maintain that development. Both male and female anlagen are present together, irrespective of genetic sex; and establishment of gonadal sex depends on the growth of one and suppression of the other. The decision as to which shall grow may depend on environment, on hormones, or on the action of sex determining genes. Since the last is commonest, there is usually a concordance between germline sex and somatic sex, but this is not inevitable. As proposed for plants, the genetic sex mechanism may involve closely linked genes for the activation of one pathway and the inactivation of the other. The argument that one sex is a null state, and the other alone is positively activated (as suggested above for mammals) is far from proven at the genetic level.

In the vertebrates, development of the somatic gonad initiates the hormone cascade which then regulates the differentiation of the sex phenotypes. It is still uncertain if the initial stimulus for gonad development comes from the activation of the sex determining genes directly, or indirectly via the H-Y antigen. If it is the latter, the universal presence of this antigen in the heterogametic sex of vertebrates argues strongly for a simple, common sex system. Certainly there is no case yet for giving up such a view. Sex phenotypes then follow from the development of the somatic gonad, and the obvious approach to their analysis is to follow the effects of hormone manipulation on gene expression. The *tfm* mouse mutant shows that the specificity of hormone receptors is an important element in this analysis.

The cell autonomy of sex expression in *Drosophila* and other insects presents quite another pattern. Manipulation of genes like $tra^{+}$ and $dsx^{+}$

shows that their products are required throughout development, just as hormones are. This indicates that the imaginal cells require instruction, cell cycle by cell cycle, as they multiply in the growing larva. Thus, the few genes we know about are clearly important, since their products regulate which genes are transcribed according to alternative sex repertoires. We have resolved the problem of alternatives by giving the mutant, double-sex, two possible states (which is a formal model not a proof) but ignored the repertoires. It is a puzzle to know how one gene product can act in all the various organs and structures which differ between the sexes, and even more to guess how the repertoires might be organized. In these respects, sex genes resemble homoeotic genes.

Perhaps it is unfortunate that the *Drosophila* Y chromosome is atypical in lacking a sex-determining gene, or gene complex, and that the X:A balance best explains the data of Table 11.1. The model in Fig. 11.6, which provisionally allocates functions to particular genes, is a first step towards the genetic dissection of the balance hypothesis. As the actions of these genes are explored, and as more relevant mutants are identified, we should progress towards an understanding of their role in organ differentiation, which Bridges (1939) rightly saw as the objective behind the study of sex differences.

# VI *Model systems and future developments*

Before we look at how far we have progressed we must consider gene regulation in single-cell eukaryotes, some of which are studied as 'model systems'. By and large, they are models which necessarily ignore determination and tissue differentiation, but some of their gene regulation systems are found in multicellular eukaryotes. They are models in this practical sense, but also because they may suggest solutions to some of the problems which are difficult to sort out within the complexities of multicellularity. We must also examine cultured somatic cells, but for a different reason. Techniques have now been developed which allow known cloned genes to be transferred into these cells where the sequences necessary for gene function can be determined, and where the role of its active product may be explored. This leads us directly to the last chapter which looks at some of the problems confronting future research.

While simple eukaryotes show many economies of gene organization, the most striking arrangement known is in the opposite direction: the presence of both mating-type genes in a yeast cell, organized in a complex gene conversion arrangement so that only one is expressed, and that at a third locus. This system of 'genes in cassettes' might account for sex-type switching in protozoa and possibly also for the complexity of the double-sex locus in *Drosophila*, though that has still to be proven. Protozoa also provide models for the non-nuclear inheritance of cortical structures of the cytoplasm, and some of these may prove relevant to clonal differences which arise when cells divide unequally (e.g. neuroblasts). However, the most important model may be the slime mould, *Dictyostelium*, where there is a simple differentiation of cells into two types: stalk and spore. Simple molecules (ammonia and cyclic AMP) are apparently involved in this 'state' decision, but as posterior prespore cells will reorganize to restore the spore–stalk balance of the slug when prestalk cells are removed from it, there are also determinative cell interactions which are now being actively studied.

Cultured somatic cells can be fused to form heterokaryons in which, since there is chromosome loss, the role of individual chromosomes may be followed. This is a relatively crude technique, except for assigning genes to chromosomes, and the most useful fusion has been for the preparation of hybridomas. These are myeloma cells fused to plasma cells which have been prestimulated to make antibody, and clones of these hybridomas secrete specific, or monoclonal, antibodies. Such antibodies have proved of great value by their matchless ability to identify specific antigens, and their use for studying 'invisible' developmental changes is sure to grow. The transfer of cloned genes into cultured cells provides a further novel tool for elucidating the sequences required for gene transcription and, possibly, the function of structural genes. Gene transfer is likely to be widely used, particularly the *Xenopus* oocyte system, which has unique advantages. The question then is: what are the priorities for geneticists, given these many new approaches?

We can expect the sequence aspects of DNA organization to be rapidly resolved, and we should soon be able to define a gene as a transcription unit in the narrow sense. It may be more difficult to work out the whole sequence organization concerned with the developmental regulation of cell-specific genes for two reasons: we are still uncertain of the role of chromatin proteins and we lack information about 'activator' molecules. The assay of maternal-effect, and of early acting, genes may provide an entry to this last problem, but that is by no means certain. The difficulties in the way of following the sequence of gene activities which specify the very precise bristle patterns of the *Drosophila* thorax are used to illustrate some of the problems ahead. In this case, previous genetic analysis has identified the major genes involved and suggested how and when they act, but what they do at the molecular level is another matter. However, as with other examples considered earlier, the hierarchy of developmental events is well defined in genetic terms, and this should provide a firm basis for future work. It may nevertheless be necessary to explore simpler systems which avoid the complexities of these terminal phenotypes. We have still much to learn about gene action during development, but progress is now so rapid that we face the future with a new confidence.

# 12 Cell systems

To suppose that the mechanisms discovered in bacteria will suffice to account for cellular differentiation in higher forms is an extrapolation which, implicitly, endows bacteria with the memory of the queen in 'Alice in Wonderland', the memory which enabled her to 'remember things before they happened'. There are many good reasons for believing that, in differentiation, other, and still unknown mechanisms must operate as well. The genetic paradox of differentiation cannot be solved on paper or on bacteria. It must be solved on somatic cells of higher organisms, for the biochemical and genetic analysis of which we are beginning to have the necessary tools.

B. Ephrussi, 1962

It is tempting to escape from the complexities of organs and organisms by studying homogeneous populations of single cells, either unicellular eukaryotes or monotypic cell cultures derived from multicellular eukaryotes. Both kinds of population have the advantage that they can be grown in quantity for molecular analysis, and they are usually easy to manipulate genetically. But both have inherent limitations. The unicells lack the division of labour found among the various cell types of the higher eukaryotes, and it is just the organization of this specialization of function during development which is a central problem. Unicells are, generally, multifunctional prototrophs adapted to particular, restricted environments. As Ephrussi emphasizes, their genetic mechanisms, specialized for their survival, are unlikely to have anticipated the 'tricks' required to establish multicellularity, let alone the genetic organization which must underpin novelty, and the generation of products and structures not found in unicellular eukaryotes. On the other hand, only a limited range of cells can be grown from multicellular eukaryotes, and these usually have a restricted cell-cycle life, dying off after ~70 cell cycles in the case of mouse fibroblasts, for example. Permanent cell lines, on the other hand, are often transformed and karyotypically abnormal. They are almost invariably differentiated cells which have already passed through the developments which interest us.

While we must remember these limitations, we can also appreciate the merits of studying single cells. Single cell eukaryotes may have evolved genetic systems of enduring merit, superior to prokaryote systems. While some of these systems may turn out to be unique to unicells, but of intrinsic interest, they may also throw light on processes which we can only dimly discern among the complexities of multicellular development. We shall consider some of these situations first, and then look at cultured cells. As Ephrussi emphasized, we can now study the characteristics of these cell lines in many ways but, more important, we can manipulate their genotypes. We can place known cloned genes in cells and ask what parts of their organization are necessary for their normal functions. That technology takes us back, full circle, to the problems raised at the beginning of this book.

## Unicellular eukaryotes

Our concern under this heading is to look only at the organization of some genes and gene systems relevant to the topics already considered. We shall first consider the regulation and organization of metabolic pathways which may throw light on some of the gene/enzyme systems already considered. Thereafter we shall describe the unexpected genetic arrangement of 'sex genes' in yeast, and in some protozoa. Protozoa also allow us to consider some interesting properties of cytoplasm and its organelles.

### *Organization of metabolic pathways*

Fungal nuclear DNA is mostly organized as unique sequences, and in *Neurospora crassa* only 10–20 percent is repetitive (Brooks and Huang, 1972). This DNA is structured into nucleosomes made up of the usual four histones, each with properties similar to those described for higher forms. The length of DNA between the nucleosomes is smaller, however, at around 30 bp, and histone H1 is consequently represented by a family of smaller proteins. In these respects fungi share characteristics of prokaryote and of higher eukaryote DNA organization, and we can ask if this progression to greater complexity is also reflected in any elaboration of gene organization?

Mutational analysis of fungi originally suggested that genes for some synthetic pathways were grouped together in clusters. More recent data show that, instead, these loci are transcription units for large polypeptide chains having multienzyme functions. A most striking example is the group of sites involved in the synthesis of chorismic acid, the precursor of the aromatic amino acids, in *N. crassa*. Initially, mutants were isolated which lacked just one of the five functions, and the fine structure map divided the region into five non-overlapping sites which complemented *inter se*, as if

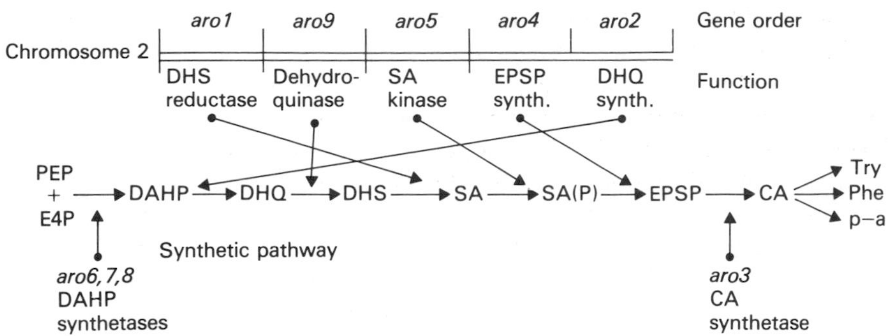


**Fig. 12.1** The synthesis of chorismic acid (CA) in *Neurospora crassa* involves nine genes. The first three synthesize DAHP (deoxyheptulosonic acid phosphate) from PEP (phosphoenolpyruvate) and E4P (erythrose-4-phosphate). DAHP is then the substrate for the products of a five gene cluster, which is transcribed as a single polypeptide, and forms EPSP (enolpyruvylshikimic acid phosphate); which is then converted into CA by the product of another gene (*aro3*). Note that the clustered gene order is not the same as the synthetic sequence, suggesting that the functional enzyme complex, thought to be a dimer, is folded. DHQ, dehydroquinic acid; DHS, dehydroshikimic acid; SA(P), shikimic acid phosphate; Try, Phe and p-a are tryptophan, phenylalanine and *p*-aminobenzoic acid (see Gaertner and Cole, 1975).

they were separate genes (Fig. 12.1). Later mutations were found which did not complement at all, or complemented with only one or two other mutants. These anomalies were explained when it was found that the 'cluster' coded for a single polypeptide (150,000 molecular weight), which forms a dimeric protein folded to give five enzymic domains (Lumsden and Coggins, 1977). There are many other similar examples, and we have seen that this economy of organization is also found in *Drosophila*, as judged by the *r* locus. What is not clear, through lack of information on their biochemical functions, is if other complex loci in higher eukaryotes can be explained in the same way.

Three further elaborations of prokaryote systems are worth noting. First, there is evidence that a structural gene product may function both as an enzyme and as a regulatory protein, as is the case for the threonine deaminase of yeast where the *ilv-1* (deaminase structural gene) mutant blocks the starvation-induced derepression of the two subsequent enzymes common to the isoleucine–valine synthesis pathway (Bollon, 1975). And there are other examples where the protein has similarly evolved this double function (autogenous regulation). Second, in yeast, and to a limited extent also in *Neurospora*, starvation of the appropriate auxotrophs for any *one* of the amino acids, arginine, lysine, histidine or tryptophan, results in derepression of the enzymes for *all* of these synthetic pathways: one repressor is a multipathway regulator (Wolfner *et al.*, 1975). A mutant at the *tra* locus is derepressed with respect to all of these enzymes even in the presence of the end-product amino acids, suggesting that the $tra^+$ product is the repressor molecule. Third, not all regulatory gene products are necessarily proteins;

they may be nuclear RNAs active only within the nucleus where they are formed. The *scon* locus in *Neurospora* is unlinked to the structural genes involved in sulphur metabolism, and a constitutive mutant (*scon*$^c$) has been found which results in the constitutive production of aryl sulphatase (and other enzymes) even in the presence of excess sulphate which would normally cause its repression. Electrophoretic variants of this enzyme are available (ars$^N$ and ars$^S$), and nuclei can be made with *scon ars* combinations. These can then be combined in the common cytoplasm of heterokaryons and exposed to repressing and derepressing environments when it is found that *scon*$^c$ affects only the *ars* allele in its own nucleus, and *scon*$^+$, in turn, allows the repression of the allele in its nucleus (Burton and Metzenberg, 1972). These data suggest that the regulatory gene product is a nuclear RNA, although such a molecule has not yet been identified.

These examples show that the classic operon arrangement of prokaryotes is elaborated in higher forms to give, in one way or another, simplified control of multiple pathways. We may expect these economies of organization to be present in higher eukaryotes, too; but proving their existence is another matter. The difficulties of handling the very large numbers required for the resolution of such complex problems of gene organization needs no emphasizing. The second important point is that these cases, and many others, provide evidence of substrate induction (or derepression) and of end-product inhibition among anabolic and catabolic pathways in fungi, and experiments with inhibitors suggest that most of this regulation is at the transcriptional level. It is no surprise, then, that constitutive mutants have been isolated, and that good *cis*-dominant, promoter–operator mutants have been found. However, in no case has the complete complement of an operon been unambiguously defined, and it is possible that operons, *sensu stricto*, do not exist in eukaryotes. If they do, they will probably be for regulating metabolic systems, and we are left without any certain answer concerning the role of operons in differentiation and the synthesis of differentiation-specific proteins.

Mutants have been found which affect the differentiation of fungi such that cell wall structures are abnormal, colonial form altered, conidial differentiation impeded or distorted, the cell cycle disturbed, and so on. Some of these changes may be, and often are, considered as simple models for differentiation in higher forms, but the data do not yet provide us with any new information relevant to the classical problems of determination and differentiation. The fact that, for example, altered hyphal branching and colonial forms in *N. crassa* can be traced to mutant forms of glucose-6-phosphate dehydrogenase and 6-phosphogluconate dehydrogenase and, via the pentose shunt, to changed levels of polyunsaturated fatty acid in cell membranes (Scott and Mahoney, 1976) alerts us to the importance of normal metabolism for normal development (cf. p. 149). This secondary consequence of a metabolic defect proves only that differentiation cannot proceed properly in the absence of the necessary building bricks. Many morphological mutants may be of this class, and we shall see that cell form may be changed for this reason (p. 335).

### Genes in cassettes: yeast mating types

The most intriguing unicell organization recently discovered involves the arrangement of the mating-type genes of *Saccharomyces cerevisiae*. In heterothallic strains there are two mating types, *a* and α, which are relatively stable, switching from one type to the other once every $10^5$ or $10^6$ budding cycles. Each mating type produces a diffusible 'sex factor'; polypeptides of different compositions. The α-factor arrests a haploid *a* cell at the G1 stage of its cell cycle prior to DNA synthesis, and makes it competent to agglutinate with an α cell to form a vegetative diploid, which subsequently sporulates giving the expected 2:2 ratio, α and *a* therefore appear to be alleles, and they map at the mating type locus (*MAT*α or *a*) on chromosome III. The α-factor has no effect on α cells and it can be used to assay the mating type of a strain (and reciprocally for the *a* factor, though less has been done with this). The situation is quite different for homothallic strains which show a high frequency of switching of mating types, hence their name. Using the α-factor assay, Hicks and Herskowitz (1976) showed that such an α ascospore usually makes an α bud, but that the next bud has a more than even

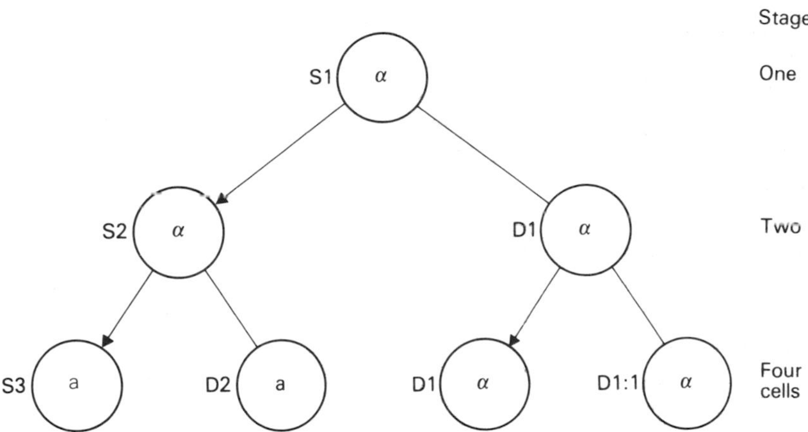


**Fig. 12.2** The pattern of mating-type switches in a homothallic yeast strain, after Strathern and Herskowitz (1979). The spore (S1) buds off one daughter (D1) and then another (D2), and the first daughter buds off a third (D1:1). When these four cells are challenged with the α-factor, two are found to have changed to *a* mating type, and two remain α as shown. This is a stable change (in the absence of HO) as can be shown by mating the individual cells to the appropriate heterothallic strain (*MAT*α *ho*), and is therefore a change of the mating-type locus. The S1 spore must carry the information for both α and *a*. Only cells which have budded once or more (experienced cells) can switch mating types, and both division products are always of the same mating type.

chance of being of *a* mating type (and so arrested by the α-factor). In each budding cycle of a homothallic yeast there is a high probability of a mating-type switch (Fig. 12.2). Study of this high 'mutation rate' has exposed a surprising new aspect of chromosome organization (reviewed by Leupold (1980)) which may have important implications for some aspects of differentiation. It is a mechanism that produces alternative phenotypes, since *a* and α cells have a number of different physiological properties.

Genetic analysis shows that other genes are involved in addition to *MAT*. A dominant (*HO*) determines homothallism, and heterothallism is determined by its recessive (*ho*). Mutants of α and *a* map to sites 30 or so units on either side of *MAT*, in a curious symmetry, with α on the left at *HML*, and *a* on the right at *HMR* (although the positions are reversed in some strains). These genes are expressed at the *MAT* locus, not at their own sites, and appear to be alleles for that reason (Fig. 12.3). Further, in the presence of *HO*, a more or less sterile α mutant (*matα*$^{ster}$) can switch to *MATa*$^{+}$, which can then make a *MATα*$^{+}$ (fertile) cell within three cell cycles. Hence, the homothallic strain must carry reserve copies of *HMLα* in its genome, and the same is found for *HMRa*. *HML* and *HMR* genes are, therefore, stored in a 'silent' form, as in a cassette, waiting to be played at the *MAT* locus. This evidence suggests that the exchange involves a recombinational event, but it differs from the similar activity of transposable elements (p. 64) in its regularity and precision. True recombination can occur between homologues (*trans* configuration) in diploids, and using the different strains noted above, haploids can be made carrying α (or *a*) at all three sites, and these are stable since they are *HMLα*, *MATα* and *HMRα*.

This 'cassette model' (Fig. 12.3) predicts that: an α gene at *MATα* will have its counterpart at *HMLα*; there will be a physical difference between *MATα* and *MATa*; and there will be a similar difference between *HMLα* and *HMRa*. These molecular predictions have been tested by using a cloned

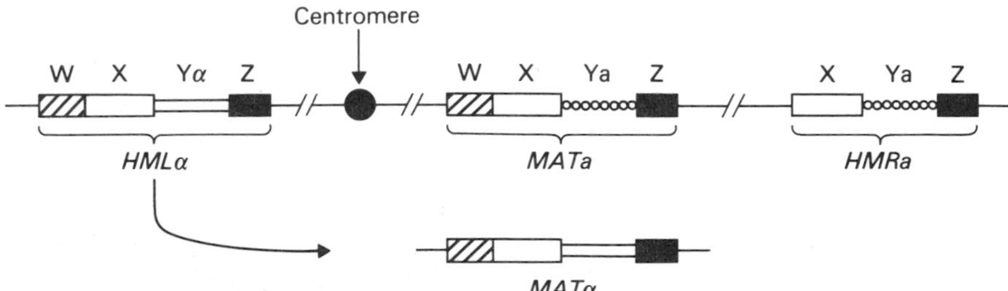


**Fig. 12.3** The structure of the mating type cassette system in *Saccharomyces*, after Strathern *et al*. (1980). The mating type locus (*MAT*) carries a copy of one (here *a*) of the genes from the unexpressed loci (HMLα and *HMRa*). The sequences common to these loci, as determined from restriction enzyme maps, are shown by letter and by the shading of the blocks. The 700 or so base pairs of the Y regions which differ between α and *a* are flanked by common sequences, and these allow the pairing of *HML* and *HMR* with *MAT* homologous sequences (see Fig. 12.5).

*MATα* sequence to probe the yeast DNA organization (Hicks *et al.*, 1979). A combined yeast–coli plasmid (YEp13) was used since it transforms both *E. coli* and yeast. The *E. coli* genome fragment carried ampicillin and tetracycline resistance, and the yeast fragment contained *leu2*$^+$, which complements the *leu2*$^-$ mutant of yeast and its equivalent, *leuB6*, in *E. coli*. Selection in both plasmid carriers was therefore possible. *Bam*H1 restriction fragments of DNA from an α yeast were cloned, and the clones screened for colonies in transformed *E. coli* (i.e. *Amp*$^R$, *Leu*$^+$, *tet*$^s$). The yeast containing clones were then screened against a *mata*$^-$, *leu2*$^-$ haploid yeast first for survival on leucine-free media, and then for the expression of α mating type, for *MATα*/*MATa* is α. Subsequent fractionation of the DNA which gave this double transformation provided a 6.6 kb fragment containing the α function in its central part. The plasmid containing this α gene was then used to probe restriction enzyme digested DNA fragments from the meiotic segregants of known *a* and α diploids, suitably marked. The restriction endonuclease digested DNA was separated on agarose gels and transferred by the Southern blot technique to nitrocellulose, where it was probed by radioactively labelled plasmid DNA. Sample autoradiographs are shown in Fig. 12.4.

In a cross which segregates *a* and α we might expect to find only two bands, *HMLα* and *MATα*, with this α probe, whereas four are found. There are therefore some DNA homologies between *a* and α, and the probe identifies both. Band C is dimorphic, and segregates consistently in a mendelian fashion with mating type. The C bands must therefore be the *MAT* locus, and the polymorphism reflects the greater size (~ 200 bp) of the α gene. Two of the additional bands should be *HML* and *HMR*, and this can also be explored using segregants of *HMLa* and *HMLα*, and *HMRa* and *HMRα*, with the other loci held the same. Band A is then found to segregate for *HMLa* and α, with the previously found size difference, and band B similarly for *HMRa* and α. The mating-type gene must be partly homologous at all three loci, and there is a real size difference between *a* and α. Finally, since strains lacking *a* (or α) can be made it is possible to test the central 'cassette switching' hypothesis. *HO HMLα MATa HMRα* will give two cell types on division, the original genotype and *HMLα MATα HMRα* arising from the only possible replacement at the *MAT* locus. If the latter cell type is isolated from its sibs, it will be a stable α cell and have one *MAT* band at C, whereas the parental type can mate with α cells and form a stable *a*/α diploid which has two bands at A, B and C. This is what is found. Oshima and Takano (1980) provide a genetic proof of the cassette transposition.

A number of experiments suggest that replacement of *a* by α, and α by *a* is a gene conversion event (Klar *et al.*, 1980a, b). An experiment by Haber and Rogers (1982) strongly suggests that it is an intrachromosomal mitotic gene conversion. By making a tandem duplication at *HMR* (*HMRa::HMRa*) they could ask if this would replace the *MATα in toto*, assuming that the homologous sites of *HMR* and *MAT* would pair (Fig. 12.5). Southern blots

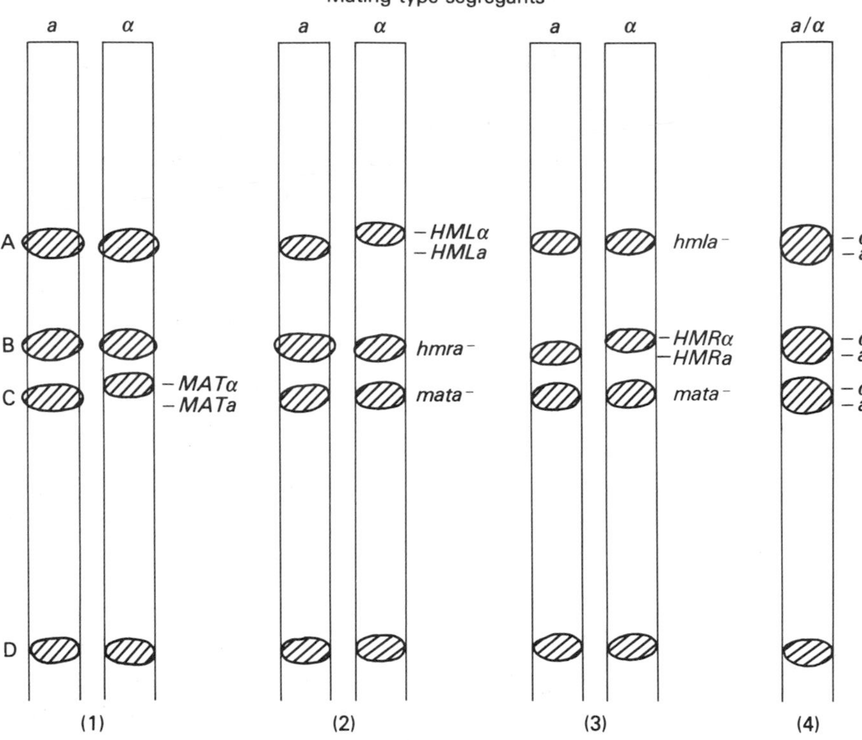


**Fig. 12.4** Diagram of Southern blots of individual meiotic segregants of *Saccharomyces cerevisiae a* and α mating types probed with an α-carrying plasmid. Four bands are found instead of the expected two; hence there must be homology between *a* and α, picked up by the α probe. The tracks in 1 show a polymorphism at the C band which segregates with mating type, thus identifying the *MAT* fragment which is ~200 bp smaller in *MATa* than in *MAT*α. By using four special strains in which *a* and α are expressed at *HML* or *HMR* (tracks 2 and 3, respectively), the same size difference should segregate with mating type, and the autoradiography polymorphisms show that band A corresponds to *HML* and band B to *HMR*. Band D is a small fragment cut by the *Hin*dIII enzyme, and is irrelevant. Track 4 shows that both *a* and α are expressed at A, B and C in an *a*/α diploid strain (see text for further details).

of digested fragments of the DNA show that this occurs, and that there is no loss from the donor locus. Conversely, a *MAT*α duplication was also replaced by a single *MATa* locus. The conversion system is therefore dependent on pairing of homologous sequences in the *HMR, HML* and *MAT* sites, and on DNA mismatch repair favouring the incoming strand.

One can also ask why only the cassette at the *MAT* site is transcribed and not those at *HML* and *HMR*. A number of non-allelic mutants are known which allow transcription at all three sites (e.g. Silent Information Regulators or *SIR* mutants). The mature transcripts appear identical to those from the *MAT* site, so it is argued that *HML* and *HMR* transcription is

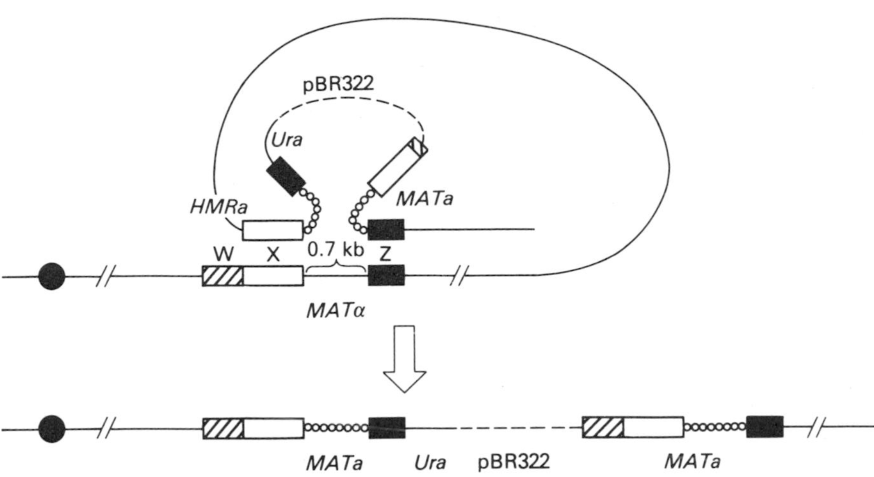


**Fig. 12.5** A model of the transposition of a tandem duplication of *HMRa* to replace *MATα*, simplified from Haber and Rogers (1982). The duplication was constructed by inserting *HMRa* and its adjacent *Ura* sequence into the plasmid pBR322, and then joining this to *MAT/HMRa* to give a linear tandem duplication. The model assumes homologous pairing of the outside sequences X and Z, and instead of replacement of 0.7 kb non-homologous region of *MATα*, a large 7.9 kb region is inserted instead of the *MATa* unique sequence. The converse is also found: a *MATα* duplication can be replaced by a single *MATa* locus.

usually prevented by *trans*-acting, negative regulators. The cassette or *MAT* would then be transcribed constitutively (Klar *et al.*, 1981) in *sir*$^-$, and the evidence suggests that *SIR*$^+$ causes a difference in chromatin structure at the *HM* loci rather than at *MAT* (Nasmyth, 1982).

Much of the detail of this system has yet to be worked out (the role of *HO*; why replacement of a cassette is 70 percent of the time by its alternative (of opposite mating type); what the molecular functions of the *HO*, *SIR* and other genes are; and so on), and it is still uncertain if it is unique to yeasts. We have seen that the *Drosophila* double-sex mutants have some of the earmarks of such an arrangement, but that is far from proving that cassettes exist in higher eukaryotes. One speculation (Herskowitz *et al.*, 1980) is that cassette switching could account economically for the complex cell lineages of higher organisms by the movement (and activation) of regulatory genes. This is a version of the binary choice model (p. 222) with this difference: each of the alternatives is a regulatory unit. Proof will therefore depend on showing that key regulatory loci are rearranged. None of the morphological higher eukaryote mutants we have considered suggests this possibility.

### *Protozoa: nuclear–cytoplasmic interactions*

Nuclear transplantation is the test for nuclear-cytoplasmic interactions, and for assessing the role of cytoplasmic signals in gene activation. The first successful transplants were done by fusing a nucleus-containing rhizoid fragment to the enucleate stalk of the unicellular alga, *Acetabularia* (Hämmerling, 1963). This showed that the morphology of the cap was dependent on the genotype of the nucleus, but only after some delay due to the long life of the mRNAs of this organism. True transplants of nuclei between species of *Amoeba* quickly followed (see Danielli and Di Barardino, 1979). The compatibility of nucleus and cytoplasm was assessed, and responses to such general insults as irradiation, chemicals and temperature elevation were measured to see if they corresponded to the patterns of the cytoplasmic recipient, or of the nuclear donor. In all cases there was some regulation between the two, and no example of a purely cytoplasmic inheritance. As Morgan (1934) pointed out, this is what we should expect if nuclei are to respond to cytoplasms, and such diversified nuclei are to change cytoplasms in their turn.

The more advanced ciliates, on the other hand, provide very clear evidence for cytoplasmic inheritance, and we shall look at one example which typifies many (see Sonneborn, 1970, Aufderheide *et al.*, 1980). The cell cortex of *Paramecium primaurelia* is sufficiently firm to anchor its special components (cilia, trichocysts, anal pore, contractile vacuoles and ingestatory apparatus) in fixed positions, and in regular anterior–posterior, left–right orientations, and these are reproduced with the same microgeography when the cell divides across its middle. This equatorial division transects the ingestatory apparatus through the vestibule which leads into the gullet, and a new gullet arises from an anlage which forms at the junction of the right vestibular wall with the gullet, and nowhere else. This gullet anlage can be transplanted to the right of the oral region of another *P. primaurelia* to produce a cell line with paired ingestatory apparatuses, or doublet paramecia can be made with two vestibules and gullets 180° apart (Fig. 12.6), and this doublet also perpetuates the condition after division. If one of the anlage positions of a doublet paramecium is irradiated with ultraviolet, its progeny lack both vestibule and gullet on the irradiated side, but not on the other side.

In these experiments the cells carry the normal nuclear complement which, presumably, makes the usual gene products. Hence, where the relevant products are assembled depends on the position of the junction of vestibule and gullet, even when this is doubled and in the wrong place. Assembly fails when something in this area is damaged, even if the surroundings are intact, and this 'receptor' cannot be replaced or repaired by the nucleus. The organization of the cell cortex in *Paramecium*, and in related species such as *Tetrahymena* and *Stentor*, is the essential and indispensable basis of the cell morphology. One *Tetrahymena* mutant, *janus*, expresses a similar mirror-image doublet organization (Frankel and Jenkins, 1979), but its molecular basis is unknown.

This highly specialized cortical organization has evolved as the characteristic feature of ciliates, and we cannot expect it to be so elaborated else-

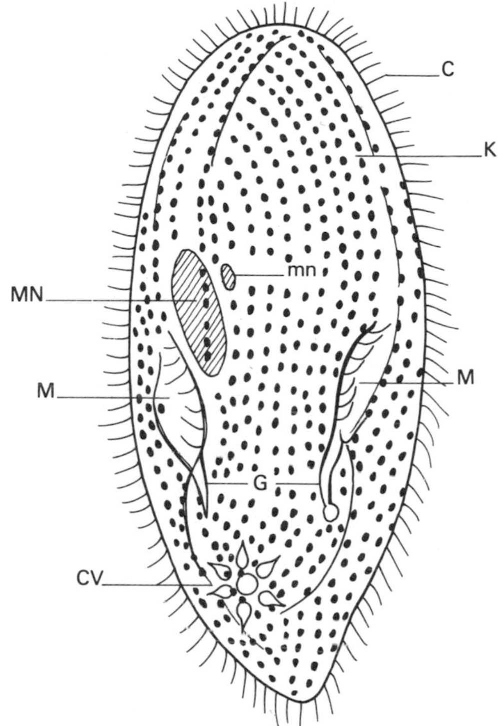


**Fig. 12.6** *Paramecium* with double mouth and gullet, after Sonneborn (1970). The paired mouth–gullet (M and G) is regularly inherited as described in the text. There is one macronucleus (MN) and one micronucleus (mn) and the excretory contractile vacuole (CV) lies towards the posterior. The cilia (C) are shown diagrammatically together with their basal kinetosomes which are interconnected to give the kinety rows (K).

where, especially in multicellular forms with a cellular division of labour. However, as we saw (p. 20), some spatial organization is expressed in enucleated eggs which may undergo partial embryogenesis, and division products of differentiated cells may form mirror-images, as is the case with cultured neuroblastoma cells (Solomon, 1979). There is therefore some cell patterning even in higher forms, and study of these notable ciliate systems may suggest the elements underlying such patterns in general.

Ciliates also have systems of mating type determination which parallel that just described for yeasts in a remarkable way, but whether they also involve cassette genes is not yet known. *Paramecium* has two kinds of nuclei, a small, paired diploid germinal nucleus which undergoes meiosis to make haploid products for fertilization, and a single 'somatic' macronucleus which derives from the zygote and is ~ 800-ploid. Prior to fertilization, one of the haploid nuclei divides mitotically to make two genetically identical gamete nuclei. These nuclei may fuse (autogamy) to form a diploid synkaryon, a single cell homozygous at all loci. If, instead, two cells of opposite mating type conjugate, one 'male' gamete nucleus is reciprocally exchanged to make two genotypically identical, but not homozygous, synkaryons in the two conjugants of the pair, whose mitotic descendants form a synclone. After autogamy or conjugation, the synkaryon nucleus divides twice mitotically and two of these products remain as micronuclei and two form macronuclei according to their position in the cell (Sonneborn, 1954). At the first cell division the micronuclei again divide and a pair passes to each cell, but the macronuclei do not divide and, instead, segregate one to each cell. At each subsequent division, both micronuclei and macronuclei divide to give two sister clones, or karyonides, each descended from a carrier of one of the new macronuclei. The macronuclei existing prior to fertilization are lost, except under special circumstances.

Mating types were originally designated by numbers, odd (O) or even (E), and each breeds true during asexual reproduction, with only rare reversals. The products of autogamy would also be expected to be true breeding, but they are not. Even when selection for one type is practised during successive autogamies, both mating types are produced: O or E is alternately silent. The unit of this inheritance is not the synclone but the karyonide, since sister karyonides, from either autogamous or conjugant cells, are often of different mating types. Haploids also show this difference between sister karyonides. Mating type is therefore determined by the macronucleus (which also determines other somatic characters). Mating type determination is influenced by temperature (higher temperature increases the frequency of E), but only when the treatment is applied just prior to the first post-zygotic cell division as the new macronuclei are forming; hence its inheritance in the karyonide. Temperature has no effect at other times, or on the mitotic descendants of the determined macronucleus. Some exceptional karyonides contain both O and E cells, and any isolated individual cell has progeny of both mating types. A mutant ($mt^o$) makes karyonides of only O mating type; but its reciprocal, only E type, has not been found.

It will certainly be profitable to see if the yeast model provides the key to ciliate mating types, but the ciliate system also illustrates two additional points. First, a stable chromosomal state is established at a very specific stage during the making of the macronucleus and, second, the cytoplasmic determination of micro- and macronuclei is also very exact. That is, we have stable nuclear diversity established within the same cell. This implies a more specific cytostructure than one might anticipate, although it is in line with the regularity of the cortical organization. But then the macronucleus is itself a surprising structure, for the DNA within it is found fragmented into gene-sized pieces, organized in nucleosomes but otherwise atypical (Lipps *et al.*, 1982). It seems unlikely that these characteristics will be found in higher organisms.

## Cellular slime moulds

The slime mould, *Dictyostelium discoideum*, exists as free amoeba-like cells feeding on soil bacteria. The haploid amoebae multiply mitotically until they consume their food supply, and then they aggregate into a multicellular 'slug' which, after migration, forms a fruiting body (Fig. 12.7). This simple morphogenesis has been taken as a model of differentiation generally (Cappuccinelli and Ashworth, 1977), but so far most work has been done on the interesting chemotaxis and cell cohesion of the aggregation process, which is now well understood at the molecular level (Loomis, 1979). Fifty or so mutants have been isolated and allocated to six linkage groups (there are seven chromosomes), and they are all recessives (Newell, 1978). There is a diploid phase and macrocyst formation, so genetic understanding of *D. discoideum* organization should develop quickly; but this will not concern us here.

### *Cell aggregation and differentiation*

The signal for differentiation of the amoebae is amino acid starvation, and there is evidence that some secretion accumulated at high population density is also involved (Margolskee *et al.*, 1980). Aggregation starts after 8 h when this differentiation is complete, and it is mediated by pulsed and relayed secretions of cyclic AMP (cAMP). Adenyl cyclase increases by an order of magnitude or more during differentiation, and the cellular cAMP by about two orders of magnitude. Its production cycles every 5–8 min so that a pulsed secretion is added to the medium. Since this would, nevertheless, saturate the medium, a phosphodiesterase which converts cAMP → 5′-AMP is also secreted, and the activity level of this enzyme is regulated by a protein inhibitor, which is feedback controlled by cAMP. That is, high levels of cAMP reduce inhibitor production which permits the phosphodiesterase to lower the cAMP level. The amoebae also have a larger number of cAMP binding sites on the cell surface so that they can accept $\sim 5 \times 10^5$ molecules per cell, and they differentiate two kinds of contact sites which cause them to cohere on collision, instead of moving away from one another as they do when feeding. Thus, when a differentiated cell in the starved population secretes a pulse of cAMP into the medium it is received as a signal by adjacent cells, which in their turn secrete a pulse of cAMP and then become refractory for some minutes. In this way, waves of chemotactic signals are propagated through the population and result in the orientated, inward movement of the cells in radial streams. Once in contact, the cells join head to tail and move together to form an irregular aggregation mound (Fig. 12.7). Two mutants, one which fails to make the phosphodiesterase and the other with a faulty contact site, fail to aggregate. The first can be 'cured' by adding the enzyme to the medium.

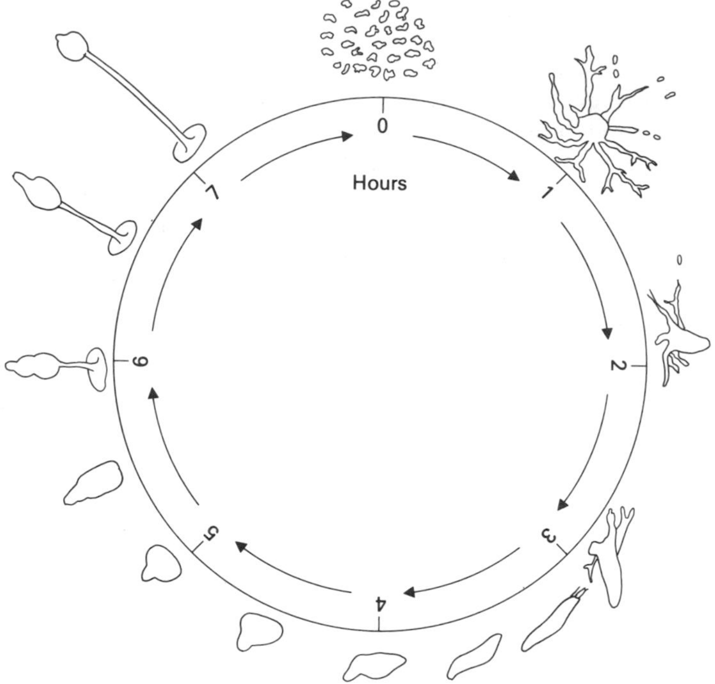


**Fig. 12.7** The differentiation of *Dictyostelium*. The zero time population of free living amoebae respond to starvation by streaming to aggregation points (1) which eventually accumulate thousands of cells. The centre of the unified mass rises up (2) and falls over to form the migrating slug (3–4) which eventually settles (4–5) to form the fruiting body (6–7). The tip cells turn into stalk cells, and the remainder form spores which are raised into the air by addition of cells to the tip of the growing stalk. The time scale is approximate.

The aggregation mound piles up a raised nipple which acts like an organizer region, and if it is removed there is a reorganization of the mound to form another. The accumulated cell mass eventually elongates and falls over with the nipple at the tip, to form a slime-coated slug, which moves over the surface for some hours until it finally differentiates. At this slug stage, the cells of the tip are histologically distinguishable as prestalk cells, and the posterior has rearguard cells which will eventually make the basal disc. Prestalk cells of the slug will differentiate if exposed to cAMP, but this morphogenesis is suppressed in the slug by the $NH_3$ excreted by these protein-metabolizing cells. Pulses of cAMP, nonetheless, seem to organize cell movement within the slug, since prestalk cells move directionally towards a cAMP source (Matsukuma and Durston, 1979). cAMP alone is not sufficient for the differentiation of stalk cells, and a diffusible oligosaccharide factor also plays a density-dependent part (Town and Stanford, 1979). When differentiation occurs, a cellulose stalk is secreted within the cell mass and is extended from its top end as stalk cells move in there, vacuolate and die. Prespore cells are carried up the stalk, where they acquire a thick wall and become a resistant spore. Each spore, under the right conditions, hatches into an amoeba to restart the cycle.

While the role of cAMP in bringing about cell aggregation is well understood, the main interest of this model system lies in the events which occur subsequently and cause the differentiation of stalk cells and spore cells. There are three arguments which favour the view that cell–cell contacts are the essential stimuli for gene activation during this differentiation. First, if cells in the post-aggregation phase are dispersed they dedifferentiate, and after 1½ h they undergo an 'erasure event' (loss of mRNAs?) following which they repeat their initial differentiation pattern if reaggregated: cell separation blocks differentiation (Soll and Mitchell, 1982). Second, cohesion depends on at least two contact sites; A sites which mediate end-to-end joining, and B sites which allow side-to-side cohesion, and only the latter are EDTA sensitive. Mutants which fail to make these tight contacts fail to synthesize the 2000–3000 co-regulated mRNAs (11 percent of the single copy genome) which typify the change from the vegetative to aggregated state. The essential contact seems to be between A sites, since mutants which associate loosely but make no EDTA resistant contacts fail to transcribe significant levels of the mRNAs, whereas a mutant which makes such contacts does, even if it subsequently fails in normal differentiation (Blumberg *et al.*, 1982). Third, monovalent antibodies made against aggregated cells block reaggregation when mixed with dissociated cells, presumably by combining with their aggregation sites, and prevent the synthesis of UDP-glucose pyrophosphorylase, an enzyme indicator of the change of cell state. However, if the antibody is pretreated with an EDTA–high salt extract of cells (to remove B-site and other specificities) aggregation is not blocked but enzyme synthesis is. This suggests to Kaleko and Rothman (1982) that there are other components on the surfaces required for enzyme induction which may be activated by contact.

There is, however, a contrary argument. It has been known for some time that cells developing at high density release a diffusible factor (DIF) which induces stalk cell differentiation in the presence of cAMP. DIF is absent from vegetative cells, but increases in amount during slug migration and accumulates in cells in parallel with the prespore-specific UDP-galactose polysaccharide transferase and other enzymes (Brookman *et al.*, 1982). There are therefore diffusible factors involved in cell differentiation, and the importance of cell contacts might be to facilitate the short-range accumulation of these molecules. Kay and Trevan (1981) therefore asked if *high* levels (5 mM) of cAMP might induce spore formation in cells so diluted that they made no cell–cell contacts. And this it did for a sporogenous mutant strain. Of course, it may be the peculiarity of this mutant that it no longer requires contact to differentiate, but it seems more likely that contacts facilitate the local diffusion of morphogenetic factors, and the intracellular accumulation of cAMP (Parish, 1979).

The importance of cell–cell contacts for *Dictyostelium* differentiation, and the difficulties of analysing their significance in detail, illustrates, as a model system should, what we are likely to be confronted with in like situations in higher organisms. As yet, it is not clear which characteristics are unique to the model, as the cAMP aggregation signal probably is, and which are more generally important (as the cAMP induction of A contact sites and DIF synthesis may be). Indeed, the many morphogenetic roles of cAMP suggest it acts more like a hormone than in its usual role as a second messenger, but whether its main activity is to phosphorylate (and activate) pre-existing regulatory proteins is unknown as, indeed, are the comparable activities of this molecule in higher cells. The coordinate induction of so many mRNAs may relate to the peculiarity of *Dictyostelium*'s DNA organization. Seventy percent of the sequences are unique, but they are flanked on one side by repetitive sequences (250–400 bp) and poly$(dT)_{25}$ sequences on the other. Both sets of repeats are transcribed with the single copies, and in so far as the repeats are common to different unique sequences they might qualify as fitting the Britten–Davidson model (Firtel and Jacobson, 1977). Clearly, this organism is a model for many things (Loomis, 1982).

## Cultured somatic cells

Culture media have come a long way since Ross Harrison (1907) studied nerve outgrowth from a fragment of frog medullary tube explanted into a lymph clot, and most experimental media are now chemically defined, except for a supplement of foetal calf, or horse, serum which is generally necessary to give more or less normal cell division rates and cell survival. It is now possible to grow cells from most embryonic and adult tissues (Cox and King, 1975) and those which proliferate in the adult, like mammalian leukocytes and fibroblasts, divide and retain their phenotype for many cell cycles before becoming senescent and dying out. Multiplying cells from terminally differentiated, non-growing tissues like muscle and epithelium, derive from populations of determined, non-cycling cells (Gelfant, 1978) and they usually differentiate after a few cell cycles and cease further growth. These cells make their typical products under good conditions (e.g. adrenal cortex cells synthesize steroid hormones, cartilage cells make chondroitin suphate, lens cells synthesize crystallins, etc.), but under poor medium conditions they may dedifferentiate. If they are returned to supplemented media (e.g. containing embryo extract, or other complex additions) they 'remember' their fate, and differentiate again. In short, differentiation, or the maintenance of differentiation, depends on culture conditions, but we have no knowledge of the exact factors involved. The role of the components of effective supplements is now being actively explored.

Most of the cells which will concern us derive from tumours, like the HeLa cells already described (p. 11), or mouse fibroblast L cells which were made by treating cultured connective tissue with the carcinogen 20-methylcholanthrene, which, after 4 months, changed some cells so that they grew rapidly and without becoming senescent. Many vigorous, permanent cell lines have since been established (Table 12.1) and they can frequently be cultured in unsupplemented media. Auxotrophs, and drug-resistant and temperature-sensitive mutants can be isolated from them using conventional selection techniques (Siminovitch, 1976) and these provide genetic markers which can be exploited. Permanent cell lines have limitations—their karyotype is frequently abnormal and their differentiation may be partial and not under the control of the normal mechanisms—but they provide experimental material which can be regularly expanded in any required amount. Cultured somatic cells can be handled, to a degree, like unicellular eukaryotes; but since they cannot be bred, genetic analysis depends on our ability to fuse genetically different cells together.

### Cell fusion

Spontaneous fusion occurs very infrequently among cells grown *in vitro*, and if two different strains are cultured together they will form a heterokaryon;

**Table 12.1** Some differentiated cultured cells

| Cell type | Properties |
|---|---|
| **Normal cells** | |
| Cartilage | Synthesis of chondroitin sulphate and specific collagen |
| Heart muscle | Muscle proteins; spontaneous contraction |
| Lens | Synthesis of lens crystallins |
| Melanocytes | Pigmented melanosomes |
| Myoblasts | Muscle actins and myosin, fusion to form myotubes |
| Adrenal cortex | Synthesis of steroid hormones |
| Thyroid | Synthesis of thyroxin |
| **Tumour cells** | |
| Chorionic carcinoma | Synthesis of chorionic gonadotropin |
| Erythroleukaemia | Synthesis of haemoglobin |
| Hepatoma | Synthesis of serum albumin, induction of TAT by glucocorticoids |
| Mammary tumour | Synthesis of milk proteins |
| Myeloma | Synthesis of immunoglobulins |
| Teratocarcinoma | Multiple pathways of differentiation |
| Neuroblastoma | Formation of neurone-like cells, synthesis of acetylcholinesterase, neurotransmitters |

two genetically different nuclei in a common cytoplasm (Barski *et al.*, 1960). The frequency of fusion is greatly increased if the cells are exposed to UV-inactivated Sendai virus (SV) which brings the membranes of the cells together by a complex reaction (Okada, 1958); or more commonly today, by adding polyethylene glycol (PEG) to the medium to cause cell fusion (Pontecorvo, 1976). Intra- and interspecific heterokaryons can be constructed, and if their nuclei undergo mitosis a mononucleate hybrid cell (*synkaryon*) is formed containing chromosomes from both parent cells (Fig. 12.8). Some of these chromosomes, usually the chromosomes of one parent in interspecific hybrids, are lost during subsequent divisions. This chromosome elimination has been used very profitably with mouse–human synkaryons, where the human chromosomes are lost, for allocating over 200 human genes to their chromosomes (Ruddle, 1981; Puck and Kao, 1982). The interactions between nucleus and cytoplasm in cell hybrids also tells us a little about the regulation of gene activity.

The chick erythrocyte nucleus is small, containing condensed chromosomes which are only minimally transcribed, if at all. If this cell is fused to an active cell, such as a mouse fibroblast, very little chick cytoplasm is transferred to the heterokaryon since most of it leaks from the erythrocyte during the virus fusion process. The erythrocyte nucleus responds to its new environment with a complex sequence of changes. The condensed chromatin

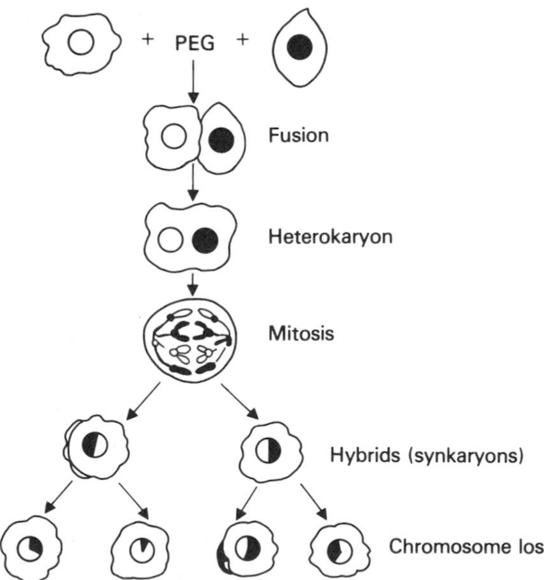


**Fig. 12.8** The formation of hybrid cells after fusion with polyethylene glycol (PEG). Viable heterokaryons divide mitotically to give a variety of hybrid cells with different chromosomal complements derived from the parent cells. As these hybrids divide, there is chromosomal loss, particularly in hybrids between species (see text for details).

becomes dispersed and the nucleus enlarges (cf. p. 330), nucleospecific mouse proteins migrate into the nucleus, hnRNAs and RNA polymerase increase in amount, a nucleolus forms and rRNA is synthesized, and chick specific proteins appear in the cell as enzymes, surface antigens and receptors etc. The erythrocyte nucleus is reactivated and again becomes partly functional, although it does not now make haemoglobin which was its previous major product. DNA synthesis takes place, too, and a synkaryon forms which rapidly loses the chick chromosomes during subsequent divisions (Harris, 1967; Ringertz and Bolund, 1974). A similar sequence of events occurs when macrophages or other more or less inactive cells are fused to active cells, and in each case there is clear evidence for the passage of effector molecules from cytoplasm to nucleus, and a reprogramming of nuclear activities. At least some of these molecules cannot be species specific in their action, though others will be.

As we might expect, many heterokaryons show properties of both parental cells (e.g. the spontaneous fusion of co-cultured rat and chick myoblasts produces myotubes containing both mouse and chick myosin (Carlsson *et al.*, 1974)): the genes are co-expressed. In other cases, like the chick erythrocyte–mouse fibroblast heterokaryon just described, the characteristic product of a parent cell (haemoglobin production) may be *extinguished*. This implies that the non-erythroid cytoplasm switches off either the transcription or the translation of haemoglobin genes. This activation/inactivation can also be studied by fusing enucleated cytoplasm fragments to cells (to form *cybrids*), but as both heterokaryons and cybrids have to be examined as single cells with a limited life, it is easier to study co-expression and extinction in synkaryons which can be multiplied to provide usable numbers.

Only about 1 in 10,000 heterokaryons go on to form synkaryons, and mutant strains can be used to select these more efficiently than by picking off and testing colonies from the culture (Fig. 12.9). Some chromosome loss may occur with both methods so it is not always certain that a particular required gene is present in the new compound nucleus: extinction is therefore less meaningful than co-expression unless subsequent progeny show reactivation of the gene. Since chromosome loss is more common in interspecific cell hybrids, the data from intraspecific hybrids are more reliable, and these hybrids also avoid the problems of species specific activators which would further confuse the results from interspecific hybrids. Few experiments meet these and other requirements (Davis and Adelberg, 1973), and we have to be content with the rather general indications of gene control mechanisms suggested by synkaryons.

The first possibility is that the parental cells have no effect on one another, and all genes are expressed autonomously or remain inactive. This seems to be the case for the enzymes of intermediary metabolism, as judged by the continued production of isozyme markers from both parents in intra- and interspecific hybrids. The $HGPRT^+$ gene on the inactive X chromosome

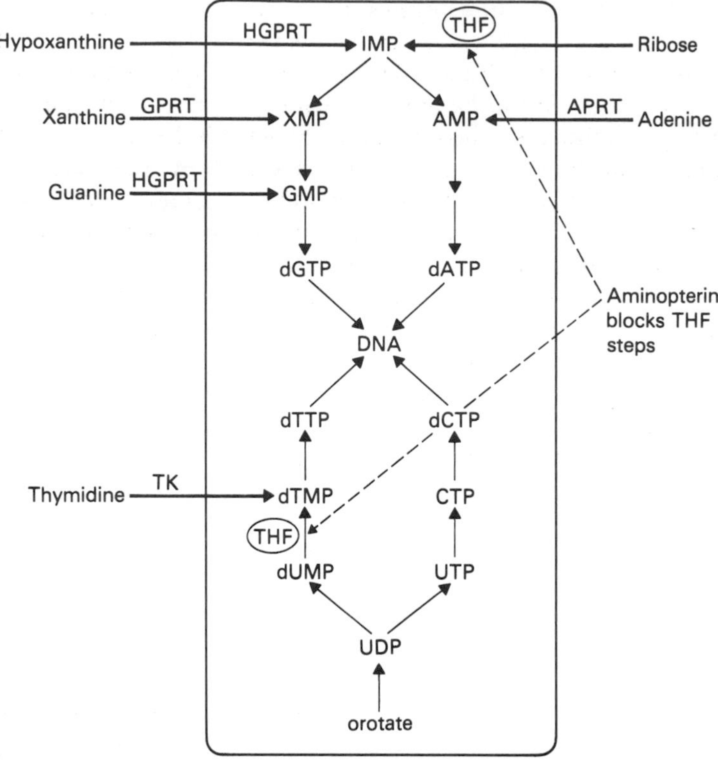


**Fig. 12.9** The HAT selection technique (Littlefield, 1964) exploits the property of the folic acid (THF) antagonist aminopterin (methotrexate) to block the synthesis *de novo* of purines and the biosynthesis of the methyl group of thymine. DNA and RNA synthesis are therefore blocked by the drug. This block is overcome in cells which can salvage hypoxanthine (and are hypoxanthine phosphoribosyltransferase (HGPRT) positive), and thymidine (and are thymidine kinase (TK) positive). Cells which are either HGPRT$^-$ or TK$^-$ will not grow in a medium which contains aminopterin even when provided with hypoxanthine and thymidine. However, if such cells are fused, the TK$^+$ and HGPRT$^+$ genes which they carry reciprocally will permit them to salvage the thymidine and the hypoxanthine, and to grow. Thus, in a medium containing hypoxanthine, aminopterin and thymidine, HGPRT$^-$:TK$^+$ and HGPRT$^+$:TK$^-$ parent cells will not survive, but the HGPRT$^-$:TK$^+$/HGPRT$^+$:TK$^-$ synkaryon will alone form colonies. Mutations blocking GPRT (guanine phosphoribosyltransferase) or APRT (adenine phosphoribosyltransferase) can be similarly exploited in place of HGPRT. The pathways are simplified.

of a female mouse cell likewise remains inactive, unless the synkaryon is subject to the selective pressure of the HAT medium of Fig. 12.9 when its activation sometimes occurs (Kahan and DeMars, 1975). X-inactivation is apparently stable in hybrid cells, and there is no new inactivation in 3X cells (i.e., two active and one inactive X are found when ♂ and ♀ cells are fused) as would be the case during embryogenesis (p. 125).

As we have seen for heterokaryons, luxury proteins (facultative markers) are often also lost in synkaryons. Syrian hamster melanoma cells fused to unpigmented mouse fibroblasts make unpigmented hybrids even when they appear to contain both chromosome complements. And the relevant enzyme, dopa oxidase, which is required for melanin synthesis, is absent—not just inhibited. However, if a near tetraploid melanoma cells is used to make a $2n$ melanoma: $1n$ fibroblast hybrid, half the clones are pigmented, and these pigmented clones generate further pigmented subclones as well as unpigmented ones which may have lost the dopa oxidase gene (Davidson, 1972; 1974). The amount of the regulatory gene product is important, one might conclude. Chromosome loss has been used to examine extinction in rat hepatoma (differentiated liver tumour) cells carrying the HGPRT$^-$ marker. When these are fused to human HGPRT$^+$ fibroblasts they no longer make tyrosine aminotransferase (TAT) and it cannot be induced by steroid hormones as it was in the hepatoma. TAT synthesis and inducibility are extinguished. However, if the clone is selected in a medium containing 8-azaguanine the human HGPRT$^+$ sex chromosome is lost because those cells that carry it are selected against, and the TAT inducibility is regained (Croce *et al.*, 1973). The human X chromosome must carry the gene for a factor which blocks TAT inducibility, and the equivalent of this gene must be inactive in the hepatoma cell.

Other examples support the points just made (Ringertz and Savage, 1976; Bernhard, 1976), but their interpretation is difficult. The melanoma hybrids suggest that something made by fibroblast genes blocks pigment production and that the relative amount of this substance is insufficient to do so in some $2n$:$1n$ hybrids. Similarly, the human X chromosome appears to make something which prevents expression of the TAT gene. But this is some way from proving that these are repressor molecules, as previously defined (p. 103), and all we can legitimately conclude is that cell hybrids do show that the products of one genotype may affect the activities of the other genotype. Unfortunately there is no clear example of the coordinate activation or inactivation of a group of structural genes, and little has yet been done to exploit the system for analysis of the determined state of embryonic cells.

### Monoclonal antibodies

Cell fusion has a most interesting use in refining the antigen–antibody reaction, which, as we saw above, allows us to investigate cell components, often without knowing what they are or what they do. Friedlander (1979) reviews this method of following changes in common and unique antigens during development. However, conventional immunological techniques are often unsatisfactory due to the antigenic complexity of, for example, cell surface proteins and the like, and the development of a technique which provides antibodies specific to single components of a heterogeneous mixture has overcome this difficulty. Indeed, the technique has wider ranging applications, from the purification of biological molecules such as interferon and individual membrane proteins, through to direct therapy where passive immunization (injection of antibody) may disable the parasite or tumour affecting a patient (Milstein, 1980). Here we shall outline only the principle of the method.

Myeloma (plasmacytoma) cells make immunoglobulins (antibody) in culture, and they secrete them into the medium. Myeloma cells are an immortal clone making a particular, abnormal antibody, and fusion of different clones results in co-expression of antibodies in the hybrid. If a plasma cell, derived from a B lymphocyte stimulated by an antigen, is fused to a myeloma cell we should therefore expect to be able to perpetuate the plasma cell's specific antibody secretion. The procedure is to immunize a mouse by injecting the required antigen (isolated protein, cell fraction or whatever) and then to fuse some of its spleen cells to cultured cells from a mouse myeloma line (Fig. 12.10). After selecting the hybrids we have a group of cells secreting a range of antibodies, depending on the specific activity of the spleen cell partner in each hybrid, and these have to be cloned. Since chromosome loss occurs in these clones, they can be further selected and recloned to establish monoclones secreting only a specific antibody, a monoclonal antibody (Köhler and Milstein, 1975). In some cases a mutant myeloma line which makes no antibody is used as the tumour cell partner in the fusion, since this allows immediate selection of clones secreting the spleen cell antibodies.

As with other cell cultures, the required clones can be multiplied to give large amounts of monoclonal antibodies which can be used to find, say, if the same antigen is present on different cells, or when the antigen appears during development, and so on (Spragg *et al.*, 1982). This is facilitated by coupling a fluorescent dye to the antibody, which can then be used to examine the distribution of the antigen on the cell surface with a precision not previously possible. But this precision brings its own problems for, while we can now enlarge our understanding of the likenesses and differences between cells, it is not always certain that the antigen we have identified is of importance in development. The establishment of monoclonal antibody 'banks', which will permit wider comparisons, may overcome this difficulty.

### Cancer cells

Myelomas are tumours, and the tumour phenotype is expressed in hybrids with normal cells. We can therefore ask if this malignancy is always expressed in other tumour/normal hybrids since, if so, it would imply the 'dominance' of whatever factors cause cancer, which would be an important generalization. Many kinds of tumour cell have been used to test this proposition—cells from spontaneous cancers, from irradiation or chemically

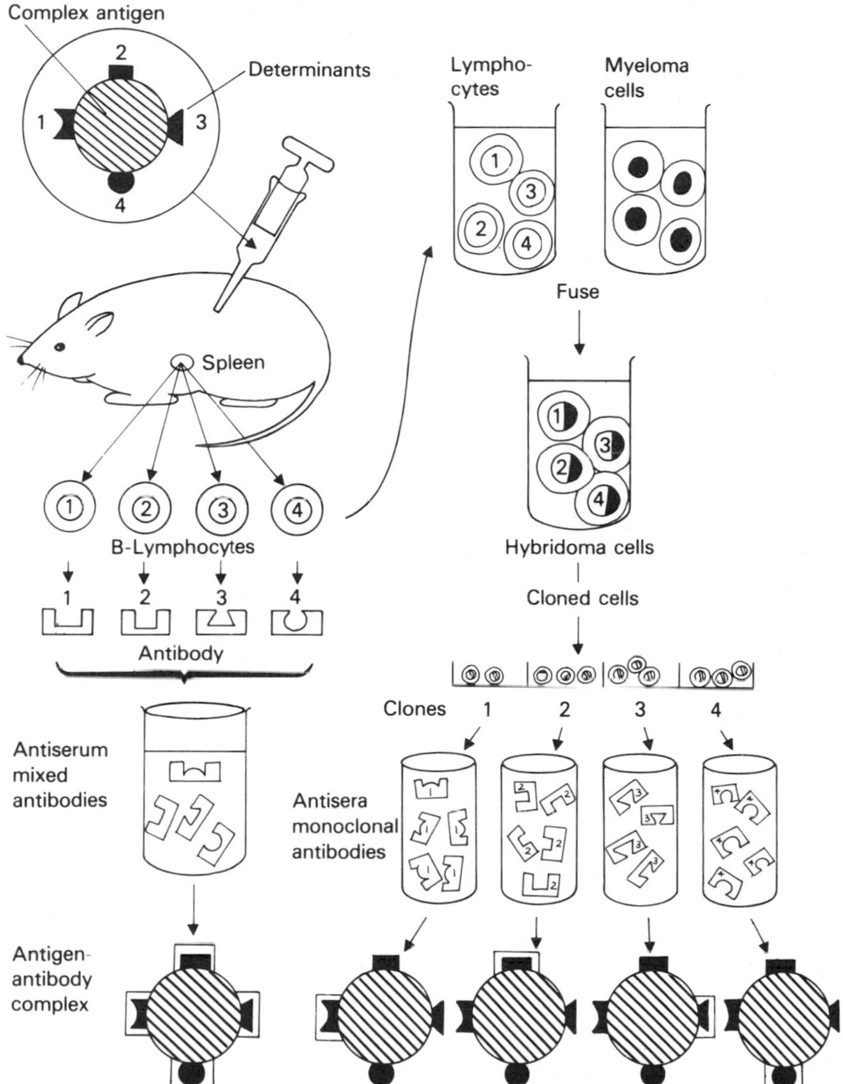


**Fig. 12.10** The production of monoclonal antibodies, after Milstein (1980). Most antigens carry a complex array of different antigenic determinants. When they are injected into a vertebrate, the immune system proliferates lines of lymphocytes, each secreting an immunoglobulin specific to a single (or part) determinant. An antiserum therefore contains a mixture of antibodies, which will combine with all the sites on the antigen, as shown on the left of the diagram. If the spleen cells are separated and fused to cultured myeloma cells, each hybrid will carry one type of antibody and the hybridomas can be cloned to produce monoclonal antibodies, each specifically binding to one determinant of the antigen. The practical routine for making these monoclonal antibodies is more complex than diagrammed above. Usually an $HGPRT^+$ myeloma cell line is used and the hybrids selected in HAT medium, and the clones are tested for antibody production and recloned as necessary. Hybridomas, like most cell lines, can be stored frozen, and kept 'banked' for use.

induced tumours, or from cells transformed by oncogenic viruses like SV40. On culture, all such cells show a loss of dependence on anchorage to a substratum for multiplication, no inhibition of growth when the culture becomes confluent (lack of contact inhibition), and an ability to grow in suspension culture or in semi-solid agar media, unlike normal cells. There is a high correlation between these characteristics and malignancy when the cells are transplanted into syngenic, or into immunosuppressed hosts. One can therefore use these culture characteristics as a preliminary assessment of the expression of the tumour phenotype in hybrids.

Early experiments (Scaletta and Ephrussi, 1965) showed that malignancy was dominant but later work showed it was not, or behaved as an intermediate characteristic (Harris and Klein, 1969). And the same hybrid combinations gave different results in different laboratories. One possible explanation of these contradictory results is chromosome loss, which is a regular feature of hybrid cells, particularly of interspecific hybrids. Stanbridge *et al.* (1982) attempted to overcome this problem by fusing HeLa cells to normal human fibroblasts (intraspecific hybrids) since these retain a high chromosome complement through many cell cycles. Such hybrids are morphologically intermediate between their parents, are non-malignant, but show the growth characteristics of the HeLa line (Table 12.2). Occasionally, as the clones grow, colonies appear which are tumorous, and these have lost chromosomes 11 or 14. This may argue for a chromosomal tumour-suppressing element, but the difficulty of carrying out the necessary chromosomal analysis illustrates one serious limitation of the cell fusion technique. Similarly, attempts to see if various tumour cell lines carried the same defect by making hybrids between them showed that some combinations complemented (had different causes of tumorigenicity) while others did not (Wiener *et al.*, 1974). However, since we do not know if the latter involve dominance, we cannot conclude that these cell lines are malignant through the same cause. We shall return to this point under the next heading.

## DNA-mediated gene transfer

The ambiguities of the cell fusion technique can be overcome when we transfer specific genes into cells. This was first done by Szybalska and Szybalski (1962) who exposed human cells deficient for the enzyme hypoxanthine phosphoribosyltransferase ($HGPRT^-$) to polymerized DNA from normal cells ($HGPRT^+$), and isolated the $HGPRT^+$ transformants using the HAT medium (Fig. 12.9). Subsequently, Munyon *et al.* (1971), showed that UV-irradiated herpes simplex virus (HSV) could rescue mouse L cells which lacked the thymidine kinase gene ($tk^-$), again using the HAT medium to select the $tk^+$ transformants; and this was confirmed by Wigler *et al.* (1977), who further showed that the transforming activity resided in a 3.4 kb frag-

**Table 12.2** Phenotypic characteristics of tumorous and non-tumorous hybrids between HeLa cells and human fibroblasts

| **Character** | **Parent fibroblast** | **Non-tumour hybrid** | **Tumour segregants** | **Parent HeLa** |
|---|---|---|---|---|
| Morphology | Fibroblastic | Intermediate | Epithelial | Epithelial |
| Microfilaments | Organized | Organized | Poor organization | Poor organization |
| Density dependent growth inhibition | Yes | No | No | No |
| Growth in agar | No | Yes | Yes | Yes |
| Serum requirement | High | Reduced | Reduced | Reduced |
| Lectin agglutination | +/– | +++ | +++ | +++ |
| Gonadotropin synthesis | Nil | Nil | Present | Present |
| Alkaline phosphatase (placental) synthesis | Low | High | High | High |

Data from Stanbridge *et al.* (1982).

ment of the viral genome which was separated after treatment with *Bam*H1 restriction enzyme. When available, individual genes can therefore be transferred into cells. Technically, this is done by co-precipitating the DNA with calcium phosphate, and the complex is added directly to the cells; but DNA can also be injected into nuclei, or transferred to the cells in fine lipid vesicles (liposomes), or whole chromosomes may be used (see Scangos and Ruddle (1981) for references).

A number of other dominant genes (those for adenine phosphoribosyltransferase, dihydrofolate reductase etc.) can be used like $tk^-$ and it is usually possible, as with HGPRT, to show that the transferred interspecific gene has the biochemical characteristics of the donor which distinguish it from the host type. Apparently not all interspecific transfers are successful, since Wigler *et al.* (1978) report that $tk^-$ mouse L cells can be transformed with mouse, human, hamster, calf and chicken DNA, but not with DNA from salmon, *Drosophila* or *Dictyostelium*. This argues for using intraspecfic

DNA transfers in studies of gene organization and activity, just as we saw for cell fusion experiments.

The obvious application of DNA-mediated gene transfer (DMGT) is to link a dominant selector gene, like HSV $tk^+$, with another known gene, like the ovalbumin gene, to make a chimeric plasmid which can then be used to transform mouse L $tk^-$ cells. When this is done (Lai *et al.*, 1980), the ovalbumin gene is found to be incorporated into the mouse chromosomes in a proportion of selected clones, but in other cases only fragments of the gene are present. Surprisingly, the complete ovalbumin gene is transcribed, the mRNA properly spliced (the mouse β-globin gene has similar junction sequences to ovalbumin) and between 1000 and 100,000 molecules of chick ovalbumin are produced per mouse cell. Not all DMGT experiments are so successful; for example, the rabbit β-globin gene in transformed mouse cells is transcribed and properly spliced but not translated (Mantei *et al.*, 1979). How far this is a consequence of the organization of the gene in the transforming plasmid and how far it results from changes in the DNA during its passage into the nucleus is now being studied.

As just described, the ovalbumin gene is constitutively transcribed apparently from its own promoter, since the $tk^+$ gene was linked to its 3′ end. Some DMGTs result in the integration of the gene, but without transcription. An instructive example is the integration of the *Drosophila* heat shock gene, *hsp70*, into mouse L $tk^+$ cells using the preceding method (Corces *et al.*, 1981). In one transformed line, as many as 20 copies of *hsp70* and its flanking sequences were integrated into different sites in the mouse genome, but not transcribed. Exposure of these cells to heat shock caused a rapid production of *hsp70* mRNA, mostly with correct or nearly correct 5′ and 3′ ends: the *Drosophila* gene responded to some mouse regulatory signal, as we might expect from the generality of the heat shock system. This shows that the signal, and the DNA sequences with which it reacts, must be highly conserved (Schlesinger *et al.*, 1982).

Again, to take quite a different kind of example, Mantei and Weismann (1982) introduced a human interferon-α gene into mouse L $tk^-$ cells, and found that this gene, and the mouse interferon genes, were transcribed when the cells were challenged with Newcastle disease virus. The integrated human gene was responsive to the viral infection. Co-transformation does not require that the gene be previously ligated to the selectable gene. Using the HSV $tk^+$ system, Wigler *et al.* (1979) found that the phage øX174, the plasmid pBR322 and rabbit β-globin sequences, provided in 1000-fold excess over the selectable $tk^+$ DNA, were present in around 80 percent of the HAT-viable clones. Selection was necessary for establishment of clones carrying the unselected genes—they could not be detected in $tk^-$ clones—and it seems likely that early in the process of DNA transfer within the cell there is some reorganization of the DNA to form a 'transgenome' which links the selectable to the unselectable gene (Huttner *et al.*, 1979). This transgenome is stable, and most selected clones show that many copies, not necessarily identical at the molecular level, may be incorporated into the cell's chromosomes. In some cases, relaxation of selection results in loss of the selected (and any associated) character, and the genes then seem to be in the subchromosomal fractions of the DNA, and therefore are not integrated and so become lost.

We have still much to learn about the events which occur during DMGT, but it is already a technique which allows known genes to be placed in cells where their activity can be studied. Genetic engineering of these DNAs will enable the analysis of the sequences involved in the control of gene expression (e.g. the 3.6 kb segment of the *Drosophila hsp70* must contain its regulatory sequences) and, conversely, the use of mutant cells permits the identification of the relevant wild-type DNA segment carrying the gene, as in the case of HSV $tk^+$. What is now required, however, are systems which will allow us to assay for developmental stage-specific gene expression. This may involve using larger DNA units (~ 50 kb) if the flanking controls are distant, and certainly requires a greater understanding of cell-specific DNA modifications and of the organization of the DNA in chromosomes.

Cells need not be genetically defective to be identified as transformed, and the tumorous change to anchorage independent growth is an example of such a selectable transformation. If DNA from a number of mouse tumour cell lines is used to transform a standard cell line grown to contact inhibition in a monolayer, the transformed cells will overgrow to form an identifiable clump. Shih *et al.* (1979) found that some DNAs from different cell lines, sheared to 30 kb, did transform, while others did not. The DNA from the transformed lines, in its turn, was competent to transform other normal cells, which formed tumours when transplanted *in vivo*. It was therefore concluded, since the transforming DNA fragments were relatively small, that a dominant mutant gene, or gene cluster, caused the original cancers and was transmissible by DMGT. We might now dispute this genetic conclusion for the following reasons.

The discovery that oncogenic RNA retroviruses carry host genes, or more exactly the properly spliced transcript from genes of the host genome, and that these host genes, called *onc* genes as a general class, are then responsible for changing cells to a tumorous state, challenges the proposition that cancers are due to simple gene mutations (see Cairns (1981) and Rigby (1982) for discussion). Instead, we have to assume that proto-*onc* genes have a normal function in cells, and that an imbalance of the gene product, either induced by carcinogenic treatments or by the elevated translation of the *onc* gene in the virus particles, causes the abnormal developments which we recognize as cancerous. This 'balance' generalization is supported by the evidence that murine leukaemia results from chromosome-15 trisomy and by other chromosome rearrangements which cause increased expression of normal cellular genes (Klein, 1981). Hence the Shih *et al.* (1979) result may be due either to duplication of *onc* genes in the integrated transforming DNA or, more likely, to the integration of multiple copies of that DNA into the genome. It seems likely, too, that these ubiquitous *onc* genes play an important part in normal development, but this has still to be explored. Since homologous sequences have been found in *Drosophila* (Shilo and Weinberg, 1981) it may be easier to examine this possibility there.

Not all tumour transformations are due to quantitative changes in the level of *onc* gene products, and a most surprising result has come from identification of the DNA sequences from a human bladder carcinoma oncogene which can transform mouse NIH 3T3 cells. Careful study of fragments of the bladder cancer cell DNA shows that specific sequences are involved; and when these are cloned and used as a probe, they correspond to a normal human gene, a proto-oncogene. The proto-oncogene had no transforming competence, ruling out a quantitative effect. Indeed, the level of the oncogene product is normal in tumours, showing that its effect is not due directly to malregulation of the gene. Instead, sequence analysis shows that the 12th amino acid of the first exon had been changed from glycine to valine (Tabin *et al.*, 1982; Reddy *et al.*, 1982). The change is a qualitative one, presumably causing a conformational change in the protein coded by the *onc* gene.

Two retroviruses, the Harvey mouse sarcoma virus and the Kirsten mouse sarcoma virus, carry the same host-derived oncogenes, again mutated at the identical site (Gly → Arg and Gly → Ser, respectively). All code for a 21,000 molecular weight protein, confirming that their tumorigenicity is due to structural gene mutations and not to regulatory gene mutations. However, this dominant bladder carcinoma oncogene has been shown to transform 3T3 cells only, not other cells. It is not a universal carcinogen, and we still have much to learn about its metabolic action. Not all *onc* genes, or retroviruses, carry these particular sequences, of course, and there must be many proto-oncogenes in the normal genome. Since they affect development, some of these loci may play an important role during embryogenesis, as was foreshadowed long ago when it was recognized that tumours carried unique antigens common to embryos. It will be an unexpected bonus if cancer research repays its debt to embryological studies by focusing our attention on important developmental genes and their products.

## *Drosophila* cells

*Drosophila* cell culture developed late, for technical reasons, but it is now easy to establish cultures from gently homogenized 5–15 h embryos (reviewed by Sang, 1981). Most cells in such suspensions are programmed to differentiate into larval tissues or, after exposure to 20-hydroxyecdysone, into adult tissues (Dübendorfer and Eichenberger-Glinz, 1980), but a small class multiplies after a few months of culture and some of these become established cell lines. Over 100 lines, some from mutant stocks, are now available, and they usually retain a normal karyotype for many cell generations, possibly indefinitely. *Drosophila* cell lines also differ from the vertebrate lines previously described by not deriving from tumours, but their normal role is unknown. Many of the cells look alike (fibroblastic, round, etc.), but enzyme assays and study of their surface antigens show that this similarity is often superficial and that like cells express different gene complements. The cells fall into two broad classes: those which are killed by exposure to 20-hydroxyecdysone and are presumably larval cells, and those which differentiate and cease to grow in the presence of this hormone. Members of this last class are convenient material for following the molecular changes induced by hormones (Cherbas and Cherbas, 1981).

Responding cells carry ecdysone receptor proteins on their surface, and the receptor–hormone complex moves to the chromosomes where, by some unknown mechanism, it induces specific transcription patterns (Schaltmann and Pongs, 1982). A number of cell lines respond by a change of form (Fig. 12.11), by a qualitative alteration of the pattern of protein synthesis and the induction of specific enzymes. One enzyme which is easily followed is acetylcholine esterase (AChE), and Berger and Wyss (1980) have fused cells which separately respond by a notable AChE induction or by little change when hormone treated. The phenotype of these hybrids is intermediate, but the AChE response is extinguished. The intermediate phenotype, which shows that the hormone is still effective, may reflect an active gene-dosage effect, but the AChE response implies the presence of a *trans*-acting repressor molecule. If this deduction is correct, we have two levels of gene control; one by specific repressor molecules and the other by the receptor–hormone complex. Such a system would explain the specificity of hormone responses—different repressor patterns—when cells apparently have the same receptors.

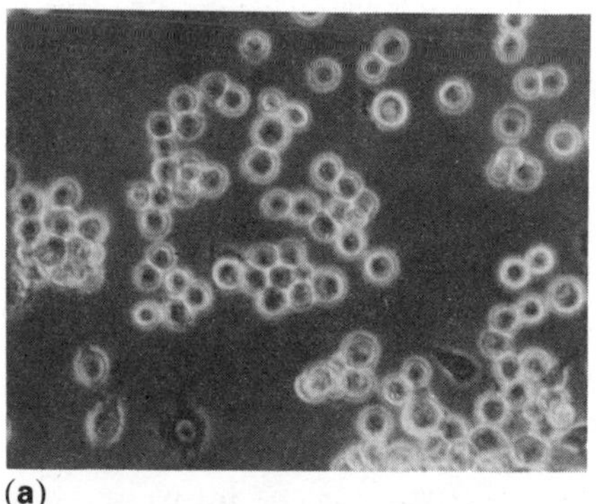
(**a**)

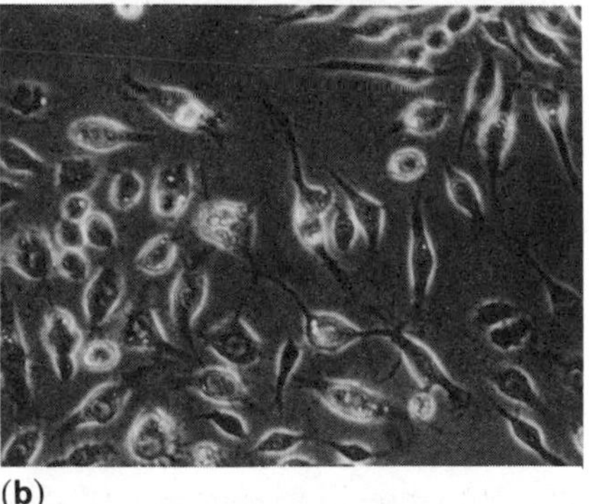
(**b**)

**Fig. 12.11** Some, but not all, *Drosophila* cell lines respond to 20-hydroxyecdysone by a change of phenotype. In this case, the Dm1 line (**a**) makes extensive processes and ceases its normal rate of multiplication (**b**). The Kc and D1 lines behave similarly and have been classed as neural-like since they show a 10 to 30-fold increase of acetylcholinesterase activity. Six or so major proteins also increase in amount (Berger *et al.*, 1978). These cells provide a useful, homogeneous, system for studying insect hormone action at the molecular level. Photographs by Dr. M. Simcox.

Primary embryonic cells do not respond to hormone, and when they are fused to the responsive cell line the hybrid proliferates and is ecdysone resistant. This change is not due to mutation and is stable; it is apparently an epigenetic event (cf. teratoma cell fusion). This first attempt to use cell fusion to probe the determined state of primary embryonic cells (Wyss, 1980) awaits further elaboration.

Cultured cells are also useful for studying the molecular changes induced by heat shock. Lindquist (1981) has so used them to make an instructive comparison between the translational control of events in *Drosophila* and in yeast (*Saccharomyces*). In both cases the overt phenomena are the same: pre-existing translation is turned off and the heat shock proteins (hsps) are synthesized. In yeast, where the mRNAs have a life of minutes, there is no selective translation of hsp mRNAs and the pre-existing mRNAs merely fade out of the protein pattern. With *Drosophila* cells, on the other hand, where the pre-existing mRNAs have a long life, there is translational control in favour of hsp sequences. This is clearly shown by the rapid recovery of the synthesis of pre-existing messages when the cells are returned to normal temperatures, even in the presence of actinomycin which blocks new RNA synthesis. There is therefore some system which sequesters the non-hsp mRNAs during heat shock, and prevents their translation. The generality of the heat shock response must mask a variety of different mechanisms for achieving the same end, and this is true of developmental systems also. The heat shock response, by blocking gene action, may account for the temperature phenocopies described earlier (Mitchell and Lipps, 1978).

Transformation of *Drosophila* cells has recently been accomplished (Sinclair *et al.*, 1983), first, using whole cell or fly DNA to rescue a shibire cell line which could not grow at 29 °C; and, second, by using a plasmid carrying a bacterial xanthine: guanine phosphoribosyltransferase (XGPRT) gene and a variant of the HAT selective medium (Fig. 12.9). Co-transformation with wild-type DNA and the plasmid was successful and, although not selected for, the plasmid circle was found to be free in the nucleus. Co-transformation of the *XGPRT*-carrying plasmid and a plasmid carrying *hsp70* genes, on the other hand, showed multiple copy integration, as previously described for mammalian cells.

The hybrid dysgenesis P element (p. 120) has been skilfully exploited in another transformation system (Rubin and Spradling, 1982). Here, the cloned element, linked to a rosy$^+$ gene, was injected into the pole plasm of rosy$^-$ eggs. Some germ cells of the hatching flies carried the rosy$^+$ gene as shown by breeding tests, and the P element was found to be integrated into the fly chromosomes, though in different sites there. Thus, it should now be possible to study particular genes, and genes which have had parts of their sequences either removed or substituted, in both cells and flies. The sequences required for the coordinate induction of, say, the yolk protein genes by 20-hydroxyecdysone, should be amenable to analysis by transformation, as should be the regulatory elements of the rosy locus (p. 139).

## Plant cells

We have seen that plant cells can be cultured and made to differentiate into plantlets by adding cytokinins and auxins to the culture medium, and this has been used commercially for the rapid propagation of plant clones and for the selection of resistant strains of various kinds (Brettell and Ingram, 1979). Anther cultures have also been developed to make haploids, but these are often mitotically unstable and become aneuploid, as do most cultured plant cells. Such haploid lines have been used to select auxotrophs and resistant lines, but their use for hybrid cell selection has not been exploited. Of course heterokaryons can be made only after removal of the cell walls by cellulase/pectinase enzymes, and these protoplasts can be fused using polyethylene glycol, or other treatments (Vasil *et al.*, 1980). Only rare heterokaryons go on to make hybrids, and these usually show selective chromosome loss, just like mammalian cells. The only hybrids which have been grown to plants are *Datura*, *Nicotiana* and *Petunia* species hybrids which are sexually compatible. They have not yet been used for developmental studies.

The transformation of plant cells has also progressed slowly through lack of a suitable vector. The tumour-inducing (Ti) plasmid of the crown gall bacterium, *Agrobacterium tumefaciens*, affects only dicotyledonous plants, and there it prevents the differentiation of infected, and immediately adjacent, cells. At least five contiguous genes of the plasmid are involved in this inhibition of differentiation, and a sixth gene, opine synthetase, which is not found in plants, provides an enzyme marker for cells carrying the integrated plasmid. Of course, plasmid-transformed cells which do not differentiate are no use for gene transmission or developmental studies. Ti is now available, however, with the five genes mutationally inactivated (Inze *et al.*, 1981) and this has been used to transfer opine synthesis to tobacco cells, and thence to plants (Ream and Gordon, 1982). The opine synthetase gene is then transmitted through pollen and egg without loss. It should now be possible, therefore, to transfer cloned genes to dicotyledonous plants and to study their developmental action. The diffusible molecules, made by the differentiation-inhibiting genes of the unmutated plasmid itself, should provide a new approach to the problems of plant differentiation since the effects of the separate mutants can now be studied.

## Conclusions

These studies show that eukaryotic unicells have evolved more complex gene control systems than prokaryotes, as Ephrussi anticipated. Some of these arrangements may still be found, or elaborated, in multicellular eukaryotes, and the importance of experiments with unicells is that they suggest what we might look for at this more sophisticated level. At present, however, we often lack critical information which would allow us to reject some models in favour of others. For instance, there is good evidence for constitutive mutations in fungi, but a dearth of similar examples among multicells. This may be due to the difficulty of recognizing and isolating such mutants, or it may reflect a different form of regulation which allows a gene to be expressed in some tissues but not in others, or both.

The importance of this partitioning of function between cell groups is readily seen in the slime moulds where cell–cell contacts and the local diffusion of regulatory molecules is necessary for normal development. The difficulty here is to know whether these molecules, such as cAMP, have similar functions in higher forms, or if these functions have been taken over by other, more complex molecules. At a different level, the remarkable regulation of macro- and micronuclei in *Paramecium* implies a very precise organization of the cellular cytoskeleton and a very local level of gene control (cf. the *Caenorhabditis* egg). Thus, unicell model systems identify important phenomena in an exaggerated form, and which are often more amenable to analysis than they would be in complex organisms. This is certainly the case for the cassette genes of yeast which exploit transposition to regulate mating type in a quite unexpected fashion. It will therefore be very interesting to find if the double-sex locus of *Drosophila* has a similar arrangement for the cell-by-cell regulation of sex. The cassette system is, as yet, a unique example of the development programme built into the organization of the DNA itself. We need more such information from unicells to guide us through the greater sequence complexities of higher forms.

Cell fusion studies have paid too little attention to changes of phenotype, both overt and molecular. The extinction phenomenon, however, suggests that 'repressor' molecules exist, but whether they affect single genes, or gene groups with coordinated functions, or whole areas of chromosomes, is quite uncertain. Conversely, the activation of *Drosophila hsp* genes in mouse cells shows that there are activator molecules, but this may indicate only a common, inherited characteristic of a physiological system of great antiquity. The equally ancient *onc* genes apparently act as suppressors of differentiation when they cause tumours, but it is not clear that this is their normal function. Both these systems may be used for the bioassay of genes, as can the fusion of mature to embryonic *Drosophila* cells (loss of hormone response), but little advantage has yet been taken of this possibility.

The transformation of cells with known, cloned, DNA sequences is likely to be a more powerful technique than cell fusion. First, isolating the sequences from a gene bank which will rescue, say, a ts lethal, should allow identification of many genes we know nothing about. Second, by removing or otherwise altering flanking sequences it should be possible to define those necessary for gene expression, including sequences common to co-induced genes, like the three hormone-induced yolk proteins of the *Drosophila* egg.

Third, transformation may allow us, with luck, to locate the structural genes for regulatory molecules which change the phenotype of the cells themselves. Our present uncertainty over the nature of these molecules is the major gap in current knowledge.

# 13 Genetics and the problems of development

Rather than search for probably non-existent uniformities such as *the* mechanism of cell differentiation or *the* mechanism of control of genic action or *the* mechanism of development, it seems to me much more realistic and fruitful to recognize the existence of a variety of these mechanisms and that they are part of a grand repertoire of biologically tested options which are potentially available for use by any organism. Each organism actually employs a certain combination of the potentially available options in its own particular programme of development. The task ahead is to discover the combination used by the organism under investigation. . . .

T. M. Sonneborn, 1970

We started by defining the problem of development as: the elucidation of the chain of mechanisms which, in orderly progression from egg to adult, results in the activation of gene sets specifying differences between cells. We must now consider how far we have progressed towards that end and what has still to be done; for most has still to be done despite the extraordinary advances of the past decade. These advances have come through the development of techniques, especially molecular genetic techniques, and largely without the benefit of theory, as Brown (1981) remarks. The methods have been employed, quite properly, to explore the organization of the more accessible genes like hormone-induced genes, histone sequences, 5S RNA, heat shock genes and the like. As we should expect, this empirical approach does not yet give us a coherent picture of development of any organism. Further, as Sonneborn (1970) emphasized, we must not expect to discover *a* mechanism of development, but rather to unravel the many mechanisms exploited by different organisms. It is too soon to decide if meaningful generalizations can be drawn from features common to a variety of organisms, but until we identify some such general principles there can be no useful theory—models for particular mechanisms are a different matter, as we shall see below. Thus, we shall adopt the same Baconian principle as the experimentalists and try to give some empirical order to the results considered earlier.

Before we do this we must emphasize a disadvantage of the pragmatic experimental approach; it may obscure the distinction between genes which are important for development, and genes which are not. In the Caplan–Ordahl model (Fig. 4.6) we made the customary distinction between housekeeping genes and differentiation-state specific genes. Among the housekeeping genes we have the subclass of universally-required genes necessary for DNA, RNA and protein synthesis and for cell division. But among the metabolic genes there may be a partitioning of function between cells and tissues of higher forms such that, for example, pyrimidine deoxyribosyltransferase is found only in the liver of the rat and not in other tissues (Knox, 1976). Or, again, mammals have two (or more) pyruvate kinase genes, one of which functions in erythrocytes/liver, and the other in the remaining tissues. Metabolic genes may therefore be inactivated in some tissues but not in others, and can therefore be classed as state specific genes. On the other hand, the state specific genes making products unique to a cell type are usually activated at a particular stage of development, although they may need a particular hormone (or other) stimulus for their expression.

As was clearly identified in the Britten–Davidson model (Fig. 4.2), morphogenetic complexity depends on the activation/inactivation of gene *sets*, not single genes; possibly in a hierarchical sequence of determinative steps. An essential concern, if the model is correct, is therefore with the genes which regulate the cell-state specific genes. This is the gene class we know virtually nothing about. In Garcia-Bellido's (1981a) terms, they should be the genes which define the 'grammar' of development, in contrast to the 'phonetics' of molecular studies. The nearest we have come to identifying such a grammar has been in the studies of transdetermination, of teratomas, of homoeotic genes and of sex. There is no reason to assume that this gene class (activation genes) will be regulated in the same ways as the gene classes described in the preceding paragraph, or that these classes are similarly regulated among themselves. Nevertheless, if we are to progress from phonetics to a grammar, it is instructive to start by considering the molecular requirements for gene activation and then proceed to the greater complexities of morphogenetic genes.

## Active genes

Transcriptional activation of a gene depends on the initial decondensation of a domain of the chromosome to be transcribed; most obviously seen in lampbrushes and puffs. Genes packaged in nucleosomes are apparently inactive. However, the DNase I test, which shows actively transcribed genes to be preferentially susceptible to digestion, is not absolute, for the ovalbumin gene is still sensitive in the hormone-withdrawn oviduct (Palmiter *et al.*, 1977). Decondensation indicates a potential to be transcribed (competence?) and is the first change in the chromatin which we can recognize. It is a primary step in gene activation. The non-histone proteins released by this DNase I digestion are predominantly of the high mobility group (HMG), and chick erythrocyte chromatin depleted of them is no longer sensitive to DNase I digestion at the globin gene (Weisbrod and Weintraub, 1979). This, and other experiments reviewed by Weisbrod (1982), strongly suggest that chromatin decondensation depends on assocation of non-histone proteins with chromatin which, in the case of the HMGs, causes the displacement of histone H1. This is not by itself a tissue specific process since the same HMGs are found in different cell types.

Specificity in decondensation could arise in two ways. Sequence-specific DNA-binding proteins are known (Jack *et al.*, 1981), but they are difficult to identify since many proteins bind to non-specific DNA sequences and coincidental binding is hard to disprove. We have no certain evidence as to how such proteins might act. Similarly, we do not yet know how hormone receptor–protein complexes activate genes with which they associate, although it is clear that the ecdysterone–receptor complex binds to the *Drosophila* salivary gland puffs which it induces (Schaltmann and Pongs, 1982). On the other hand, specificity may reside in the many post-synthetic modifications (acetylation, phosphorylation, methylation or poly-ADP-ribosylation) which can alter the properties of nuclear proteins or of the DNA. These particular possibilities have still to be explored.

We can also ask if genes can be inactivated, which is more difficult to discern than activation. If many somatic cell nuclei are injected into the large nucleus of the *Xenopus* oocyte, they enlarge and rapidly synthesize RNAs (cf. p. 313), but do not divide. Some, but not all, of the mRNAs are efficiently translated so one can ask if new genes are activated and others inactivated, as shown by the new patterns of cytoplasmic proteins. Generally, the results suggest that nuclei adjust to the pattern of transcription of the oocyte, suggesting that this cell contains components which inactivate, as well as activate, genes. This change occurs even when the donor nuclei come from unrelated species, since HeLa nuclei change their pattern of protein synthesis when so transplanted (DeRobertis *et al.*, 1977), and there must be some generality of these controls. Of course, these results do not disprove that the oocyte nucleus activates all genes, for it may be that post-transcriptional events determine which proteins are synthesized in the cytoplasm (see Darnell (1982) for a general review). The oocyte nucleus has its own commitment to protein products, which may set limits to the utility of this transplantation technique for exploring gene activation/inactivation.

The *Xenopus* oocyte nuclear injection procedure, which leaves the genes in their normal chromosomal environment, can be exploited in two further ways. The *Xenopus* oocyte-type 5S RNA is not transcribed in somatic cells (p. 54). If erythrocyte nuclei are injected into eggs they are transcribed as already described, and we get oocyte-type gene activation without nuclear division. However, there is a class of female which makes non-activating oocytes, and this transcription does not occur in these. If the erythrocyte nuclei are pretreated with 0.35 M NaCl, which removes a minority of chromosomal proteins prior to injection into the non-activating eggs, the oocyte-type 5S RNA is transcribed (Korn and Gurdon, 1981). The chromosomal proteins determine the transcription pattern, and by using this system for bioassay it should be possible to identify the activating component in normal eggs (by injection into non-activating eggs) and the chromosomal proteins removed by the salt extraction.

The same *Xenopus* oocyte system (Fig. 13.1) can be used to explore the DNA organization necessary for transcription, since injected circular molecules quickly form into nucleosome-organized mini-chromosomes, a proportion of which are transcribed within 1 hour. This transcription is not regulated, is generally, but not invariably, from the correct coding strand, but may be inefficiently spliced and terminated (Gurdon and Melton, 1981). There is an exception to these generalities: the *Drosophila hsp70* sequences spliced into phage λ are not transcribed until the oocyte is heat shocked. Isolated, cloned *hsp* sequences are irregularly transcribed at 20 °C, on the other hand (Voellmy and Rungger, 1982). There are therefore some gene organization arrangements which should permit the study of the regulation of cloned genes inserted into the nucleus.

In this system RNA polymerase I transcribes rDNA, polymerase II transcribes histones and other structural genes, and polymerase III transcribes 5S and tRNAs, in the normal fashion. As we have seen (p. 54), there is

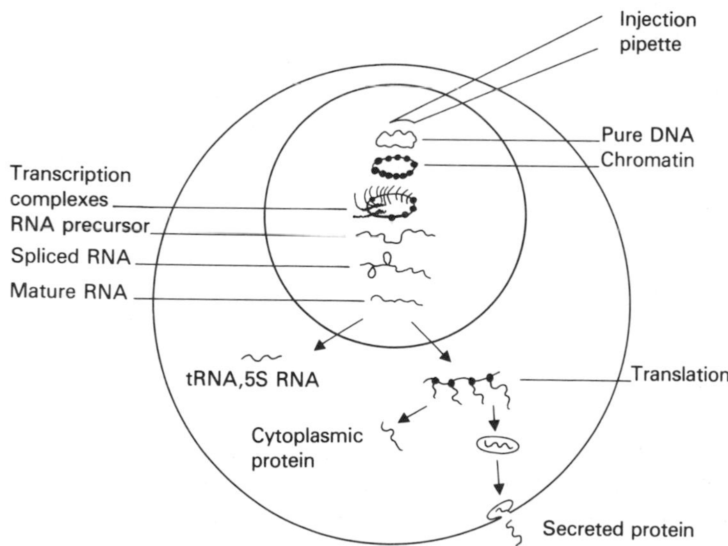


**Fig. 13.1** Diagram of the sequence of events which occurs when purified DNA or RNA is injected into a *Xenopus* oocyte, after Gurdon and Melton (1981). The egg is 1.2 mm in diameter, and the nucleus one third of this and located nearer the upper, pigmented pole. Injection is facilitated by centrifuging the eggs to bring the nucleus into sight on the surface (Kressman and Birnstiel, 1980). See text for further details.

an internal control region for the transcription of 5S RNA, and a further transcription factor IIIA also binds to this central region of the gene (Engelke *et al.*, 1980). Similarly, there is an internal promoter for tRNA transcription; and transcription terminates at a row of T residues in both classes. The sequences required for RNA polymerase II transcription are more complex since deletions as distant as 200 bases upstream from the start site of histone H1 reduce the RNA formed to one fourth (Grosschedl and Birnstiel, 1981).

The role of these 'prelude sequences', upstream from the initiation site, has been carefully explored using the Herpes simplex *tk* gene, which is the gene most faithfully transcribed in the *Xenopus* oocyte system. The method is to substitute *Bam*H1 linker sequences (10 bases) for normal sequences, as shown in Fig. 13.2. This has been done at 18 sites along the DNA, from the initiation site up to –105 bases. Changes to the normal sequence at, and around, the initiation site do not block transcription, although it usually starts somewhat higher in the substituted sequence. As might be expected, alterations to the TATA box reduce, or nearly eliminate, transcription. The exciting result is that substitution between –47 and –61, and again between –80 and –105 lowers transcription to a tenth and a twentieth, respectively. Alterations between these sequences are without effect on transcription, as are deletions above –105 or within the structural gene itself (McKnight and Kingsbury, 1982). Unlike RNA polymerase III, polymerase II is regulated by DNA sequences distant from the structural gene, which may facilitate its entry to the gene by allowing contact with the TATA box which directs initiation of transcription.

It seems likely that this study defines the general relationship between RNA polymerase II and the DNA sequence organization generally necessary for the transcription of structural genes since a similar upstream promoter element has been found at –47 to –66 residues for the *Drosophila hsp70* gene (Pelham, 1982). Further, a glance at the relevant *tk* upstream sequences (Fig. 13.2) suggests that the two regulatory regions might come together to form a hairpin loop. If this is found, DNA conformation may be important for transcription regulation, too.

Our understanding of the sequence signals for termination is less precise. Hofer and Darnell's (1981) discovery that transcription of the mouse β-globin gene runs some 1300 bases beyond the 3′ end of the gene confirms the general conclusion that the average hnRNA is very much larger than gene-sized, even allowing for introns. It therefore seems that RNA polymerase II terminates at some, as yet uncertain, sequence distant from the structural gene, that an endonuclease cleaves this to 15 bases 3′ to the AAUAAA sequence, and this new 3′-end is polyadenylated (Proudfoot, 1982). Until we are more certain about the function of this 'extra' 3′-flanking DNA, if it has any function, we cannot redefine a gene as the unit of transcription, although such a definition would bring together the molecular and genetic approaches.

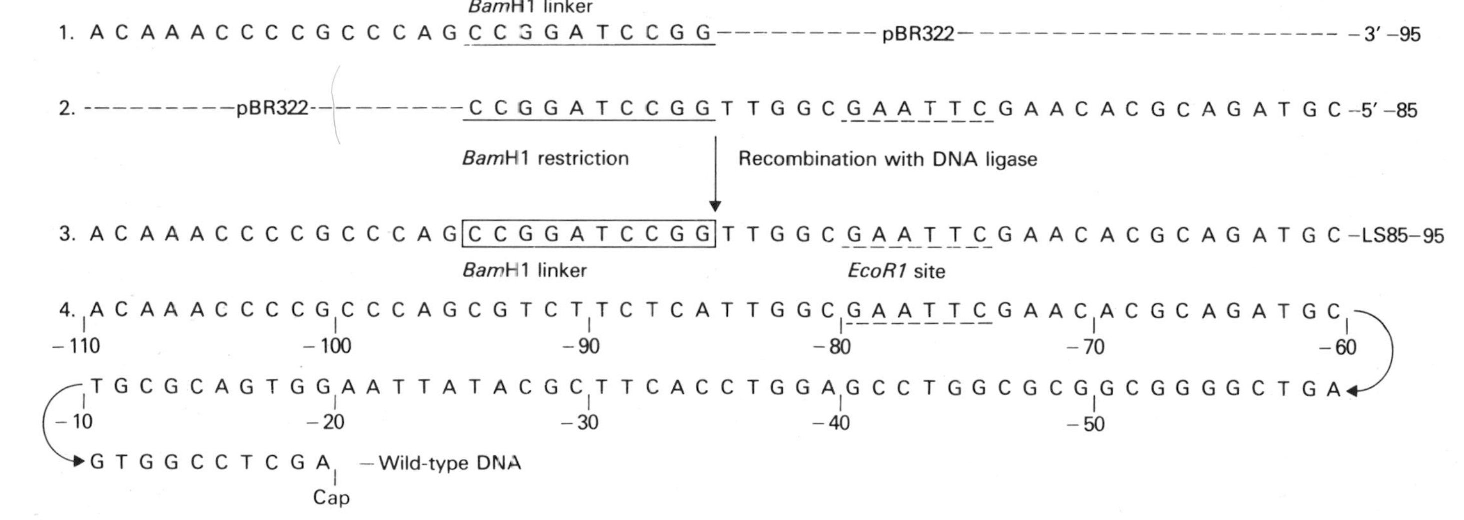


**Fig. 13.2** The linker scanning technique devised by McKnight and Kingsbury (1982) to produce mutants at specific sites in the thymidine kinase gene. A series of deletion mutants is made using exonuclease III and S1 nuclease, and a synthetic *Bam*H1 restriction site is added to the ends of the 3′ and 5′ deletions before they are incorporated into the plasmid pBR322. The sequences of the cloned *tk* DNAs are determined, and matching 5′ and 3′ deletion mutants are identified whose deletion termini are separated by 10 nucleotides. Such a matching pair is shown in the upper sequences, where the 3′ deletion is at −95 (1) and the 5′ deletion at −85 (2). In (3) restriction with *Bam*H1, and recombination with DNA ligase then substitutes the *Bam*H1 linker sequence for the wild-type sequence (see 4) between −85 and −95. An *Eco*R1 site is indicated and was used for end labelling. The mutant and natural sequences were compared by the chemical method of nucleotide sequence determination thus confirming the site-directed mutagenesis. The process is repeated for other matching pairs, and the role of flanking sequences can be determined, as explained in the text. The entire wild-type upstream sequence is given in (4).

Since the *tk* gene is not regulated in the *Xenopus* system, we cannot identify the DNA sequences which may be necessary for gene activation/inactivation, and hence the sequences which regulate cell-state specific genes. We need cloned genes which can be activated/inactivated in one or other of the transformation systems before the linker-scanning technique, or other methods of directed mutation (Shortle *et al.*, 1981), can be applied to this problem. With such systems it should be possible to define the essential DNA organization required for cell-specific gene transcription, and then to pursue further the problems of chromatin organization. These are major topics now on the molecular geneticist's agenda. However, if many loci have the organizational complexity of bithorax, with introns and inserted sequences extending the gene over many kilobases of DNA, there will be formidable problems to overcome.

## Gene activation

The unequal distribution of maternal gene products during the first divisions of the egg is usually taken as the first step towards differential embryonic gene activation (Gurdon, 1981). We can now elaborate on this. In mosaic eggs (e.g. *Caenorhabditis*), the initial differences are stable and comprehensive, and many morphogenetic maternal-effect mutations can be identified (p. 184). In regulative eggs on the other hand, a few initial differences are reinforced and elaborated by subsequent cell interactions of various undefined sorts during embryogenesis. And in the extreme case (e.g. mammals), initial differences are hard to discern, morphogenetic maternal-effect mutations have yet to be described, and stable cell differences first appear during early embryonic stages. The factors which activate/inactivate cell-specific genes are laid down in some eggs, whereas their elaboration in other eggs depends on interactions among the early division products. Or in some cases both strategies are practised. These different classes present us with different practical problems, but they all present us with a major difficulty: we do not know what we are looking for. Nor can we be sure that the same mechanisms operate at these different developmental stages. This is a major aspect of Sonneborn's programmatic diversity.

The products of maternal-effect genes with activation/inactivation effects will be difficult to identify since most of the highly determinate eggs are small, and most of the larger eggs have extensive RNA and protein populations. Inhibition of RNA synthesis in the *Xenopus* oocyte has no effect on the first 8–9 h of development, for example, by which time the embryo has 30,000 or more cells; and maternally formed mRNAs are still being used by late sea urchin embryos (Davidson *et al.*, 1982). There are exceptions to these rules, but they are not very helpful. For example, new mRNAs are synthesized by *Drosophila* blastula cells (Anderson and Lengyel, 1979), but we have seen that these cells, or some of them, are determined by this stage. Two approaches are being used to overcome these difficulties. At the molecular level, cDNA clones are being made from polysomal mRNAs formed during early defined stages of development (Bruskin *et al.*, 1982; Sina and Pellegrini, 1982), in the expectation that stage-unique clones may have identifiable effects and may relate, in the case of *Drosophila*, to known genes marked in the salivary gland chromosomes. Alternatively, maternal-effect mutations are being examined to ensure that their apparent effects are not the indirect consequences of mutant housekeeping genes; the 'good' ones are being cloned, and their RNAs and proteins studied. This seems the more profitable approach, but it ignores important aspects of this central problem.

Maternal-effect gene products (synthesized by the oocyte?) must themselves be regulated; and they must become spatially organized in the egg if they are to be precisely positioned in its division products, although there is little evidence of this (Carpenter and Klein, 1982). We have cause to wonder if our assumptions are correct when we have to ask: what regulates the regulators?

The problems of identifying and defining the determinative events of early embryogenesis are also difficult to approach. Embryological studies show that most changes involve small groups of cells, their division patterns, and the contacts they make with each other as separate clones or polyclones. Except for plant hormones and the cAMP of slime moulds, we have no information about the effector molecules involved, and we lack the ultra-microtechnology which might identify them. Here, too, there is an additional problem, since the generality of the effects of the plant hormones (and cAMP) suggests that responsive cells have already progressed to a state of competence to respond. In short, the embryologists' old problems of evocators and inducers and of responsive tissues have not yet yielded to molecular-genetic analysis, although there must be many lethal and semi-lethal mutations, yet to be identified, which affect these systems.

A variety of extrinsic agents have been shown to induce differentiation in cloned cells, ranging from the switch to erythroid differentiation of Friend cells exposed to dimethylsulphoxide and the differentiation of keratinocytes plated in methylcellulose-containing media, to the induction of prolactin synthesis of pituitary tumour cells exposed to 5-bromodeoxyuridine (Biswas *et al.*, 1979). In all cases, only a proportion of the cells responds to the treatment, apparently at random, and they always change from one state to the other differentiated state. The cells are programmed and the agent activates the programme only in some cells, it would appear (but see Levenson and Housman (1981) for a more general discussion). There is the case where differentiation of a mammary cell line is induced by a physiological regulator, the saturated fatty acids, butyric and myristic acids (Dulbecco *et al.*, 1980). These acids cause this carcinoma line to undergo a normal morphological differentiation and to form domes of tissue; and the fatty acids are normal, accumulating, cell products. However, as with all differentiated cell lines, we are probably observing here only the ultimate event in a sequence of changes which has already led to cell commitment.

A more startling manipulation has been reported which changes the post-embryonic development of the brine shrimp, *Artemia salina* (Hernandorena, 1980). These shrimps can be cultured axenically in an artificial medium, where they grow normally. However, if the purine/pyrimidine balance of the diet is manipulated to cause a relative purine deficit, supernumerary limbs (gonopodes) develop on the abdomen which is naturally always devoid of limbs (Fig. 13.3). It is not yet known if this induced homoeosis is due directly to a deficit of adenylic or guanylic acids, or of a product such as a pteridine; but as with other homoeoses we have here an atavistic change which shows that the development of abdominal limbs is usually suppressed (cf. bithoraxoid).

These two exampies suggest that normal metabolic products may sometimes be involved in the regulation, both positive and negative, of gene expression. We do not know whether they act directly, or indirectly by

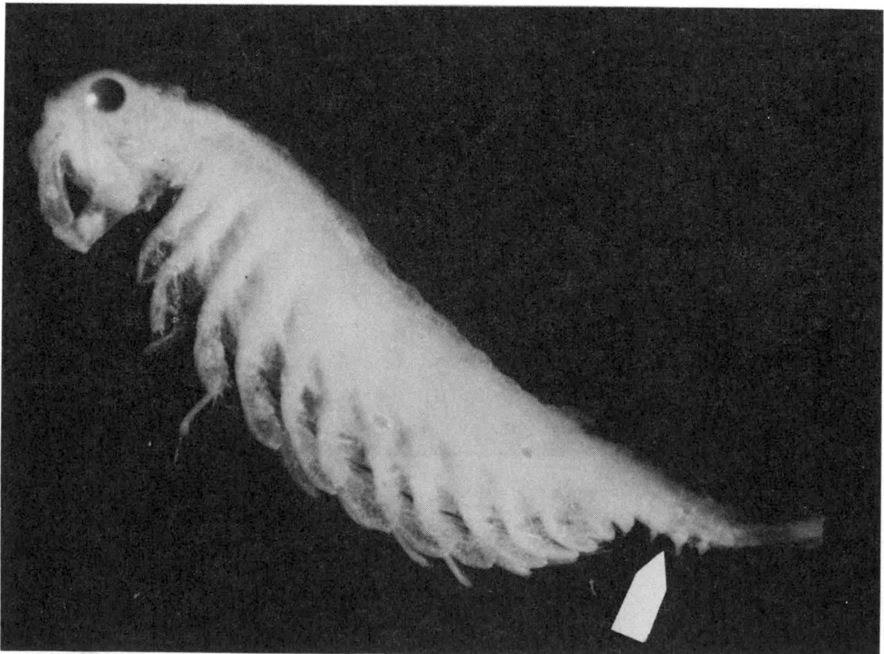

**Fig. 13.3** The brine shrimp, *Artemia*, has no abdominal legs, but when larvae are grown on diets relatively deficient in purines supernumerary gonopodes develop on the abdominal segments (arrowed). The Utah strain used cannot synthesize the purine ring *de novo*, and a dietary deficiency of adenylic acid during early larval stages induces this homoeosis. A pyrimidine deficiency does not; but excess pyrimidine does unless balanced by an increase of dietary adenylic acid (Hernandorena, 1980).

activating other regulatory genes after complexing with say, the sensor of the Britten–Davidson model. But we obviously need more information about activator molecules which will allow us to manipulate systems, as the *Artemia* system can be manipulated. Without this we cannot satisfactorily test any of our models of gene organization, or how genes are regulated in the egg or in the early embryo. This is a most significant gap in present knowledge.

## Genetic analysis

Our models of gene control (Ch. 4) are hierarchical and postulate master-switch genes which activate developmental modules, such as eye or limb modules. Brenner (1981) has pointed out that these subsystems may not be independent, but may have common components and may interact with one another. In that event, development may proceed from a rough form to refined structures, just as an artist first blocks in the main features of a portrait and then remodels them. The embryologist's theory of 'fields' suggests such a sequence. The same genes might then be involved in the elaboration of different fields and also in their subsequent development. Also, we do not at all understand the rules which relate gene sequences to the phenotypic form of their product, or as Brenner puts it, how to map "genetic space on to organismic space". An interesting, if theoretical, example of this second problem is provided by Oster *et al.* (1980; 1983), who have explored the properties of a viscoelastic model of the cell (Fig. 13.4). In a blastula-like ball of such cells, a change in shape of one cell (possibly due to the effect of $Ca^{2+}$ on the actin gel) causes a successive firing of shape changes in neighbouring cells by mechanical action alone, and the formation of a 'gastrula'. All that is required is the activation of the initial cell and 'development' proceeds without positional information programmes or clocks, at least in the computer model! This was not predictable from conventional descriptions of the cell, and it is salutory to remember that cells have intrinsic properties affecting their shape and motility (Weeds, 1982), not readily predictable from their active genes.

In principle, genetic analysis should allow us to resolve some of these difficulties, and our final business is to explore this proposition. The rapid evolution of techniques will provide more tools for this task, but at present, we still have problems. Except for mutants with obvious morphogenetic consequences, like the homoeotic mutants, it is not clear what priority we should give to the pursuit of some genes rather than others, when our concern is with development. A mutation like rudimentary which has a secondary morphogenetic effect via pyrimidine metabolism, or a structural gene mutation that perhaps disturbs development by blocking the synthesis of a muscle myosin, and others like them, would fall in the low priority

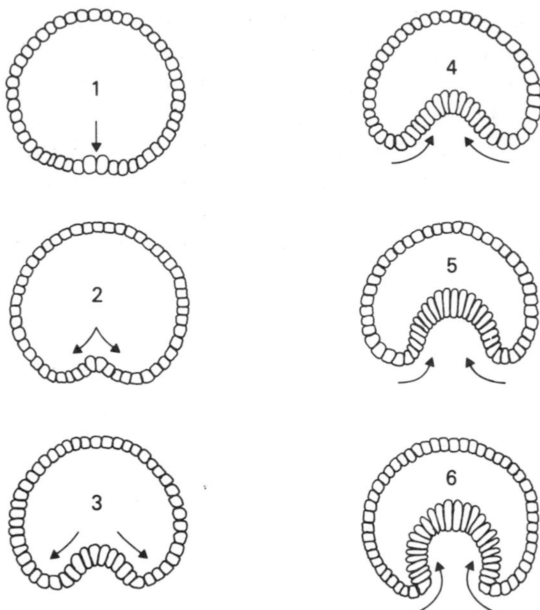


**Fig. 13.4** The computer-model sequence of changes in a ball of cells having the viscoelastic properties of an actomyosin gel, after Oster (1982). When the bottom cell is triggered (1) a wave of contraction spreads outward (2 and 3) creating a circle of bending movements which invaginate the epithelium layer (4–6) mimicking the gastrulation movement seen in a sea urchin blastula. Note that the cells which initiate gastrulation in the sea urchin are specified at the end of cleavage. The model assumes such an 'active' cell, and that the remainder are passive and respond only as a consequence of the mechanochemical properties of their cytoskeleton.

class; but we have to recognize that we understand their indirect role only after they have been studied. The attraction of the homoeotics is that they seem obvious candidates for developmental regulatory genes, but the *Artemia* phenocopy warns us that this has still to be proven. It is therefore not surprising that Brenner (1981) concludes that "... answers to questions about the relations between genomes and complex organisms will come from detailed knowledge of the structure and expression of individual genes and from an insight of how their products participate in the biochemical and cellular processes underlying development". Must we take this conservative view?

Meinhardt (1982), after surveying mathematical models of pattern formation, suggests an alternative proposition which recognizes how development progresses from very general to special local events, as we saw in Ch. 1. The first step in the sequence is establishment of the embryonic axes, followed by the separation of somites or segments etc., and then by the segregation of some cells as the primordia of particular structures and, within these, a further, and finer, cell identification as precursors of ultimately differentiated hairs, sensilla, stomata or other constructs. New patterns are built on old patterns and are the foundations for further subpatterns, and this has only to be said to be obvious. Where we have genetic data (see e.g. Ch. 10), the evidence is that particular gene functions are necessary for the implementation of each step of this developmental progression, just as each phase can be described by a different mathematical model. Before we anticipate difficulties, we must therefore first explore the total genetic context of each stage, or phase, of development. For the common experimental organisms, this means filling the gaps in present knowledge.

Conventional genetics starts from the identification of mutant phenotypes, and the earliest and most general of these phenotypes are embryonic. We would argue that priority should be given to the study of early embryonic mutants, and first priority to morphogenetic maternal-effect mutants. We have seen that perhaps a hundred or more genes control maternal effects in *Caenorhabditis*, only a dozen or two in *Drosophila*, and possibly none in mammals. In the first case, virtually all the structure of the organism is determined by maternally-acting genes, although later-acting mutations may cause disturbed division patterns which superimpose altered developmental pattern details (Sulston and Horvitz, 1981). In the second, maternal controls determine body polarity (shown by the mutants bicaudal, dorsal etc.), germ cell determination (Illmensee *et al.*, 1976), segment number (Newman and Schubiger, 1980) and the competence of segments to differentiate (extra sex combs), matching the small number of genes involved. This inadequate sample suggests an evolution of embryological controls from maternal determination to the zygotic gene determination of higher forms.

For technical reasons, we can expect the *Drosophila* maternal-effect mutants to be understood first, especially as some of the genes are being cloned and a bioassay is available for the dorsal group (p. 169). Some mutants come into Brenner's (1981) category of genes which affect a set of kernel processes, like the dorsalization of cells, which are followed by a refinement process involving the activity of zygotic genes. But this is not the only kind of maternal effect, for the mutant, fused, regulates segment polarity and polar granules control the immediate differentiation of germ cells. The *Drosophila* maternal-effect mutants provide an entry to a number of different processes which underpin determination, and while we cannot yet say that they are switch genes, in the strict sense of regulating gene batteries, they at least delimit the potential of genome activity.

It is more difficult to identify morphogenetic mutants among the mass of early embryonic (zygotic) lethals, many of which must be the consequence of metabolic defects. Again homoeotics provide an obvious class of choice, and they have already given one answer to an important question. Polycomb has an apparently identical action to extra sex combs, but mitotic recombination shows that its activity is required as an activator/repressor of BX-C throughout development. So also is the opposite action of *Rg-bx* (p. 258). It follows that expression of *bx* genes depends on the continuous action of this postulated repressor/activator system and not on some permanent *cis*-change in the chromatin structure of the genes. The cell's 'memory' is not built into its chromatin and cell-inherited in this way, but depends on the continuous function of putative regulatory genes. The identification of the products of *esc, Pc* and *Rg-bx* is therefore of great interest, and should soon follow the cloning of these genes. We shall then see if the postulated gradients of these *trans*-active products are real or not.

It may be that continuous control is a more general phenomenon than, for example, the Caplan–Ordahl model of gene organization (p. 112) assumes, since the sex phenotype genes are also required throughout development in *Drosophila*, and hormone balance can change sex phenotypes even in adult vertebrates. However, some hormones may act as switches, for while male *Drosophila* make yolk proteins only if continuously exposed to 20-hydroxyecdysone, females require only a single pulse of the hormone. And the morphogenetic changes of pupation similarly require only one pulse. Thus, in some situations hormones cause permanent changes in transcription while in others they do not, and *Drosophila* yolk synthesis may provide a minimal genetic difference system for exploring this dichotomy.

Other early-acting mutants include the segmentation group (p. 172) in *Drosophila*, which owe their repetitive nature to that characteristic of the embryo, and Notch and almondex which alter early determination in favour of nerve (ectoderm) at the expense of mesoderm (muscle). Precisely how far these changes depend on alterations of primary determination and how far on subsequent changes in proliferation patterns is not yet clear, but the latter may be important in some cases. For example, in *Caenorhabditis* some larval mutants repeat parental and grandparental division patterns (Fig. 13.5), with complex consequences for neuroblast lineages or for larval development (Chalfie *et al.*, 1981). They may also have pleiotropic effects and alter the number and position of a variety of body cells. Thus there are genes which prevent reiteration of cell lineages. If arthropods have evolved by segment duplication (and elaboration) as Lewis (1978) suggests, and neurones by cell duplication (Goodman, 1977), the repeat division patterns exposed by these mutations must be important. Similar dependence on a single active gene of the stem cell pattern of asymmetric division is illustrated by the continuous proliferative divisions of brain neuroblasts in the lethal giant larva mutation, and by the tumorous growth of the classic female sterile mutant ovaries. These two fundamental mechanisms for segregating developmental potential (symmetric and asymmetric division) are common to all multicellular organisms. The mutants show they are under gene control, but whether or not the genes act as master-switch genes or merely precipitate (or block) a cascade of sequential gene action will be difficult to sort out since most of these genes are cell autonomous.

The difficulty of identifying control genes, and their action, increases as developmental complexity progresses. What genetic analysis suggests, as we have seen, is that most of the important control genes are maternal-effect

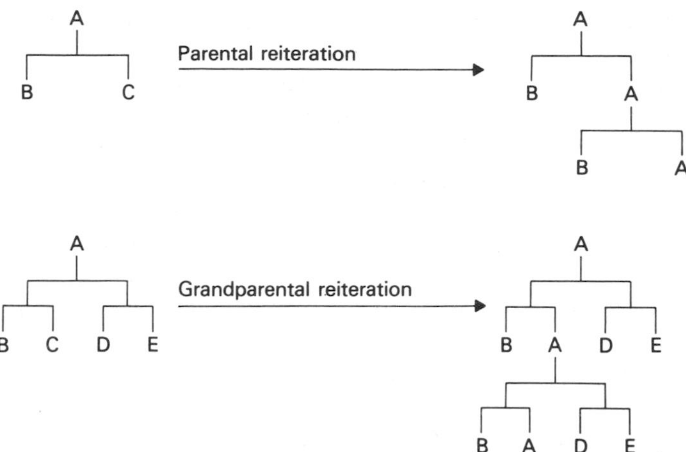


**Fig. 13.5** In a number of situations a parental cell, A, may divide to produce two differently determined cells, B and C, which may proliferate or differentiate. In the *unc86* mutants of *C.elegans* there is, instead, a parental reiteration which disturbs the lineages of post-embryonic neuroblasts. Note that *Drosophila* neuroblasts normally undergo this parental reiteration. The second situation is where four cells are formed after two divisions to give different products, as in the case of the insect bristle precursor, and mutation may cause the grandparental reiteration shown, or some variant thereof. There are also examples where the B, C, D, E determined or differentiated stages are not achieved, and the progeny of A proliferate in tumour fashion (based on Chalfie *et al.*, 1981).

loci in simple forms like *C. elegans*, whereas in *Drosophila* these are few and the homoeotic loci take over the control of adult differentiation. We lack comparable mutants that might tell us the temporal and spatial sequence of determinative events in vertebrates. This general statement conceals the fact that we do not know in any detail what these apparent control loci do: how they regulate cell number in a tissue or organ; how they create tissue and cell variety of the kind seen, say, in a *Drosophila* limb. At least for the homoeotic loci, there seems to be no ambiguity about their function as switch genes—bithoraxoid$^+$ blocks leg disc differentiation and causes histoblasts to differentiate instead—but this, and Lewis' model (Fig. 10.7), leave all the steps between to be explained. Similarly, while we know that *Drosophila* blastoderm cells, and parts from mouse primitive-streak embryos, develop autonomously, we do not know how this happens. We need some systems where these missing steps can be followed in simple terms.

These systems need not be the simple ones described in the preceding chapter, although these may be useful in this context. The important point is that they should be open to genetic analysis and our final example is a complex one which illustrates this point. One long-studied genetic system is the bristle (chaeta) patterns of the Drosophilidae, which are so invariant that they are used as taxonomic characters. These patterns are not altered by extremes of culture temperature or by malnutrition, and departures from the standard are all due to gene mutations (including modifier mutations, which we shall ignore here). In *D. melanogaster* the pattern on the dorsal mesothorax (notum) is made up of about 100 microchaetes (hairs) and 11 precisely positioned macrochaetes (bristles) among some 10,000 epidermal cells (Fig. 13.6). There are the same differentiated structures on tergites and legs, but we need not consider them. These chaetae are sensory organs, with a shaft sited in a socket, and with two internal nerve components, each separately differentiated from the four division products of a single cell (Lawrence, 1966). We shall look first at the genetics of this regular pattern and then at the development of the chaeta unit.

Although many genes affect chaeta patterns the most important are the achaeta-scute complex at the tip of the X chromosome (Fig. 13.7), hairy on the third chromosome and extramacrochaetae, also on the third chromosome. Garcia-Bellido (1979) has recently provided a detailed analysis of the achaeta-scute region, using point mutation, deletion and inversion to explore its structure. Centrally, there is a lethal of scute (*l-sc*) region, and a tandem reverse repeat of independent scute elements on either side, mutants, or

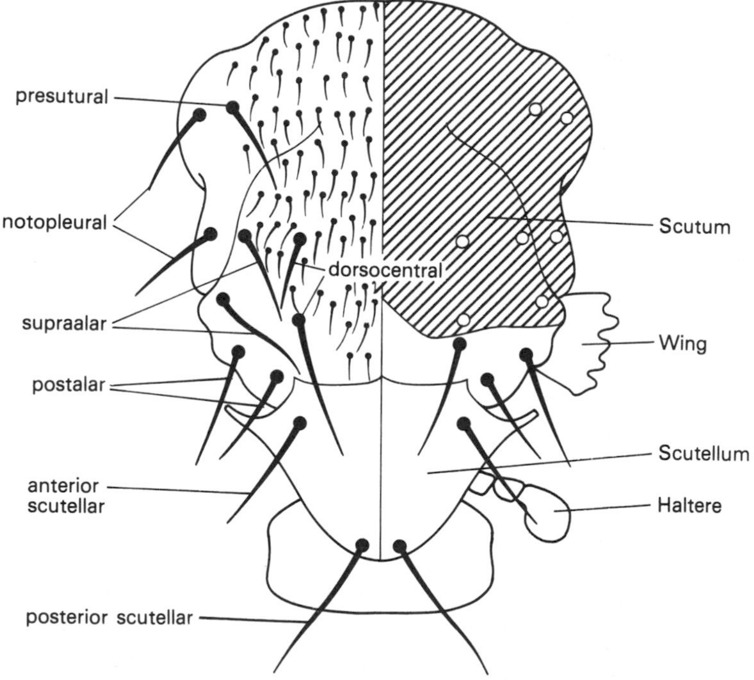


**Fig. 13.6** The 11 macrochaetes found on the hemi-mesothorax are shown on the left half of the diagram, together with the smaller microchaetes. The hemi-thorax shown on the right is mosaic for the *achaeta* mutation anterior to the posterior dorsocentral bristle (shaded area), this sector only just includes this bristle, and a bristle is formed in the wild-type tissue adjacent to the normal bristle site.

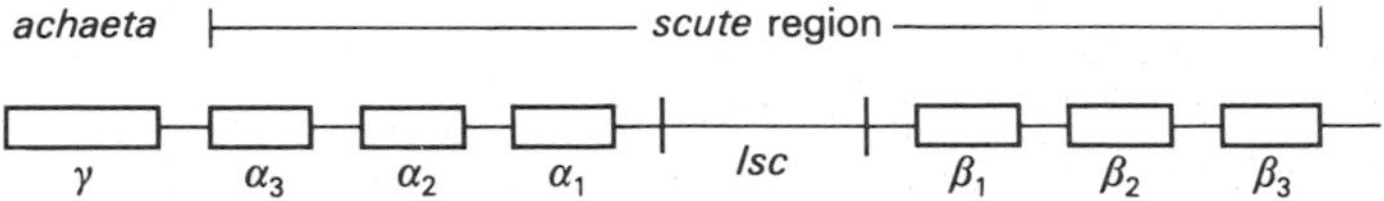


**Fig. 13.7** The *achaeta-scute* region is located between chromosome bands 1B1 and 1B4–5. Duplications and deficiencies for this region identify a centrally located lethal site (lethal of scute *lsc*), with tandem reversed repeat *scute* elements ($\alpha_{1-3}$ and $\beta_{1-3}$) on either side. The *achaeta* element (γ) lies on the left. 62 kb of the *scute* β-region has been cloned (Carramolino *et al.*, 1982). As with BX-C, known 'point' mutants are found to involve large deletions and insertions (gipsy transposon-like elements or such) and there is no evidence of repetitive elements. This similarity suggests that the region should now be named the *achaeta-scute* complex locus (AS-C) for we again have very varied sequence changes causing the same phenotypic effects.

deletions, or double heterozygotes of which remove macrochaetes. There is no evidence that the elements of the scute (*sc*) group are responsible for the differentiation of particular macrochaetes, and the fact that some combinations show weak complementation suggests that their various products may form a complex multimeric protein, though this is uncertain. The achaeta (*ac*) locus is distal to the scute region, and mutations or deletions remove mainly microchaetes. Duplications within the region (and a putative point mutation) result in a phenotype first described as Hairy-wing (*Hw*), and which is also expressed as extra microchaetes on the notum. This is dominant in the presence of multiple copies of the wild-type locus, and thus behaves as an overproducer or operator-constitutive mutation (p. 125). Deletion of the entire locus is, of course, lethal in homozygotes, and focus mapping (p. 203) pinpoints a primary action on the embryonic central nervous system, which is degenerate (Jiminez and Campos-Ortega, 1979).

Like *Hw*, hairy (*h*) mutations produce extra microchaetes which are similarly distributed over the notum. In flies homozygous for *h*, heterozygous *ac* reduces the number of extra hairs, homozygous *ac* eliminates them and extra copies of *ac-sc* increase the number. *sc/sc* is without effect. This suggests that $ac^+$ is the structural gene and $h^+$ its regulator (repressor), as suggested by Falk (1963). The extramacrochaetae (*emc*) locus is quite separate from *h*; its mutation increases the number of macrochaetes, as its name implies. Deletions of *sc* (but not *ac*) eliminate the mutant activity, and extra doses of $sc^+$ in $emc^1/emc^2$ non-complementing mutants increase the number of macrochaetes. The data are compatible with the assumption that $emc^+$ represses the $sc^+$ structural gene. It is particularly interesting that the maximum number of macrochaetes produced by the last mutant combination appear in the pattern found in the genus *Leptocera*, which may be the archetypal pattern (see Garcia-Bellido, 1981b). If this is correct, evolution is associated with gene repression, as with homoeotics.

We can now ask how chaetae develop. Gynandromorph analysis shows that neighbouring chaetae may derive from different lineages, and that the pattern does not arise from a spatial redistribution of cells of the same origin, but as a particular event within a clone of epidermal cells (Sturtevant, 1929). Clonal analysis using the *ac-sc* deletion, which eliminates notal chaetae, shows that mutant clones established 24 h prior to pupation contain microchaetes (i.e. the cells are irreversibly determined as microchaetes somewhat earlier) and that clones made 48 h before pupation contain macrochaetes, but not microchaetes (Garcia-Bellido and Santamaria, 1978). The 100 microchaetes are therefore determined when the notum region contains about 1000 cells, and the 11 macrochaetes when there are about 250 cells in the anlage.

Garcia-Bellido (1981b) argues from this calculation that 1 epidermal cell in 10 becomes determined as a microchaete and about 1 in 20 as a macrochaete. Since these determined cells would be radially separated from one another by only two or three cells, it is unlikely that positional information is involved. So how do epidermal cells decide which will form chaetae? The classic studies of Stern (1954a, b) show that when an early induced clone of achaeta mutant cells covers a bristle site no bristle is formed, but when the border of such a clone included the bristle site only marginally, a non-mutant bristle forms some distance from the bristle site (Fig. 13.6). This suggests that any chaeta position is first defined as a small region and then refined to a single cell. Since wing epidermal cells divide once every 8 h, the region may be a clone of determined cells which compete for some diffusible morphogen, the chaetogen of Ghysen and Richelle (1979), (or in the destruction of the *h* or *emc* repressors) such that only one member of the clone is ultimately activated.

Although many details have still to be worked out, this system is important as it shows that cells earlier determined as wing, and probably subsequently identified as notal epidermis, become further restricted as potential macrochaetes or, later, as potential microchaetes. This last step, which depends on the pupation hormone for its final expression, results from some local gene action within small groups of cells. In Garcia-Bellido's (1981) phrase, 'cells talk to cells' at this detailed level of sequential differentiation; but what language they use, and what stimulates its expression at that particular time in development, is unknown. What matters, however, is that we have all the descriptive and genetic information needed to pursue what is a very general developmental phenomenon, particularly important for studies of pattern elements. We need more systems of this sort for the following reason.

This last example completes the circle, so to speak, by again asking the question we started with: what molecules cause the *ac* and *sc* genes to be released from their respective repressors? Although this question may imply a simple prokaryote-type of control system, much more is involved. A detailed pattern is laid down in the determined chaetal precursor cell such that each of the four terminal cells has a different fate: a bristle, a socket, a nerve and a glial-like cell. Thus there is a final level of problem which is very difficult to comprehend. How do the chromosomes from two successive cell divisions come to have different active genes? Perhaps the answer lies in the complexity of the *ac-sc* locus; and the accumulated mutants there should allow this to be explored. However this may turn out, the obvious deficiency in our knowledge concerns the non-DNA molecules which successively determine blastoderm cells to be wing, to be notal cells, to be epidermis, or to be macrochaete or microchaete precursors and which give these precursors their final instructions as to how they differentiate.

The question of identifying the inducing agents (which react with the sensor in the Britten–Davidson model), and then following the consequences of their actions brings us to the next generation of problems. It is idle to speculate about these when we are still uncertain that we have the right experimental tools to hand. A simple, recent experiment illustrates the complexity that may confront us. Striated muscle and endoderm tissue isolated from the anthomedusa *Podocoryne carnea* survive as pure cell types in sea water, without regeneration or transdifferentiation. However, if the endoderm is combined with muscle pretreated with collagenase the regeneration thus initiated forms the manubrium of the medusa, the feeding organ containing the gonads. Altogether seven new cell types appear, including sex cells; and radioactive labelling and separate inhibition of division of the two initial cell types shows that each can transdifferentiate to the full set (Schmidt *et al.*, 1982). It is difficult to envisage the kind of cellular event caused by bringing the two cell types into apposition, and how it results in such a complex outcome; and it is even more difficult to see how this system might be experimentally analysed in terms of gene regulation without mutants to guide our work. The corollary to Sonneborn's statement which heads this chapter is that we must identify systems which are most amenable to experimental analysis and, we have argued, these must be molecular-genetic systems. *Omnis ex* DNA.

Genetic analysis, then, allows us to consider both sides of the coin. On the obverse, we have the phenotypic changes caused by a mutation. These may be trivial from our point of view, like the white eye mutations in *Drosophila*, but may nevertheless serve as a tool for exploring gene organization and interactions; or they may cause developmental anomalies, like the homoeotic mutants which apparently expose genes which have different classes of function (Table 10.4). It seems unlikely that we have discovered all the genes which will direct our attention to the regulation of the crucial events preceding determination and differentiation, but further genetic analysis from the top down will no doubt identify them. On the reverse of the coin, we have the flood of new work taking advantage of these older studies and proceeding from the bottom up, from the detailed analysis of DNA organization, in all its unexpected complexity, towards an understanding of the action of the gene product in development. It is the combination of these two approaches, focused on favourable systems, that will make the future studies of developmental systems even more exciting than those we have described.

# Bibliography

**Abbott, L. C., Karpen, G. H.** and **Schubiger, G.** (1981) Compartment restrictions and blastema formation during pattern regulation in *Drosophila* imaginal leg discs, *Devl. Biol.* **87**, 64–75.

**Abraham, I.** and **Doane, W. W.** (1978) Genetic regulation of tissue-specific expression and Amylase structural genes in *Drosophila melanogaster*, *Proc. Natl. Acad. Sci. U.S.A.* **75**, 4446–50.

**Abrahamson, S., Wurgler, F. E., De Jongh, C.** and **Meyer, H. U.** (1980) How many loci on the X-chromosome of *Drosophila melanogaster* can mutate to recessive lethals? *Envir. Mutag.* **2**, 447–53.

**Adler, P.** (1978) Mutants of the bithorax complex and determinative states in the thorax of *Drosophila melanogaster*, *Devl. Biol.* **65**, 447–61.

**Air, G. M.** (1979) Rapid DNA sequence analysis, *Crit. Rev. Biochem.* **6**, 1–33.

**Akam, M. E.** (1983) Decoding the *Drosophila* complexes, *Trends Biochem. Sci.* **8**, 173–7.

**Akam, M. E., Roberts, D. B., Richards, G. P.** and **Ashburner, M.** (1978) *Drosophila*: the genetics of two larval proteins, *Cell* **13**, 215–26.

**Allan, J., Hartman, P. G., Crane-Robinson, C.** and **Aviles, F. X.** (1980) The structure of histone H1 and its location in chromatin, *Nature* **288**, 675–9.

**Alwine, J. C., Kemp, D. J.** and **Stark, G. R.** (1977) Method for detection of specific RNAs in agarose gels by transfer to diazobenzyloxymethyl-paper and hybridization with DNA probes, *Proc. Natl. Acad. Sci. U.S.A.* **74**, 5350–54.

**Anderson, K.** and **Nüsslein-Volhard, C.** (1982) Genetic control of dorsoventral polarity in the *Drosophila* embryo. Report at the 3rd EMBO *Drosophila* Conference, Crete.

**Anderson, K. V.** and **Lengyel, J. A.** (1979) Rates of synthesis of major classes of RNA in *Drosophila* embryos, *Devl. Biol.* **70**, 217–31.

**Andres, G.** (1953) Experiments on the fate of dissociated embryonic cells (chick) disseminated by the vascular route, *J. Exp. Zool.* **122**, 507–40.

**Angelier, N.** and **Lacroix, J. C.** (1975) Complexes de transcription d'origine nucleolaire et chromosomique d'oocytes de *Pleurodeles waltii* et *P.poireti*, *Chromosoma* **51**, 323–35.

**Arking, R.** (1975) Temperature-sensitive cell-lethal mutants of *Drosophila*: isolation and characterization, *Genetics* **80**, 519–37.

**Arst, H. N.** (1976) Integrator gene in *Aspergillus nidulans*, *Nature* **262**, 231–4.

**Artavanis-Tsakonas, S., Schedl, P., Tschudi, C., Pirrotta, V., Steward, R.** and **Gehring, S.** (1977) The 5S genes of *Drosophila melanogaster*. *Cell* **12**, 1057–67.

**Ashburner, M.** (1971) Induction of puffs in polytene chromosomes of *in vitro* cultured salivary glands in *Drosophila melanogaster* by ecdysone analogues, *Nature New Biol.* **230**, 222–4.

**Ashburner, M.** (1973) Temporal control of puffing activity in polytene chromosomes, *Cold Spring Harbor Symp. Quant. Biol.* **38**, 655–62.

**Ashburner, M.** (1974) Sequential gene activation by ecdysone in polytene chromosomes of *Drosophila*. II. Effects of inhibitors of protein synthesis, *Devl. Biol.* **39**, 141–57.

**Ashburner, M.** (1975) The puffing activities of salivary gland chromosomes. In *Handbook of Genetics. Vol. 3. Invertebrates of Genetic Interest*. Ed. R. C. King. Plenum, New York. 793–811.

**Ashburner, M.** (1980) Drosophila at Kolymbari, *Nature* **288**, 538–40.

**Ashburner, M.** (1982) Transformation and gene organization in *Drosophila*, *Nature* **300**, 15–16.
**Ashburner, M.** and **Berendes, H. D.** (1978) Puffing of polytene chromosomes. In *Genetics and Biology of* Drosophila. Vol. 2b. Eds. M. Ashburner and T. R. F. Wright. Academic Press, New York. 316–95.
**Ashburner, M.** and **Bonner, J. J.** (1979) The induction of gene activity in *Drosophila* by heat shock, *Cell* **17**, 241–54.
**Ashburner, M.** and **Richards, G. P.** (1976) Puffing patterns in *Drosophila*. In *Insect Development*. Ed. P. A. Lawrence. Blackwell, Oxford. 203–25.
**Ashburner, M., Chihara, C., Meltzer, P.** and **Richards, G.** (1974) Temporal control of puffing activity in polytene chromosomes, *Cold Spring Harbor Symp. Quant. Biol.* **38**, 655–62.
**Astauroff, B. L.** (1929) Studien über die erbliche Veränderung der Halteren bei *Drosophila melanogaster*, *Arch. Entw. Mech. Org.* **115**, 448–63.
**Auerbach, C.** (1936) The development of legs, wings and halteres in wild type and some mutant strains of *Drosophila melanogaster*, *Trans. R. Soc. Edinb.* **58**, 787–815.
**Aufderheide, K. J., Frankel, J.** and **Williams, N. E.** (1980) Formation and positioning of surface-related structures in Protozoa, *Microbiol. Rev.* **44**, 252–302.

**Bacon, L. D.** (1970) Immunological tolerance to female wattle isografts in male chickens, *Transplantation* **10**, 124–6.
**Bahn, E.** (1971) Position-effect variegation for an isoamylase in *Drosophila melanogaster*, *Hereditas* **67**, 79–82.
**Bahn, E.** (1973) A suppressor locus for the pyrimidine requiring mutant: *rudimentary*, *Drosophila Inform. Serv.* **49**, 98.
**Baker, B. S., Carpenter, A. T. C., Esposito, M. S., Esposito, R. E.** and **Sandler, L.** (1976) The genetic control of meiosis, *Ann. Rev. Genet.* **10**, 53–134.
**Baker, B. S.** and **Ridge, K. A.** (1980) Sex and the single cell. 1. On the action of major loci affecting sex determination in *Drosophila melanogaster*, *Genetics* **94**, 383–423.
**Baker, R. H.** and **Sakai, R. K.** (1976) Male determining factor in the mosquito *Culex tritaeniorhynchus*, *J. Hered.* **67**, 289–94.
**Baker, W. K.** (1968) Position-effect variegation, *Adv. Genet.* **14**, 133–69.
**Baker, W. K.** (1978a) A fine-structure gynandromorph fate map of the *Drosophila* head, *Genetics* **88**, 743–54.
**Baker, W. K.** (1978b) A clonal analysis reveals early development restrictions in the *Drosophila* head. *Devl. Biol.* **62**, 447–63.
**Baker, W. K.** and **Tsai, L. J.** (1977) Malformed, a mutation of *Drosophila melanogaster* producing mirror-image duplication of a portion of the orbit, *Devl. Biol.* **57**, 221–5.
**Balkaschina, E. I.** (1929) Em Fall der Erbhomoosis (die Genovariation 'aristopedia') bei *Drosophila melanogaster*, *Arch. Entw. Mech. Org.* **115**, 448–63.
**Barski, G., Sorieul, S.** and **Cornefert, F.** (1960) Production dans des cultures *in vitro* de deux souches cellulaire en assocation, de cellules de caractere 'hybride', *C. R. Hebd. Seances Acad. Sci. Paris* **251**, 1825–27.
**Bateson, W.** (1894) '*Materials for the Study of Variation*'. Macmillan, London.
**Bautz, A. M.** (1971) Chronologie de la mise en place de l'hypoderine imaginale de l'abdomen de *Calliphora erythrocephala* Meigen, *Arch. Zool. Exp. Genet.* **112**, 157–78.
**Beadle, G. W.** and **Tatum, E. L.** (1941) Genetic control of biochemical reactions in *Neurospora*. *Proc. Natl. Acad. Sci. U.S.A.* **27**, 499–506.
**Beadle, G. W.** (1958) Genes and Chemical reactions in *Neurospora*—Nobel Lecture. Supplement VI in I. H. Herskowitz. *Genetics*. 2nd. edn. Little, Brown and Co., Boston.
**Beatty, R. A** (1949) Studies on reproduction in wild type and female sterile mutants of *Drosophila melanogaster*, *Proc. R. Soc. Edinb.* **B63**, 249–70.
**Becker, H. J.** (1957) Über Röntgenmosaitflecken und Defectmutationen au Auge von *Drosophila* und die Entwicklungsphysiologie des Auges, *Z. Ind. Abst. Vererb.* **88**, 333–73.
**Becker, H. J.** (1959) Die Puffs der Speichedrusenchromosomen von *Drosophila melanogaster*. 1. Beobachtungen zum verhalten des Puffmusters un Normalstamm und bei zwei Mutanten, *giant* und *lethal-giant-larvae*, *Chromosoma* **10**, 654–78.
**Beckingham-Smith, K.** and **Tata, J. R.** (1976) The hormonal control of amphibian metamorphosis. In *The developmental biology of plants and animals*. Eds. C. F. Graham and P. F Wareing. Blackwell, Oxford. 232–45.
**Beckwith, J.** and **Rossow, P.** (1974) Analysis of Genetic Regulatory Mechanisms, *Ann. Rev. Genet.* **8**, 1–13.

**Bedicheck-Pipkin, S.** (1959) Sex balance in *Drosophila melanogaster*. Aneuploidy of short regions of chromosome 3, using the triploid method, *Univ. Texas Publ.* **5915**, 69–88.

**Beerman, W.** (1952) Chromometenkoustanz und spezifische Modifitatione der Chromosomenstruktur un der Entwicklung und Organdifferenzierung von *Chironomus tentans, Chromosoma* **5**, 139–98.

**Beerman, W.** (1955) Cytologische Analyse eines Camptochironomus-Artbastards. 1. Kruzungsergebnisse und die Evolution des Karyotypus, *Chromosoma* **7**, 198–259.

**Beerman, W.** (1956) Nuclear differentiation and functional morphology of chromosomes, *Cold Spring Harbor Symp. Quant. Biol.* **21**, 217–30.

**Beerman, W.** (1972) Chromomeres and genes. In *Results and Problems in Cell Differentiation. Vol. 4. Development Studies on Giant Chromosomes*. Springer, Berlin. 1–31.

**Beerman, W.** (1973) Directed changes in the pattern of Balbiani ring puffing in *Chironomus*: effects of sugar treatment, *Chromosoma* **41**, 297–326.

**Bekendorf, S. K.** and **Kafatos, F. C.** (1976) Differentiation in the salivary glands of *Drosophila melanogaster*: Characterization of the glue proteins and their developmental appearance, *Cell* **9**, 365–73.

**Belote, J. M.** and **Baker, B. S.** (1982) Sex determination in *Drosophila melanogaster*: Analysis of transformer-2, a sex transforming locus, *Proc. Natl. Acad. Sci. U.S.A.* **79**, 1568–72.

**Belote, J. M.** and **Lucchesi, J. C.** (1980) Control of X chromosome transcription by the maleless gene in *Drosophila*, *Nature* **285**, 573–5.

**Bennett, D.** (1975) The T-locus of the mouse, *Cell* **6**, 441–54.

**Bentley, M. M.** and **Williamson, J. H.** (1979) The control of aldehyde oxidase and xanthine dehydrogenase activities by the *cinnamon* gene in *Drosophila melanogaster*, *Can. J. Genet. Cytol.* **21**, 457–71.

**Berendes, H. D.** (1967) The hormone ecdysone as effector of specific changes in the pattern of gene activity of *Drosophila hydei*, *Chromosoma* **22**, 274–93.

**Berendes, H. D.** (1970) Polytene chromosome structure at the sub-microscope level. 1. A map of the region X, 1-4E of *D. melanogaster*, *Chromosoma* **29**, 118–30.

**Berger, E., Ringler, R., Alahiotis, S.** and **Frank, M.** (1978) Ecdysone-induced changes in morphology and protein synthesis in *Drosophila* cell cultures, *Devl. Biol.* **62**, 498–511.

**Berger, E.** and **Wyss, C.** (1980) Acetylcholinesterase induction by β-ecdysone in *Drosophila* cell lines and their hybrids, *Somatic Cell Genet.* **6**, 631–40.

**Bernhard, H. P.** (1976) The control of gene expression in somatic cell hybrids, *Int. Rev. Cytol.* **47**, 289–325.

**Bigelis, R., Keesey, J.** and **Fink, G. R.** (1977) The *His*4 fungal gene cluster is not polycistronic. In *Molecular Approaches to Eukaryotic Genetic Systems*. Eds. G. Wilcox, J. Abelson and C. F. Fox. Academic Press, New York. 179–87.

**Bingham, P.** (1980) The regulation of *white* locus expression: a dominant mutant allele at the *white* locus of *Drosophila melanogaster*, *Genetics* **95**, 341–53.

**Bingham, P. M., Kidwell, M. G.** and **Rubin, G. M.** (1982) The molecular basis of P–M hybrid dysgenesis: The role of the P element, a P-strain-specific transposon family, *Cell* **29**, 995–1004.

**Bingham, P. M., Levis, R.** and **Rubin, G. M.** (1981) Cloning of DNA sequences from the *white* locus of *D. melanogaster* by a novel and general method, *Cell* **25**, 693–704.

**Bird, A. P.** (1980) Gene reiteration and gene amplification. In *Cell Biology*. Vol. 3. Eds. L. Goldstein and D. M. Prescott. Academic Press, New York. 62–111.

**Bishop, J. O., Morton, J. C., Rosbach, M.** and **Richardson, M.** (1974) Three abundance classes in HeLa cell mRNA, *Nature* **250**, 199–203.

**Biswas, D. K., Abdullah, K. T.** and **Brennessel, B. A.** (1979) On the mechanism of 5-bromodeoxyuridine induction of prolactin synthesis in rat pituitary tumor cells, *J. Cell Biol.* **81**, 1–9.

**Blumberg, D. D., Margolskee, J. P., Barklis, E., Chung, S. N., Cohen, N. S.** and **Lodish, H. F** (1982) Specific cell–cell contacts are essential for induction of gene expression during differentiation of *Dictyostelium discoideum*, *Proc. Natl. Acad. Sci. U.S.A.* **79**, 127–31.

**Bodenstein, D.** (1943) Hormones and tissue competence in the development of *Drosophila*, *Biol. Bull. Woods Hole* **84**, 34–58.

**Bodenstein, D.** (1955) Contributions to the problem of regeneration in insects, *J. Exp. Zool.* **129**, 209–24.

**Bogenhagen, D. F., Sakonju, S.** and **Brown, D. D.** (1980) A control region in the centre of the 5S RNA gene directs specific

initiation of transcription: II. The 3′ border of the region, *Cell* **19**, 27–35.
**Bohn, H.** (1976) Tissue interactions in the regenerating cockroach leg, *Symp. R. Entomol. Soc. Lond.* **8**, 170–85.
**Bollon, A. P.** (1975) Regulation of the *ilvI* multifunctional gene in *Saccharomyces cerevisiae*, *Mol. Gen. Genet.* **142**, 1–12.
**Bonner, J. J.** and **Pardue, M. L.** (1976) The effect of heat shock on RNA synthesis in *Drosophila* tissues, *Cell* **8**, 43–50.
**Bonner, J. J.** and **Pardue, M. L.** (1977) Polytene chromosome puffing and in situ hybridization measure different aspects of RNA metabolism, *Cell* **12**, 227–34.
**Botchan, P., Reeder, R. H.** and **Dawid, I. B.** (1977) Restriction analysis of the nontranscribed spacers of *X. laevis* rDNA, *Cell* **11**, 599–607.
**Bournias-Vardiabasis, N.** and **Bownes, M.** (1978) Developmental analysis of the *tumorous head* mutation in *Drosophila melanogaster*, *J. Embryol. Exp. Morph.* **44**, 227–41.
**Boveri, T.** (1887) Uber Differenzierung der Zellkern wahrend der Furchung des Eies von *Ascaris megalocephala*, *Anat. Anz.* **2**, 688–93.
**Boveri, T.** (1899) Die Entwicklung von *Ascaris megalocephala* mit besonderer Rucksicht auf die kernverhaltnisse. In *Festschr. Fur C.v. Kupffer*. G. Fischer, Jena. 383–430.
**Bownes, M.** (1982) Hormonal and genetic regulation of vitellogenesis in *Drosophila*, *Q. Rev. Biol.* **57**, 247–74.
**Bownes, M., Cline, T. W.** and **Schneiderman, H. A.** (1976) Daughters from daughterless mothers—rescuing a female-lethal maternal effect by cytoplasmic transplantation in *Drosophila* embryos, *Roux's Arch.* **181**, 279–84.
**Bownes, M.** and **Roberts, S.** (1979) Acquisition of differentiative capacity in imaginal wing discs of *Drosophila melanogaster*, *J. Embryol. Exp. Morph.* **49**, 103–13.
**Bownes, M.** and **Seiler, M.** (1977) Developmental effects of exposing *Drosophila* embryos to ether vapour, *J. Exp. Zool.* **199**, 9–24.
**Breathnach, R., Mandel, J. L.** and **Chambon, P.** (1977) Ovalbumin gene is split in chicken DNA, *Nature* **270**, 314–9.
**Bregliano, J. C., Picard, G., Bucheton, A., Pelisson, A., Lavige, J. M.** and **L'Heritier, P.** (1980) Hybrid dysgenesis in *Drosophila melanogaster*, *Science* **207**, 606–11.
**Brehme, K. S.** (1939) A study of the effects on development of "Minute" mutations in *Drosophila melanogaster*, *Genetics* **24**, 131–61.
**Brenner, S.** (1974) The genetics of *Caenorhabditis elegans*, *Genetics* **77**, 71–94.
**Brenner, S.** (1981) Genes and Development. In *Cellular Controls in Differentiation*. Eds. C. W. Lloyd and D. A. Rees. Academic Press, New York. 3–7.
**Brettell, R. I. S.** and **Ingram, D. S.** (1979) Tissue culture in the production of novel disease-resistant plants, *Biol. Rev.* **54**, 329–45.
**Bridges, C. B.** (1921) Triploid intersexes in *Drosophila melanogaster*, *Science* **54**, 252–4.
**Bridges, C. B.** (1935) Salivary chromosome maps with a key to the banding of the chromosomes of *D. melanogaster*, *J. Hered.* **26**, 60–4.
**Bridges, C. B.** (1939) Cytological and genetic basis of sex. In *Sex and Internal Secretions*. 2nd edn. Ed. E. Allen. Baillière, London. 15–63.
**Bridges, C. B.** and **Dobzhansky, Th.** (1933) The mutant "proboscipedia" in *Drosophila melanogaster*: a case of hereditary homoeosis, *Roux's Arch.* **127**, 575–90.
**Bridges, C. B.** and **Morgan, T. H.** (1923) The third chromosome group of mutant characters of *Drosophila melanogaster*, *Carnegie Inst. Wash. Publ.* **327**, 251.
**Briggs, R.** (1973) Developmental genetics of the Axolotl. In *Genetic Mechanisms of Development*, Ed. F. H. Ruddle. Academic Press, New York. 164–201.
**Briggs, R.** and **King, T. J.** (1952) Transplantation of living nuclei from blastula cells into enucleated frog's eggs, *Proc. Natl. Acad. Sci. U.S.A.* **38**, 455–63.
**Briggs, R., Signoret, J.** and **Humphrey, R. R.** (1964) Transplantation of nuclei of various cell types from neurulae of the Mexican axolotl (*Ambystoma mexicanum*), *Devl. Biol.* **10**, 233–46.
**Britten, R. J.** and **Davidson, E. H.** (1969) Gene regulation for higher cells: a theory, *Science* **165**, 349–58.
**Britten, R. J.** and **Kohne, D. E.** (1968) Repeated sequences in DNA, *Science* **161**, 529–40.
**Brookman, J. J., Town, D. C., Jermyn, K. A.** and **Kay, R. R.** (1982) Developmental regulation of a stalk cell differentiation-inducing factor in

*Dictyostelium discoideum, Devl. Biol.* **91**, 191–6.
**Brooks, R. R.** and **Huang, P. C.** (1972) Redundant DNA of *Neurospora crassa, Biochem. Genet.* **6**, 41–9.
**Brothers, V. M., Tsubota, S., Germeraad, S.** and **Fristrom, J. W.** (1978) The rudimentary locus of *Drosophila melanogaster*: partial purification of a carbamyl phosphate synthetase-aspartate transcarbamylase-dihydroorotase complex, *Biochem. Genet.* **16**, 321–32.
**Brower, D. L., Lawrence, P. A.** and **Wilcox, M.** (1981) Clonal analysis of the undifferentiated wing disc of *Drosophila, Devl. Biol.* **86**, 448–55.
**Brown, D. D.** (1981) Gene expression in eukaryotes, *Science* **211**, 667–74.
**Brown, D. D.** and **Gurdon, J. B.** (1978) Cloned single repeating units of 5S DNA direct accurate transcription of 5S RNA when injected in *Xenopus* oocytes, *Proc. Natl. Acad. Sci. U.S.A.* **75**, 2849–53.
**Brown, D. D.** and **Jordan, E.** (1978) A variety of 5S RNA genes are transcribed accurately by an extract of oocyte nuclei, *Carnegie Inst. Wash. Yearb.* **77**, 139–40.
**Brown, S. W.** and **Chandra, H. S.** (1977) Chromosome imprinting and the differential regulation of homologous chromosomes. In *Genetic Mechanisms of Cells. Vol. 1. Cell Biology*. Ed. L. Goldstein and D. M. Prescott. Academic Press, New York. 110–89.
**Brown, S. W.** and **Nelson Rees, W. A.** (1961) Radiation analysis of a lecanoid genetic system, *Genetics* **46**, 983–1007.
**Brun, G.** and **Plus, N.** (1980) The viruses of *Drosophila*. In *The Genetics and Biology of* Drosophila. Vol. 2d. Ed. M. Ashburner and T. R. F. Wright. Academic Press, New York. 625–702.
**Bruskin, A. M., Bedard, P. A., Tyner, A. C., Snowman, R. M., Brandhorst, B. P.** and **Klein, W. H.** (1982) A family of proteins accumulating in ectoderm of sea urchin embryos specified by two related cDNA clones, *Devl. Biol.* **91**, 317–24.
**Bryant, P. J.** (1971) Regeneration and duplication following operations *in situ* on the imaginal discs of *Drosophila melanogaster, Devl. Biol.* **26**, 637–51.
**Bryant, P. J.** (1975) Pattern formation in the imaginal wing disc of *Drosophila melanogaster*. Fate map, regeneration and duplication, *J. Exp. Zool.* **193**, 49–78.
**Bryant, P. J.** (1978) Pattern formation in imaginal discs. In *The Genetics and Biology of* Drosophila. Vol. 2c. Ed. M. Ashburner and T. R. F. Wright. Academic Press, New York. 230–335.
**Bryant, P. J., Adler, P. N., Duranceau, C., Fam, M. J., Glenn, S., Hsei, B., James, A. A., Littlefield, C. L., Reinhardt, C. A., Strub, S.** and **Schneiderman, H. A.** (1978) Regulative interactions between cells from different imaginal discs of *Drosophila melanogaster, Science* **201**, 928–30.
**Bryant, P. J.** and **Schneiderman, H. A.** (1969) Cell lineage, growth, and determination in the imaginal leg discs of *Drosophila melanogaster, Devl. Biol.* **20**, 263–90.
**Bryant, S. V., French, V.** and **Bryant, P.** (1981) Distal regeneration and symmetry, *Science* **212**, 993–1002.
**Bryant, S. V.** and **Iten, L. E.** (1976) Supernumerary limbs in amphibians: Experimental production in *Notophthalmus viridescens* and a new interpretation of their formation, *Devl. Biol.* **50**, 212–34.
**Bulyzhenkov, V. E.** and **Ivanov, V. I.** (1978) Interaction of homoeotic genes of *Antennapedia, aristapedia* and *Polycomb* in *Drosophila melanogaster, Biol. Zbl.* **97**, 527–33.
**Bukhari, A. I., Shapiro, J. A.** and **Adhya, S. L.** (Eds.) (1977) *DNA Insertion Elements, Plasmids and Episomes*, Cold Spring Harbor Laboratory, New York.
**Bull, A.** (1956) Developmental effects of the $v^{gB}$ deficiency in *Drosophila melanogaster, J. Exp. Zool.* **132**, 467–507.
**Bull, A.** (1966) *Bicaudal*, a genetic factor which affects the polarity of the embryo in *Drosophila melanogaster, J. Exp. Zool.* **161**, 221–41.
**Bull, J. J.** (1980) Sex determination in reptiles, *Q. Rev. Biol.* **55**, 3–21.
**Burton, E. G.** and **Metzenberg, R. L.** (1972) Novel mutation causing derepression of several enzymes of sulfur metabolism in *Neurospora crassa, J. Bacteriol.* **109**, 140–51.

**Cairns, J.** (1981) The origin of human cancers, *Nature* **289**, 353–7.
**Callan, H. G.** (1963) The nature of lampbrush chromosomes, *Int. Rev. Cytol.* **15**, 1–34.
**Callan, H. G.** (1981) Croonian Lecture: Lampbrush chromosomes, *Proc. R. Soc. Lond.* **214**, 417–48.
**Callan, H. G.** and **Lloyd, L.** (1960) Lampbrush chromosomes of crested newts *Triturus cristatus* (Laurenti), *Phil.*

*Trans. R. Soc. Lond.* **B243**, 135–219.
**Calos, M. P.** and **Miller, J. H.** (1980) Transposable elements, *Cell* **20**, 579–95.
**Campbell, J. G.** (1962) A retro-ocular teratoma containing pinealomatous tissue in a young chick, *Brit. J. Cancer* **16**, 258–66.
**Campos-Ortega, J. A** and **Waitz, M.** (1978) Cell clones and pattern formation: developmental restrictions in the compound eye of *Drosophila*, *Roux's Arch.* **184**, 155–70.
**Capdevila, M. P.** and **Garcia-Bellido, A.** (1981) Genes involved in the activation of the bithorax complex of *Drosophila*, *Roux's Arch.* **190**, 339–50.
**Caplan, A. I.** and **Ordahl, C. P.** (1978) Irreversible gene repression model for control of development, *Science* **201**, 120–30.
**Cappuccinelli, P.** and **Ashworth, J. M.** (1977) *Development and Differentiation in the Cellular Slime Moulds*. Elsevier/North-Holland, Amsterdam.
**Carlon, N.** and **Erickson, G. F.** (1978) Fine structure of prefollicular and developing germ cells in the male and female left embryonic chick gonads *in vitro* with and without androgenic steroids, *Biochem. Biophys.* **18**, 335–49.
**Carlson, P. S.** (1971) A genetic analysis of the *rudimentary* locus of *Drosophila melanogaster*, *Genet. Res.* **17**, 53–81.
**Carlsson, S. A., Luger, O., Ringertz, N. R.** and **Savage, R. E.** (1974) Phenotypic expression in chick erythrocyte × rat myoblast hybrids and in chick myoblast × rat myoblast hybrids, *Expl. Cell Res.* **84**, 47–55.
**Carpenter, C. D.** and **Klein, W. H.** (1982) A gradient of poly ($A^+$) RNA sequences in *Xenopus laevis* eggs and embryos, *Devl. Biol.* **91**, 43–9.
**Carramolino, L., Ruiz-Gomez, M., Guerrero, M. C.** and **Modelell, J. C.** (1982) DNA map of mutations at the *scute* locus of *Drosophila* melanogaster, *EMBO J.* **1**, 1185–91.
**Case, S. T.** and **Daneholt, B.** (1978) The size of the transcription unit in Balbiani ring 2 of *Chironomus tentans* as derived from analysis of the primary transcript and 75S RNA, *J. Mol. Biol.* **124**, 223–41.
**Cassada, R., Isnenghi, E., Culotti, M.** and **von Ehrenstein, G.** (1981) Genetic analysis of temperature-sensitive embryogenesis mutants in *Caenorhabditis elegans*, *Devl. Biol.* **84**, 193–205.
**Cather, J. N.** (1971) Cellular interactions in the regulation of development in annelids and molluscs, *Adv. Morphogenet.* **9**, 67–125.
**Cattanach, B. M., Evans, E. P., Burtenshaw, M. D.** and **Barlow, J.** (1982) Male, female and intersex development in mice of identical chromosome constitution, *Nature* **300**, 445–6.
**Cattanach, B. M., Pollard, C. E.** and **Hawkes, S. G.** (1971) Sex reversed mice: XX and XO males, *Cytogenetics* **10**, 318–37.
**Chalfie, M., Horvitz, H. R.** and **Sulston, J. E. (1981)** Mutations that lead to reiterations in the cell lineages of *C.elegans*, *Cell* **24**, 59–69.
**Chamberlin, M. E., Britten, R. J.** and **Davidson, E. H.** (1975) Sequence organization in Xenopus DNA studied by the electron microscope, *J. Mol. Biol.* **96**, 317–34.
**Chambon, P., Benoist, C., Breathnach, R., Cochet, M., Gannon, F., Gerlinger, P., Krust, A., Le Meur, M., Le Pennec, J. P., Mandel, J. L., O'Hare, K.** and **Perrin, F.** (1979) Structural organization and expression of ovalbumin and related chicken genes. In *From Gene to Protein: Information Transfer in Normal and Abnormal Cells*. Ed. T. R. Russell, K. Brew, H. Faber and J. Schultz. Academic Press, New York. 55–78.
**Chan, L. N** and **Gehring, W.** (1971) Determination of blastoderm cells in *Drosophila melanogaster*, *Proc. Natl. Acad. Sci. U.S.A.* **68**, 2217–21.
**Charlesworth, B.** and **Charlesworth, D.** (1978) A model for the evolution of dioecy and gynodioecy, *Amer. Nat.* **112**, 975–97.
**Chen, P. S.** and **Hadorn, E.** (1954) Vergleichende Untersuchusigen über die freien Aminosäuren in der larvalen Haemolymphe von *Drosophila*, *Ephestia* und *Corethra*, *Rev. Suisse Zool.* **61**, 437–51.
**Cherbas, L.** and **Cherbas, P.** (1981) The effects of ecdysteroid hormones on *Drosophila melanogaster* cell lines, *Adv. Cell Culture* **1**, 91–124.
**Chovnick, A., Gelbart, W.** and **McCarron, M.** (1977) Organization of the *rosy* locus in *Drosophila melanogaster*, *Cell* **11**, 1–10.
**Chovnick, A., McCarron, M., Hilliker, A., O'Donnell, J., Gelbart, W.** and **Clark, S.** (1978) Gene organization in *Drosophila*, *Cold Spring Harbor Symp. Quant. Biol.* **42**, 1011–21.
**Clever, U.** (1964) Actinomycin and

Puromycin: Effects on sequential gene activation by ecdysone, *Science* **146**, 794–95.
**Cline, T. W.** (1978) Two closely linked mutations in *Drosophila melanogaster* that are lethal to opposite sexes and interact with daughterless, *Genetics* **90**, 683–98.
**Cline, T. W.** (1979) A male-specific lethal mutation in *Drosophila melanogaster* that transforms sex, *Devl. Biol.* **72**, 266–75.
**Cochet, M., Gannon, E., Hen, R., Maroteaux, L., Perrin, F.** and **Chambon, P.** (1979) Organization and sequence studies of the 17-piece chicken conalbumin gene, *Nature* **282**, 567.
**Cock, A. G.** (1964) Dosage compensation and sex chromatin in non-mammals, *Genet. Res.* **5**, 354–65.
**Coe, E. H.** and **Neuffer, M. G.** (1978) Embryo cells and their destinies in the corn plant. In *Clonal Basis of Development*. Eds. S. Súbtelny, and I. M. Sussex. Academic Press, New York. 113–30.
**Cohn, R. J.** and **Kedes, L. H.** (1979) Non allelic histone gene clusters of individual sea urchins (*Lytechinus pictus*): mapping of homologies in coding and spacer DNA, *Cell* **18**, 855–64.
**Compton, J. L.** and **McCarthy, B. J.** (1978) Induction of the Drosophila heat shock response in isolated polytene nuclei, *Cell* **14**, 191–202.
**Conklin, E. G.** (1905) The organization and cell lineage of the ascidian egg, *J. Acad. Nat. Sci. Philadelphia* **13**, 1–119.
**Cook, P. R.** and **Brazell, I. A.** (1980) Mapping sequences in loops of nuclear DNA by their progressive detachment from the nuclear cage, *Nucleic Acids Res.* **18**, 2895–906.
**Corces, V. Holingren, R., Freund, R., Morimoto, R.** and **Meselson, M.** (1980) Four heat shock proteins of *Drosophila melanogaster* coded within a 12-kilobase region in chromosome sub-division 67B, *Proc. Natl. Acad. Sci. U.S.A.* **77**, 5390–93.
**Corces, V., Pellicer, A., Axel, R.** and **Meselson, M.** (1981) Integration, transcription and control of a *Drosophila* heat shock gene in mouse cells, *Proc. Natl. Acad. Sci. U.S.A.* **78**, 7038–42.
**Corden, J., Wasylyk, B., Buchwalder, A., Sassone-Corsi, P., Kedinger, C.** and **Chambon, P.** (1980) Promoter sequences of eukaryotic protein-coding genes, *Science* **209**, 1406–14.
**Correns, C.** (1907) *Die Bestimmung und Vererbund des Geschlechtes nach neuen Versuchen mit hohereu Pflanzen*. Berlin.
**Cox, R. P.** and **King, J. C.** (1975) Gene expression in cultured mammalian cells, *Int. Rev. Cytol.* **43**, 281–351.
**Craig, E. A., McCarthy, B. J.** and **Wadsworth, S. C.** (1979) Sequence organization of two recombinant plasmids containing genes for the major heat shock induced protein of *D. melanogaster*, *Cell* **16**, 575–88.
**Crain, W. R., Eden, F. C., Pearson, W. R., Davidson, E. H.** and **Britten, R. J.** (1976) Absence of short period interspersion of repetitive sequences in the DNA of *Drosophila melanogaster*, *Chromosoma* **56**, 309–26.
**Crick, F. H. C.** (1971a) General model for the chromosomes of higher organisms, *Nature* **234**, 25–7.
**Crick, F. H. C.** (1971b) The scale of pattern formation. In *Control Mechanisms of Growth and Differentiation. Symp. Soc. Exp. Biol.* **25**, Cambridge University Press, Cambridge. 429–38.
**Crick, F. H. C.** (1979) Split genes and RNA splicing, *Science* **204**, 264–271.
**Crick, F. H C.** and **Lawrence, P. A.** (1975) Compartments and polyclones in insect development, *Science* **189**, 340–7.
**Croce, C. M., Litwack, G.** and **Koprowski, H.** (1973) Human regulatory gene for inducible tyrosine aminotransferase in rat-human hybrids, *Proc. Natl. Acad. Sci. U.S.A.* **70**, 1268–72.
**Cross, D. P.** and **Sang, J. H.** (1978) Cell culture of individual *Drosophila* embryos: II. Culture of X-linked embryonic lethals, *J. Embryol. Exp. Morph.* **45**, 173–87.
**Crouse, H. V.** (1960) The nature of the influence of X-translocations on sex of progeny in *Sciara coprophilia*, *Chromosoma* **11**, 146–66.
**Cummings, M. R.** and **King, R. C.** (1970) The cytology of the vitellogenic stages of oogenesis in *Drosophila melanogaster*. 1. General staging characteristics, *J. Morphol.* **128**, 427–42.

**Dale, L.** and **Bownes, M.** (1980) Is regeneration in *Drosophila* the result of epimorphic regulation? *Roux's Arch.* **189**, 91–6.
**Daneholt, B.** (1972) Giant RNA transcript in a Balbiani ring, *Nature New Biol.* **240**, 229–32.
**Daneholt, B., Case, S. T., Lam, M. M., Nelson, L.** and **Wieslander, L.** (1978) The 75S RNA transcription unit in Balbiani ring 2 and its relation to chromosome structure, *Phil. Trans. R. Soc. Lond.* **B283**, 383–9.

**Danielli, J. F.** and **DiBarardino, M. A.** (1979) Overview. In *Nuclear Transplantation. Int. Rev. Cytol., Suppl.* **9**, 4–9.

**Darnborough, C.** and **Ford, P. J.** (1976) Cell free translation of messenger RNA from oocytes of *Xenopus laevis*, *Devl. Biol.* **50**, 285.

**Darnell, J. E.** (1982) Variety in the level of gene control in eukaryotic cells, *Nature* **297**, 365–71.

**Darnell, J. E.** (1979) Transcription units for mRNA production in eukaryotic cells and their DNA viruses, *Prog. Nucleic Acid Res. Mol. Biol.* **22**, 327–53.

**Davidson, E. H.** (1976) *Gene Activity in Early Development*. 2nd edn. Academic Press, New York.

**Davidson, E. H.** and **Britten, R. J.** (1973) Organization, transcription and regulation in the animal genome, *Q. Rev. Biol.* **48**, 565–613.

**Davidson, E. H.** and **Britten, R. J.** (1979) Regulation of gene expression: a new look at the possible role of interspersed sequences, *Science* **204**,1052–9.

**Davidson, E. H., Hough-Evans, B. R.** and **Britten, R. J.** (1982) Molecular biology of the sea urchin embryo, *Science* **217**, 17–26.

**Davidson, E. H., Hough-Evans, B. R., Amenson, C. S.** and **Britten, R. J.** (1973) General interspersion of repetitive and nonrepetitive elements in the DNA of *Xenopus*, *J. Mol. Biol.* **77**, 1–24.

**Davidson, R. L.** (1972) Regulation of melanin synthesis in mammalian cells: Effects of gene dosage on the expression of differentiation, **69**, 951–5.

**Davidson, R. L.** (1974) Gene expression in somatic cell hybrids, *Ann. Rev. Genet.* **8**, 195–218.

**Davis, F. M** and **Adelberg, E. A.** (1973) Use of somatic cell hybrids for analysis of the differentiated state, *Bacteriol. Rev.* **37**, 197–214.

**Davis, K. T.** and **Shearn, A.** (1977) *In vitro* growth of imaginal discs from *Drosophila melanogaster*, *Science* **196**, 438–9.

**Dawid, I. B.** and **Botchan, P.** (1977) Sequences homologous to ribosomal insertions occur in the *Drosophila* genome outside the nucleolus organizer, *Proc. Natl. Acad. Sci. U.S.A.* **74**, 4233–7.

**Dawid, I. B.** and **Wellauer, P. K.** (1978) Ribosomal DNA and related sequences in *Drosophila melanogaster*, *Cold Spring Harbor Symp. Quant. Biol.* **42**, 1185–200.

**Demerec, M.** (1936) Frequency of "cell-lethals" among lethals obtained at random in the X-chromosome of *Drosophila melanogaster*, *Proc. Natl. Acad. Sci. U.S.A.* **22**, 350–4.

**Denell, R. E.** (1973) Homoeosis in *Drosophila*: 1. Complementation studies with revertants of *Nasobemia*, *Genetics* **75**, 279–97.

**Denell, R. E.** (1982) Homoeosis in *Drosophila*: Evidence for a maternal effect of the *polycomb* locus, *Devl. Genet.* **3**, 103–13.

**Denell, R. E., Hummels, K. R, Wakimoto, G. T** and **Kaufman, T. C.** (1981) Developmental studies of lethality associated with the *Antennapedia* gene complex in *Drosophila melanogaster*, *Devl. Biol.* **81**, 43–50.

**Deppe, U., Schierenberg, E., Cole, T., Krieg, C., Schmitt, O., Yoder, B.** and **von Erenstein, G.** (1978) Cell lineages of the embryo of the nematode, *Caenorhabditis elegans*, *Proc. Natl. Acad. Sci. U.S.A.* **75**, 376–80.

**DeRobertis, E. M.** and **Gurdon, J. B.** (1977) Gene activation in somatic nuclei after injection into amphibian oocytes, *Proc. Natl. Acad. Sci. U.S.A.* **74**, 2470–4.

**DeRobertis, E. M.** and **Olson, M. W.** (1979) Transcription and processing of cloned yeast tyrosine tRNA genes microinjected into frog oocytes, *Nature* **278**, 137–43.

**DeRobertis, E. M., Partington, G. A., Longthorne, R. F.** and **Gurdon, J. B.** (1977) Somatic nuclei in amphibian oocytes: evidence for selective gene expression, *J. Embryol. Exp. Morph.* **40**, 199–214.

**Devlin, R. B.** and **Emerson, C. P.** (1978) Coordinate regulation of contractile protein synthesis during muscle differentiation, *Cell* **13**, 599–611.

**Dewey, M. J., Martin, D. W., Martin, G. R.** and **Mintz, B.** (1977) Mosaic mice with teratocarcinoma-derived mutant cells deficient in hypoxanthine phosphoribosyltransferase, *Proc. Natl. Acad. Sci. U.S.A.* **74**, 5564–8.

**Diaz, M. O., Barsacchi-Pilone, G., Mohon, K. A.** and **Gall, J. G.** (1981) Transcripts from both strands of a satellite DNA occur on lampbrush chromosome loops of the newt (*Notophthalmus*), *Cell* **24**, 649–59.

**DiBarardino, M. A.** (1980) Genetic stability and modulation of metazoan nuclei transplanted into eggs and oocytes, *Differentiation* **17**, 17–30.

**Dickinson, W. J.** (1975) A genetic locus affecting the developmental expression of an enzyme in *Drosophila melanogaster*,

*Devl. Biol.* **42**, 131–40.

**Dickinson, W. J.** (1980) Complex *cis*-acting regulatory genes demonstrated in *Drosophila* hybrids, *Devl. Genet.* **1**, 229–40.

**Diez, J. L., Santa-Cruz, M. C., Villanueva, A.** and **Aller, P.** (1980) Dependance of Balbiani ring puffing on protein synthesis, *Chromosoma* **81**, 263–9.

**Doane, W. W.** (1973) Role of hormones in insect development. In *Developmental Systems: Insects*. Vol. 2. Eds. S. J. Counce and C. H. Waddington. Academic Press, New York. 291–497.

**Doane, W. W.** (1977) Further evidence for temporal gene controlling *Amy* expression in *Drosophila melanogaster*, *Genetics* **86**, 515–16.

**Doane, W. W., Abraham, I., Kolar, M. M., Martenson, R. E.** and **Deibler, G. E.** (1975) Purified *Drosophila* α-amylase isozymes: Genetical, biochemical and molecular characterisation. In *Isozymes IV*. Ed. C. Markert. Academic Press, New York. 585–607.

**Dobzhansky, T.** and **Schultz, J.** (1934) The distribution of sex factors in the X-chromosome of *Drosophila melanogaster*, *J. Genet.* **28**, 349–86.

**Dohmen, M. R.** and **Verdonk, N. H.** (1979) The ultrastructure and role of the polar lobe in development of molluscs. In *Determinants of Spatial Organization*. Eds. S. Subtelny and I. R. Konigsberg. Academic Press, New York. 3–27.

**Dübendorfer, A.** and **Eichenberger-Glinz, S.** (1980) Development and metamorphosis of larval and adult tissues of *Drosophila in vitro*. In *Invertebrate Systems In Vitro*. Eds. E. Kurstak, and K. Marmorsch and A. Dübendorfer. Elsevier/North-Holland, Amsterdam. 169–86.

**Dübendorfer, A., Shields, G.** and **Sang, J. H.** (1975) Development and differentiation *in vitro* of *Drosophila* imaginal disc cells from dissociated early embryos, *J. Embryol. Exp. Morph.* **33**, 487–98.

**Dübendorfer, K.** and **Nöthiger, R.** (1982) A clonal analysis of cell lineage and growth in the male and female genital disc of *Drosophila melanogaster*, *Roux's Arch.* **191**, 42–55.

**Duck, P.** and **Chovnick, A.** (1975) Resolution of an equivocal genetic element in *Drosophila melanogaster*: organization of the *maroon-like* locus, *Genetics* **79**, 459–66.

**Dudler, R., Egg, A. H., Kubli, E., Artavanis-Tsakonas, S., Gehring, W., Steward, R.** and **Schedl, P.** (1980) Transfer RNA genes of *Drosophila melanogaster*, *Nucleic Acids Res.* **8**, 2921–37.

**Dugaiczyk, A., Woo, S. L. C., Colbert, D. A., Lai, E. C., Mace, M. L.** and O'Malley, B. W. (1979) The ovalbumin gene: cloning and molecular organization of the entire natural gene, *Proc. Natl. Acad. Sci. U.S.A.* **76**, 2253–7.

**Dulbecco, R., Bologna, M.** and **Unger, M.** (1980) Control of a mammary cell line by lipids. *Proc. Natl. Acad. Sci. U.S.A.* **77**, 1551–5.

**Duncan, I.** and **Lewis, E. B.** (1982) Genetic control of body segment differentiation in Drosophila. In *Development and Order; Its Origin and Regulation*. Ed. S. Subtelny. Liss, New York. 533–44.

**Dunsmuir, P., Brorein, W. J., Simon, M. A.** and **Rubin, G. M.** (1980) Insertion of the *Drosophila* transposable element *copia* generates a 5 base pair duplication, *Cell* **21**, 575–9.

**Du Praw, E. J.** (1970) *DNA and Chromosomes*, Holt, Rinehart and Winston, New York.

**Edwards, J. W.** and **Gardner, E. J** (1962) Genetics of the *eye-reduced* mutant of *Drosophila melanogaster*, with special reference to homoeosis and eyelessness, *Genetics* **53**, 785–98.

**Edwards, T. C. R., Candido, E. P. M.** and **Chovnick, A.** (1977) Xanthine dehydrogenase from *Drosophila melanogaster*, *Mol. Gen. Genet.* **154**, 1–6.

**Egyhazi, E.** (1976) Quantitation of turnover and export to the cytoplasm of hnRNA transcribed in the Balbiani rings, *Cell* **7**, 507–15.

**Engelke, D. R., Ng, S.** and **Shastry, B. S.** (1980) Specific interaction of a purified transcription factor with an internal control region of 5S RNA genes, *Cell* **19**, 717–28.

**Engels, W. R.** (1981) Germline hypermutability in *Drosophila* and its relation to hybrid dysgenesis and cytotype, *Genetics* **98**, 565–87.

**Ephrussi, B.** (1962, but published 1979) Mendelism and the new genetics, *Somatic Cell Genet.* **5**, 681–95.

**Epper, F.** (1981) Morphological analysis and fate map of the intersexual genital disc of the mutant doublesex-Dominant in *Drosophila melanogaster*, *Devl. Biol.* **88**, 104–14.

**Epper, F.** and **Nöthiger, R.** (1982) Genetic and developmental evidence for a repressed genital primordium in *Drosophila melanogaster*, *Devl. Biol.* **94**, 163–75.

**Ernst, G. S., Hough-Evans, B. R., Britten, R. J.** and **Davidson, E. H.** (1980) Limited complexity of the RNA in micromeres of 16-cell sea urchin embryos, *Devl. Biol.* **79**, 119–27.

**Esau, K.** (1965) *Plant Anatomy*. Wiley, New York.

**Evans, E. P., Burtenshaw, M. D.** and **Cattanach, B. M.** (1982) Meiotic crossing-over between X and Y chromosomes of male mice carrying the sex-reversing (*Sxr*) factor, *Nature* **300**, 443–5.

**Fain, M. J.** and **Schneiderman, H. A.** (1979) Wound healing and regenerative response of fragments of the *Drosophila* wing imaginal disc cultured *in vitro*, *J. Insect Physiol.* **25**, 913–24.

**Falk, D. R., Bellamy, R., el Kouni, M.** and **Naguib, F.** (1980) A genetic and dietary study of the physiology of pyrimidine synthesis in *Drosophila melanogaster*, *J. Insect Physiol.* **26**, 735–40.

**Falk, D. R.** and **Nash, D.** (1974) Sex-linked auxotrophic and putative auxotrophic mutants of *Drosophila melanogaster*, *Genetics* **76**, 755–66.

**Falk, R.** (1963) A search for a gene control system in *Drosophila*, *Amer. Nat.* **97**, 129–32.

**Falkenthal, S., Graham, M. L., Korn, E. L.** and **Lengyel, J. A.** (1982) Transcription, processing and turnover of RNA from the *Drosophila* mobile genetic element *copia*, *Devl. Biol.* **92**, 294–305.

**Falkenthal S.** and **Lengyel, J. A.** (1980) Structure, translation and metabolism of the cytoplasmic copia ribonucleic acid of *Drosophila melanogaster*, *Biochemistry* **19**, 5842–50.

**Fausto-Sterling, A.** (1980) Studies on the female sterile mutant *rudimentary* of *Drosophila melanogaster*. III. Cell death in rudimentary wing imaginal disc, *J. Exp. Zool.* **213**, 383–90.

**Fell, H. B.** and **Mellanby, E.** (1953) Metaplasia produced in cultures of chick ectoderm by high vitamin A, *J. Physiol.* **119**, 470–88.

**Ferguson, M. W. J.** and **Joanen, T.** (1982) Temperature of egg incubation determines sex in *Alligator mississippiensis*, *Nature* **296**, 850–3.

**Ferrus, A.** and **Kankel, D. R.** (1981) Cell lineage relationships in *Drosophila melanogaster*. The relationship of cuticular to internal tissues, *Devl. Biol.* **85**, 485–504.

**Fielding, C. J.** (1967) Developmental genetics of the mutant *grandchildless* of *Drosophila subobscura*, *J. Embryol. Exp. Morph.* **17**, 375–84.

**Fincham, J. R. S.** and **Day P. R.** (1965) *Fungal Genetics*. Blackwell, Oxford.

**Fincham, J. R. S., Day, P. R.** and **Radford, A.** (1979) *Fungal Genetics*. 4th edn. Blackwell, Oxford.

**Fincham, J. R. S.** and **Sastry, G. R. K.** (1974) Controlling elements in maize, *Ann. Rev. Genet.* **8**, 15–50.

**Findly, R. C.** and **Pederson, T.** (1981) Regulated transcription of the genes for actin and heat-shock proteins in cultured *Drosophila* cells, *J. Cell Biol.* **88**, 323–8.

**Finnegan, D. J., Rubin, G. M., Young, M. W.** and **Hogness, D. S.** (1978) Repeated gene families in *Drosophila melanogaster*, *Cold Spring Harbor Symp. Quant. Biol.* **42**, 1053–63.

**Finnerty, V.** and **Johnson, G.** (1979) Post translational modification as a potential explanation of high levels of enzyme polymorphism: xanthine dehydrogenase and aldehyde oxidase in *Drosophila melanogaster*, *Genetics* **91**, 695–722.

**Firtel, R. A.** and **Jacobson, A.** (1977) Structural organization and transcription of the genome of *Dictyostelium discoideum*, *Int. Rev. Biochem.* **15**, 377–86.

**Fischberg, M.** and **Blackler, A. W.** (1961) How cells specialize, *Scient. Am.* **205**, 124–41.

**Flavell, A. J., Ruby, S. W., Toole, J. J., Roberts, B. E.** and **Rubin, G. M.** (1980) Translation and developmental regulation of RNA encoded by the eukaryotic transposable element *copia*, *Proc. Natl. Acad. Sci. U.S.A.* **77**, 7107–11.

**Ford, C. E.** (1970) Cytogenetics and sex discrimination in man and mammals, *J. Biosoc. Sci. Suppl.* **2**, 7–30.

**Ford, C. E., Evans, E. P.** and **Gardner, R. L.** (1975) Marker chromosome analysis of two mouse chimeras, *J. Embryol. Exp. Morph.* **33**, 447–57.

**Ford, P. J.** and **Southern, E. M.** (1973) Different sequences for 5S RNA in kidney cells and ovaries of *Xenopus laevis*, *Nature New Biol.* **241**, 7–12.

**Doolittle, Ford W.** and **Sapienza, C.** (1980) Selfish genes, the phenotype paradigm and genome evolution, *Nature* **284**, 601–3.

**Franco, M. G., Rubini, P. G.** and **Vecchi, M.** (1982) Sex determinants and their distribution in various populations of *Musca domestica* L. of Western Europe, *Genet. Res.* **40**, 279–93.

**Franke, W. W., Sheer, U., Trendelenburg, M., Zentgraf, H.** and **Spring, H.** (1978) Morphology of transcriptionally active chromatin, *Cold Spring Harbor Symp. Quant. Biol.* **42**, 755–60.
**Frankel, J.** and **Jenkins, L. M.** (1979) A mutant of *Tetrahymena thermophila* with a partial mirror-image duplication of cell surface pattern. II. Nature of genic control, *J. Embryol. Exp. Morph.* **49**, 203–27.
**French, V.** (1978) Intercalary regeneration around the circumference of the cockroach leg, *J. Embryol. Exp. Morph.* **47**, 53–84.
**French V., Bryant, P. J.** and **Bryant, S. V.** (1976) Pattern regulation in epimorphic fields, *Science* **193**, 966–81.
**Friedlander, M.** (1979) Immunological approaches to embryonic development and differentiation. In *Current Topics in Developmental Biology*, **13**, (Part I). Academic Press, New York.
**Friesen, H.** (1936) Spermatogoniales crossing-over bie *Drosophila*, *Z. Indukt. Abstamm. Vererbungsl.* **71**, 501–26.
**Fristrom, D.** (1969) Cellular degeneration in the production of some mutant phenotypes in *Drosophila melanogaster*, *Mol. Gen. Genet.* **103**, 363–79.
**Fritsch, E. F., Lawn, R. M.** and **Maniatis, T.** (1979) Characterisation of deletions which affect the expression of foetal globin genes in man, *Nature* **279**, 598–603.
**Fullilove, S. L.** and **Jacobson, A. G.** (1978) Embryonic Development—descriptive. In *The Genetics and Biology of* Drosophila. Vol. 2c. Eds. M. Ashburner and T. R. F. Wright. Academic Press, New York. 105–227.
**Fyrberg, E. A., Kindle, K. L.** and **Davidson, N.** (1980) The actin genes of *Drosophila*: a dispersed multigene family, *Cell* **19**, 365–78.

**Gaertner, F. H.** and **Cole, K. W.** (1975) A gene cluster: evidence for one gene, one polypeptide, five enzymes, *Biochem. Biophys.* Res. Commun. **75**, 259–64.
**Gafner, J.** and **Philippsen, P.** (1980) Common features of transposition—a yeast transposon also generates duplications of the target sequence, *Nature* **286**, 414–8.
**Galau, G. A., Klein, W. H., Davis, M. M., Wold, B. J., Britten, R. J.** and **Davidson, E. H.** (1976) Structural gene sets active in embryos and adult tissues of the sea urchin, *Cell* **7**, 487–505.
**Gall, J. G.** and **Pardue, M. L.** (1969) Formation and detection of RNA–DNA hybrid molecules in cytological preparations, *Proc. Natl. Acad. Sci. U.S.A.* **63**, 378–83.
**Gans, M.** (1953) Etude genetique et physiologique du mutant *z* de *Drosophila melanogaster*, *Bull. Biol. France Belg. (suppl.)* **38**, 1–90.
**Gans, M., Audit, C.** and **Masson, M.** (1975) Isolation and characterization of sex-linked female-sterile mutants in *Drosophila melanogaster*, *Genetics* **81**, 683–704.
**Garcia-Bellido, A.** (1975) Genetic control of wing disc development in *Drosophila*. In *Cell Patterning*. Ciba Foundation Symp. **29**, Elsevier, Amsterdam. 161–82.
**Garcia-Bellido, A.** (1977) Homoeotic and atavic mutations in insects, *Amer. Zool.* **17**, 613–29.
**Garcia-Bellido, A.** (1979) Genetic analysis of the achaeta-scute system of *Drosophila melanogaster*, *Genetics* **91**, 491–520.
**Garcia-Bellido, A.** (1981a) Opening remarks. In *Development and Neurobiology of* Drosophila. Ed. O. Siddiqui. Plenum, New York. 1–2.
**Garcia-Bellido, A.** (1981b) From the gene to the pattern: chaeta differentiation. In *Cellular Controls in Differentiation*. Ed. C. W. Lloyd and D. A. Rees. Academic Press, New York. 281–304.
**Garcia-Bellido, A.** and **Capdevila, M. P.** (1978) The initiation and maintenance of gene activity in a development pathway of *Drosophila*. In *Clonal Basis of Development*. Eds. S. Subtelny and I. M. Sussex. Academic Press, New York. 3–21.
**Garcia-Bellido, A.** and **del Prado, J. M.** (1979) Genetical analysis of maternal information in *Drosophila*, *Nature* **278**, 346–8.
**Garcia-Bellido, A.** and **Merriam, J. R.** (1969) Cell lineage of the imaginal disc in *Drosophila* gynandromorph, *J. Exp. Zool.* **170**, 61–76.
**Garcia-Bellido, A.** and **Merriam, J. R.** (1971) Genetic analysis of cell heredity in imaginal discs of *Drosophila melanogaster*, *Proc. Natl. Acad. Sci. U.S.A.* **68**, 2222–6.
**Garcia-Bellido, A., Ripoll, P.** and **Morata, G.** (1973) Developmental compartmentalisation of the wing disc of *Drosophila*, *Nature New Biol.* **245**, 251–3.
**Garcia-Bellido, A.** and **Santamaria, P.** (1972) Developmental analysis of the wing disc in the mutant *engrailed* of *Drosophila melanogaster*, *Genetics* **72**, 87–104.
**Garcia-Bellido, A.** and **Santamaria, P.** (1978) Developmental analysis of the achaeta-scute system of *Drosophila melanogaster*, *Genetics* **88**, 469–86.

**Gardner, R. L.** (1968) Mouse chimeras obtained by the injection of cells into the blastocyst, *Nature* **220**, 596–7.
**Gardner, R. L.** (1981) *In vivo* and *in vitro* studies on cell lineage and determination in the early mouse embryo. In *Cellular Controls in Differentiation*; Eds. C. W. Lloyd and D. A. Rees. Academic Press, London. 257–78.
**Gardner, R. L.** and **Rossant, J.** (1976) Determination during embryogenesis. In *Embryogenesis in Mammals*. Ciba Foundation Symp. **40**. Elsevier, Amsterdam. 5–25.
**Garen, A.** and **Gehring, W.** (1972) Repair of the lethal developmental defect in *deep orange* embryos of *Drosophila* by injection of normal egg cytoplasm, *Proc. Natl. Acad. Sci. U.S.A.* **69**, 2982–5.
**Garen, A., Kanvar, L.** and **Lepesant, J.-A.** (1977) Roles of ecdysone in *Drosophila* development, *Proc. Natl. Acad. Sci. U.S.A.* **74**, 5099–103.
**Gasc, J. M.** (1978) Growth and sexual differentiation in the gonads of chick and duck embryos, *J. Embryol. Exp. Morph.* **44**, 1–13.
**Gateff, E.** (1978a) The genetics and epigenetics of neoplasms in *Drosophila*, *Biol. Rev.* **53**, 123–68.
**Gateff, E.** (1978b) Malignant neoplasms of genetic origin in *Drosophila melanogaster*, *Science* **200**, 1448–59.
**Gateff, E.** and **Schneiderman, H. A.** (1974) Developmental capacities of benign and malignant neoplasms of *Drosophila*, *Roux's Arch.* **176**, 23–65.
**Gearhart, J. D.** and **Mintz, B.** (1972) Clonal origins of somites and their muscle derivatives: evidence from allophenic mice, *Devl. Biol.* **29**,27–37.
**Gehring, W. J.** (1967) Clonal analysis of determination dynamics in cultures of imaginal discs in *Drosophila melanogaster*, *Devl. Biol.* **16**, 438–56.
**Gehring, W. J.** (1970) A recessive lethal (l(4)29) with a homoeotic effect in *Drosophila melanogaster. Drosophila Inform. Serv.* **45**, 103.
**Gehring, W. J.** (1972) The stability of the determined state in cultures of imaginal discs in *Drosophila*. In *The Biology of Imaginal Discs*. Eds. H. Ursprung and R. Nöthiger. Springer–Verlag, New York. 35–58.
**Gehring, W. J.** and **Paro, R.** (1980) Isolation of a hybrid plasmid with homologous sequences to a transposing element of *Drosophila melanogaster. Cell* **79**, 897–904.
**Gelbart, W. M.** (1974) A new mutant controlling mitotic chromosome disjunction in *Drosophila melanogaster. Genetics* **76**, 51–63.
**Gelbart, W. M., McCarron, M., Pandey, J.** and **Chovnick, A.** (1974) Genetic limits of the xanthine dehydrogenase structural element within the *rosy* locus in *Drosophila melanogaster. Genetics* **78**, 869–86.
**Gelfant, S.** (1978) Tissue and tumor cell proliferation. In *Cell Reproduction*. Eds. **E. R. Durksen, D. M. Prescott** and **C. F. Fox**. Academic Press, New York. 689–99.
**Ghysen, A.** and **Richelle, J.** (1979) Determination of sensory bristles pattern formation in *Drosophila*. II. The Achaete-Scute Locus. *Devl. Biol.* **70**, 438–52.
**Gilbert, W.** (1978) Why genes in pieces? *Nature* **271**, 501.
**Girton, J. R.** (1981) Pattern triplications produced by a cell-lethal mutation in *Drosophila, Devl. Biol.* **84**, 164–72.
**Girton, J. R.** and **Bryant, P. J.** (1980) The use of cell lethal mutations in the study of *Drosophila* development, *Devl. Biol.* **77**, 233–43.
**Glass, B.** (1957) In pursuit of a gene, *Science* **126**, 683–9.
**Glassman, E.** (1965) Genetic regulation of xanthine dehydrogenase in *Drosophila melanogaster, Fed. Proc.* **24**,1243–51.
**Gloor, H.** (1947) Phänokopie-Versuche mit Ather an *Drosophila, Rev. Suisse Zool.* **54**, 637–712.
**Gloor, H.** (1954) Phänotypus der Heterozygoten bei der unvollstandig dominaten, homozygot lethalen mutante *kr* (= krüppel) von *Drosophila melanogaster, Arch. Klaus-Stift. VererbForsch.* **29**, 277–87.
**Glover, D. M.** (1980) *Genetic Engineering—Cloning DNA*. Chapman and Hall, London.
**Glover, D. M.** and **Hogness, S.** (1977) A novel arrangement of the 18S and 28S sequences in a repeating unit of *Drosophila melanogaster* rDNA, *Cell* **10**, 167–76.
**Glover, D. M., White, R. L., Finnegan, D. J.** and **Hogness, D. S.** (1975) Characterization of six cloned DNAs from *Drosophila melanogaster*, including one that contains the genes for rRNA, *Cell* **5**, 149–57.
**Goldschmidt, E.** and **Lederman-Klein, A.** (1958) Recurrence of a forgotten homoeotic mutant in *Drosophila*, *J. Hered.* **49**, 262–6.
**Goldschmidt, R.** (1938) *Physiological Genetics*. McGraw-Hill, New York.

**Goldschmidt, R.** (1940) *The Material Basis of Evolution.* Yale University Press, New Haven.

**Goldschmidt, R. B.** (1955) *Theoretical Genetics.* University of California Press, Berkeley.

**Goodman, C. S.** (1977) Neuron duplications and deletions in locust clones and clutches, *Science* **197**, 1384–6.

**Gordon, C.** (1936) The frequency of heterozygosis in free living populations of *Drosophila melanogaster* and *Drosophila subobscura*, *J. Genet.* **33**, 25–60.

**Gowen, M. S.** and **Gowen, J. W.** (1922) Complete linkage in *Drosophila melanogaster*, *Amer. Nat.* **56**, 286–8.

**Graham, C. F.** (1977) Teratocarcinoma cells and normal mouse embryogenesis. In *Concepts in Mammalian Embryogenesis.* Eds. M. I. Sherman. MIT Press, Cambridge, MA. 315–94.

**Graham, D. E., Neufeld, B. R., Davidson, E. H.** and **Britten, R. J.** (1974) Interspersion of repetitive and nonrepetitive DNA sequences in the sea urchin genome, *Cell* **1**, 127–38.

**Graves, J. A. M.** (1982) 5-Azacytidine-induced reexpression of alleles on the inactive X chromosome in a hybrid mouse cell line, *Expl. Cell Res.* **141**, 99–105.

**Green, M. M.** (1980) Transposable elements in *Drosophila* and other Diptera, *Ann. Rev. Genet.* **14**, 109–20.

**Green, M. M.** and **Green, K. C.** (1949) Crossing over between alleles at the *lozenge* locus in *Drosophila melanogaster, Proc. Natl. Acad. Sci. U.S.A.* **35**, 586–91.

**Griffin-Shea, R., Thireos, G.** and **Kafatos, F. C.** (1980) Chorion cDNA clones of *D. melanogaster* and their use in studies of sequence homology and chromosomal location of chorion genes, *Cell* **19**, 915–22.

**Gross, R. J.** (1957) The relation of skin to defect regulation in regenerating half limbs, *J. Morphol.* **100**, 547–64.

**Grosschedl, R.** and **Birnstiel, M. L.** (1980) Identification of regulatory sequences in the prelude sequences of an H2A histone gene by the study of specific deletion mutants *in vivo*, *Proc. Natl. Acad. Sci. U.S.A.* **77**, 1432–6.

**Grüneberg, H.** (1938) An analysis of the "pleiotropic" effects of a new lethal mutation in the rat (*Mus norvegicus*), *Proc. R. Soc. Lond.* **B125**, 123–44.

**Grunstein, M.** and **Hogness, D. T.** (1975) Colony hybridization: a method for the isolation of cloned DNAs that contain a specific gene, *Proc. Natl. Acad. Sci. U.S.A.* **72**, 3961–5.

**Gurdon, J. B.** (1974) *The Control of Gene Expression in Animal Development.* Oxford University Press, Oxford.

**Gurdon, J. B.** (1981) Concepts of gene control in development. In *Developmental Biology Using Purified Genes.* ICN–UCLA Vol. 23; Ed. **D. D. Brown.** Academic Press, New York, 1–10.

**Gurdon, J. B., Laskey, R. A.** and **Reeves, O. R.** (1975) The developmental capacity of nuclei transplanted from keratinised skin cells of adult frogs, *J. Embryol. Exp. Morph.* **34**, 93–112.

**Gurdon, J. B.** and **Melton, D. A.** (1981) Gene transfer in amphibian eggs and oocytes, *Ann. Rev. Genet.* **15**, 189–218.

**Gvozdev, V. A., Gerasimova, T. I., Birstein, V. Y.** (1973) Inactivation of the structural gene of 6-phosphogluconate dehydrogenase when transferred to heterochromatin in *Drosophila melanogaster*, *Genetika* **9**, 64–72.

**Haber, J. E.** and **Rogers, D. T.** (1982) Transposition of a tandem duplication of yeast mating-type genes. *Nature* **296**, 768–90.

**Hadlaczky, G., Summer, A. T.** and **Ross, A.** (1981) Protein depleted chromosomes. II. Experiments concerning the reality of chromosome scaffolds. *Chromosoma* **81**, 557–67.

**Hadorn, E.** (1945) Zur Pleiotropie der Genwirkurg. *Arch. Jul. Klaus-Stiftg.* **20**, 82–95.

**Hadorn, E.** (1960) *Developmental Genetics and Lethal Factors.* Methuen, London.

**Hadorn, E.** (1963) Differenzierungsleitungen wiederholt fragmentierter Teilstücke mänlicker Genitalscheiben von *Drosophila melanogaster* nach Kultur *in vivo. Devl. Biol.* **7**, 617–29.

**Hadorn, E.** (1965) Problems of determination and transdetermination. In *Genetic Control of Differentiation*, Brookhaven National Laboratory, New York.

**Hadorn, E.** (1978) Transdetermination. In *Genetics and Biology of* Drosophila. Vol. 2c. Eds. M. Ashburner and T. R. F. Wright. Academic Press, New York. 556–617.

**Hall, J. C., Gelbart, W. M.** and **Kankel, D. R.** (1976) Mosaic Systems. In *The Genetics and Biology of* Drosophila. Vol. 1a. Eds. M. Ashburner and E. Novitski. Academic Press, New York. 265–314.

**Hämmerling, J.** (1963) Nucleo-cytoplasmic interactions in *Acetabularia* and other cells, *Ann. Rev. Plant Physiol.* **14**, 65–92.
**Ham, R. G.** and **Veomett, M. J.** (1980) *Mechanisms of Development*, C.V. Mosby Co., St. Louis.
**Handler, A. M.** (1982) Ecdysteroid titres during pupal and adult development in *Drosophila melanogaster*, *Devl. Biol.* **93**, 73–82.
**Hannah-Alava, A.** and **Stern, C.** (1957) The sex-combs in males and intersexes of *Drosophila melanogaster*, *J. Exp. Zool.* **134**, 533–56.
**Harper, M. I., Fosten, M.** and **Monk, M.** (1982) Preferential paternal X-inactivation in extra-embryonic tissues of early mouse embryos, *J. Embryol. Exp. Morph.* **67**, 127–35.
**Harris, H.** (1967) The reactivation of the red cell nucleus, *J. Cell. Sci.* **2**, 23–32.
**Harris, H.** and **Klein, G.** (1969) Malignancy of somatic cell hybrids, *Nature* **224**, 1315–6.
**Harrison, R. G.** (1907) Observations on the living developing nerve fiber, *Proc. Soc. Exp. Biol.* **4**, 140–3.
**Hartman, P. E.** and **Roth, J. R.** (1973) Mechanisms of suppression, *Adv. Genet.* **17**, 1–105.
**Hayashi, S., Gillam, I. C., Delaney, A. D., Dunn, R., Tener, G. M., Grigliatti, T. A.** and **Suzuki, D. T.** (1980) Hybridization of tRNAs of *Drosophila melanogaster* to polytene chromosomes, *Chromosoma* **76**, 65–84.
**Haynie, J. L.** and **Bryant, P. J.** (1976) Intercalary regeneration in the imaginal wing disc of *Drosophila melanogaster*, *Nature* **259**, 659–62.
**Henikoff, S.** and **Meselson, M.** (1977) Transcription at two heat shock loci in *Drosophila*, *Cell* **12**, 441–51.
**Hernandorena, A.** (1980) Programmation of postembryonic development in *Artemia* by dietary supplies of purine and pyrimidine. In *The Brine shrimp* Artemia. Vol. 2. Eds. G. Persoone, P. Sorgeloos, O. Roels and E. Jaspers. Universa Press, Belgium. 209–18.
**Herskowitz, I. H.** (1949) *Hexaptera*, a homoeotic mutant in *Drosophila melanogaster*, *Genetics* **34**, 10–25.
**Herskowitz, I., Blair, L., Forbes, D., Hicks, J., Kassir, Y., Kuskuer, P., Rine, J., Sprague, G.** and **Strathern, J.** (1980) Control of cell type in the yeast *Saccharomyces cerevisiae* and a hypothesis for development in higher eukaryotes. In *The Molecular Genetics of Development.* Eds. T. Leighton and W. F. Loomis. Academic Press, New York. 79–118.
**Hess, O.** (1975) Y-linked factors affecting male fertility in *Drosophila melanogaster* and *Drosophila hydei*. In *Handbook of Genetics*. Vol. 3. Ed. R. C. King. Plenum, New York. 747–56.
**Hicks, J. B.** and **Herskowitz, I.** (1976) Interconversion of yeast mating types. I.Direct observation of the action of the homothallism gene, *Genetics* **83**, 245–8.
**Hicks, J., Strathern, J.** and **Klar, A. J. S.** (1979) Transposable mating type genes in *Saccharomyces cerevisiae*, *Nature* **282**, 478–83.
**Hilliker, A. J., Clark, S. H.** and **Chovnick, A.** (1980) Cytogenetic analysis of the chromosomal region immediately adjacent to the rosy locus in *Drosophila melanogaster*, *Genetics* **95**, 95–110.
**Hillman, N.** and **Hillman, R. J.** (1975) Ultra-structural studies of $t^{w32}/t^{w32}$ mouse embryos, *J. Embryol. Exp. Morph.* **33**, 685–95.
**Hillman, N., Sherman, M. I.** and **Graham, C.** (1972) The effect of spatial arrangement on cell determination during mouse development, *J. Embryol. Exp. Morph.* **28**, 253–78.
**Hillman, R.** and **Wright, T. R. F.** (1977) Perspective—*Drosophila* and development, *Amer. Zool.* **17**, 519–20.
**Hiraizumi, Y.** (1971) Spontaneous recombination in *Drosophila melanogaster* males, *Proc. Natl. Acad. Sci. U.S.A.* **60**, 268–70.
**Hirsh, J.** and **Davidson, N.** (1981) Isolation and characterization of the dopa decarboxylase gene of *Drosophila melanogaster*, *Mol. Cell Biol.* **1**, 475–85.
**Hochman, B.** (1976) The fourth chromosome of *Drosophila melanogaster*. In *The Genetics and Biology of* Drosophila. Vol. 1b. Eds. M. Ashburner and E. Novitski. Academic Press, New York. 903–28.
**Hodgetts, R. B.** (1975) The response of dopa decarboxylase activity to variations in gene dosage in *Drosophila*: a possible location of the structural gene, *Genetics* **79**, 45–54.
**Hodgkin, J. A.** (1980) More sex-determination mutants of *Caenorhabditis elegans*, *Genetics* **96**, 649–64.
**Hodgkin, J. A.** and **Brenner, S.** (1977) Mutations causing transformation of the sexual phenotype in the nematode, *Caenorhabditis elegans*, *Genetics* **86**, 275–87.
**Hofer, E.** and **Darnell, J. E.** (1981) The

primary transcription unit of the mouse beta-major globin gene, *Cell* **23**, 585–93.

**Holiday, R.** and **Pugh, J. E.** (1975) DNA modification mechanisms and gene activity during development, *Science* **187**, 226–32.

**Holmgren, R., Livak, J., Morimoto, R., Freund, R.** and **Meselson, M.** (1979) Studies of cloned sequences from four *Drosphila* heat shock loci, *Cell* **18**, 1359–70.

**Holtfreter, J.** and **Hamburger, V.** (1956) Embryogenesis: progressive differentiation. Amphibians. In *Analysis of Development*. Eds. B. H. Willier, P. A. Weiss and V. Hamburger. Saunders, Philadelphia.

**Holtzer, H., Rubinstein, N., Fellini, S., Yeoh, G., Chi, J., Birnbaum, J.** and **Okayama, M.** (1975) Lineages, quantal cell cycles, and the generation of cell diversity, *Q. Rev. Biophys.* **8**, 523–57.

**Holzworth, K. W., Gottlieb, F. J.** and **Spector, C.** (1974) A unique case of female sterility in *Drosophila melanogaster*, *Roux's Arch.* **174**, 267–75.

**Horstadius, S.** (1935) Über die Determination im Verlaufe der Eiachse bei Seeigeln, *Publ. Staz. Zool. Napoli.* **14**, 251–79.

**Horstadius, S.** (1939) The mechanics of sea urchin development studied by operative methods; *Biol. Rev. Camb. Phil. Soc.* **14**, 132–79.

**Hotta, Y.** and **Benzer, S.** (1972) Mapping of behaviour in *Drosophila* mosaics, *Nature* **240**, 527–35.

**Howard, G. C., Abmayr, S. M., Shinefeld, L. A., Sato, V. L.** and **Elgin, S. C. R.** (1981) Monoclonal antibodies against a specific nonhistone chromosomal protein of *Drosophila* associated with active genes, *J. Cell. Biol.* **88**, 219–25.

**Hovemann, B.** and **Galler, R.** (1982) Vitellogenin in *Drosophila*: a comparison of the YPI and YPII genes and their transcription products, *Nucleic Acids Res.* **10**, 2261–74.

**Hung, M.-C.** and **Wensink, P. C.** (1981) The sequence of the *Drosophila melanogaster* gene for yolk protein 1, *Nucleic Acids Res.* **9**, 6407–19.

**Huttner, K. M., Scangos, G.** and **Ruddle, F. H.** (1979) DNA-mediated gene transfer of a circular plasmid into murine cells, *Proc. Natl. Acad. Sci. U.S.A.* **76**, 5820–4.

**Illmensee, K.** (1976) Nuclear and cytoplasmic transplantation in *Drosophila*. In *Insect Development*, Symp. R. Entomol. Soc. Lond., No. 8. Ed. P. E. Lawrence, Wiley, New York. 76–96.

**Illmensee, K.** and **Hoppe, P. C.** (1981) Nuclear transplantation in *Mus musculus*: Developmental protential of nuclei from preimplantation embryos, *Cell* **23**, 9–18.

**Illmensee, K., Mahowald, A. P.** and **Loomis, M. R.** (1976) The ontogeny of germ plasm during oogenesis in *Drosophila*, *Devl. Biol.* **49**, 40–65.

**Illmensee, K.** and **Mintz, B.** (1976) Totipotency and normal differentiation of single teratocarcinoma cells cloned by injection into blastocysts. *Proc. Natl. Acad. Sci. U.S.A.* **73**, 549–53.

**Ilyin, Y. V., Chemeliauskaite, V. G., Ananiev, E. V.** and **Georgiev, G. P.** (1980) Isolation and characterisation of a new family of mobile dispersed genetic elements, mdg 3, in *Drosophila melanogaster. Chromosoma* **81**, 27–53.

**Ingham, P. W.** (1981) *Trithorax*: a new homoeotic mutation of *Drosophila melanogaster*. II. The role of $trx^+$ after embryogenesis, *Roux's Arch.* **190**, 365–9.

**Ingham, P. W.** and **Whittle, J. R. S.** (1980) *Trithorax*: a new homoeotic mutation of *Drosophila melanogaster* causing transformations of abdominal and thoracic imaginal segments, *Mol. Gen. Genet.* **179**, 607–14.

**Ingolia, T. D.** and **Craig, E. A.** (1981) Primary sequence of the 5′ flanking regions of *Drosophila* heat shock genes in chromosome subdivision 67B, *Nucleic Acids Res.* **9**, 1627–42.

**Inze, D., Gheysen, G., Van Montague, M.** and **Schell, J.** (1981) Region specific transposon mutagenesis of a Ti plasmid of *Agrobacterium tumefaciens*, *Arch Int. Phys. Biochem.* **89**, B176.

**Ish-Horowicz, D., Holden, J. J.** and **Gehring, W. J.** (1977) Deletions of two heat activated loci in *D. melanogaster* and their effects on heat induced protein synthesis, *Cell* **12**, 643–52.

**Ish-Horowicz, D.** and **Pinchin, S. M.** (1980) Genomic organisation of the 87A7 and 87C1 heat-induced loci of *Drosophila melanogaster*, *J. Mol. Biol.* **142**, 231–45.

**Ising, G.** and **Block, K.** (1981) Derivation-dependant distribution of insertion sites for a *Drosophila* transposon, *Cold Spring Harbor Symp. Quant. Biol.* **45**, 527–44.

**Ising, G.** and **Ramel, C.** (1976) The behaviour of a transposing element in *Drosophila melanogaster*. In *The Genetics and Biology of* Drosophila. Vol. 1b. Eds. M. Ashburner and E. Novitski. Academic Press, New York. 947–54.

**Jack, J. W.** and **Judd, B. H.** (1979) Allelic pairing and gene regulation: a model for the *zeste-white* interaction in *Drosophila melanogaster*, *Proc. Natl. Acad. Sci. U.S.A.* **76**, 1368–72.
**Jack, R. S., Gehring, W. J.** and **Brack, C.** (1981) Protein component from *Drosophila* larval nuclei showing sequence specificity for a short region near a major heat-shock protein gene, *Cell* **24**, 321–31.
**Jackle, H.** and **Kalthoff, K.** (1981) Proteins foretelling head or abdomen development in the embryo of *Smittia* spec, *Devl. Biol.* **85**, 287–98.
**Jacobsen, J. V.** (1977) Regulation of ribonucleic acid metabolisms by plant hormones, *Ann. Rev. Plant Physiol.* **28**, 537–64.
**James, A. A.** and **Bryant, P. J.** (1981) Mutations causing pattern deficiencies and duplications in the imaginal wing disc of *Drosophila melanogaster*, *Devl. Biol.* **85**, 39–54.
**Jamrich, M., Greenleaf, A. L.** and **Bautz, E. K. F.** (1977) Localization of RNA polymerase in polytene chromosomes of *Drosophila melanogaster*, *Proc. Natl. Acad. Sci. U.S.A.* **74**, 2079–83.
**Janning, W.** (1974) Entwicklungsgenetische Untersuchungen am Gynandern von *Drosophila melanogaster* 1. Die inern Organe der Imago, *Roux's Arch.* **174**, 313–32.
**Janning, W.** (1978) Gynandromorph fate maps in *Drosophila*. In *Results and Problems in Cell Differentiation, Genetic Mosaics and Cell Differentiation*. Vol. 9. Ed. W. J. Gehring. Springer–Verlag, Berlin. 1–28.
**Jarry, B P.** (1976) Isolation of a multifunctional complex containing the first three enzymes of pyrimidine biosynthesis in *Drosophila melanogaster*, *FEBS Lett.* **70**, 71–5.
**Jarry, B. P.** (1979) Genetical and cytological location of the structural parts coding for the first three steps of pyrimidine biosynthesis in *Drosophila melanogaster*, *Mol. Gen. Genet.* **172**, 199–202.
**Jarry, B.** and **Falk, D. R.** (1974) Functional diversity within the *rudimentary* locus of *Drosophila melanogaster*, *Mol. Gen. Genet.* **135**, 113–22.
**Jelinek, W.** and **Leinwand, J.** (1978) Low molecular weight RNAs hydrogen-bonded to nuclear and cytoplasmic poly(A)-terminated RNA from cultured Chinese hamster ovary cells, *Cell* **15**, 205–14.
**Jelinek, W. G., Toomey, T. P., Leinwand, L., Duncan, C. H., Biro, P. A., Choudary, P. W., Weismann, S. M., Rubin, C. M., Houck, C. M., Deininger, P. L.** and **Schmidt, C. N.** (1980) Ubiquitous, interspersed repeated sequences in mammalian genomes, *Proc. Natl. Acad. Sci. U.S.A.* **77**, 1398–402.
**Jimenez, F.** and **Campos-Ortega, J.** (1979) A region of the *Drosophila* genome necessary for central nervous development, *Nature* **282**, 310–12.
**John, H. A., Birnstiel, M. L.** and **Jones, K. W.** (1969) RNA–DNA hybrids at the cytological level, *Nature* **223**, 582–7.
**Johnson, D. R.** (1974) Hairpintail: A case of post-reductional gene action in the mouse egg? *Genetics* **76**, 795–805.
**Johnson, E. M.** and **Allfrey, V. G.** (1978) Postsynthetic modifications of histone primary structure. In *Biochemical Actions of Hormones*. Ed. G. Litwack. Academic Press, New York. 2–51.
**Johnson, J. H.** and **King, R. C.** (1972) Studies on *fes*, a mutation affecting cytokinesis, in *Drosophila melanogaster*, *Biol. Bull Woods Hole* **143**, 525–47.
**Johnson, M. H.** and **Ziomek, C.** (1981) The foundation of two distinct cell lineages within the mouse morula, *Cell* **24**, 71–80.
**Jost, A.** (1965) Gonadal hormones in the sex differentiation of the mammalian foetus. In *Organogenesis*. Ed. R. L de Haan and H. Ursprung. Holt, New York. 611–28.
**Jowett, T.** and **Sang, J. H.** (1979) Nutritional regulation of antennal/leg homoeotic mutants in *Drosophila melanogaster*, *Genet. Res.* **34**, 143–61.
**Judd, B. H.** (1974) Genes and chromosomes of *Drosophila*. In *The Eukaryotic Chromosome*. Eds. W. J. Peacock and R. D. Brock. Australian National University Press, Canberra. 169–84.
**Judd, B. H.** (1976) Genetic Units of *Drosophila*: Complex loci. In *The Genetics and Biology of* Drosophila. Vol. 1b. Eds. M. Ashburner and E. Novitski, Academic Press, New York. 169–84.
**Judd, B. H.** and **Young, M. W.** (1973) An examination of the one cistron–one chromosome concept, *Cold Spring Harbor Symp. Quant. Biol.* **38**, 573–9.

**Kacser, H.** and **Burns, J. A.** (1981) The molecular basis of dominance, *Genetics* **97**, 639–66.

**Kafatos, F. C., Maniatis, T., Efstratiadis, A., Kee, S. G., Regier, J. C.** and **Nadel, M.** (1977) The moth chorion as a system for studying the structure of developmentally regulated gene sets. In *The Organization and Expression of the Eukaryotic Genome*. Eds. E. M. Bradbury and K. Javaherian. Academic Press, New York. 393–420.

**Kahan, B.** and **DeMars, R.** (1975) Localized derepression on the human inactive X-chromosome in mouse–human cell hybrids, *Proc. Natl. Acad. Sci. U.S.A.* **72**, 1510–4.

**Kaleko, M.** and **Rothman, F. G.** (1982) Membrane sites regulating developmental gene expression in *Dictyostelium discoideum*, *Cell* **28**. 801–11.

**Kalthoff, K.** (1979) Analysis of a morphogenetic determinant in an insect embryo (*Smittia* spec., Chironomidae, Diptera). In *Determinants of Spatial Organization*. Eds. S. Subtelny and I. Konigsberg. Academic Press, New York. 97–126.

**Kamalay, J. C.** and **Goldberg, R. B.** (1980) Regulation of structural gene expression in tobacco, *Cell* **19**, 935–46.

**Kankel, D. R.–1 and Hall, J. C.** (1976) Fate mapping of the nervous system and other internal tissues in genetic mosaics of *Drosophila melanogaster*, *Devl. Biol.* **48**, 1–24.

**Karlsson, J.** (1983) Morphogenesis and compartments in *Drosophila*. In *Pattern Formation*. Macmillan, London (in press).

**Karpen, G. H.** and **Schubiger, G.** (1981) Extensive regulatory capabilities of a *Drosophila* imaginal disc blastema, *Nature* **294**, 744–7.

**Kauffman, S. A.** (1973) Control circuits for determination and transdetermination, *Science* **181**, 310–18.

**Kauffman, S. A.** (1980) Heterotopic transplantation in the syncytial blastoderm of *Drosophila*: Evidence for anterior and posterior commitments, *Roux's Arch.* **189**, 135–45.

**Kauffman, S. A.** (1981) Patterns of temperature sensitivity in *Contrabithorax/Ultrabithorax* heterozygotes of *Drosophila*, *Devl. Biol.* **88**, 341–51.

**Kauffman, S. A.** and **Ling, E.** (1980) Timing and heritability of the *Nasobemia* transformation in *Drosophila*, *Roux's Arch.* **189**, 147–53.

**Kauffman, S. A., Shymko, R. M** and **Tarbert, K.** (1978) Control of sequential compartment formation in *Drosophila*, *Science* **199**, 259–70.

**Kaufman, T. C.** (1978) Cytogenetic analysis of chromosome 3 in *Drosophila melanogaster*: isolation and characterization of four new alleles of the *Proboscipedia* (pb) locus, *Genetics* **90**, 579–96.

**Kaufman, T. C., Lewis, R.** and **Wakimoto, B.** (1980) Cytogenetic analysis of chromosome 3 in *Drosophila melanogaster*: the homoeotic gene complex in polytene chromosome interval 84A–B, *Genetics* **94**, 115–33.

**Kaufman, T. C., Tasaka, S. E.** and **Suzuki, D. T.** (1973) The interaction of two complex loci, *zeste* and *bithorax* in *Drosophila melanogaster*, *Genetics* **75**, 299–321.

**Kavenoff, R.** and **Zimm, B. H.** (1973) Chromosome sized DNA molecules from *Drosophila*, *Chromosoma* **41**, 1–28.

**Kay, R. R.** and **Trevan, D. J.** (1981) *Dictyostelium* amoebae can differentiate into spores without cell-to-cell contact, *J. Embryol. Exp. Morph.* **62**, 369–78.

**Keene, M. A., Corces, V., Lowenhaupt, K.** and **Elgin, S. C. R.** (1981) DNase I hypersensitive sites in *Drosophila* chromatin occur at the 5′ ends of regions of transcription, *Proc. Natl. Acad. Sci. U.S.A.* **78**, 143–6.

**Kerridge, S.** (1981) Distal into Proximal (*Dipr*): a homoeotic mutation of *Drosophila melanogaster*, *Mol. Gen. Genet.* **184**, 519–25.

**Kerridge, S.** and **Sang, J. H.** (1981) Developmental analysis of the homoeotic mutation *bithoraxoid* of *Drosophila melanogaster*, **61**, 69–86.

**Kidd, S. J.** and **Glover, D. M.** (1980) A DNA segment from *D. melanogaster* which contains five tandemly repeating units homologous to the major rDNA insertion, *Cell* **19**, 103–19.

**Kidwell, M. G., Kidwell, J. F.** and **Sved, J. A.** (1977) Hybrid dysgenesis in *Drosophila melanogaster*. A syndrome of aberrant traits including mutation, sterility and male recombination, *Genetics* **86**, 813–33.

**Kimble, J.** (1981) Alterations in the cell lineage following laser ablation of cells in the somatic gonad of *Caenorhabditis elegans*, *Devl. Biol.* **87**, 286–300.

**King, R. C.** (1970) *Ovarian development in* Drosophila melanogaster. Academic Press, New York.

**King, R. C., Bahns, M., Horowitz, R.** and **Larramendi, P.** (1978) A mutation that affects female and male germ cells differentially in *Drosophila melanogaster*, *Int. J. Insect Morphol. Embryol.* **7**, 359–75.

**King, R. C.** and **Mohler, J. D.** (1975) The genetic analysis of oogenesis in *Drosophila melanogaster*. In *Handbook of Genetics*. Vol. 3. Ed. R. C. King. Plenum, New York. 757–92.

**King, R. C., Riley, S. F., Cassidy, J. D.** and **White, P. E.** (1981) Giant polytene chromosomes from the ovaries of a *Drosophila* mutant, *Science* **212**, 441–3.

**King, T. J.** (1979) Prologue: nuclear transplantation in amphibia, *Int. Rev. Cytol. Suppl.* **9**, 101–8.

**Klar, A. J. S., McIndoo, J., Strathern,** and **Strathern, J. N.** (1980) Precise mapping of the homothallism genes HML and HMR in *Saccharomyces cerevisiae*, *Genetics* **96**, 315–20.

**Klar, A. J. S., McIndoo, J., Strathern, J. N.** and **Hicks, J. B.** (1980) Evidence for a physical interaction between the transposed and substituted sequences during mating type gene transposition in yeasts, *Cell* **22**, 291–8.

**Klar, A. J. S., Strathern, J. N., Broach, J. B.** and **Hicks, J. B.** (1981) Regulation of transcription in expressed and unexpressed mating type cassettes in yeast, *Nature* **289**, 293.

**Klein, G.** (1981) The role of gene dosage and genetic transpositions in carcinogenesis, *Nature* **294**, 313–8.

**Klug, A., Rhodes, D., Smith, J., Finch, J. T.** and **Thomas, J. O.** (1980) A low resolution structure for the histone core of the nucleosome, *Nature* **287**, 509–16.

**Knox, W. E.** (1976) *Enzyme Patterns in Fetal, Adult and Neoplastic Rat Tissue*. Karger, Basel.

**Koch, E. A., Smith, P. A.** and **King, R. C.** (1967) The division and differentiation of *Drosophila* cystocytes, *J. Morphol.* **121**, 55–70.

**Köhler, G.** and **Milstein, C.** (1975) Continuous cultures of fused cells secreting antibody of predefined specificity, *Nature* **256**, 495–7.

**Komma, D. J.** (1966) Effect of sex transformation genes on glucose-6-phosphate dehydrogenase activity in *Drosophila melanogaster*, *Genetics* **54**, 497–503.

**Konopka, R. J.** and **Benzer, S.** (1971) Clock mutants of *Drosophila melanogaster*, *Proc. Natl. Acad. Sci. U.S.A.* **68**, 2112–6.

**Koo, G. C., Wachtel, S. S., Krupen-Brown, K., Mittl, L. R., Breg, W. R., Genel, M., Rosenthal, I. M., Borgaonkar, D. S., Miller, D. A., Tantravahi, R., Schreck, R. R., Erlanger, B. F.** and **Miller, O. J.** (1977) Mapping the locus of the H-Y gene on the human Y chromosome, *Science* **198**, 940–2.

**Korge, G.** (1977a) Larval saliva in *Drosophila melanogaster*: Production, composition and relationship to chromosome puffs, *Devl. Biol.* **58**, 339–55.

**Korge, G.** (1977b) Direct correlation between a chromosome puff and the synthesis of a larval saliva protein in *Drosophila melanogaster*, *Chromosoma* **62**, 155–74.

**Korge, G.** (1980) Gene activities in larval salivary glands of insects, *Verh. Dtsch. Zool. Ges.* 94–110.

**Korge, G.** (1981) Genetic analysis of the larval secretion gene *sgs-4* and its regulatory chromosome sites in *Drosophila melanogaster*, *Chromosoma* **84**, 373–90.

**Korn, L. J.** and **Gurdon, J. B.** (1981) The reactivation of developmentally inert 5S genes in somatic nuclei injected into *Xenopus* oocytes, *Nature* **289**, 461–5.

**Kornberg, T.** (1981) Compartments in the abdomen of *Drosophila* and the role of the *engrailed* locus, *Devl. Biol.* **86**, 363–72.

**Kraminsky, G. P., Clark, W. C., Estelle, M. A., Gietz, R. D., Sage, B. A., O'Connor, J. D.** and **Hodgetts, R. B** (1980) Induction of translatable mRNA for dopa decarboxylase in *Drosophila*: An early response to ecdysterone, *Proc. Natl. Acad. Sci. U.S.A.* **77**, 4175–9.

**Kress, H.** (1981) Ecdysone-induced puffing in *Drosophila*: a model, *Naturwissenschaften* **68**, 28–33.

**Kressman, A.** and **Birnstiel, M. L.** (1980) Surrogate genetics in the frog oocyte. In *Transfer of Cell Constituents into Eukaryotic Cells*. Eds. J. E. Celis, A. Grassmann and A. Loyter. Plenum, New York. 383–407.

**Kuhn, D. T.** and **Cunningham, G. N.** (1977) Aldehyde oxidase compartmentalization in *Drosophila melanogaster* wing imaginal discs, *Science* **196**, 875–877.

**Kuhn, D. T., Woods, D. F.** and **Andrew, D. J.** (1981a) Deletion analysis of the tumorous-head (*tuh-3*) gene in *Drosophila melanogaster*, *Genetics* **99**, 99–107.

**Kuhn, D. T., Woods, D. F.** and **Cook, J. L.** (1981b) Analysis of a new homoeotic mutation *iab-2* within the bithorax complex in *Drosophila melanogaster*, *Mol. Gen. Genet.* **181**, 82–6.

**Kuhn, D. T., Züst, B.** and **Illmensee, K.** (1979) Autonomous differentiation of the Tumorous-Head phenotype in *Drosophila melanogaster*, *Mol. Gen. Genet.* **168**,117–24.

**Lai, E. C., Woo, S. L. C., Bordelon-Riser, M. E., Fraser, T. H.** and **O'Malley, B. W.** (1980) Ovalbumin is synthesized in mouse cells transformed with the natural chicken ovalbumin gene, *Proc. Natl. Acad. Sci. U.S.A.* **77**, 244–8.

**Lakhotia, S. C.** (1971) Benzamide as a tool for gene activity studies in *Drosophila*, *Proc. IVth Cell Biol. Conf.*, New Delhi.

**Lakhotia, S. C.** and **Mukherjee, A. S.** (1969) Chromosomal basis of dosage compensation in *Drosophila*: Cellular autonomy of hyperactivity of the male-X chromosome in salivary gland and sex differentiation, *Genet. Res.* **14**, 137–50.

**Laird, C. D.** (1971) Chromatid structure: relationship between DNA content and nucleotide sequence diversity, *Chromosoma* **32**, 378–406.

**Laird, C. D.** (1980) Structural paradox of polytene chromosomes, *Cell* **22**, 869–74.

**Lamb, M. M.** and **Daneholt, B.** (1979) Characterization of active transcription units in Balbiani rings of *Chironomus tentans*, *Cell* **7**, 835–48.

**Lamb, M. M.** and **Laird, C. D.** (1976) Increase in nuclear poly(A)-containing RNA at syncytial blastoderm in *Drosophila melanogaster* embryos, *Devl. Biol.* **52**, 31–42.

**Lambert, B.** (1973) Tracing of RNA from a puff in a polytene chromosomes to the cytoplasm in *Chironomus tentans* salivary gland cells, *Nature* **242**, 51.

**Lambert, B.** and **Edstrom, J.-E.** (1974) Balbiani ring nucleotide sequences in cytoplasmic 75S RNA of *Chironomus tentans* salivary gland cells, *Mol. Biol. Rep.* **1**, 457–64.

**Landstrom, U.** and **Løvtrup, S.** (1979) Fatemaps and cell differentiation in the amphibian embryo—an experimental study, *J. Embryol. Exp. Morph.* **54**, 113–30.

**Lastowski, D. M.** and **Falk, D. R.** (1980) Characterization of an autosomal rudimentary-shaped wing mutation in *Drosophila melanogaster* that affects pyrimidine synthesis, *Genetics* **96**, 471a–9.

**Laufer, J. S., Bazzicalupo, P.** and **Wood, W. B.** (1980) Segregation of developmental potential in early embryos of *Caenorhabditis elegans*, *Cell* **19**, 569–77.

**Lawrence, P. A.** (1966) Development of hairs and bristles in the milkweed bug, *Oncopeltus fasciatus*, *J. Cell Sci.* **1**, 475–98.

**Lawrence, P. A.** (1981a) A general cell marker for clonal analysis of *Drosophila* development, *J. Embryol. Exp. Morph.* **64**, 321–32.

**Lawrence, P. A.** (1981b) The cellular basis of segmentation in insects, *Cell* **26**, 3–10.

**Lawrence, P. A.** and **Brower, D. L.** (1982) Myoblasts from *Drosophila* wing disc can contribute to developing muscle throughout the fly, *Nature* **295**, 55–7.

**Lawrence, P. A.** and **Green, S. M.** (1979) Cell lineage in the developing retina of *Drosophila*, *Devl. Biol.* **71**, 142–52.

**Lawrence, P. A.** and **Morata, G.** (1977) The early development of mesothoracic compartments in *Drosophila*, *Devl. Biol.* **56**, 40–51.

**Lawrence, P. A.** and **Struhl, G.** (1982) Further studies of the *engrailed* phenotype in *Drosophila*, *EMBO J.* **1**, 827–33.

**Leder, P.** (1980) The organization and expression of cloned globin genes, *The Harvey Lectures* **74**, 81–100.

**Leder, A.** and **Leder, P.** (1975) Butyric acid, a potent inducer of erythroid differentiation in cultured erythroleukemic cells, *Cell* **5** 319–22.

**Lee, L. W.** and **Gerhart, J. C.** (1973) Dependance of transdetermination frequency on the developmental stage of cultured imaginal discs of *Drosophila melanogaster*, *Devl. Biol.* **35**, 62–82.

**LeFever, H. M.** (1973) Analysis of three *white* mutants resulting in two new recombination sites at the white locus in *Drosophila melanogaster*. *Drosophila Inform. Serv.* **50**, 109–10.

**LeFevre, G.** (1973) The one band–one gene hypothesis: evidence from a cytogenetic analysis of mutant and non-mutant rearrangement break point in *Drosophila melanogaster*, *Cold Spring Harbor Symp. Quant. Biol.* **38**, 591–9.

**LeFevre, G.** (1976) A photographic representation and interpretation of the polytene chromosomes of *Drosophila melanogaster* salivary glands. In *The Genetics and Biology of* Drosophila. Vol. la. Eds. M. Ashburner and E. Novitski, Academic Press, New York. 31–66.

**Leupold, U.** (1980) Transposable mating-type genes in yeasts, *Nature* **283**, 811–2.

**Levenson, R.** and **Housman, D.** (1981) Commitment: How do cells make the decision to differentiate, *Cell* **25**, 5–6.

**Levis, R., Bingham, P. M.** and **Rubin, G. M.** (1982a) Physical map of the *white* locus of *Drosophila melanogaster*, *Proc. Natl. Acad. Sci. U.S.A.* **79**, 564–8.

**Levis, R., Collins, M.** and **Rubin, G. M.** (1982b) FB elements are the common basis for the instability of the $W^{D2L}$ and $W^{C}$ *Drosophila* mutations, *Cell* **30**, 551–65.

**Levis, R., Dunsmuir, P.** and **Rubin, G. M** (1980) Terminal repeats of the *Drosophila* transposable element *copia*: nucleotide sequence and genomic organization, *Cell* **21**, 581–8.
**Lewin, B.** (1980a) Alternatives for splicing: recognizing the ends of introns, *Cell* **22**, 324–6.
**Lewin, B.** (1980b) Alternatives for splicing: an intron-coded protein, *Cell*, **22**, 645–6.
**Lewin, B.**(1980c) *Gene expression*. Vol. 2 Eukaryotic Chromosomes 2nd edn. Wiley, New York.
**Lewis, E. B.** (1950) The phenomenon of position effect, *Adv. Genet.* **3**, 73–115.
**Lewis, E. B.** (1955) Some aspects of position pseudoallelism *Amer. Nat.* **89**, 73–89.
**Lewis, E. B.** (1964) Genetic control and regulation of developmental pathways. In *The Role of Chromosomes in Development*. Ed. M. Locke. Academic Press, New York. 231–52.
**Lewis, E. B.** (1968) Genetic control of developmental pathways in *Drosophila melanogaster*. *Proc. XII. Int. Congr. Genet., Tokyo* **2**, 96–7.
**Lewis, E. B.** (1978) A gene complex controlling segmentation in *Drosophila*, *Nature* **276**, 565–70.
**Lewis, E. B.** (1981) Developmental genetics of the bithorax complex in *Drosophila*. In *Developmental Biology Using Purified Genes, ICN–UCLA Symposia on Molecular and Cellular Biology*, Vol. XXIII. Eds. D.D. Brown and C. F. Fox. Academic Press, New York. 189–208.
**Lewis, E. B.** (1982) Control of body segment differentiation in *Drosophila* by the bithorax gene complex. In *Embryonic Development, Part A Genetic Aspects* Eds. M. M. Burger and R. Weber. Liss, New York. 269–87.
**Lewis, K. R.** and **John, B.** (1963) *Chromosome Marker*, Churchill, London.
**Lewis, K. R.** and **John, B.** (1968) The chromosomal basis of sex determination, *Int. Rev. Cytol.* **23**, 277–376.
**Lewis, M., Helmsing, P. J.** and **Ashburner, M.** (1975) Parallel changes in puffing activity and patterns of protein synthesis in salivary glands of *Drosophila*, *Proc. Natl. Acad. Sci. U.S.A.* **72**, 3604–8.
**Lewis, P. H.** (1947) New mutants. *Drosophila Inform. Serv.* **21**,69.
**Lewis, R. A., Wakimoto, B. T., Denell, R. E.** and **Kaufman, T. C.** (1980) Genetic analysis of the Antennapedia gene complex (ANT-C) and adjacent chromosomal regions of *Drosophila melanogaster*. II.Polytene chromosome segments 84A–84B1,2. *Genetics* **95**, 383–97.
**Lifschytz, E.** and **Green, M. M.** (1979) Genetic identification of dominant overproducing mutations: the *Beadex* gene, *Molec. Gen. Genet.* **171**, 153–9.
**Lifschytz, E.** and **Hareven, D.** (1977) Gene expression and control of spermatid morphogenesis in *Drosophila melanogaster*, *Devl. Biol.* **58**, 276–94.
**Lifton, R. P., Goldberg, M. L., Karp, R. W.** and **Hogness, D.** (1977) The organization of the histone genes in *D. melanogaster*: functional and evolutionary implications, *Cold Spring Harbor Symp. Quant. Biol.* **42**, 1047–51.
**Lillie, F. R.** (1902) Differentiation without cleavage in the egg of the annelid *Chaetopterus pergamentaceus*, *Arch. Entwicklungsmech. Org.* **14**, 477–99.
**Lindquist, S.** (1981) Regulation of protein synthesis during heat shock, *Nature* **293**, 311–4.
**Lindsley, D. L.** and **Grell, E. H.** (1968) *Genetic variations of* Drosophila melanogaster. Carnegie Institute of Washington, Washington, DC.
**Lindsley, D., Sandler, L., Baker, B. S. Carpenter, A., Denell, R. E., Hall, J. C., Jacobs, P. A., Miklos, G. L. G., Davis, B. K., Gethmann, R. C., Hardy, R. W., Hessler, A., Miller, S. M., Nozawa, H., Perry, D. M.** and **Gould-Somero, M.** (1972) Segmental aneuploidy and the genetic gross structure of the *Drosophila* genome, *Genetics* **71**, 157–84.
**Lipps, H. J., Gruissem, W.** and **Prescott, D. M.** (1982) Higher order DNA structure in macronuclear chromatin of the hypotrichous ciliate *Oxytrichia nova*, *Proc. Natl. Acad. Sci. U.S.A.* **79**, 2495–9.
**Lis, J., Prestidge, L.** and **Hogness, D. S.** (1978) A novel arrangement of tandemly repeated genes at a major heat shock site in *D. melanogaster*, *Cell* **14**, 901–19.
**Littlefield, J. W.** (1964) Selection of hybrids from matings of fibroblasts *in vitro* and their presumed recombinants, *Science* **145**, 709–10.
**Liu, C. P.** and **Lim, J. K.** (1975) Complementation analysis of methyl methanesulfonate induced recessive lethal mutations in the zeste-white region of the X-chromosome of *Drosophila melanogaster*, *Genetics* **79**, 601–11.
**Livak, K. J., Freund, R., Schweber, M., Wensink, P. C.** and **Meselson, M.** (1978) Sequence organization and transcription at two heat shock loci of

*Drosophila*, *Proc. Natl. Acad. Sci. U.S.A.* **75**, 5613–7.

**Lloyd, D. G.** and **Webb, C. J.** (1977) Secondary sex characters in seed plants, *Bot. Rev.* **43**, 177–216.

**Lohs-Schardin, M.** (1982) *Dicephalic*—A *Drosophila* mutant affecting polarity in follicle organization and embryonic patterning, *Roux's Arch.* **191**, 28–36.

**Lohs-Schardin, M., Sander, K., Cremer, C., Cremer T.** and **Zorn, C.** (1979) Localized UV laser microbeam irradiation of early *Drosophila* embryos. Fate maps based on location and frequency of adult defects, *Devl. Biol.* **68**, 533–45.

**Loomis, W. F.** (1979) Biochemistry of aggregation in *Dictyostelium*, *Devl. Biol.* **70**, 1–12.

**Loomis. W. F.** (1982) The spatial pattern of cell-type differentiation in *Dictyostelium*, *Devl. Biol.* **93**, 279–84.

**Løvtrup, S., Landström, U.** and **Løvtrup-Rein, H.** (1978) Polarities, cell differentiation and primary induction in the amphibian embryo, *Biol. Rev.* **53**, 1–42.

**Lucchesi, J. C.** and **Skripsky, T.** (1981) The link between dosage compensation and sex differentiation in *Drosophila melanogaster*, *Chromosoma* **82**, 217–28.

**Lumsden, J.** and **Coggins, J. R.** (1977) The subunit structure of the *arom* multi-enzyme complex of *Neurospora crassa*. A possible pentafunctional polypeptide chain, *Biochem. J.* **161**, 599–607.

**Lyon, M.** (1972) X-chromosome inactivation and developmental patterns in mammals. *Biol. Rev.* **47**, 1–35.

**McCarrey, J. R.** and **Abbott, U. K.** (1979) Mechanisms of sex determination, gonadal sex differentiation and germ-cell development in animals, *Adv. Genet.* **20**, 217–90.

**McCarron, M., O'Donnell, J., Chovnick, A., Bhullar, B. S., Hewitt, J.** and **Candido, E. P. M.** (1979) Organization of the *rosy* locus in *Drosophila melanogaster*: further evidence in support of a *cis*-acting control element adjacent to the xanthine dehydrogenase structural element, *Genetics* **91**, 275–93.

**McClintock, B.** (1949) Mutable loci in maize, *Carnegie Inst. Wash. Yearb.* **48**, 157–67.

**McClintock, B.** (1951) Chromosome organization and gene expression, *Cold Spring Harbor Symp. Quant. Biol.* **16**, 13–47.

**McClintock, B.** (1962) Topographical relations between elements of control systems in maize, *Carnegie Inst. Wash. Yearb.* **61**, 448–61.

**McCready, S. J., Godwin, J., Mason, D. W., Brazell, I. A.** and **Cooke, P. R.** (1980) DNA is replicated at the nuclear cage, *J. Cell Sci.* **46**, 365–86.

**MacGillivray, A. J.** (1977) The analysis of chromatin non-histone proteins. In *The Organization and Expression of the Eukaryotic Genome*. Eds. E. M. Bradbury and K. Javaherian. Academic Press, New York. 21–41.

**McGinnis, W., Farrell, J. R.** and **Beckendorf, S. K.** (1980) Molecular limits on the size of a genetic locus in *Drosophila melanogaster*, *Proc. Natl. Acad. Sci. U.S.A.* **77**, 7367–71.

**Macgregor, H. C.** (1980) Recent developments in the study of lampbrush chromosomes, *Heredity* **44**, 3–35.

**McKenzie, S. L.** and **Meselson, M.** (1977) Translation *in vitro* of *Drosophila* heat shock messages, *J. Mol. Biol.* **117**,279–83.

**McKnight, G. S., Hager, L.** and **Palmiter, R. D.** (1980) Butyrate and related inhibitors of histone deacetylation block the induction of egg white genes by steroid hormones, *Cell* **22**, 469–77.

**McKnight, S. L.** and **Kingsbury, R.** (1982) Transcriptional control signals of a eukaryotic protein-coding gene, *Science* **217**, 316–24.

**McLaren, A.** (1976a) Genetics of the early mouse embryo, *Ann. Rev. Genet.* **10**, 361–88.

**McLaren, A.** (1976b) *Mammalian Chimeras*. Cambridge University Press, Cambridge.

**McLaren, A.** and **Monk, M.** (1982) Fertile females produced by inactivation of an X-chromosome of "*sex-reversed*" mice, *Nature* **300**, 446–8.

**Maden, M.** (1982) Vitamin A and pattern formation in the regenerating limb, *Nature* **295**, 672–5.

**Madhavan, M. M.** and **Schneiderman, H. A.** (1977) Histological analysis of the dynamics of growth of imaginal discs and histoblast nests during the larval development of *Drosophila melanogaster*, *Roux's Arch.* **83**, 269–305.

**Mahowald, A. P., Caulton, J. H.** and **Gehring, W. J.** (1979) Ultra structural studies of oocytes and embryos derived from female flies carrying the *grandchildless* mutation in *Drosophila subobscura*, *Devl. Biol.* **69**, 118–32.

**Mandaron, P.** (1973) The effects of α-ecdysone, β-ecdysone and inokosterone on the *in vitro* evagination of *Drosophila* leg disc and subsequent differentiation of imaginal integumentary structures, *Devl. Biol.* **31**, 101–13.

**Mange, A. P.** and **Sandler, L.**(1973) A note on maternal effect mutants *daughterless* and *abnormal oocyte* in *Drosophila melanogaster*, *Genetics* **73**, 73–86.

**Mangold, O.** (1923) Transplantationsversache zur Frage der Spezifität und der bildung der Keimblätter bei *Triton, Arch. Micr. Anat. w. Entwicklungsmech.* **100**, 198–301.

**Mangold, O.** (1931) Das Determinationsproblem. III. Das Wirbeltierauge in der Entwicklung und Regeneration, *Ergebn. Biol.* **7**, 193–403.

**Maniatis, T., Fritsch, E. F.** and **Sambrook, J.** (1982) *Molecular Cloning*. Cold Spring Harbor Laboratory, New York.

**Maniatis, T., Hardison, R. C., Lacy, E., Lauer, J., O'Connell, C., Quon, D., Sim, G. K.** and **Efstratiadis, A.** (1978) The isolation of structural genes from libraries of eukaryotic DNA, *Cell* **15**, 687–701.

**Manning, J. E., Schmid, C. W.** and **Davidson, N.** (1975) Interspersion of repetitive and non-repetitive DNA sequences in the *Drosophila melanogaster* genome, *Cell* **4**, 141–55.

**Mantei, N.** and **Weissmann, C.** (1982) Controlled transcription of a human α-interferon gene introduced into mouse cells, *Nature* **297**, 128–32.

**Mantei, N., Boll, W.** and **Weissmann, C.** (1979) Rabbit β-globin mRNA production in mouse L cells transformed with cloned rabbit β-globin chromosomal DNA, *Nature* **281**,, 40–6.

**Margolskee, J. P., Froshauer, S., Skomska R.** and **Lodish, H. F.** (1980) The effects of cell density and starvation on early developmental events in *Dictyostelium discoideum*, *J. Embryol. Exp. Morph.* **74**, 409–21.

**Markert, C. L.** (1975) *Isozymes. Vol. 3. Developmental Biology*. Academic Press, New York.

**Markert, C. L.** and **Ursprung, H.** (1962) The ontogeny of isozyme patterns of lactate dehydrogenase in the mouse, *Devl. Biol.* **5**, 363–81.

**Marsh, J. L.** and **Wright, T. R. F.** (1979) Control of dopa decarboxylase expression during development in *Drosophila*. In *Eukaryotic Gene Regulation* Eds. B. Axel, T. Maniatis and C. F. Fox. Academic Press, New York. 183–94.

**Marsh, J. L.** and **Wright, T. R. F.** (1980) Developmental relationship between dopa decarboxylase, dopamine acetyltransferase and ecdysone in *Drosophila, Devl. Biol.* **80**, 379–87.

**Marshall, T.** and **Whittle, J. R. S.** (1978) Genetic analysis of the mutation *Female-lethal* in *Drosophila melanogaster, Genet. Res.* **32**, 103–11.

**Martin, G. R., Epstein, C. J., Travis, B., Tucker, G., Yatzio, S., Martin, D. W., Cleft, S.** and **Cohen, S.** (1978) X-chromosome inactivation during differentiation of female teratocarcinoma stem cells *in vitro, Nature* **271**, 329–33.

**Marx, J. L.** (1981) Genes that control development, *Science* **213**, 1485–8.

**Matsukuma, S.** and **Durston, A. J.** (1979) Chemotactic cell sorting in *Dictyostelium discoideum*, *J. Embryol. Exp. Morph.* **50**, 243–51.

**Mehl, Y.** and **Jarry, B. P.** (1978) Developmental regulation of the first three enzymes of pyrimidine biosynthesis in *Drosophila melanogaster, Devl. Biol.* **67**, 1–10.

**Meinhardt, H.** (1982) *Models of Biological Pattern Formation*. Academic Press, London.

**Meinke, D. W.** and **Sussex, I. M.** (1979) Isolation and characterization of six embryo lethal mutants of *Arabidopsis thaliana, Devl. Biol.* **72**, 62–72.

**Melvold, R. W.** (1977) Evidence suggesting two H-Y antigens in the mouse, *Immunogenetics* **5**, 33–41.

**Merriam, J. R.** (1978) Estimating primordial cell numbers in *Drosophila* imaginal discs and histoblasts. In *Genetic Mosaics and Cell Differentiation*. Ed. W. J. Gehring. Springer-Verlag, Berlin. 71–96.

**Metz, C. W.** (1938) Chromosome behaviour, inheritance and sex determination in *Sciara, Amer. Nat.* **72**, 485–520.

**Meyerowitz, E. M.** and **Hogness, D. S.** (1982) Molecular organization of a *Drosophila* puff site that responds to ecdysone, *Cell* **28**, 165–76.

**Mikamo, K.** and **Witschi, E.** (1964) Masculinization and breeding of WW *Xenopus, Experientia* **20**, 622–3.

**Milani, R.** (1975) The housefly *Musca domestica*. In *Handbook of Genetics*. Vol. 3. Ed. R. C. King. Plenum, New York. 377–99.

**Miller, O. L.** and **Beatty, B. R.** (1969) Visualization of nucleolar genes, *Science* **164**, 955–7.

**Milner, M. J.** and **Dübendorfer, A.**

(1982) Tissue-specific effects of the juvenile hormone analogue ZR515 during metamorphosis in *Drosophila* cell cultures, *J. Insect Physiol.* **28**, 661–6.
**Milner, M. J.** and **Sang, J. H.** (1974) Relative activities of α-ecdysone and β-ecdysone for the differentiation *in vitro* of *Drosophila melanogaster* imaginal discs, *Cell* **3**, 141–3.
**Milstein, C.** (1980) Monoclonal antibodies, *Sci. Am.* **243**, 56–64.
**Mintz, B.** (1978) Mutant mice from mutagenized terato-carcinoma cells. In *Differentiation and Development*; Eds. F. Ahmed, J. Schultz, T. A. Russell and R. Werner. Miami Winter Symp. Vol. 15. Academic Press, New York. 441–56.
**Mintz, B.** and **Baker, W. W.** (1967) Normal mammalian muscle differentiation and gene control of isocitrate dehydrogenase synthesis, *Proc. Natl. Acad. Sci. U.S.A.* **58**, 592–8.
**Mirault, M. E., Goldschmidt-Clermont, M., Artavanis-Tsakonas, S.** and **Schedl, P.** (1979) Organization of the multiple genes for the 70,000 dalton heat-shock protein in *Drosophila melanogaster, Proc. Natl. Acad. Sci. U.S.A.* **76**, 5254–8.
**Mirault, M. E., Goldschmidt-Clermont, M., Moran, L., Arrigo, A. P.** and **Tissières, A.** (1978) The effect of heat shock on gene expression in *D. melanogaster, Cold Spring Harbor Symp. Quant. Biol.* **42**, 819–27.
**Mirzabekov, A. D.** (1980) Nucleosome structure and its dynamic transitions, *Q. Rev. Biophys.* **13**, 255–95.
**Mitchell, H. K.** and **Lipps, L. S.** (1978) Heat shock and phenocopy induction in *Drosophila, Cell* **15**, 907–18.
**Mittenthal, J. E.** (1981) The rule of normal neighbours: A hypothesis for morphogenetic pattern regulation, *Devl. Biol.* **88**, 15–26.
**Mittwoch, U.** (1971) Sex determination in birds and mammals, *Nature* **231**, 432–4.
**Mittwoch, U.** (1973) *Genetics of Sex Differentiation*. Academic Press, New York.
**Miwa, J., Schierenberg, E., Miwa, S.** and **von Ehrenstein, G.** (1980) Genetics and mode of expression of temperature-sensitive mutations arresting embryonic development in *Caenorhabditis elegans, Devl. Biol.* **76**, 160–74.
**Mizuno, S.** and **Macgregor, H. C.** (1974) Chromosomes, DNA sequences, and evolution in salamanders of the genus *Plethodon, Chromosoma* **48**, 239–96.
**Moen, R. C.** and **Palmiter, R. D.** (1980) Changes in hormone responsiveness of chick oviduct during primary stimulation with estrogen, *Devl. Biol.* **78**, 450–63.
**Mohan, J.** (1976) Ribosomal DNA and its expression in *Drosophila melanogaster* during growth and development, *Mol. Gen. Genet.* **147**, 217–23.
**Monk, M.** (1978) Biochemical studies of mammalian X chromosome activity. In *Development in Mammals III*. Ed. M. H. Johnson. North-Holland, Amsterdam. 189–224.
**Monk, M.** (1981) A stem-line model for cellular and chromosomal differentiation in early mouse development, *Differentiation* **19**, 71–6.
**Monod, J., Jacob, F.** and **Gross, F.** (1962) Structural and rate-determining factors in the biosynthesis of adaptive enzymes, *Biochem. Soc. Symp.* **21**, 104–32.
**Morata, G.** and **Garcia-Bellido, A.** (1976) Developmental analysis of some mutants of the bithorax system of *Drosophila, Roux's Arch.* **179**, 125–43.
**Morata, G.** and **Kerridge, S.** (1981) Sequential functions of the bithorax complex of *Drosophila, Nature* **290**, 778–81.
**Morata, G.** and **Kerridge, S.** (1982) The role of position in determining homoeotic gene function in *Drosophila, Nature* **300**, 191–2.
**Morata, G.** and **Lawrence, P. A.** (1975) Control of compartment development by the *engrailed* gene in *Drosophila, Nature* **255**, 614–7.
**Morata, G.** and **Lawrence, P. A.** (1978) Anterior and posterior compartments in the head of *Drosophila, Nature* **274**, 473–4.
**Morata, G.** and **Lawrence, P. A.** (1979) Development of the eye–antenna imaginal disc of *Drosophila, Devl. Biol.* **70**, 355–71.
**Morgan, T. H.** (1910) Sex limited inheritance in *Drosophila, Science* **32**, 120–2.
**Morgan, T. H.** (1914) Mosaics and gynandromorphs in *Drosophila, Proc. Soc. Exp. Biol.* **11**, 171–2.
**Morgan, T. H.** (1929) *The Theory of the Gene*. 1964 reprint. Hafner Publishing Co., New York.
**Morgan, T. H.** (1934) *Embryology and Genetics*. Columbia University Press, New York.
**Mortin, M. A.** and **Kaufman, T. C.** (1982) Developmental genetics of a temperature sensitive RNA polymerase II mutation in *Drosophila melanogaster, Mol. Gen. Genet.* **187**, 120–5.

**Mortin, M. A.** and **LeFevre, G.** (1981) An RNA polymerase II mutation in *Drosophila melanogaster* that mimics *Ultrabithorax, Chromosoma* **82**, 237–47.
**Mullen, R. J.** (1978) Mosaicism in the central nervous system of mouse chimeras. In *The Clonal Basis of Development*; Eds. S. Subtelny and I. M. Sussex. Academic Press, New York. 83–102.
**Muller, H. J.** (1930) Types of visible variations induced by X-rays in *Drosophila, J. Genet.* **22**, 299–334.
**Muller, H. J.** (1932) Further studies on the nature and causes of gene mutations, *Proc. 6th Int. Congr. Genet. (Ithaca)* **1**, 213–55.
**Muller, H. J.** (1966) The gene material as the initiator and the organizing basis of life, *Amer. Nat.* **100**, 493–517.
**Muller, U., Guichard, A., Reyss-Brun, M.** and **Scheib, D.** (1980) Induction of H-Y antigen in the gonads of male quail embryos by diethylstilbestrol, *Differentia* **16**, 129–33.
**Munyon, W., Kraiselburd, E., Davis, D.** and **Mann, J.** (1971) Transfer of thymidine kinase to thymidine kinaseless L cells by infection with ultraviolet-irradiated *Herpes simplex* virus, *J. Virol.* **7**, 813–20.
**Murphy, C.** (1974) Cell death and autonomous gene action in lethals affecting imaginal discs in *Drosophila melanogaster, Devl. Biol.* **39**, 23–36.
**Muskavitch, M. A. T.** and **Hogness, D. S.** (1980) Molecular analysis of a gene in a developmentally regulated puff of *Drosophila melanogaster. Proc. Natl. Acad. Sci. U.S.A.* **77**, 7362–6.

**Nagl, W.** (1981) Polytene chromosomes in plants, *Int. Rev. Cytol.* **73**, 21–54.
**Nakamura, O.** and **Toivonen, S.** (1978) *Organizer—a milestone of a half century from Spemann*. Elsevier North-Holland, Amsterdam.
**Nasmyth, K. A.** (1982) The regulation of yeast mating-type chromatin structure by SIR: An action at a distance affecting both transcription and transposition, *Cell* **30**, 567–78.
**Nelson, O. E.** (1968) The *waxy* locus in maize. II. The location of the controlling element alleles, *Genetics* **60**, 475–91.
**Newby, W. W.** (1949) Abnormal growths on the head of *Drosophila melanogaster, J. Morphol.* **85**, 177–95.
**Newell, P. C.** (1978) Genetics of the cellular slime moulds, *Ann. Rev. Genet.* **12**, 69–93.
**Newman, S. M.** and **Schubiger, G.** (1980) A morphological and developmental study of *Drosophila* embryos ligated during nuclear multiplication, *Devl. Biol.* **79**, 128–38.
**Nishioka, Y.** and **Leder, P.** (1979) The complete sequence of a mouse chromosomal α-globin gene reveals elements conserved throughout vertebrate evolution, *Cell* **18**, 875–82.
**Niu, M. C.** (1956) New approaches to the problem of embryonic induction. In *Cellular Mechanisms in Differentiation and Growth*. Ed. D. Rudnick. University Press, Princeton. 155–71.
**Norby, S.** (1970) A specific nutritional requirement for pyrimidines in *rudimentary* mutants of *Drosophila melanogaster, Hereditas* **66**, 205–14.
**Nordheim, A., Pardue, M. L., Lafer, E. M., Moller, A., Stollar, B. D.** and **Rich, A.** (1981) Antibodies to left-handed Z-DNA bind to interband regions of *Drosophila* polytene chromosomes, *Nature* **294**, 417–21.
**Nöthiger, R., Dübendorfer, A.** and **Epper, F.** (1977) Gynandromorphs reveal two separate primordia for male and female genitalia in *Drosophila melanogaster, Roux's Arch.* **181**, 367–73.
**Nur, U.** (1967) Reversal of heterochromatization and the activity of the paternal chromosome set in the male mealy bug, *Genetics* **56**, 375–89.
**Nüsslein-Volhard, C.** (1977) Genetic analysis of pattern-formation in the embryo of *Drosophila melanogaster*. Characterisation of the maternal effect mutant *bicaudal, Roux's Arch.* **183**, 249–68.
**Nüsslein-Volhard, C.** (1979) Maternal effect mutations that alter the spatial coordinates of the embryo of *Drosophila melanogaster*, In *Determinants of Spatial Organization*. Eds. S. Subtelny and I. Konigsberg. Academic Press, New York. 185–211.
**Nüsslein-Volhard, C., Lohs-Schardin, M., Sander, K.** and **Cremer, C.** (1980) A dorso-ventral shift of embryonic primordia in a new maternal-effect mutant of *Drosophila, Nature* **283**, 474–6.
**Nüsslein-Volhard, C.** and **Wieschaus, E.** (1980) Mutations affecting segment number and polarity in *Drosophila, Nature* **287**, 795–801.

**O'Brien, S. J.** (1973) On estimating functional gene number in eucaryotes, *Nature New Biol.* **242**, 52–4.
**O'Brien, S. J.** and **MacIntyre, R. J.** (1978) Genetics and biochemistry of

enzymes and specific proteins of *Drosophila*. In *The Genetics and Biology of* Drosophila. Vol. 2a. Eds. M. Ashburner and T. R. F. Wright. Academic Press, New York. 396–551.

**Ohno, S.** (1976) Major regulatory genes for mammalian sexual development, *Cell* **7**, 315–21.

**Ohno, S.** (1977) Hormone-like action of H-Y antigen and gonadal development of XY-XX mosaic males and hermaphrodites, *Human Genet.* **35**, 21–5.

**Ohno, S.** (1979) *Major sex-determining genes*. Monographs on Endocrinology. Vol. II. Springer-Verlag, Berlin.

**Okada, M., Kleinman, I. A.** and **Schneiderman, H. A.** (1974) Repair of a genetically caused defect in oogenesis in *Drosophila melanogaster* by transplantation of cytoplasm from wild-type eggs and by injection of pyrimidine nucleosides, *Devl. Biol.* **37**, 55–62.

**Okada, Y.** (1958) The fusion of Erlich's tumor cells caused by HVJ virus *in vitro, Biken's J.* **1**, 103–10.

**Old, R. W., Callan, H. G.** and **Gross, K. W.** (1977) Localisation of histone gene transcripts in newt lamp brush chromosomes by *in situ* hybridisation, *J. Cell Sci.* **27**, 57–79.

**Old, R. W.** and **Primrose, S. B.** (1980) *Principles of Gene Manipulation*. Blackwell, Oxford.

**O'Malley, B. W., Stein, J. P., Wao, S. L. C., Catterall, J. F., Tsai, M.-J.** and **Means, A. R.** (1979) The ovomucoid gene: organization, structure and regulation. In *From Gene to Protein: Information Transfer in Normal and Abnormal Cells*. Eds. T. R. Russell, K. Brew, H. Faber and J. Schultz. Academic Press, New York. 15–22.

**Oppenheimer, J. M.** (1967) *Essays in the History of Embryology and Biology*. MIT Press, Cambridge, MA.

**Orgel, L. E.** and **Crick, F. H.** (1980) Selfish DNA: the ultimate parasite, *Nature* **284**, 604–7.

**Oshima, T.** and **Takano, I.** (1980) Duplicated gene producing transposable controlling elements for the mating-type differentiation in *Saccharomyces cerevisiae, Genetics* **94**, 859–70.

**Oster, G. F.** (1982) Mechanochemistry and Morphogenesis. In *Aharon Katzur-Katchalsky Memorial Symposium*. June 6–11, 1982, Rehovot, Israel.

**Oster, G., Odell, G.** and **Alberch, P.** (1980) Mechanics, Morphogenesis and Evolution, *Lectures on Maths in the Life Sciences* **13**, 165–255.

**Oster, G., Odell, G.** and **Burnside, B.** (1983) The mechanical basis of morphogenesis II: The mechanochemistry of cytogel contractility. *Devl. Biol.* (In press).

**Ouweneel, W. J.** (1969) Influence of environmental factors on the homoeotic effect of *loboid-ophthalmoptera* in *Drosophila melanogaster, Roux's Arch.* **164**, 15–36.

**Ouweneel, W. J.** (1970a) New mutants, *Drosophila Inform. Serv.* **45**, 35.

**Ouweneel, W. J.** (1970b) Genetic analysis of *loboid-ophthalmoptera*, a homoeotic strain in *Drosophila melanogaster, Genetica* **41**, 1–20.

**Ouweneel, W. J.** (1970c) Developmental capacities of young and mature wild type and *opht* eye imaginal discs in *Drosophila melanogaster, Roux's Arch.* **166**, 76–88.

**Ouweneel, W. J.** (1976) Developmental genetics of homoeosis, *Adv. Genet.* **18**, 179–248.

**Painter, T. S.** (1933) A new method for the study of chromosome rearrangements and the plotting of chromosome maps, *Science* **78**, 585–6.

**Painter, T. S.** (1934) Salivary chromosomes and the attack on the gene, *J. Hered.* **25**, 465–76.

**Palmiter, R., Mulvihill, E., McKnight, S.** and **Senear, A.** (1977) Regulation of gene expression in the chick oviduct by steroid hormones, *Cold Spring Harbor Symp. Quant. Biol.* **42**, 639–47.

**Papaioannou, V. E., Gardner, R. L., McBurney, M. W., Babinet, C.** and **Evans, M. J.** (1978) Participation of cultured teratocarcinoma cells in mouse embryogenesis, *J. Embryol. Exp. Morph.* **44**, 93–104.

**Pardue, M. L., Weinberg, E., Kedes, L.** and **Birnstiel, M.** (1972) Localization of sequences coding for histone messenger RNA in the chromosomes of *Drosophila melanogaster, J. Cell Biol.* **55**, 199.

**Pardue, M. L.** and **J. G. Gall.** (1972) Chromosome structure studied by nucleic acid hybridization in cytological preparations. In *Chromosomes Today*, **3**. Eds. C. D. Darlington and K. R. Lewis. Hafner Publishing Co., New York. 47–52.

**Parish, R. W.** (1979) Cyclic AMP induces the synthesis of developmentally regulated plasma membrane proteins in *Dictyostelium, Biochim. Biophys. Acta* **553**, 179–82.

**Parks, H. B.** (1936) Cleavage patterns in *Drosophila* and mosaic formation, *Ann. Entomol. Soc. Am.* **29**, 350–92.
**Paterson, B. M.**, **Roberts, B. E.** and **Kuff, E. L.** (1977) Structural gene identification and mapping by DNA/mRNA hybrid arrested cell-free translation, *Proc. Natl. Acad. Sci. U.S.A.* **74**, 4370–4.
**Paul, J.** (1972) General theory of chromosome structure and gene activation in eucaryotes, *Nature* **238**, 444–6.
**Paul, J.**, **Zollner, E. J.**, **Gilmour, R. S.** and **Birnie, G. D.** (1978) Properties of transcriptionally active chromatin, *Cold Spring Harbor Symp. Quant. Biol.* **42**, 597–603.
**Paulson, J. R.** and **Laemmli, U. K.** (1977) The structure of histone-depleted metaphase chromosomes, *Cell* **12**, 817–28.
**Pavan, C.** and **DaCunha, A. B.** (1969) Gene amplification in ontogeny and phylogeny of animals, *Genetics Suppl.* **61**, 289–304.
**Peacock, W. J.**, **Lohe, A. R.**, **Gerlach, W. L.**, **Dunsmuir, P.**, **Dennis, E. S.** and **Appels, R.** (1978) Fine structure and evolution of DNA in heterochromatin, *Cold Spring Harbor Symp. Quant. Biol.* **42**, 1121–35.
**Pelham, H. R. B.** (1982) A regulatory upstream promoter element in the *Drosophila* Hsp 70 heat-shock gene, *Cell* **30**, 517–28.
**Perlman, S. M.**, **Ford, P. J.** and **Rosbash, M. M.** (1977) Presence of tadpole and adult globin RNA sequences in oocytes of *Xenopus laevis*, *Proc. Natl. Acad. Sci. U.S.A.* **74**, 3835–9.
**Peterson, A. C.** (1974) Chimera mouse study shows absence of disease in genetically dystrophic muscle, *Nature* **248**, 561–4.
**Peterson, N. S.**, **Moller, G.** and **Mitchell, H. K.** (1979) Genetic mapping of the coding regions for three heat-shock proteins in *Drosophila melanogaster*, *Genetics* **92**, 891–902.
**Philips, J. P.** and **Forrest, H. S.** (1980) Ommochromes and pteridines, In *The Genetics and Biology of* Drosophila. Vol. 2d. Eds. M. Ashburner and T. R. F. Wright. Academic Press, New York. 542–623.
**Picard, G.** (1976) Non-mendelian female sterility in *Drosophila melanogaster*: Hereditary transmission of I factor, *Genetics* **83**, 107–23.
**Pontecorvo, G.** (1976) Production of indefinitely multiplying mammalian somatic cell hybrids by polyethylene glycol (PEG) treatment, *Somatic Cell Genet.* **1**, 397–400.
**Poodry, C. A.**, **Hall, L.** and **Suzuki, D. T.** (1973) Developmental properties of *shibire*$^{ts1}$, a pleiotropic mutation affecting larval adult locomotion and development, *Devl. Biol.* **32**, 373.
**Poodry, C. A.** and **Schneiderman, H. A.** (1976) Pattern formation in *Drosophila melanogaster*: The effects of mutations on polarity in the developing leg, *Roux's Arch.* **180**, 175–88.
**Portin, P.** (1977) Effect of temperature on interaction of *Abruptex* and recessive alleles of the *Notch* complex locus of *Drosophila melanogaster, Hereditas* **87**, 77–84.
**Portin, P.** (1981) The antimorphic mode of action of lethal *Abruptex* alleles of the *Notch* locus in *Drosophila melanogaster, Hereditas* **95**, 247–51.
**Postlethwait, J. H.** (1974) Development of the temperative-sensitive homoeotic mutant *ophthalmoptera* of *Drosophila melanogaster, Devl. Biol.* **36**, 212–7.
**Postlethwait, J. H.**, **Bownes, M.** and **Jowett, T.** (1980) Sexual phenotype and vitellogenin systhesis in *Drosophila melanogaster, Devl. Biol.* **79**, 379–87.
**Postlethwait, J. H.**, **Bryant, P. J.** and **Schubiger, G.** (1972) The homoeotic effect of 'tumorous head' in *Drosophila melanogaster, Devl. Biol.* **29**, 337–42.
**Postlethwait, J. H.** and **Handler, A. M.** (1978) Nonvitellogenic female sterile mutants and the regulation of vitellogenesis in *Drosophila melanogaster, Devl. Biol.* **67**, 202–13.
**Postlethwait, J. H.**, **Poodry, C. A.** and **Schneiderman, H. A.** (1971) Cellular dynamics of pattern duplication in imaginal discs of *Drosophila melanogaster, Devl. Biol.* **26**, 125–32.
**Postlethwait, J. H.** and **Schneiderman, H. A.** (1971a) A clonal analysis of development in *Drosophila melanogaster*: morphogenesis, determination, and growth in the wild-type antenna, *Devl. Biol.* **24**, 447–519.
**Postlethwait, J. H.** and **Schneiderman, H. A.** (1971b) Pattern formation and determination in the antenna of the homoeotic mutant *Antennapedia* of *Drosophila melanogaster, Devl. Biol.* **25**, 606–40.
**Postlethwait, J. H.** and **Weiser, K.** (1973) Vitellogenesis induced by juvenile hormone in the female sterile mutant apterous-four in *Drosophila melanogaster, Nature New Biol.* **244**, 284–5.
**Potter, S. S.**, **Brorein, W. J.**, **Dunsmuir, P.**

and **Rubin, G. M.** (1979) Transposition of elements of the 412, *copia* and 297 dispersed repeated gene families in *Drosophila*; *Cell* **17**, 415–27.

**Potter, S., Truett, M., Phillips, M.** and **Maher, A.** (1980) Eukaryotic transposable genetic elements with inverted terminal repeats, *Cell* **20**, 639–47.

**Poulson, D. F.** (1940) The effects of certain X-chromosome deficiencies on the embryonic development of *Drosophila melanogaster, J. Exp. Zool.* **83**, 271–325.

**Poulson, D. F.** (1945) Chromosomal control of embryogenesis in *Drosophila, Amer. Nat.* **79**, 340–63.

**Poulson, D. F.** (1950) Histogenesis, organogenesis and differentiation in the embryo of *Drosophila melanogaster* Meigen. In *Biology of* Drosophila. Ed. M. Demerec. Wiley, New York. 168–270.

**Proudfoot, N.** (1982) The end of the message, *Nature* **298**, 516–7.

**Proudfoot, N. J., Shander, M. H. M., Manley, J. L., Gefter, M. L.** and **Maniatis, T.** (1980) Structure and *in vitro* transcription of human globin genes, *Science* **209**, 1329–36.

**Prunell, A.** and **Kornberg, R. D.** (1978) Relation of nucleosomes to DNA sequences, *Cold Spring Harbor Symp. Quant. Biol.* **42**, 103–8.

**Puck, T. T.** and **Kao, F. T.** (1982) Somatic cell genetics and its application to medicine, *Ann. Rev. Genet.* **16**, 225–71.

**Pukkila, P. J.** (1975) Identification of lampbrush chromosome loops which transcribe 5S ribosomal RNA in *Notophthalmus* (*Triturus*) *viridescens, Chromosoma* **53**, 71–89.

**Ramenofsky, M., Faulkner, D. J.** and **Ireland, C.** (1974) The effects of juvenile hormone on cirriped metamorphosis, *Biochem. Biophys. Res. Commun.* **60**, 172–8.

**Rawls, J. M.** (1980) Identification of a small genetic region that encodes orotate phosphoribosyltransferase and orotidylate decarboxylase in *Drosophila melanogaster, Mol. Gen. Genet.* **178**, 43–9.

**Rawls, J. M., Chambers, C. L.** and **Cohen, W. S.** (1981) A small genetic region that controls dihydroorotate dehydrogenase in *Drosophila melanogaster, Biochem. Genet.* **19**, 115–27.

**Rawls, J.** and **Fristrom, J.** (1975) A complex genetic locus that controls the first three steps of pyrimidine biosynthesis in *Drosophila, Nature* **255**, 738–40.

**Rayle, R. E.** (1968) New mutants, *Drosophila Inform. Serv.* **43**, 62.

**Razin, A.** and **Riggs, A. D.** (1980) DNA methylation and gene function, *Science* **210**, 604–10.

**Ready, D. F., Hanson, T. E.** and **Benzer, S.** (1976) Development of the *Drosophila* retina, a neurocrystalline lattice, *Devl. Biol.* **53**, 217–40.

**Ream, K. W.** and **Gordon, M. P.** (1982) Crown gall disease and prospects for genetic manipulation of plants, *Science* **218**, 854–9.

**Reddy, E. P., Reynolds, R. K., Santos, E.** and **Barbacid, N.** (1982) A point mutation is responsible for the acquisition of transforming properties by the T24 human bladder carcinoma oncogene, *Nature* **300**, 149–52.

**Regenass, U.** and **Bernhard, H. P.** (1978) Analysis of the *Drosophila* mutant *mat(3)1* by pole cell transplantation experiments, *Mol. Gen. Genet.* **164**, 85–91.

**Reinert, J.** and **Bajaj, Y. P. S.** (1977) *Plant Cell, Tissue and Organ Culture*. Springer-Verlag, Berlin.

**Ribbert, D.** (1979) Chromomeres and puffing in experimentally induced polytene chromosomes of *Calliphora erythrocephala, Chromosoma* **74**, 269–98.

**Rice, T. B.** and **Garen, A.** (1975) Localized defects of blastoderm formation in maternal-effect mutants of *Drosophila, Devl. Biol.* **43**, 277–86.

**Rice, T. B., Rice, F. A.** and **Garen, A.** (1979) Adult abnormalities resulting from a localized blastoderm defect in a maternal effect mutant of *Drosophila, Devl. Biol.* **69**, 194–201.

**Richards, G. P.** (1976) The control of prepupal puffing patterns *in vitro*: Implications for prepupal ecdysone titres in *Drosophila melanogaster, Devl. Biol.* **48**, 191–5.

**Richards, G.** (1980a) Ecdysteroids and puffing in *Drosophila melanogaster*. In *Progress in Ecdysone Research*. Ed. J. A. Hoffman. Elsevier/North-Holland, Amsterdam.

**Richards. G.** (1980b) The polytene chromosomes in the fat body nuclei of *Drosophila melanogaster, Chromosoma* **79**, 241–50.

**Richards, G.** (1981) Insect hormones in development, *Biol. Rev.* **56**, 501–49.

**Rickoll, W. L.** and **Counce, S. J.** (1981) Morphogenesis in the embryo of *Drosophila melanogaster*—germ band extension in the maternal-effect lethal *mat(3)6, Roux's Arch.* **190**, 245–51.

**Rigby, P. W. J.** (1982) The oncogenic

circle closes, *Nature* **297**, 451–3.
**Ringertz, N. R.** and **Bolund, L.** (1974) Reactivation of chick erythrocyte nuclei by somatic cell hybridization, *Int. Rev. Exp. Path.* **8**, 83–116.
**Ringertz, N. R.** and **Savage, R. D.** (1976) *Cell Hybrids*. Academic Press, New York.
**Ripley, S.** and **Kalthoff,** (1981) Double abdomen induction with low UV-doses in *Smittia* spec. (Chironomidae, Diptera): sensitive period and complete photoreversibility, *Roux's Arch.* **190**,49–54.
**Ripoll, P.** and **Garcia-Bellido, A.** (1973) Cell autonomous lethals in *Drosophila melanogaster, Roux's Arch.* **169**, 200–15.
**Ritossa, F.** (1962) A new puffing pattern induced by temperature shock and DNA in *Drosophila, Experientia* **18**, 571–3.
**Roberts, D. B.** and **Brock, H. W.** (1981) The major serum proteins of Dipteran larvae, *Experientia* **37**, 103–10.
**Rodgers, W. H.** and **Gross, P. R.** (1978) Inhomogeneous distribution of egg RNA sequences in the early embryo, *Cell* **14**, 279–88.
**Roop, D. R., Nordstrom, J. L., Tsai, S. Y., Tsai, M-J,** and **O'Malley, B. W.** (1978) Transcription of structural and intervening sequences in the ovalbumin gene and identification of potential ovalbumin mRNA precursors, *Cell* **15**, 671–85.
**Rose, R. W.** and **Hillman, R.** (1974) Increased aminoacylation of tRNA as a function of the *Abnormal Abdomen* genotype of *Drosophila melanogaster, Mol. Gen. Genet.* **133**, 213–24.
**Rose, S. M.** (1962) Tissue-arc control of regeneration in the amphibian limb, *Symp. Soc. Study Dev. Growth* **20**, 153–76.
**Roseland, C. R.** and **Schneiderman, H. A.** (1979) Regulation and metamorphosis of the abdominal histoblasts of *Drosophila melanogaster, Roux's Arch.* **186**, 235–65.
**Rossant, J.** and **Lis, W. T.** (1979) Possible dual origin of the ectoderm of the chorion in the mouse embryo, *Devl. Biol.* **70**, 249–54.
**Rothfels, K.** and **Nambiar, R.** (1981) A cytological study of natural hybrids between *Prosimulium multidentatum* and *P. magnum*. With notes on sex determination in the Simuliidae (Diptera), *Chromosoma* **82**, 673–92.
**Rotmann, E.** (1935) Der Anteil von Induktor und reagierenden Gewebe an der Entwicklung des Haltfadens, *Roux's Arch.* **133**, 193–24.
**Rubin, G. M., Kidwell, M. G.** and **Bingham, P. M.** (1982) The molecular basis of P-M hybrid dysgenesis: The nature of induced mutations, *Cell* **29**, 987–94.
**Rubin, G. M.** and **Spradling, A. C.** (1982) Genetic transformation of *Drosophila* with transposable element vectors, *Science* **218**, 348–53.
**Rubini, P. G.** (1967) Ulteriori osservazioni sui determinati sessuali di *Musca domestica* L., *Genet. Agrar.* **21**, 363–84.
**Ruddle, F. H.** (1981) A new era in mammalian gene mapping: somatic cell genetics and recombinant DNA methodologies, *Nature* **294**, 115–20.
**Runnström, J.** (1928) Plasmabau und Determination bei den Ei von *Paracentrotus lividus* Lik, *Roux's Arch.* **113**, 556–81.
**Russell, M. A.** (1974) Pattern formation in the imaginal discs of a temperature-sensitive cell-lethal mutant of *Drosophila melanogaster, Devl. Biol.* **40**, 24–39.
**Rylander, L., Pigon, A.** and **Edström, J.-E.** (1980) Sequences translated by Balbiani ring 75S RNA *in vitro* are present in giant secretory protein from *Chironomus tentans, Chromosoma* **81**, 101–13.

**Saint, R. B.** and **Egan, J. B.** (1979) A method which facilitates the ordering of DNA restriction fragments. *Mol. Gen. Genet.* **171**, 103–6.
**Sakonju, S., Bogenhagen, D. F.** and **Brown, D. D.** (1980) A control region in the centre of the 5S RNA gene directs specific initiation of transcription: 1. The 5′ border of the region, *Cell* **19**, 13–25.
**Samols, D.** and **Swift, H.** (1979) Genomic organisation in the Flesh Fly, *Sarcophaga bullata, Chromosoma* **75**, 129–34.
**Sander, K.** (1976) Specification of the basic body pattern in insect embryogenesis, *Adv. Insect Physiol.* **12**, 125–238.
**Sander, K., Löhs-Schardin, M.** and **Baumann, M.** (1980) Embryogenesis in a *Drosophila* mutant expressing half the normal segment number, *Nature* **287**, 841–3.
**Sang, J. H.** (1963) Penetrance, expressivity and thresholds, *J. Hered.* **54**, 143–51.
**Sang, J. H.** (1969) Biochemical basis of hereditary melanotic tumors in *Drosophila*. In *Neoplasms and Related Disorders of Invertebrate and Lower Vertebrate Animals*. Nat. Cancer Inst. Monograph **31**, National Cancer Institute, Bethesda, MD 291–301.
**Sang, J. H.** (1978) The nutritional requirements of *Drosophila*. In *The*

*Genetics and Biology of* Drosophila. Vol. 2a. Eds. M. Ashburner and T. R. F. Wright. Academic Press, New York. 159–92.

**Sang, J. H.** (1981) *Drosophila* cells and cell lines, *Adv. Cell Culture* **1**, 125–92.

**Sang, J. H.** and **McDonald, J. M.** (1954) Production of phenocopies in *Drosophila* using salts, particularly sodium metaborate, *J. Genet.* **52**, 392–412.

**Santamaria, P.** (1979) Heat shock induced phenocopies of dominant mutants of the bithorax complex in *Drosophila melanogaster, Mol. Gen. Genet.* **172**, 161–3.

**Sass, H.** and **Bautz, E. K. F.** (1982) Immunoelectron microscopic localisation of RNA polymerase B on isolated chromosomes of *Chironomus tentans, Chromosoma* **85**, 633–42.

**Sauer-Locher, E.** (1954) Keimblätterbildung und Differenzierungsleistungen in isolierten Eiteilen der Honigbiene, *Roux's Arch.* **147**, 302–54.

**Saura, A. O.** and **Sorsa, V.** (1979) Electron microscope analysis of the banding pattern in the salivary gland chromosomes of *Drosophila melanogaster, Hereditas* **90**, 39–49.

**Scaletta, L. J.** and **Ephrussi, B.** (1965) Hybridisation of normal and neoplastic cells *in vitro, Nature* **205**, 1169–71.

**Scalenghe, F., Turco, E., Edström, J. E., Pirrotta, V.** and **Melli, M.** (1981) Microdissection and cloning of DNA from a specific region of *Drosophila melanogaster* polytene chromosomes, *Chromosoma* **82**, 205–16.

**Scandalios, J. G., Chang, D-Y., McMillin, D. E., Tsaftaris, A.** and **Moll, R. H.** (1980) Genetic regulation of the catalase developmental programme in maize scutellum: Identification of a temporal regulatory gene, *Proc. Natl. Acad. Sci. U.S.A.* **77**, 5360–4.

**Scangos, G.** and **Ruddle, F. J.** (1981) Mechanisms and applications of DNA-mediated gene transfer in mammalian cells—a review, *Gene* **14**, 1–10.

**Scazzocchio, C.** and **Arst, H. N.** (1978) The nature of an initiator constitutive mutation in *Aspergillus nidulans, Nature* **274**, 177–9.

**Scazzocchio, C.** and **Gorton, D.** (1977) Regulation of purine breakdown. In *Genetics and Physiology of* Aspergillus. Eds. J. E. Smith and J. A. Pateman. Academic Press, New York. 255–65.

**Scazzocchio, C.** and **Sealy-Lewis, H. M.** (1978) A mutation of xanthine dehydrogenase (purine hydroxylase 1) of *Aspergillus nidulans* resulting in altered specificity, *J. Biochem.* **91**, 99–109.

**Schaffner, W., Kunz, G., Daetwyler, H., Telford, J., Smith, H. O.** and **Birnstiel, M. L.** (1978) Genes and spacers of cloned sea urchin DNA analysed by sequencing, *Cell* **14**, 655–71.

**Schaltmann, K.** and **Pongs, O.** (1982) Identification and characterization of the ecdysterone receptor in *Drosophila melanogaster* by photoaffinity labeling, *Proc. Natl. Acad. Sci. U.S.A.* **79**, 6–10.

**Scheller, R. H., Constantini, F. D., Kozlowski, M. R., Britten, R. J.** and **Davidson, E. H.** (1978) Specific representation of cloned repetitive DNA sequences in sea urchin RNAs, *Cell* **15**, 189–203.

**Schierenberg, E., Miwa, J.** and **von Ehrenstein, G.** (1980) Cell lineages and developmental defects of temperature-sensitive embryonic arrest mutants in *Caenorhabditis elegans, Devl. Biol.* **76**, 141–59.

**Schlesinger, M. J., Aliperti, G.** and **Kelley, P. M.** (1982) The response of cells to heat shock, *Trends Biochem. Sci.* **7**, 222–5.

**Schlesinger, M. J., Ashburner, M.** and **Tissières, A.** (1982) *Heat Shock: From Bacteria to Man.* Cold Spring Harbor Laboratory, New York.

**Schmidt, V., Wydler, M.** and **Alder, H.** (1982) Transdifferentiation and regeneration *in vitro, Devl. Biol.* **92**, 476–88.

**Schubiger, G.** (1968) Anlageplan, Determinationszustand und Transdetermination-leistungen der männlichen Vorderbeinscheibe von *Drosophila melanogaster, Roux's Arch.* **160**, 9–40.

**Schubiger, G.** (1971) Regeneration, duplication and transdetermination in fragments of the leg disc of *Drosophila melanogaster, Devl. Biol.* **26**, 277–95.

**Schubiger, G.** (1974) Acquisition of differentiative competence in the imaginal leg disc of *Drosophila melanogaster, Roux's Arch.* **174**, 303–11.

**Schubiger, G.** and **Alpert, G. D.** (1975) Regeneration and duplication events in a temperature sensitive homoeotic mutant of *Drosophila melanogaster, Devl. Biol.* **42**, 292–304.

**Schubiger, G., Moseley, R. C.** and **Wood, W. J.** (1977) Interaction of different egg parts in determination of various body regions in *Drosophila melanogaster, Proc.*

*Natl. Acad. Sci. U.S.A.* **74**, 2050–3.

**Schüpbach, T.** (1982) Autosomal mutations that interfere with sex determination in somatic cells of *Drosophila* have no direct effect on the germline, *Devl. Biol.* **89**, 117–27.

**Schüpbach, T., Wieschaus, E.** and **Nöthiger, R.** (1978) A study of the female germ line in mosaics of *Drosophila, Roux's Arch.* **184**, 41–56.

**Scott, M. P.** and **Pardue, M. L.** (1981) Translational control in lysates of *Drosophila melanogaster* cells, *Proc. Natl. Acad. Sci. U.S.A.* **78**, 3353–7.

**Scott, W. A.** and **Mahoney, E.** (1976) Defects of glucose-6-phosphate and 6-phosphogluconate dehydrogenase in *Neurospora, Current Topics in Cell Regulation* **10**, 205–36.

**Shannon, M. P., Kaufman, T. C., Shen, M. W.** and **Judd, B. H.** (1972) Lethality patterns and morphology of selected lethal and semi-lethal mutants in the *Zeste-white* region of *Drosophila melanogaster, Genetics* **72**, 615–38.

**Shearn, A.** (1979) What is the normal function of genes which give rise to homoeotic mutations. In *Development and Neurobiology of* Drosophila. Eds. O. Siddiqi, P. Babu, L. M. Hall and J. C. Hall. Plenum, New York. 155–62.

**Shearn, A., Davis, K. T.** and **Hersperger, E.** (1978) Transdetermination of *Drosophila* imaginal discs cultured *in vitro, Devl. Biol.* **65**, 536–40.

**Shearn, A.** and **Garen, A.** (1974) Genetic control of imaginal disc development in *Drosophila, Proc. Natl. Acad. Sci. U.S.A.* **71**, 1393–7.

**Shearn, A., Rice, T., Garen, A.** and **Gehring, W.** (1971) Imaginal disc abnormalities in lethal mutants of *Drosophila, Proc. Natl. Acad. Sci. U.S.A.* **68**, 2594–8.

**Shellenbarger, D. L.** and **Mohler, J. D.** (1978a) Temperature sensitive periods and autonomy of pleiotropic effects of $l(1)N^{ts1}$, a conditional Notch lethal in *Drosophila, Devl. Biol.* **62**, 432–46.

**Shellenbarger, D. L.** and **Mohler, J. D.** (1978b) Temperature-sensitive mutations of the *Notch* locus in *Drosophila melanogaster, Genetics* **81**, 143–62.

**Shen, C-K.,** and **Maniatis, T.** (1980) Tissue specific DNA methylation in a cluster of rabbit β-like globin genes, *Proc. Natl. Acad. Sci. U.S.A.* **77**, 6634–8.

**Sheridan, W. F.** and **Neuffer, M. G.** (1982) Maize developmental mutants, *J. Hered.* **73**, 318–29.

**Sherman, M. I.** and **Wudl, L. R.** (1977) T-complex mutations and their effects. In *Concepts in Mammalian Embryogenesis.* Ed. M. I. Sherman. MIT Press, Cambridge, MA.

**Shermoen, A. W.** and **Bekendorf, S. K.** (1982) A complex of interacting DNase I-hypersensitive sites near the *Drosophila* glue protein gene, *Sgs4, Cell* **29**, 601–7.

**Shiba, T.** and **Saigo, K.** (1983) Origin of eukaryotic movable genetic elements: Identification of retrovirus-like particles containing *copia* RNA in *Drosophila melanogaster, Nature* **302**, 119–24.

**Shields, G., Dübendorfer, A.** and **Sang, H. J.** (1975) Differentiation *in vitro* of larval cell types from early embryonic cells of *Drosophila melanogaster, J. Embryol. Exp. Morph.* **33**, 159–75.

**Shih, C., Shilo, B-Z., Goldfarb, M. P., Dannenberg, A.** and **Weinberg, R. A.** (1979) Passage of phenotypes of chemically transformed cells via transfection of DNA and chromatin, *Proc. Natl. Acad. Sci. U.S.A.* **76**, 5714–8.

**Shilo, B-Z.,** and **Weinberg, R. A.** (1981) DNA sequences homologous to vertebrate oncogenes are conserved in *Drosophila melanogaster, Proc. Natl. Acad. Sci. U.S.A.* **78**, 6789–92.

**Shimke, R. T., Alt, R. W., Kellems, R. E., Kaufman, R. J.** and **Bertino, J. R.** (1977) Amplification of dihydrofolate reductase genes in methotrexate-resistant cultured mouse cells, *Cold Spring Harbor Symp. Quant. Biol.* **42**, 647–57.

**Shimotohno, K., Mizutani, S.** and **Temin, H. M.** (1980) Sequence of retrovirus provirus resembles that of transposable elements, *Nature* **285**, 550.

**Shortle, D., DiMaio, D.** and **Nathans, D.** (1981) Directed mutagenesis, *Ann. Rev. Genet.* **15**, 265–94.

**Shukla, P. T.** (1980) *Vestigial wingless*: a dominant allele of *vestigial* in *D. melanogaster, Drosophila Inform. Serv.* **55**, 210.

**Shur, B. D.** and **Bennett, D.** (1979) A specific defect in galactosyltransferase regulation on sperm bearing mutant alleles of the T/+ locus, *Devl. Biol.* **71**, 243–59.

**Silver, L. M., Artzt, K.** and **Bennett, D.** (1979) A major testicular cell protein specified by a mouse *T*/+ complex gene, *Cell* **17**, 275–84.

**Silvers, W. K., Billingham, R. E.** and **Sanford, B. H.** (1968) The H-Y transplantation antigen: A Y-linked or sex influenced factor? *Nature* **220**, 401–3.

**Silvers, W. K., Gasser, D. L.** and **Eicher,**

**E. M.** (1982) H-Y antigen, serologically detectable male antigen and sex determination, *Cell* **28**, 439–40.

**Simcox, A. A.** and **Sang, J. H.** (1982) Cell determination in the *Drosophila melanogaster* embryo. In *Embryonic Development, Part A: Genetic Aspects*. Eds. M. M. Burger and R. Weber. Liss, New York. 349–61.

**Siminovitch, L.** (1976) On the nature of heritable variation in cultured somatic cells, *Cell* **7**, 1–11.

**Sina, B. J.** and **Pellegrini, M.** (1982) Genomic clones coding for some of the initial genes expressed during *Drosophila* development, *Proc. Natl. Acad. Sci. U.S.A.* **79**, 7351–5.

**Sinclair, J. J., Burke, J. F., Ish-Horowicz, D.** and **Sang, J. H.** (1983) Extrachromosomal replication of a *copia*-based vector transfected into *Drosophila* cells, *Nature* (in press).

**Singh, L.** and **Jones, K. W.** (1982) Sex reversal in the mouse (*Mus musculus*) is caused by a recurrent non-reciprocal crossover involving the X and an aberrant Y chromosome, *Cell* **28**, 205–16.

**Skripsky, T.** and **Lucchesi, J. C.** (1982) Intersexuality resulting from the interaction of sex-specific lethal mutations in *Drosophila melanogaster, Devl. Biol.* **94**, 153–62.

**Slifer, E. J.** (1943) A mutant stock of *Drosophila* with extra sex combs. *J. Exp. Zool*, **90**, 31–40.

**Smith, H. O.** and **Birnstiel, M. L.** (1976) A simple method for DNA restriction site mapping, *Nucleic Acids Res.* **3**, 2387–98.

**Smith, L. D.** and **Williams, M. A.** (1975) Germinal plasm and determination of the primordial germ cells. In *The Developmental Biology of Reproduction*. Eds. C. L. Markert and J. Papaconstantinou. Academic Press, New York. 3–24.

**Snow, M. H. L.** (1981) Autonomous development of parts isolated from primitive-streak-stage mouse embryos. Is development clonal? *J. Embryol. Exp. Morph.* **65**, (Suppl.). 269–87.

**Soll, D. R.** and **Mitchell, L. H.** (1982) Differentiation and dedifferentiation can function simultaneously and independantly in the same cells in *Dictyostelium discoideum, Devl. Biol.* **91**, 183–90.

**Solomon, F.** (1979) Detailed neuritic morphologies of sister neuro-blastoma cells are related, *Cell* **16**, 165–70.

**Sonneborn, T. M.** (1954) Patterns of nucleo-cytoplasmic integration in *Paramecium, Caryologia*, **6** (Suppl.), 307–25.

**Sonneborn, T. M.** (1970) Gene action in development, *Proc. R. Soc. Lond.*, **B176**, 347–66.

**Sonneborn, T. M.** (1977) Genetics of cellular differentiation: stable nuclear differentiation in eukaryotic unicells, *Ann. Rev. Genet.* **11**, 349–67.

**Southern, E. M.** (1975) Detection of specific sequences among DNA fragments separated by gel electrophoresis, *J. Mol. Biol.* **98**, 503–17.

**Spemann, H.** and **Mangold, H.** (1924) Über Induktion von Embryonanlagen durch Implantation artfreunder Organisatoren, *Arch. f. mikr. Anat. Entw. Mech.* **100**, 599–638.

**Spemann, H.** (1938) *Embryonic Development and Induction*. Yale University Press, New Haven.

**Spencer, F. A., Hoffmann, F. M.** and **Gelbart, W. M.** (1982) Decapentaplegic: A gene complex affecting morphogenesis in *Drosophila melanogaster, Cell* **28**, 451–61.

**Spofford, J. B.** (1976) Position-effect variegation in *Drosophila*. In *The Genetics and Biology of* Drosophila. Vol. 1c. Eds. M. Ashburner and E. Novitski. Academic Press, New York. 955–1018.

**Spradling, A. C., Digan, M. E.** and **Mahowald, A. P.** (1980) Two clusters of genes for major chorion proteins of *Drosophila melanogaster, Cell* **19**, 904–14.

**Spradling, A. C.** and **Mahowald, A. P.** (1980) Amplification of genes for chorion proteins during oogenesis in *Drosophila melanogaster, Proc. Natl. Acad. Sci. U.S.A.* **77**, 1096–100.

**Spradling, A. C., Pardue, M. L.** and **Penman, S.** (1977) Messenger RNA in heat shocked *Drosophila* cells, *J. Mol. Biol.* **109**, 559–88.

**Spradling, A. C.** and **Rubin, G. M.** (1981) *Drosophila* genome organization: conserved and dynamic aspects, *Ann. Rev. Genet.* **15**, 219–64.

**Spradling, A. C., Waring, G. L.** and **Mahowald, A. P.** (1979) *Drosophila* bearing the *ocelliless* mutation underproduces two major chorion proteins both of which map near this gene, *Cell* **16**, 609–16.

**Spragg, J. H., Bebbington, C. R.** and **Roberts, D. B.** (1982) Monoclonal antibodies recognizing cell surface antigens in *Drosophila melanogaster, Devl. Biol.* **89**, 339–52.

**Sprey, T. E.** (1971) Cell death during the development of the imaginal discs of

*Calliphora erythrocephala, Netherlands J. Zool.* **21**, 221–64.

**Stanbridge, E. J., Der, C. J., Doersen, C-J., Nishimi, R. Y., Peehl, D. M., Weissman, B. E.** and **Wilkinson, J. E.** (1982) Human cell hybrids: analysis of transformation and tumorigenicity, *Science* **215**, 252–9.

**Steffensen, D. M.** (1968) A reconstruction of cell development in the shoot apex of maize, *Ann. J. Bot.* **55**, 254–369.

**Steiner, E.** (1976) Establishment of compartments in the developing leg imaginal discs of *Drosophila melanogaster, Roux's Arch.* **180**, 9–30.

**Stepshin, V. P.** and **Ginter, E. K.** (1974) A study of the homoeotic genes of *Antennapedia* and *Nasobemia* in *Drosophila melanogaster*. III. Influence of temperature on penetrance and expressivity of the *Apx* and *Ns* genes, *Sov. Genet.* **8**, 1252–7.

**Stern, C.** (1936) Somatic crossing over and segregation in *Drosophila melanogaster, Genetics* **21**, 625–730.

**Stern, C.** (1954a) Genes and developmental patterns, *Caryologia* **6** (Suppl.), 355–69.

**Stern, C.** (1954b) Two or three bristles, *Am. Scient.* **42**, 213–47.

**Stern, C.** (1968) *Genetic Mosaics and Other Essays*. Harvard University Press, Cambridge, MA.

**Stevens, B., Alvarez, C. M., Bohman, R.** and **O'Connor, J. D.** (1980) An ecdysteroid-induced alteration in the cell cycle of cultured *Drosophila* cells, *Cell* **22**, 675–82.

**Steward, F. C.** (1970) From cultured cells to whole plants: the induction and control of their growth and morphogenesis, *Proc. R. Soc. Lond.* **B175**, 1–30.

**Stewart, B.** and **Merriam, J.** (1980) Dosage Compensation. In *The Genetics and Biology of* Drosophila. Vol. 2d. Eds. M. Ashburner and T. R. F. Wright. Academic Press, New York. 107–40.

**Stewart, R. N.** (1978) Ontogeny of the primary body in chimeral forms of higher plants. In *The Clonal Basis of Development*. Eds. S. Subtelny and I. M. Sussex. Academic Press, New York. 131–60.

**Stewart, R. N.** and **Dermen, J.** (1970) Determination of number and mitotic activity of shoot apical initial cells by analysis of mericlinal chimeras, *Am. J. Bot.* **57**, 1010–6.

**Storti, V., Scott, M. P., Rich, A.** and **Pardue, M. L.** (1980) Translational control of protein synthesis in response to heat shock in *D. melanogaster* cells, *Cell* **22**, 825–34.

**Strathern, J. N.** and **Herskowitz, I.** (1979) Asymmetry and directionality in production of new cell types during clonal growth: the switching pattern of homothallic yeast, *Cell* **17**, 371–81.

**Strathern, J. N., Spatola, E., McGill, C.** and **Hicks, J. B.** (1980) Structure and organization of transposable mating type cassettes in *Saccharomyces* yeast, *Proc. Natl. Acad. Sci. U.S.A.* **77**, 2839–43.

**Strub, S.** (1977) Localisation of cells capable of transdetermination in a specific region of the male foreleg disc of *Drosophila, Roux's Arch.* **182**, 69–74.

**Struhl, G.** (1977) Developmental compartments in the proboscis of *Drosophila, Nature* **270**, 723–5.

**Struhl, G.** (1981a) A gene product required for correct initiation of segmental determination in *Drosophila, Nature* **293**, 36–41.

**Struhl, G.** (1981b) A homoeotic mutation transforming leg to antenna in *Drosophila, Nature* **292**, 635–8.

**Struhl, G.** (1982a) Genes controlling segmental specification in the *Drosophila* thorax, *Proc. Natl. Acad. Sci. U.S.A.* **79**, 7380–4.

**Struhl, G.** (1982b) Spineless-aristapedia: A homoeotic gene that does not control the development of specific compartments in *Drosophila, Genetics* **102**, 737–49.

**Sturtevant, E. H.** (1929) The claret mutant type of *Drosophila simulans*: a study of chromosome elimination and cell lineage, *Z. Wiss. Zool.* **135**, 323–56.

**Sulston, J. E.** and **Horvitz, H. R.** (1981) Abnormal cell lineages in mutants of the nematode *Caenorhabditis elegans, Devl. Biol.* **82**, 41–55.

**Sutton, W. D.** and **McCallum, M.** (1971) Mismatching and the reassociation rate of mouse satellite DNA, *Nature New Biol.* **232**, 83–5.

**Suzuki, D. T., Kaufman, T., Falk, D.** and the **UBC Drosophila Research Group** (1976) Conditionally expressed mutations in *Drosophila melanogaster*. In *The Genetics and Biology of* Drosophila. Vol. 1a. Eds. M. Ashburner and E. Novitski. Academic Press, New York. 207–63.

**Szabad, J., Simpson, P.** and **Nöthiger, R.** (1979) Regeneration and compartments in *Drosophila, J. Embryol. Exp. Morph.* **49**, 229–41.

**Szybalska, E. H.** and **Szybalski, W.** (1962) Genetics of human cell lines. IV.

DNA-mediated heritable transformation of a biochemical trait, *Proc. Natl. Acad. Sci. U.S.A.* **48**, 2026–34.

**Tabin, C. J., Bradley, S. M., Bargmann, C. I., Weinberg, R. A., Papageorge, A. G., Scolnick, E, M., Dhar, R., Lowy, D. R.** and **Chang, E. H.,** (1982) Mechanism of activation of a human oncogene, *Nature* **300**, 143–9.
**Tandler, J.** and **Keller, K.** (1911) Über das Verhalten des chorions bie verschiedengeschlechtlicher Zwillingsgravidität des Rines und über die Morphologie des Genitales der weiblichen Tiere, welche einer solcher Gravidität entstammen, *Dtsch. Tieraerztl. Wolchenschr.* **18**, 148.
**Tank, P. W.** and **Holder, N.** (1981) Pattern regulation in the regenerating limbs of urodele amphibians, *Q. Rev. Biol.* **56**, 113–42.
**Tarkowski, A. K.** (1961) Mouse chimeras developed from fused eggs, *Nature* **190**, 857–60.
**Temin, H. M.** (1980) Origin of retroviruses from cellular moveable genetic elements, *Cell* **21**, 599–600.
**Thiery-Mieg, D.** (1982) Paralog, a control mutant in *Drosophila melanogaster, Genetics* **100**, 209–37.
**Thoma, F., Koller, T.** and **Klug, A.** (1979) Involvement of histone H1 in the organization of the nucleosome and of the salt-dependent superstructures of chromatin, *J. Cell. Biol.* **83**, 403–26.
**Thompson, W. F., Murray, M. G.** and **Cuellar, R. E.** (1980) Contrasting patterns of DNA sequence organization in plants. In *Genome Organization and Expression in Plants*. Ed. C. L. Leaver. Plenum, New York. 1–15.
**Thörig, G. W. E., Heinstra, P. W. H.** and **Scharloo, W.** (1981) The action of the *Notch* locus in *Drosophila melanogaster*. 1. Effects of the $N^8$ deficiency on mitochondrial enzymes, *Mol. Gen. Genet.* **182**, 31–8.
**Timoféeff-Ressovsky, N. W.** (1935) Über 'muetterliche Vererbung' bei *Drosophila*, Naturwissenschaften **23**, 494–6.
**Tissières, A., Mitchell, H. K.** and **Tracy, U. M.** (1974) Protein synthesis in salivary glands of *D. melanogaster* in relation to chromosome puffs, *J. Mol. Biol.* **84**, 389–98.
**Timberlake. W. E.** (1980) Developmental gene regulation in *Aspergillus nidulans, Devl. Biol.* **78**, 497–510.
**Tobler, H., Smith, K. D.** and **Ursprung, H.** (1972) Molecular aspects of chromatin elimination in *Ascaris lumbricoides, Devl. Biol.* **27**, 190–203.
**Tokunaga, C.** (1961) The differentiation of a secondary sex comb under the influence of the gene *engrailed* in *Drosophila melanogaster, Genetics* **46**, 157–76.
**Tokunaga, C.** (1978) Genetic mosaic studies of pattern formation in *Drosophila melanogaster*. In *Genetic Mosaics and Cell Differentiation*. Ed. W. J. Gehring. Springer–Verlag, Berlin. 157–204.
**Town, C.** and **Stanford, E.** (1979) An oligosaccharide-containing factor that induces cell differentiation in *Dictyostelium discoideum*, *Proc. Natl. Acad. Sci. U.S.A.* **76**, 308–12.
**Treat-Clemons, L. G.** and **Doane, W. W.** (1980) Control of amylase levels by the *trans*-acting, tissue specific *map* locus in *Drosophila melanogaster, Genetics* **94**, 5105.

**Uenoyama, T., Fukunaga, A.** and **Ioshi, K.** (1982) Studies on the sex-specific lethals of *Drosophila melanogaster*. V. Sex transformation caused by interactions between a female-specific lethal, $Sxl^{f1}$, and the male-specific lethals *mle(3)132*, and *mle*, *Genetics* **102**, 233–43.
**Underwood, E. M., Turner, F. R.** and **Mahowald, A. P.** (1980) Analysis of cell movements and fate mapping during early embryogenesis in *Drosophila melanogaster, Devl. Biol.* **74**, 286–301.

**Van der Ploeg, L. H. T.** and **Flavell, R. A.** (1980) DNA methylation in the human γδβ-globin locus in erythroid and nonerythroid tissue, *Cell* **19**, 947–58.
**Vasil, I. K.** (1980) Perspectives in Plant Cell and Tissue Culture, *Int. Rev. Cytol. Suppl.* **11B**. Academic Press, New York.
**Vlad, M.** and **Macgregor, H. C.** (1975) Chromomere number and its genetic significance in lampbrush chromosomes, *Chromosoma* **50**, 327–47.
**Voellmy, R., Goldschmidt-Clermont, M., Southgate, R., Tissières, A., Levis, R.** and **Gehring, W.** (1971) A DNA segment isolated from chromosomal site 67B in *D. melanogaster* contains four closely linked heat-shock genes, *Cell* **23**, 261–70.
**Voellmy, R.** and **Rungger, D.** (1982) Transcription of a *Drosophila* heat shock gene is heat-induced in *Xenopus* oocytes, *Proc. Natl. Acad. Sci. U.S.A.* **79**, 1776–80.
**Vogel, O.** (1977) Regionalisation of segment-forming capacities during early embryogenesis in *Drosophila melanogaster, Roux's Arch.* **182**, 9–32.

**Vogt, M.** (1946a) Zur labilen determination die Imaginal scheiben von *Drosophila*: VI. Die umwandlung presumptiven russelgewebes in Bien-oder Fuhlergewebe, *Z. Naturforsch.* **1**. 469–75.
**Vogt, M.** (1946b) Beitrag zur Determination der Imaginalscheiben bei *Drosophila, Experientia* **2**, 313–5.

**Wachtel, S. S.** (1977) H-Y antigen and the genetics of sex determination, *Science* **193**, 1134–6.
**Wachtel, S. S., Bresler, P. A.** and **Koide, S. S.** (1980) Does H-Y antigen induce the heterogametic ovary, *Cell* **20**, 859–4.
**Wachtel, S. S., Koo, G. C., Ohno, S., Gropp, A., Deu, V. G., Tantravahli, R., Miller, C. A.** and **Miller, D. J.** (1976) The H-Y antigen and the origin of XY female wood lemmings (*Myopus schisticolor*), *Nature* **264**, 638–9.
**Waddington, C. H.** (1939) Genes as evocators in development, *Growth* **1**,37–44.
**Waddington, C. H.** (1940) *Organisers and Genes*. Cambridge University Press, Cambridge.
**Waddington, C. W.** (1956) *Principles of Embryology*. Macmillan, New York.
**Waddington, C. H.** and **Pilkington, R. W.** (1943) The structure and development of four mutant eyes in *Drosophila, J. Genet.* **43**, 44–50.
**Wahl, R. C., Warner, C. K., Finnerty, V.** and **Rajagopalan, K. V.** (1982) *Drosophila melanogaster ma-l* mutants are defective in the sulfuration of desulfo-Mo-hydroxylases, *J. Biol. Chem.* **287**, 3958–62.
**Wahli, W.** and **Dawid, I. B.** (1980) Comparative analysis of the structural organization of two closely related vitellogenin genes in *X. laevis, Cell* **20**, 107–17.
**Wakimoto, B. T.** and **Kaufman, T. C.** (1981) Analysis of larval segmentation in lethal genotypes associated with the Antennapedia Gene Complex in *Drosophila melanogaster, Devl. Biol.* **81**, 51–64.
**Wakusagi, N.** (1974) A genetically determined incompatibility system between spermatazoa and eggs leading to embryonic death in mice, *J. Reprod. Fertil.* **41**, 85–96.
**Wareing, P. F.** and **Phillips, I. D. J.** (1978) *The Control of Growth and Differentiation in Plants*. 2nd Edn. Pergamon Press, Oxford.
**Waring, G. L.** and **Mahowald, A. P.** (1979) Identification and the time synthesis of chorion proteins in *Drosophila melanogaster, Cell* **16**, 599–607.
**Weeds, A.** (1982) Actin binding proteins—regulators of cell architecture and motility, *Nature* **296**, 811–5.
**Weinberg, E. S., Overton, G. C., Hendricks, M. B., Newrock, K. M.** and **Cohen, L. H.** (1977) Histone gene heterogeneity in the sea urchin, *Strongylocentrotus purpuratus, Cold Spring Harbor Symp. Quant. Biol.* **42**, 1093–100.
**Weintraub, H.** and **Groudine, M.** (1976) Chromosomal subunits in active genes have an altered conformation, *Science* **193**, 848–56.
**Weisbrod, S.** (1982) Active chromatin, *Nature* **297**, 289–95.
**Weisbrod, S.** and **Weintraub, H.** (1979) Isolation of a subclass of nuclear proteins responsible for conferring a DNase 1-sensitive structure on globin chromatin, *Proc. Natl. Acad. Sci. U.S.A.* **76**, 630–4.
**Weiss, P.** (1939) '*Principles of Development*'. Holt, New York.
**Wellauer, P. K., Dawid, I. B.** and **Tartof, K. D.** (1978) Y and X chromosomal ribosomal DNA of *Drosophila*: comparisons of spacers and insertions, *Cell* **14**, 269–78.
**Welshons, W. J.** (1971) Analysis of a gene in *Drosophila,*·*Science* **150**, 1122–9.
**Wensink, P. C., Tabata, S.** and **Pachi, C.** (1979) The clustered and scrambled arrangement of moderately repetitive elements in *Drosophila* DNA, *Cell* **18**, 1231–46.
**Westergaard, M.** (1958) The mechanism of sex determination in dioecious flowering plants, *Adv. Genet.* **9**, 217–81.
**White, B. N., Dunn, R., Gellam, I., Tener, G. M., Armstrong, D. J., Skoog, F., Frihart, C. A.** and **Leonard, N. J.** (1975) An analysis of five serine transfer ribonucleic acids from *Drosophila, J. Biol. Chem.* **250**, 515–21.
**Whitehouse, H. L. K.** (1965) *Towards an Understanding of the Mechanism of Heredity*. Arnold, London. pp. 372.
**Whittaker, J. R.** (1973) Segregation during ascidian embryogenesis of egg cytoplasmic information for tissue specific enzyme development, *Proc. Natl. Acad. Sci. U.S.A.* **70**, 2096–100.
**Whittaker, J. R.** (1977) Segregation during cleavage of a factor determining endodermal alkaline phosphatase development in ascidian embryos, *J. Exp. Zool.* **202**, 139–54.
**Whittaker, J. R.** (1979) Cytoplasmic determinants of tissue differentiation in the ascidian egg. In *Determinants of Spatial Organisation*. Eds. S. Subtelny and I. R. Konigsberg. Academic Press, New York. 29–51.

**Whittaker, J. R., Ortolani, G.** and **Farinella-Ferruzza, N.** (1977) Autonomy of acetylcholinesterase differentiation in muscle lineage cells of ascidian embryos, *Devl. Biol.* **55**, 196–200.
**Whittinghill, M.** (1950) Two crossover-selector systems: New tools in genetics, *Science* **111**, 377–8.
**Whittle, J. R. S.** (1976) Clonal analysis of a genetically caused duplication of the anterior wing in *Drosophila melanogaster, Devl. Biol.* **51**, 257–68.
**Wiener, F., Klein, G.** and **Harris, H.** (1974) The analysis of malignancy by cell fusion. VI. Hybrids between different tumour cells, *J. Cell Sci.*, **16**, 189–98.
**Wieschaus. E.** (1978) Cell lineage relationships in the *Drosophila* embryo. In *Genetic Mosaics and Cell Differentiation*. Ed. W. J. Gehring. Springer–Verlag, Berlin. 97–118.
**Wieschaus, E.** and **Gehring, W.** (1976a) Clonal analysis of primordial disc cells in the early embryo of *Drosophila melanogaster, Devl. Biol.* **50**, 249–65.
**Wieschaus, E.** and **Gehring, W. J.** (1976b) Gynandromorph analysis of the thoracic disc primoridia in *Drosophila melanogaster, Roux's Arch.* **180**, 31–46.
**Wieschaus, E., Marsh, J. L.** and **Gehring, W.** (1978) *fs(1)K10*, a germline dependant female sterile mutation causing abnormal chorion morphology in *Drosophila melanogaster, Roux's Arch.* **184**, 75–82.
**Wieschaus, E.** and **Nöthiger, R.** (1982) The role of the transformer genes in the development of genitalia and analia of *Drosophila melanogaster, Devl. Biol.* **90**, 320–34.
**Wieschaus, E.** and **Szabad, J.** (1979) The development and function of the female germ line in *Drosophila melanogaster*: a cell lineage study, *Devl. Biol* **68**, 29–46.
**Wigglesworth, V. B.** (1935) Function of the corpus allatum in insects, *Nature* **136**, 338.
**Wigler, M., Pellicer, A., Silverstein, S., Axel, R., Urlaub, G.** and **Chasin, L.** (1979) DNA-mediated transfer of the adenine phosphoribosyltransferase locus into mammalian cells, *Proc. Natl. Acad. Sci. U.S.A.* **76**, 1373–6.
**Wigler, M., Pellicer, A., Silverstein, S.** and **Axel, R.** (1978) Biochemical transfer of single-copy eukaryotic genes using total cellular DNA as donor, *Cell* **11**, 725–31.
**Wigler, M., Silverstein, S., Lee, L-S., Pellicer, A., Cheng, T.** and **Axel, R.** (1977) Transfer of purified *Herpes* virus thymidine kinase gene to cultured mouse cells, *Cell* **11**, 223–32.
**Wilcox, M.** and **Smith, R. J.** (1977) Regenerative interactions between *Drosophila* imaginal discs of different types, *Devl. Biol.* **60**, 287–97.
**Wildermuth, H. R.** (1968) Autoradiographische Untersuchungen zum Vermelurungs-muster der Zellen in proliferierenden Rüsselprimordien von *Drosophila melanogaster, Devl. Biol.* **18**, 1–13.
**Wildermuth, H. R.** (1970a). Determination and transdetermination in cells of the fruitfly, *Sci. Prog.* **58**, 329–58.
**Wildermuth, H. R.** (1970b) Differenzierungsleistungen, Mustergliederung und Transdeterminations-mechanismen in hetero-und homoplastischen Transplantaten der Rüsselprimordien von *Drosophila, Roux's Arch.* **163**, 375–90.
**Willier, B. H.** and **Oppenheimer, J. M.** (1964) *Foundations of Experimental Embryology*, Prentice-Hall, Englewood Cliffs, NJ.
**Willis, J. H.** (1981) Juvenile hormone: The status of 'status quo', *Amer. Zool.* **21**, 763–73.
**Wilson, E. B.** (1904) Experimental studies in germinal localisation. II. Experiments on the cleavage-mosaic in *Patella* and *Dentalium, J. Exp. Zool.* **1**, 197.
**Wilson E. B.** (1925) *The Cell in Development and Heredity*. 3rd Edn. Macmillan, New York.
**Wilson, T. G.** (1980) Correlation of phenotypes of the apterous mutation in *Drosophila melanogaster, Devl. Genet.* **1**, 195–204.
**Wimber, D. E.** and **Steffensen, D. M.** (1973) Localization of gene function, *Ann. Rev. Genet.* **7**, 205–23.
**Wold, B. J., Klein, W. H., Hough-Evans, B. R., Britten, R. J.** and **Davidson, E. H.** (1978) Sea urchin embryo mRNA sequences in the nuclear RNA of adult tissues, *Cell* **14**, 941–50.
**Wolff, E.** and **Wolff, E.** (1951) The effects of castration on bird embryos, *J. Exp. Zool.* **116**, 59–98.
**Wolfner, M., Yep, D., Messenguy, F.** and **Fink, G. R.** (1975) Integration of amino acid biosynthesis into the cell cycle of *Saccharomyces cerevisiae, J. Mol. Biol.* **96**, 273–90.
**Wolpert, L.** (1969) Positional information and the spatial pattern of cellular differentiation, *J. Theor. Biol.* **25**, 1–47.
**Wolpert, L.** (1971) Positional information and pattern formation, *Curr. Topics Devl. Biol.* **6**, 183–224.

**Wolpert, L.** and **Lewis, J. H.** (1975) Towards a theory of development, *Fed. Proc.* **34**, 14–20.

**Wosnick, M. A.** and **White, B. N.** (1977) A doubtful relationship between tyrosine tRNA and suppression of the *vermilion* mutant in *Drosophila, Nucleic Acids, Res.* **4**, 3919–30.

**Wright, T. R. F.** (1970) The genetics of embryogenesis in *Drosophila, Adv. Genet.* **15**, 262–395.

**Wright, T. R. F., Bewley, G. C.** and **Sherald, A. F.** (1976a) The genetics of dopa decarboxylase in *Drosophila melanogaster*. II. Isolation and characterization of dopa decarboxylase deficient mutants and their relationship to the α-methyl dopa hypersensitive mutant, *Genetics* **84**, 287–310.

**Wright, T. R. F., Hodgetts, R. B., Sherald, A. F.** (1976b) The genetics of dopa decarboxylase in *Drosophila*. I. Isolation and characterisation of deficiencies that delete the dopa-decarboxylase-dosage-sensitive region and the α-methyldopa hypersensitive region, *Genetics* **84**, 267–85.

**Wright, T. R. F., Black, B. C., Bishop, C. P., Marsh, J. L., Pentz, E. S., Steward, R.,** and **Wright, E. Y.** (1982) The genetics of dopa decarboxylase in *Drosophila melanogaster, Mol. Gen. Genet.* **188**, 18–26.

**Wu, C.** (1980) The 5′ ends of *Drosophila* heat shock genes in chromatin are hypersensitive to DNase 1, *Nature* **286**, 854–60.

**Wudl, L.** and **Chapman, V. M.** (1976) The expression of α-glucuronidase during preimplantation development of mouse embryos, *Nature* **270**, 137–40.

**Wudl, L. R.** and **Sherman, M. I.** (1976) *In vitro* studies of mouse embryos bearing mutations at the T-locus: $t^{w5}$ and $t^{12}$, *Cell* **9**, 523–31.

**Wudl, L. R., Sherman, M. I.** and **Hillman, N.** (1977) Nature of lethality of *t* mutations in embryos, *Nature* **270**, 137–40.

**Wyss, C.** (1980) Loss of ecdysterone sensitivity of a *Drosophila* cell line after hybridization with embryonic cells, *Expl. Cell Res.* **125**, 121–6.

**Yajima, H.** (1964) Studies on embryonic determination of the harlequin fly, *Chironomus dorsalis*: II. Effects of partial irradiation of the eggs by ultraviolet light, *J. Embryol. Exp. Morph.* **12**, 89–100.

**Yanders, A. F.** (1957) New mutants, *Drosophila Inform. Serv.* **31**, 85.

**Young, M. W.** (1979) Middle repetitive DNA: a fluid component of the *Drosophila* genome, *Proc. Natl. Acad. Sci. U.S.A.* **76**, 6274–8.

**Young, M. W.** and **Judd, B. H.** (1978) Nonessential sequences, genes and the polytene chromosome bands of *D. melanogaster, Genetics* **88**, 723–42.

**Young, R. A., Hagenbuchle, O.** and **Schibler, U.** (1981) A single mouse α-amylase gene specifies two different tissue-specific mRNAs, *Cell* **23**, 451–8.

**Zachar, Z.** and **Bingham, P, M.** (1982) Regulation of *white* locus expression: The structure of mutant alleles at the *white* locus of *Drosophila melanogaster*, *Cell* **30**, 529–41.

**Zalokar, M.** (1947) L'ablation des disques imaginaux chez la larve de Drosophile, *Rev. Suisse Zool.* **50**, 232–7.

**Zalokar, M.** (1976) Autoradiographic study of protein and RNA formation during early development of *Drosophila* eggs, *Devl. Biol.* **49**, 425–37.

**Zalokar, M.** (1977) Genetic analysis of differentiated cell nuclei transplanted into polar plasm of *Drosophila* eggs, *Genetics* **88**, 572.

**Zalokar, M., Audit, C.** and **Erk, I.** (1975) Developmental defects of female-sterile mutants of *Drosophila melanogaster, Devl. Biol.* **47**, 419–32.

**Zalokar, M.** and **Erk, I.** (1976) Division and migration of nuclei during early embryogenesis of *Drosophila melanogaster, J. Microsc. Biol. Cell* **25**, 97–106.

**Zalokar, M., Erk, I.** and **Santamaria, P.** (1980) Distribution of ring-X chromosomes in the blastoderm of gynandromorphic *D. melanogaster, Cell* **19**, 133–41.

**Zambryski, P., Holsters, M., Kruger, K., Depicker, A., Schell, J., Van Montague, M.** and **Goodman, H. M.** (1980) Tumor DNA structures in plant cells transformed by *A. tumefaciens, Science* **209**, 1385–91.

**Zeevaart, J. A. D.** (1976) Physiology of flower formation, *Ann. Rev. Plant Physiol.* **27**, 321–48.

**Zhimulev, I. F.** and **Kolesnikov, N. N.** (1975) Synthesis and secretion of mucoprotein glue in the salivary gland of *Drosophila melanogaster, Roux's Arch.* **178**, 12–28.

**Zwilling, E.** (1956) Interaction between limb bud ectoderm and mesoderm in the chick embryo. IV. Experiments with a wingless mutant. *J. Exp. Zool.* **132**, 241–53.

# Index